AF849824

Business Process Analytics

The fourth edition of this widely used textbook offers a new perspective. Previously titled ***Business Process Modeling, Simulation and Design,*** as the new title suggests, this book is about analytical business process modeling and design. However, this new edition introduces analytics to the title and to the presentation.

The main objective of this book is to provide students with a comprehensive understanding of the multitude of analytical tools that can be used to model, analyze, understand, and ultimately design business processes.

The most flexible and powerful of these tools, although not always the most appropriate, is discrete-event simulation. The wide range of approaches covered in this book include graphical flowcharting tools, deterministic models for cycle time analysis and capacity decisions, and analytical queuing methods, as well as machine learning.

The authors focus on business processes as opposed to just manufacturing processes or general operations management problems and emphasize on simulation modeling using state-of-the-art commercial simulation software.

Business Process Analytics: Modeling, Simulation, and Design can be thought of as a hybrid between traditional books on process management, operations management, and simulation. The growing interest in simulation-based tools suggests that an understanding of simulation modeling, its potential as well as its limitations for analyzing and designing processes, is of key importance to students looking for a future career in operations management.

Changes from the previous edition include the following:

- New section on data-driven process improvement (with data visualization)
- Added a subsection of control charts to the 6-sigma section
- Replaced business process reengineering with business process management

- Updated all text, figures, examples, and exercises to ExtendSim10 (current version)
- More coverage on design of experiments
- More coverage of machine learning and neural networks

Manuel Laguna is a MediaOne Professor of Management Science at the Leeds School of Business at the University of Colorado. He received his doctoral degree in Operations Research and Industrial Engineering from the University of Texas at Austin. He has more than 100 publications in data analytics methods and applications and is the editor-in-chief of the *Journal of Heuristics*.

Johan Marklund is a Professor of Production Management at Lund University, Faculty of Engineering in Sweden. He holds a PhD in Production Management and BSc in Business Administration from Lund University, and an MSc in Industrial Engineering and Management from Linköping University. He has published in numerous scientific journals, and his research interests include inventory theory, supply chain management, and logistics.

Textbooks in Mathematics
https://www.routledge.com/Textbooks-in-Mathematics/book-series/
CANDHTEXBOOMTH

Series editors: Al Boggess, Kenneth H. Rosen

Graphs and Digraphs, Seventh Edition
Gary Chartrand, Heather Jordon, Vincent Vatter and Ping Zhang

An Introduction to Optimization with Applications in Data Analytics and Machine Learning
Jeffrey Paul Wheeler

Encounters with Chaos and Fractals, Third Edition
Denny Gulick and Jeff Ford

Differential Calculus in Several Variables
A Learning-by-Doing Approach
Marius Ghergu

Taking the "Oof!" Out of Proofs
Alexandr Draganov

Vector Calculus
Steven G. Krantz and Harold Parks

Intuitive Axiomatic Set Theory
José Luis García

Fundamentals of Abstract Algebra
Mark J. DeBonis

A Bridge to Higher Mathematics
James R. Kirkwood and Raina S. Robeva

Advanced Linear Algebra, Second Edition
Nicholas Loehr

Mathematical Biology: Discrete and Differential Equations
Christina Alvey and Daniel Alvey

Numerical Methods and Analysis with Mathematical Modelling
William P. Fox and Richard D. West

Business Process Analytics
Modeling, Simulation, and Design
Manuel Laguna and Johan Marklund

https://www.routledge.com/Textbooks-in-Mathematics/book-series/
CANDHTEXBOOMTH

Business Process Analytics

Modeling, Simulation, and Design

Fourth Edition

Manuel Laguna and Johan Marklund

CRC Press
Taylor & Francis Group
Boca Raton London New York

CRC Press is an imprint of the
Taylor & Francis Group, an **informa** business
A CHAPMAN & HALL BOOK

Fourth edition published 2025
by CRC Press
2385 NW Executive Center Drive, Suite 320, Boca Raton FL 33431
and by CRC Press

4 Park Square, Milton Park, Abingdon, Oxon, OX14 4RN

CRC Press is an imprint of Taylor & Francis Group, LLC

© 2025 Manuel Laguna and Johan Marklund

First edition published by Prentice Hall 2004

Second edition published by Chapman and Hall/CRC 2013

Sixth edition published by Chapman and Hall/CRC 2018

Reasonable efforts have been made to publish reliable data and information, but the author and publisher cannot assume responsibility for the validity of all materials or the consequences of their use. The authors and publishers have attempted to trace the copyright holders of all material reproduced in this publication and apologize to copyright holders if permission to publish in this form has not been obtained. If any copyright material has not been acknowledged please write and let us know so we may rectify in any future reprint.

Except as permitted under U.S. Copyright Law, no part of this book may be reprinted, reproduced, transmitted, or utilized in any form by any electronic, mechanical, or other means, now known or hereafter invented, including photocopying, microfilming, and recording, or in any information storage or retrieval system, without written permission from the publishers.

For permission to photocopy or use material electronically from this work, access www.copyright.com or contact the Copyright Clearance Center, Inc. (CCC), 222 Rosewood Drive, Danvers, MA 01923, 978-750-8400. For works that are not available on CCC please contact mpkbookspermissions@tandf.co.uk

Trademark notice: Product or corporate names may be trademarks or registered trademarks and are used only for identification and explanation without intent to infringe.

ISBN: 978-1-032-59542-9 (hbk)
ISBN: 978-1-032-61724-4 (pbk)
ISBN: 978-1-032-61723-7 (ebk)

DOI: 10.1201/9781032617237

Typeset in Palatino
by KnowledgeWorks Global Ltd.

Contents

Preface

As the title suggests, this book is about analytical business process modeling and design. It is the result of several years of teaching undergraduate and graduate process design and simulation courses to business students at the Leeds School of Business (University of Colorado Boulder) and engineering students at the Department of Industrial Management and Logistics (Lund University). The main motivation for writing this textbook stems from our struggle to find a book that approaches business process design from a broad, quantitative modeling perspective. The main objective of this book is thus to provide students with a comprehensive understanding of the multitude of analytical tools that can be used to model, analyze, understand, and ultimately design business processes. We believe that the most flexible and powerful of these tools, although not always the most appropriate, is discrete-event simulation.

The wide range of approaches covered in this book include graphical flowcharting tools, deterministic models for cycle time analysis and capacity decisions, and analytical queuing methods, as well as data mining. What distinguishes this textbook from general operations management books, most of which cover many of the same topics, is its focus on business processes as opposed to just manufacturing processes or general operations management problems, and its emphasis on simulation modeling using state-of-the-art commercial simulation software. Essentially, *Business Process Analytics: Modeling, Simulation, and Design* can be thought of as a hybrid between traditional books on process management, operations management, and simulation. Although it would be desirable for all students in operations management to take several courses in each of these topics, the reality is that few business school curricula today allow that. In our experience, simulation especially tends to be shoved to the side simply because it is perceived to be too technical. However, our firm belief, manifested by our writing of this book, is that this need not and should not be the case. The rapid development of user-friendly simulation software with graphical interfaces has made the use of simulation accessible even to those lacking computer programming skills, and it provides a great medium for illustrating and understanding the implications of capacity restrictions and random variation for process performance. Furthermore, the growing interest in simulation-based tools in industry suggests that an understanding of simulation modeling, its potential as well as its limitations for analyzing and designing processes, is of key importance to students looking for a future career in operations management.

Before proceeding with a discussion of how we picture this book being used, it is worthwhile to summarize what the book is not. It is not a traditional, qualitatively oriented book on process management, although we address these important issues throughout the book. It is not a traditional book on operations management, although it covers some of the traditional tools found in most operations management books. Furthermore, it is not a traditional book on simulation, although discrete-event simulation is used extensively. It is a book that attempts to bring these topics together by placing an analytical modeling perspective on process design and particularly emphasizing the power of simulation modeling as a vehicle for analyzing and designing business processes.

This textbook is ideal for a one-semester undergraduate or MBA course within an operations management business school curriculum. The MBA course typically would cover some of the more advanced topics in greater depth, such as process analytics (which does not require that students have some previous exposure to linear programming, but it is helpful), simulation optimization, and queuing. The MBA course also could include a more involved simulation project. In addition, we envision this book being used for an undergraduate course in industrial engineering or within an engineering management program.

In terms of requirements, the textbook assumes that students have taken a basic course in business statistics. However, if students have been exposed to a basic course in operations management and have some prior knowledge of quantitative techniques, this gives an additional opportunity to explore some of the topics covered in this textbook in more detail. In terms of how to cover the material, we recommend following the chapter sequence.

The text is organized as follows:

- Chapter 1 sets the stage by defining what we mean by a business process and business process design. It also points to the importance of these issues for overall business performance and the organization's strategic positioning.
- Chapter 2 explains some fundamental principles for successful process management and also takes a closer look at two of the more influential, process-oriented improvement programs in recent years: Six Sigma and business process management. The rationale is that although the focus of the book is on analytical modeling and design rather than implementation and change management, even the best possible design is of little use if it is not implemented and managed properly.
- Chapter 3 presents a methodology for business process design projects. The approach identifies a number of steps or issues that typically need to be dealt with during a process design project from initialization to implementation. The approach can be seen as a roadmap to the remainder of the book in the sense that the tools and methods discussed in the following chapters can be used to address some of the specified issues.

- Chapter 4 deals with a range of basic tools for analyzing and designing processes that display limited variability in demand and activity times (i.e., deterministic as opposed to stochastic models). The first part of the chapter discusses several graphical tools for charting and describing processes. These tools are particularly useful for understanding existing processes. The second part of the chapter investigates seven fundamental process design principles and associated methods of analysis.
- Chapter 5 focuses on how to manage process flows, particularly with regard to cycle time and capacity analysis. As in Chapter 4, we consider only deterministic situations with perfect information regarding demand, resource availability, and activity times.
- Chapter 6 introduces queuing as a means for explicitly incorporating randomness and variability into the process analysis. The chapter discusses queuing strategies, properties of the exponential distribution, the Poisson process, traditional birth-and-death processes, and the corresponding queuing models including single and multiple servers with and without restrictions on the queuing population and/or the queue lengths. A discussion of waiting/shortage costs and applications to process design situations is included.
- Chapter 7 introduces the notion of simulation in general and, in particular, the area of discrete-event simulation. A detailed illustrative example is used to introduce the elements of this simulation technique. Practical issues associated with successful simulation projects are discussed. The important question of when not to simulate is also addressed in this chapter.
- Chapter 8 provides an introduction to the simulation software ExtendSim, which is used for simulation modeling. The focus of the first part of the chapter is on how to get a model up and running, including working with simple animation, modeling random processing times, and limited labor resources. The latter part of the chapter discusses advanced modeling techniques and how to collect data and use the built-in tools for statistical analysis. In addition, the chapter explores how to use ExtendSim to model more complex business processes and capture features such as prioritization, attribute assignment, blocking, balking and reneging queues, routing through multiple and parallel paths, batching, resource allocation, activity-based costing, and cycle time analysis.
- Chapter 9 deals with the important issue of the statistical analysis of input and output data. Topics covered include determination of input data distributions, random number generation, and how to analyze output data. Of particular importance is how to compare the performance of alternative process designs.

- Chapter 10 introduces the design of experiments as a tool for fine-tuning simulations and discusses state-of-the-art methods for optimizing design parameters.
- Chapter 11 starts with a discussion of business analytics in relation to process design and improvement. Relevant issues related to business process management systems, business rules, and the application of process mining are discussed. The chapter ends by showing how machine learning models are used to enhance the performance of business processes.

The simulation modeling is done entirely with the ExtendSim software, enabling students to work hands-on with models of their own. A potential criticism associated with this software-integrated approach is that the acquired knowledge may be thought of as being software dependent. However, our experience is that after initially investing some time in understanding the software, the main challenge and learning lie in creating abstract models of processes, a conceptual framework that is software independent. Furthermore, because the functionality of most modern simulation software is very similar, exposure to one product promotes a valuable understanding of the potential and limitations of simulation modeling in general.

The textbook includes access to the following online supplements:

ExtendSim software (https://extendsim.com/academic)

Student materials (https://app.box.com/s/82n38yjp606q6y74dx4pfgfy3mi2v21j)

Instructor materials (https://app.box.com/s/tvqkzzop8kurxyff4csr1aha284ogd7h)

There are many individuals among our colleagues, families, and friends who have contributed to this book as sources of inspiration, encouragement, and support. Our deepest gratitude goes to all of you. Others have had a more direct influence on this book, and we take this opportunity to express our appreciation for their valuable help. Special thanks go to Dave Krahl at Imagine That Inc., who worked with us on the simulation models. We would also like to thank two anonymous reviewers, who provided many valuable suggestions in shaping this textbook.

Our gratitude also goes to past and present students in the undergraduate and graduate programs at the Leeds School of Business and Lund University, who have helped shape this book into what it is today. Special thanks go to Marco Better for helping us put together the instructor's manual.

And thank you to Bob Ross at Taylor & Francis, who encouraged us to pursue this project and provided help and advice along the way, as well as to Jose Soto, Joette Lynch, and Andrew Corrigan, who capably guided our manuscript through production. Thank you for making this book a reality!

1

Introduction to Business Process Design

Booming competition in an increasingly global marketplace leaves no room for successful companies to harbor internal inefficiencies. Even more importantly, customers are becoming more demanding; if one product or service does not live up to their expectations, there are many more from which to choose. The stakes are high, and so is the penalty for not satisfying the right customers with the right products and services. The quest for internal efficiency and external effectiveness means that organizations must align their internal activities and resources with external requirements, or to put it differently, business processes must be designed appropriately. To that end, the main objective of this book is to provide the reader with a comprehensive understanding of the wide range of analytical tools that can be used for modeling, analyzing, and ultimately designing business processes. Particular emphasis is placed on discrete event simulation, as it represents one of the most flexible and powerful tools available for these purposes. However, an important message is to choose tools that are appropriate for the situation at hand, and simulation is not always the best choice.

Before investigating various analytical tools in Chapters 4 through 11, we need to understand what business processes and process design are all about. To that end, this first chapter provides a general discussion and definition of business processes and business process design. Important concepts include process hierarchies, process architecture, and incremental process improvement versus process design. Moreover, an important distinction is made between the activities of designing a process and implementing the design, the former being the focus of this book.

This introductory chapter also discusses the importance of business process design for overall business performance and the organization's business strategy. An interesting question is why inefficient and ineffective business processes come to exist in the first place.

Chapter 2 deals with the important issues of managing and improving processes, including challenges of implementation and change management. Chapter 3 then discusses a framework for structuring business process design projects. This can be construed as a road map for the remaining eight chapters of this book, which focus on tools and modeling.

DOI: 10.1201/9781032617237-1

1.1 What Is a Business Process?

From a pragmatic point of view, a business process describes how something is done in an organization. However, for a more in-depth analysis and design of business processes, the theme of this book, further examination is needed.

Let us start by dissecting the words *business* and *process*. Most people probably would say that they have a clear notion about what a business is. In broad terms, we will define a business as "an organizational entity that deploys resources to provide customers with desired products or services." This definition serves our purposes because it encompasses profit-maximizing firms and supply chains, as well as nonprofit organizations and governmental agencies.

A process is a somewhat more ambiguous concept with different meanings, depending on the context in which it is used. For example, a biologist or medical doctor refers to breathing as a life-sustaining *process*. In mathematics or engineering, the concept of random or deterministic *processes* describes event occurrences. In politics, the importance of election *processes* is obvious, in education a key concept is the learning *process*, and so on. *Merriam-Webster's Dictionary* defines *process* as (i) a natural phenomenon marked by gradual changes that lead to a particular result, (ii) a continuing natural or biological activity or function, or (iii) a series of actions or operations conducing to an end. The last of these definitions is of particular interest, as it leads into the traditional high-level definition of a process used in the operations management literature, illustrated in Figure 1.1: *A process specifies the transformation of inputs to outputs*. The transformation can be of many different forms, but a broad classification into four different types is commonly used:

- *Physical*, for instance, the transformation of raw materials into a finished product.
- *Locational*, for instance, the transportation service provided by an airline.
- *Transactional*, for instance, banking and transformation of cash into stocks by a brokerage firm.
- *Informational*, for instance, the transformation of financial data into information in the form of financial statements.

The simple transformation perspective, illustrated in Figure 1.1, forms the basis for the so-called *process view* of an organization. According to this perspective, any organizational entity or business can be characterized as a process or a network of processes.

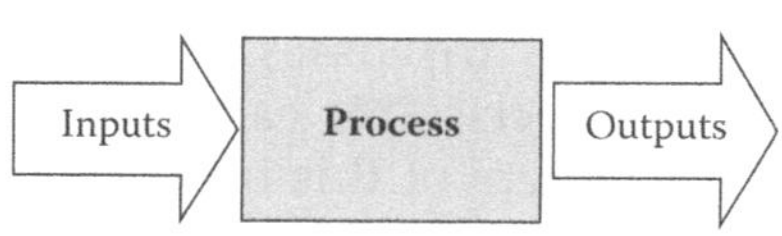

FIGURE 1.1
The transformation model of a process.

TABLE 1.1

Examples of Generic Business Processes Other Than Traditional Production and Manufacturing Processes

Accounts payable	Inventory management
Accounts receivable	Order fulfillment
Admissions (e.g., hospitals and universities)	New employee training
Auditing	Performance appraisal
Billing	Product approval (e.g., pharmaceuticals)
Budget planning	Product development
Business planning	Purchasing
Client acquisition	Receiving
Continuous improvement	Shipping
Credit approval	Vendor certification
Human resource planning and hiring	Warranty and claims processing

The process view makes no assumptions about the types of processes constituting the organization. However, often, a process is automatically thought of as a manufacturing or production process. We employ the term *business process* to emphasize that this book is not focusing on the analysis and design of manufacturing processes per se, although they will be an important subset of the entire set of business processes that define an organization. Table 1.1 provides some examples of generic processes other than traditional production/manufacturing processes that one might expect to find in many businesses. For the remainder of this book, the terms *process* and *business process* will be used interchangeably.

The simple transformation model of a process depicted in Figure 1.1 is a powerful starting point for understanding the importance of business processes. However, for purposes of detailed analysis and design of the transformation process itself, we need to go further and look behind the scenes, inside the "black box," at process types, process hierarchies, and determinants of process architecture.

1.1.1 Process Types and Hierarchies

Based on their scope within an organization, processes can be characterized into three different types: *individual processes,* which are carried out by separate individuals; *vertical* or *functional processes,* which are contained within a certain functional unit or department; and *horizontal* or *cross-functional processes,* which cut across several functional units (or, in terms of supply chains, even across different companies). See Figure 1.2 for an illustration.

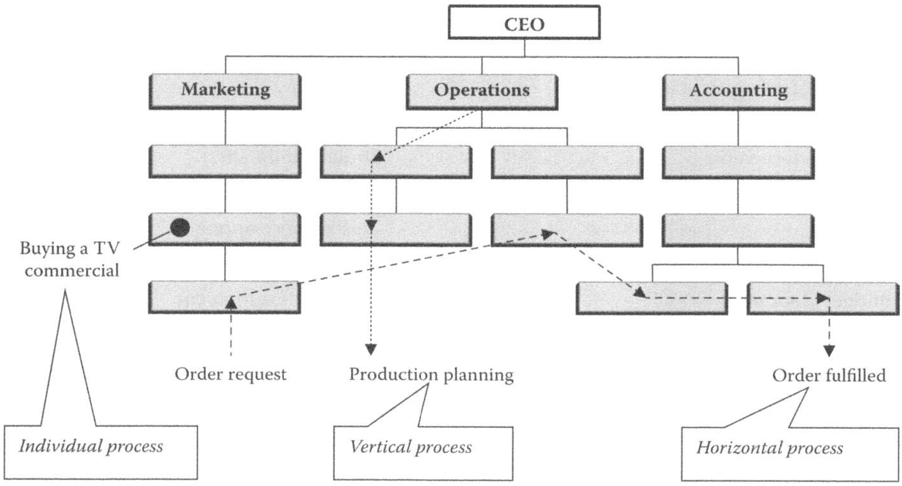

FIGURE 1.2
Illustration of individual, vertical, and horizontal processes.

It follows that a hierarchy exists between the three process types in the sense that a cross-functional process typically can be decomposed into several connected functional processes or subprocesses, which consist of several individual processes. Moving down even further in detail, any process can be broken down into one or more activities that consist of a few tasks. As an illustration, consider the order-fulfillment process in Figure 1.2, which in its entirety is cross functional. However, it consists of functional subprocesses (e.g., in the sales and marketing departments) that receive the order request by phone, process the request, and place a production order with the operations department. The order receiving itself is an activity consisting of the tasks of answering the phone, talking to the customer, and verifying that all necessary information is available to process the order. If we assume that the order-receiving activity is performed by a single employee, this constitutes an example of an individual process.

In terms of process design, cross-functional business processes that are core to the organization and include a significant number of nonmanufacturing-related activities often offer the greatest potential for improvement. Core processes are defined by Cross et al. (1994) as all the functions and the sequence of activities (regardless of where they reside in the organization), policies and procedures, and supporting systems required to meet a marketplace need through a specific strategy. A core process includes all the functions involved in the development, production, and provision of specific products or services to customers. An underlying reason why cross-functional processes often offer high improvement potential compared with functional processes in general, and production/manufacturing processes in particular, is that they are more difficult to coordinate and often suffer from suboptimization.[1] An important reason for this tendency toward suboptimization is the strong departmental interests inherent in the functional[2] organization's management structure.

Roberts (1994) provides some additional explanations for the high improvement potential of cross-functional business processes.

- Improvements in cross-functional business processes have not kept up with improvements in manufacturing processes over the years. In other words, the current margin for improvement is greater in nonmanufacturing-related business processes.
- Waste and inefficiency are more difficult to detect in cross-functional processes than in functional processes due to increased complexity.
- Cross-functional business processes often devote as little as 5 percent or less of the available process time to activities that deliver value to the customers.
- Customers are five times more likely to take their business elsewhere because of poor service-related business processes than because of poor products.

1.1.2 Determinants of the Process Architecture

The process architecture or process structure can be characterized in terms of five main components or elements (see, for example, Anupindi et al., 2012): the inputs and outputs, the flow units, the network of activities and buffers, the resources, and the information structure, some of which we have touched on briefly already. Process design and analysis primarily has to do with understanding the restrictions and opportunities these elements offer in a particular situation. We will deal with this extensively in Chapters 2 through 9.

1.1.2.1 Inputs and Outputs

The first step in understanding and modeling a process is to identify its boundaries, that is, its entry and exit points. When that is done, identifying the input needed from the environment for the process to produce the desired output is usually fairly straightforward. It is important to recognize that inputs and outputs can be tangible (raw materials, cash, and customers) or intangible (information, knowledge, energy, and time). To illustrate, consider a manufacturing process where inputs in terms of raw materials and energy enter the process and are transformed into desired products. Another example would be a transportation process where customers enter a bus station in New York as inputs and exit as outputs in Washington. To perform the desired transformation, the bus will consume gasoline as input and produce pollution as output. Similarly, a barbershop takes hairy customers as input and produces less hairy customers as output. An example in which inputs and outputs are information or data is an accounting process where unstructured financial data enter as input and a well-structured financial statement is the output. To summarize, the inputs and outputs establish the interaction between the process and its environment.

1.1.2.2 Flow Units

A *flow unit* can be defined as a "transient entity[3] that proceeds through various activities and finally exits the process as finished output." This implies that, depending on the context, the flow unit can be a unit of input (e.g., a customer or raw material), a unit of one or several intermediate products or components (e.g., the frame in a bicycle-assembly process), or a unit of output (e.g., a serviced customer or finished product). The characteristics and identity of flow units in a system, as well as the points of entry and departure, can vary significantly from process to process. Typical types of flow units include materials, orders, files or documents, requests, customers, patients, products, paper forms, cash, and transactions. A clear understanding and definition of flow units is important when modeling and designing processes because of their immediate impact on capacity and investment levels.

It is customary to refer to the generic flow unit as a *job*, and we will use this connotation extensively in Chapters 3 through 9. Moreover, an important measure of flow dynamics is the flow rate, which is the number of jobs that flow through the process per unit of time. Typically, flow rates vary over time and from point to point in the process. We will examine these issues in more detail in Chapter 5.

1.1.2.3 Network of Activities and Buffers

From our discussion of process hierarchies, we know that a process is composed of activities. In fact, an accurate way to describe a process would be as a network of activities and buffers through which the flow units or jobs must pass to be transformed from inputs to outputs. Consequently, for in-depth analysis and design of processes, we must identify all relevant activities that define the process and their precedence relationships; that is, the sequence in which the jobs will go through them. To complicate matters, many processes accommodate different types of jobs that have different paths through the network, implying that the precedence relationships typically are different for different types of jobs. As an example, consider an emergency room, where a patient in cardiac arrest clearly has a different path through the emergency room process than a walk-in patient with a sprained ankle. Moreover, most processes also include buffers between activities, allowing storage of jobs between them. Real-world examples of buffers are waiting rooms at a hospital, finished goods inventory, and lines at an airport security checkpoint. A common goal in business process design is to try to reduce the time jobs spend waiting in buffers and thereby achieve a higher flow rate for the overall process.

Activities can be thought of as micro processes consisting of a collection of tasks. Finding an appropriate level of detail to define an activity is

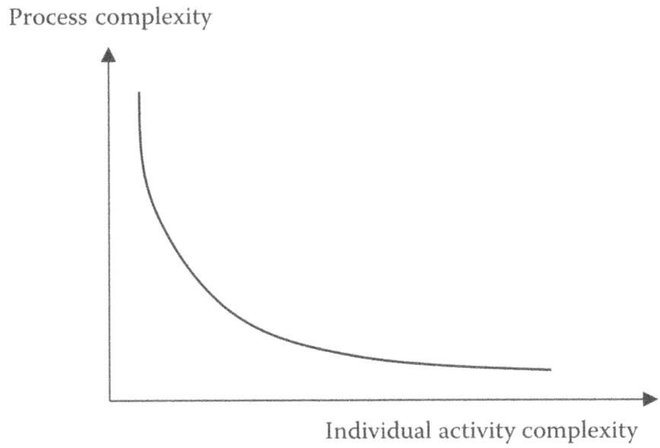

FIGURE 1.3
Process complexity versus activity complexity.

crucial in process analysis. A trade-off exists between activity and process complexity. Increasing the complexity of the individual activities by letting them include more tasks decreases the complexity of the process description (see Figure 1.3). The extreme is when the entire process is defined as one activity, and we are back to the simple "black box" input/output transformation model.

Several classifications of process activities have been offered in the literature. Two basic approaches are depicted in Figure 1.4. Both approaches classify the activities essential for the process to meet the customers' expectations as value-adding activities. They also classify handoffs, delays, and rework as non-value-adding activities. The difference is in the classification of control and policy compliance activities. One approach is based on the belief that although control activities do not directly add value to the customer, they are essential to conducting business. Therefore, these activities should be classified as *business-value-adding* (control) activities. The contrasting view is a classification based on the belief that anything that does not add value to the customer should be classified as a

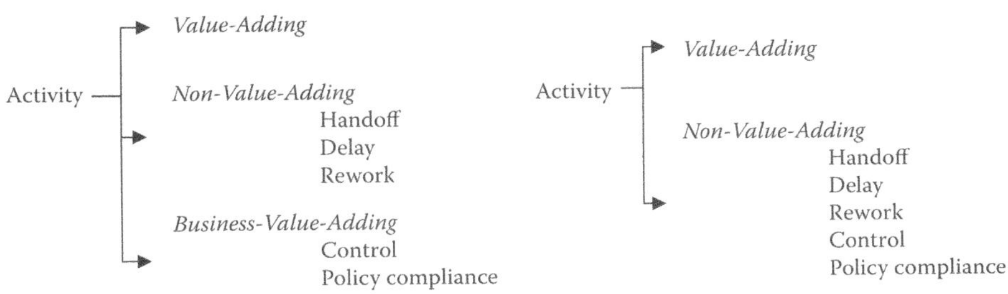

FIGURE 1.4
Classification of activities.

non-value-adding activity. This classification can be determined by asking the following question: Would your customer be willing to pay for this activity? If the answer is no, then the activity does not add value. It is then believed that control activities, such as checking the credit history of a customer in a credit-approval process, would fall into the category of non-value-adding activities because the customer (in this case, the applicant) would not be willing to pay for such an activity.

To understand the value-adding classification, the concept of value must be addressed. Although this is an elusive concept, for our purposes, it is sufficient to recognize that it involves doing the right things in the right ways. Doing the right things means providing the customers with what they want (i.e., being effective). Activities that contribute to transforming the product or service to better conform to the customer's requirements have the potential to add value. For example, installing the hard drive in a new personal computer brings the product closer to being complete and is, therefore, value adding. However, to be truly value adding, the activity also must be carried out efficiently using a minimum of resources; that is, it must be done right. Inefficient activities incur in waste that cannot be eliminated without compromising the process effectiveness. If the hard drive installation was not done properly, resulting in damaged and scrapped hard drives or the need for rework, the installation activity would not be value adding in its true sense. Even though the right activity was performed, it was not done correctly, resulting in more than the minimum amount of resources being consumed. Clearly, this extra resource consumption is something for which the customer would not be willing to pay.

The task of classifying activities should not be taken lightly, because the elimination of non-value-adding activities in a process is one of the cornerstones of designing or redesigning efficient processes. In the business process management (BPM) literature, these two steps of value classification and waste elimination are often jointly referred to as *value-added analysis*; see, for example, Dumas et al. (2013). One of the most common and straightforward strategies for eliminating non-value-adding and control activities is the integration of tasks. Task consolidation typically eliminates wasteful activities because handoffs account for a large percentage of non-value-adding time in many processes. Also, controls (or control activities) are generally in place to make sure that the work performed upstream in the process complies with policies and regulations. Task aggregation tends to eliminate handoffs and controls; therefore, it increases the ratio of value-adding activities to non-value-adding activities (and/or business-value-adding activities).

To illustrate the process of classifying activities, let us consider the famous IBM Credit example introduced by Hammer and Champy (1993). IBM Credit offered financing of computers, software, and services sold by the IBM Corporation. This was an important business because financing customers' purchases is extremely profitable. The process at the time consisted of the following sequence of activities.

1. Field sales personnel called in financing requests to a clerical group of 14 people.
2. The person taking the call logged information on a piece of paper.
3. The paper was taken upstairs to the credit department.
4. A specialist (a) entered the information into the IT system and (b) did a credit check.
5. The results of the credit check (a) were written on a piece of paper and (b) sent to the business practices department.
6. Standard loan contracts were modified to meet customer requirements.
7. The request was (a) sent to a "pricer," where (b) an interest rate was determined.
8. The interest rate (a) was written on a piece of paper and (b) sent to a clerical group.
9. A quote was developed.
10. The quote was sent to field sales via FedEx.

A possible classification of these activities into the categories of value adding, non-value adding, and business value adding is given in Table 1.2.

Note that only a small fraction of the activities are considered to add value to the customer (i.e., the sales field agent). Arguably, the only activity that adds value is Activity 9—to develop a quote; however, we also have included Activity 1, because the customer triggers the process with this activity, and Activity 6, because the customization of the contract is forced by the peculiarities of each specific request. Handoffs account for a considerable percentage of the non-value-adding activities (3, 5(b), 8(b), and 10). This description does not include all the delays that are likely to occur between handoffs because, in practice, agents in this process did not start working on a request immediately upon receipt. In fact, although the actual work took in total only 90 minutes to perform, the entire process took 6 days on average and sometimes took as long as 2 weeks. In Section 3.6.2 of Chapter 3, we will discuss how the process was redesigned to achieve average turnaround times of 4 hours.

TABLE 1.2

Activity Classifications for IBM Credit

Category	Activities
Value adding	1, 6, and 9
Nonvalue adding	2, 3, 4(a), 5, 7(a), 8, and 10
Business value adding	4(b) and 7(b)

1.1.2.4 Resources

Resources are tangible assets that are necessary to perform activities within a process. Examples of resources include machinery in a job shop, aircraft in an airline, and faculty at an educational institution. As these examples imply, resources often are divided into two categories: capital assets (e.g., real estate, machinery, equipment, and computer systems) and labor (i.e., the organization's employees and the knowledge they possess). As opposed to inputs, which flow through the process and leave, resources are used rather than consumed. For example, an airline uses resources such as aircraft and personnel to perform the transportation process day after day. At the same time, jet fuel, which is an input to the process, is not only used but also consumed.

1.1.2.5 Information Structure

The last element needed to describe the process architecture is the information structure. The information structure specifies which information is required and which is available to make the decisions necessary for performing the activities in a process. For example, consider the inventory replenishment process at a central warehouse; the information structure specifies the information needed to implement important operational policies, such as placing an order when the stock level falls below a specified reorder point.

Based on our definition of a business and the exploration of process hierarchies and architecture, a more comprehensive and adequate definition of a business process emerges.

A *business process* is a network of connected activities and buffers with well-defined boundaries and precedence relationships, which use resources to transform inputs into outputs for the purpose of satisfying customer requirements (see Figure 1.5).

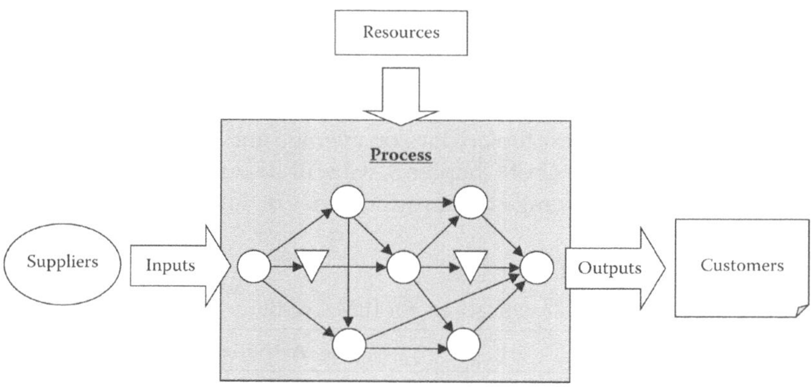

FIGURE 1.5
A process defined as a network of activities and buffers with well-defined boundaries transforming inputs to customer-required outputs using a collection of resources.

1.1.3 Workflow Management Systems

The term *workflow* is often used in connection with process management and process analysis, particularly in the field of information systems. It refers to how a job flows through the process, the activities that are performed in it, the people involved, and the information needed for its completion. The workflow is, therefore, defined by the flow units along with the network of activities and buffers, the resources, and the aforementioned information structure. Management of administrative processes is often referred to as *workflow management*, and information systems that support workflow management are called *workflow management systems*. For our purposes, we can think of workflow management as synonymous with process management.

Document processing has a fundamental place in workflow management, as business documents are the common medium for information processes (such as data flow analysis, database storage and retrieval, transaction processing, and network communications) and business processes. Workflow management controls the actions taken on documents moving through a business process. Specifically, workflow management software is used to determine and control who can access which document, what operations employees can perform on a given document, and the sequence of operations that are performed on documents by the workers in the process.

Workers have access to documents (e.g., purchase orders and travel authorizations) in a workflow management system. These individuals perform operations, such as filling and modifying fields, on the documents. For example, an employee who is planning to take a business trip typically must fill out a travel authorization form to provide information such as destination, dates of travel, and tentative budget. The document (electronic or paper) might go to a travel office where additional information is added. Then, the form probably goes to a manager, who might approve it as is, ask for additional information, or approve a modified version (for instance, by reducing the proposed travel budget). Finally, the form goes back to the individual who requested the travel funds. This simple example illustrates the control activities that are performed before the request to travel is granted. The activities are performed in sequence to guarantee that the approved budget is not changed at a later stage. Although the sequence is well established and ensures a desired level of control, workflow management software should be capable of handling exceptions (e.g., when the manager determines that it is in the best interest of the company to bypass the normal workflow sequence), reworking loops (e.g., when more information is requested from the originator of the request), and obtaining permissions to modify and update the request before the manager makes a final decision. Workflow management software developers are constantly incorporating additional flexibility into their products so that buyers can deal with the increasing complexity of modern business processes without the need for additional computer programming.

With its expanded scope and flexibility, modern workflow management software is often referred to as BPM software systems.

1.2 The Essence of Business Process Design

The essence of business process design can be described as how to do things in a good way. *Good* in this context refers to process efficiency and process effectiveness. The last statement is important; process design is about satisfying customer requirements in an efficient way. An efficient process that does not deliver customer value is useless. A well-designed process does the right things in the right way.

From a more formal perspective, business process design is concerned with configuring the process architecture (i.e., the inputs and outputs, the flow units, the network of activities and buffers, the resources, and the information structure; see Figure 1.5) so that customer requirements are satisfied in an efficient way. It should be noted that process customers can be internal or external to the organization. However, it is important that the requirements of the internal customers are aligned with the overall business goals, which ultimately means satisfying the desires of the external customers targeted in the business strategy.

Although business process design concerns all types of processes, we have already indicated that for reasons of coordination and suboptimization, it is often most valuable when dealing with complex horizontal processes that cut across the functional or departmental lines of the organization and reach all the way to the end customer. An example is the order-fulfillment process illustrated in Figure 1.2.

The roots of the functional organization date back to the late 1700s, when Adam Smith proposed the division of labor concept in *An Inquiry into the Nature and Causes of the Wealth of Nations* (1776). Referring to the 17 operations required to produce a pin, Smith argued that assigning one task to each of 17 workers would be more efficient and would produce more pins than having 17 workers each autonomously perform all 17 tasks. According to Smith, dividing work into discrete tasks provides the following benefits.

- Each worker's skill increases by repeatedly performing the same task.
- No time is lost due to switching workers from one task to another.
- Workers are well positioned to develop improved tools and techniques as a result of focusing on a single task.

What Smith's model does not address is that as products and services become more complex, and customers require more varieties of them, the need for many more activities and the coordination of the resulting basic tasks comprising these activities becomes extremely difficult. The division of labor has the goal of creating highly specialized workers who can complete basic tasks efficiently. Even if the workers become highly efficient at their tasks, the process as a whole can be inefficient. Inefficiencies in a process design with the division of labor paradigm are primarily related to the need for handing off work from one station to the next and for coordinating the flow of jobs in the process. Perhaps the most significant drawbacks associated with handing off work are the delays and errors introduced when a job is passed from one worker (or workstation) to another.

Consider, for example, the traditional approach to designing new products illustrated by Shafer and Meredith (1998). The process typically begins when the marketing department collects information about customers' needs and desires. This information is then relayed to a research and development (R&D) department, which is responsible for designing the product to meet the customers' requirements. After the design is completed, it is up to the manufacturing department to produce the product exactly as specified in the design. After the product is produced, it is the responsibility of the sales department to sell it. Finally, after a consumer purchases the product, the customer service department must provide after-sales services such as help with installation and warranty repairs. This process, although involving several departments, appears to have a logical structure, but in reality, it is not uncommon that the engineers in R&D design a feature that manufacturing cannot produce or one that can be produced only at a high cost. In this sequential approach (commonly known as *over-the-wall design*), the production problems are not discovered until after the design is finalized and handed off to manufacturing. Upon discovering the problem, manufacturing has to send the design back to R&D for modifications. Clearly, each time a design goes back and forth between R&D and manufacturing, it involves yet another handoff and increases the delay in introducing the new product.

This example illustrates an essential issue often addressed in process design—namely, that completing a set of activities sequentially, one at a time, tends to increase the time required to complete the entire process. In addition, significant delays can occur in the discovery of important information that would have influenced the quality of the output from activities already completed; this tends to extend the completion time even further. The solution is to try to do as many activities as possible in parallel. Sophisticated product development teams use the concept of concurrent engineering[4] to avoid the "over-the-wall design" problem and shorten the process time.

As discussed earlier, the road to a good process design starts with a profound understanding of what the customers want and thereby what the process is supposed to deliver. Although understanding customer preferences and purchasing behavior is a field of study in its own right, for process design purposes, four dimensions of customer requirements are particularly important: cost, quality, response time, and variety. Arriving at a good process design also requires a thorough understanding of the current process (if one exists) and any new enablers that change the rules of the game, such as information technology (IT) developments (today often referred to as digitalization), breakthrough technologies, or changes in legislation. A first approach, particularly in redesign situations, is to eliminate waste, waiting times, and non-value-adding activities to speed up the process. However, it is important not to get stuck in the old ways of doing things but to leverage the power of process design by challenging rooted perceptions of how to do things.

Throughout the book, we will use the terms *design* and *redesign* interchangeably because, from an analytical point of view, no major difference can be distinguished between designing a new process and redesigning an existing one. This statement hinges on the fact that in this book, we make a clear distinction between designing a process and implementing the design. The latter is a tremendously challenging task that requires good skills in change management and in knowing how to motivate and inspire an organization to change. Ultimately, success will depend on the buy-in from people at all levels of the organization. In implementing a new design, a big difference is evident between making changes to an existing process and establishing a process in a "greenfield organization" (or an organization that does not have any processes in place), the former usually being the more difficult situation to handle. Because the focus of this book is on the analytical modeling, analysis, and design of processes, from our perspective, the analytical approach for design and redesign is the same. We will touch briefly on the important issues of process and change management in Chapter 2. However, for a more in-depth treatment of this important area, we refer to books solely devoted to the subject of how to achieve and sustain organizational changes; these can be found in the management and human resource literature. For a comprehensive treatment of the specific field of BPM, we refer to the book by Dumas et al. (2013).

1.2.1 Incremental Process Improvement and Process Design

When exploring the essence of business process design, an important aspect to consider is the relationship between incremental improvement and design changes. Clearly, both are about how to do things in a good way, and most of the tools for analyzing processes that are covered in this book are just as useful in an incremental improvement project as in a design project. However, a subtle but important difference can be found in the objectives and the degrees of freedom available to a process design team compared with an incremental improvement team. The scope for the former usually far exceeds that of the latter.

The difference between incremental process improvement and process design parallels the difference between systems improvement and systems design that John P. Van Gigch (1978) so eloquently described many years ago.

Van Gigch's book championed the notion of a systems approach to problem-solving, a field from which process design and reengineering[5] have borrowed many concepts. In the following, we adapt Van Gigch's descriptions to the context of process design, acknowledging that if a system is defined as an assembly or set of related elements, a process is nothing else but an instance of a system.

Many of the problems arising in the practice of designing business processes stem from the inability of managers, planners, analysts, administrators, and the like to differentiate between incremental process improvement and process redesign. Incremental improvement refers to the transformation or change that brings a process closer to standard or normal operating conditions. The concept of incremental improvement carries the connotation that the design of the process is set and that norms for its operation have been established. In this context, improving a process refers to tracing the causes of departures from established operating norms or investigating how the process can be made to yield better results—results that come closer to meeting the original design objectives. Because the design concept is not questioned, the main problems to be solved are as follows:

1. The process does not meet its established goals.
2. The process does not yield predicted results.
3. The process does not operate as initially intended.

The incremental improvement process starts with a problem definition. Through analysis, one then searches for elements of components or subprocesses that might provide possible answers to the questions posed. Finally, deduction is used to draw certain tentative conclusions.

Design also involves transformation and change. However, design is a creative process that questions the assumptions on which old forms have been built. It demands a completely new outlook to generate innovative solutions with the capability of increasing process performance significantly. Process design is by nature inductive. (See Chapter 3 for a further discussion of inductive versus deductive thinking.)

1.2.2 An Illustrative Example

To illustrate the power of process design in improving the efficiency and effectiveness of business processes, we consider an example from the insurance industry.[6]

Sensing competitive pressure coupled with the need to be more responsive to its customers, a large insurance company decided to overhaul its process for handling claims regarding replacement of automobile glass. The chief executive officer (CEO) thought that if her plan was successful, she would be able to use any expertise acquired from this relatively low-risk endeavor as a springboard for undertaking even more ambitious process design projects later.

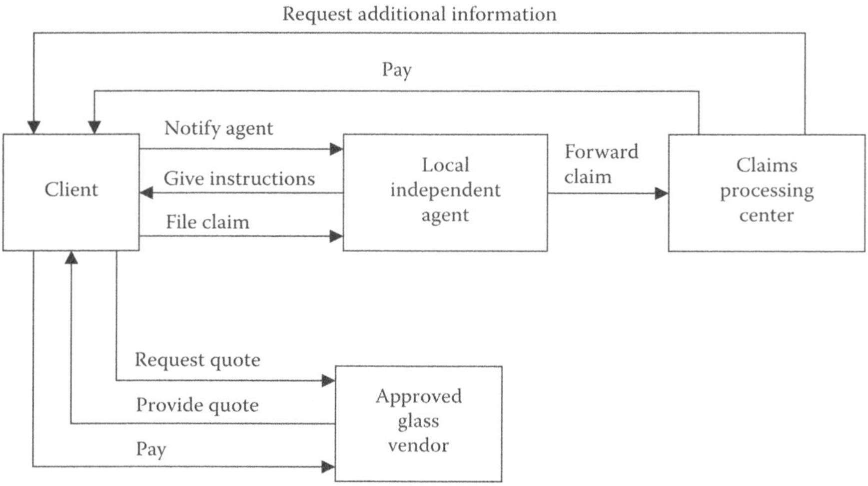

FIGURE 1.6
Original claims-handling process.

As a first step, the CEO appointed an executive sponsor to shepherd the project. Following some preliminary analysis of the potential payoff and impact, the CEO and the executive sponsor developed a charter for a process design team and collaborated to handpick its members.

Early on, the team created a chart of the existing process to aid in their understanding of things as they were. This chart is depicted in simplified form in Figure 1.6. (In Chapter 4, we will introduce methods and tools for creating several charts for process analysis.) This figure summarizes the following sequence of events for processing claims under the existing configuration.

1. The client notifies a local independent agent that she wishes to file a claim for damaged glass. The client is given a claims form and is told to obtain a replacement estimate from a local glass vendor.
2. After the client has obtained the estimate and completed the claims form, the independent agent verifies the accuracy of the information and then forwards the claim to one of the regional processing centers.
3. The processing center receives the claim and logs its date and time of arrival. A clerk registers the contents of the claim into the company's IT system (mainly for archiving purposes). Then, the form is placed in a hard copy file and routed, along with the estimate, to a claims representative.
4. If the representative is satisfied with the claim, it is passed along to several others in the processing chain, and a check eventually is

issued to the client. However, if any problems are associated with the claim, the representative attaches a form and mails a letter back to the client for the necessary corrections.

5. On receiving the check, the client can go to the local glass vendor and have the glass replaced.

In this original process, a client usually had to wait 1–2 weeks from the filing of a claim until he or she could proceed and get the automobile glass replaced.

Given the goal of coming up with a new process design with a much shorter waiting time for the clients, the team recommended the solution depicted in Figure 1.7. The team accomplished this after evaluating several process configurations. The evaluation was done by weighing the costs of each configuration against the benefits.

Structural as well as procedural changes were put into place. Some of these, especially the procedural changes, are not entirely evident simply by contrasting Figures 1.6 and 1.7. The changes in procedures involved the following.

- The claims representative was given final authority to approve the claim.
- A long-term relationship was established with a select number of glass vendors, enabling the insurance company to leverage its purchasing power and to pay the vendor directly. Furthermore, thanks to pre-negotiated prices, it is not necessary to obtain price estimates from the vendor. (Note that this procedural change is similar to the well-established total quality management [TQM] notion of *vendor certification*.)
- Rather than going through a local agent, obtaining an estimate, and filling out a form, the client now simply contacts the processing center directly by phone, or the company's web portal, to register a claim.

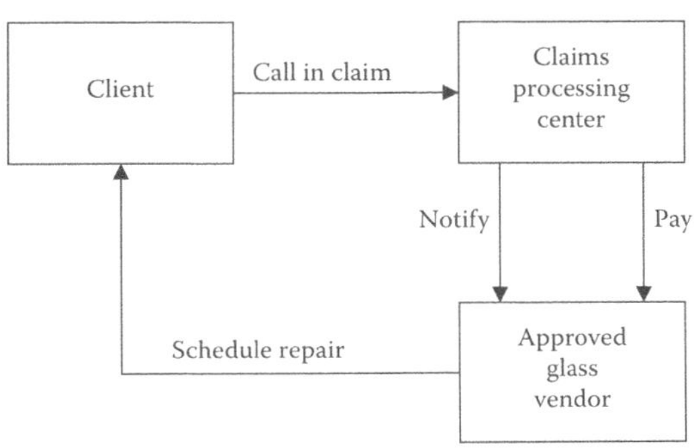

FIGURE 1.7
New improved design of the claims-handling process.

The structural changes are manifested in the following sequence of events, which describe the flow of the process.

1. Using a newly installed, 24-hour hot line, the client speaks directly with a claims representative at one of the company's regional processing centers when filing the claim. Alternatively, the client may register the claim through the company's new web portal.
2. The claims representative gathers the pertinent information over the phone, enters the data into the electronic system, and resolves any problems related to the claim on the spot. The representative then tells the client to expect a call from a certain glass vendor, who will make arrangements to repair the glass at the client's convenience.
3. Because the claim exists as an electronic file that is shared by all authorized users of the system, the accounting department can immediately begin issuing a payment to the local glass vendor.

A number of significant benefits—some more quantifiable than others—resulted from changing the process design.

- In the new process (without the web portal option), the client gets the glass repaired in as little as 24 hours, versus 10 days in the original process. This represents a 90 percent improvement in the process cycle time (i.e., the elapsed time from when the claim is called in until the glass is replaced).
- The client has less work to do because a single phone call sets the process in motion. Also, the client is no longer required to obtain a repair estimate.
- Problems are handled when the claim is made, thus preventing further delays in the process.
- Problems with lost or mishandled claims virtually disappear.
- The claim now passes through fewer hands, resulting in lower costs. By establishing a long-term relationship with a select number of glass vendors, the company was able to leverage its purchasing power to obtain a 30–40 percent savings on a paid claim.
- Because fewer glass vendors are involved, a consolidated monthly payment can be made to each approved vendor, resulting in additional savings in handling costs.
- By dealing with preapproved glass vendors, the company can be assured of consistent and reliable service.
- The claims representatives feel a greater sense of ownership in the process because they have broader responsibilities and expanded approval authority.

The new process was designed by framing the question to the project team as: How do we settle claims in a manner that will cause the least impact on the client while minimizing processing costs? If, instead, the CEO and top management had asked: How can we streamline the existing process to make it more efficient? their inquiry might not have removed the independent agent from the process. The fundamental difference between these two questions is that the first one is likely to lead to a change in the process design, because it goes to the core issue of how to best satisfy the customer requirements, and the second one is likely to lead to the introduction of technology without process change (i.e., automation without redesign). Could further efficiencies still be realized in this process? Without a doubt! This is where an incremental and continuous improvement approach should take over.

1.3 Business Process Design, Overall Business Performance, and Strategy

To put process design into a larger context, it is interesting to consider its links to overall business performance and strategy.

1.3.1 Business Process Design and Overall Business Performance

What is overall business performance, and how is it measured? A detailed answer to this question is to a large extent company specific. However, in general terms, we can assert that the performance of a business should be measured against its stated goals and objectives.

For a profit-maximizing firm, the overarching objective is usually to maximize long-term profits or shareholder value. In simple terms, this is achieved by consistently, over time, maximizing revenues and minimizing costs, which implies satisfying customer requirements in an efficient way. For nonprofit organizations, it is more difficult to articulate a common goal. However, on a basic level, presumably one objective is to survive and grow while providing customers with the best possible service or products. Consider, for example, a relief organization with the mission of providing shelter and food for people in need. To fulfill this mission, the organization must first of all survive, which means that it must maintain a balanced cash flow by raising funds and keeping operational costs at a minimum. Still, to effectively help as many people as possible in the best possible way requires more than just capital and labor resources; it requires a clear understanding of exactly what kind of aid their "customers" need the most. Furthermore, the ability to raise funds in the long run most likely will be related directly to how effectively the organization meets its goals; that is, how well it performs.

To summarize, a fundamental component in overall business performance regardless of the explicitly stated objectives is to satisfy customer requirements in an efficient way. In the long term, this requires well-designed business processes.

1.3.2 Business Process Design and Strategy

Strategy guides businesses toward their stated objectives and overall performance excellence. In general terms, *strategy* can be defined as "the unifying theme that aligns all individual decisions made within an organization."

In principle, profit-maximizing firms address their fundamental objective of earning a return on their investment that exceeds the cost of capital in two ways. Either the firm establishes itself in an industry with above-average returns or it leverages its competitive advantage over the other firms within an industry to earn a return that exceeds the industry average. These two approaches define two strategy levels that are distinguishable in most large enterprises. The corporate strategy answers the question: Which industries should we be in? The business strategy answers the question: How should we compete within a given industry?

Although both strategies are of utmost importance, the intensified competition in today's global economy requires that no matter what industry one operates in, it is necessary to be highly competitive. Consequently, a prerequisite for success is an effective business strategy. In fact, nonprofit organizations need to be competitive as well. For example, the relief organization mentioned previously most likely has to compete against other organizations of the same type for funding. Similarly, a university competes against other universities for students, faculty, and funding.

Developing a business strategy is therefore an important undertaking in any organization. Many different approaches deal with this issue in a structured, step-by-step fashion. However, developing a sound business strategy requires a profound understanding of the organization's external and internal environment combined with a set of clearly defined, long-term goals. The external environment includes all external factors (e.g., social, political, economic, and technological) that influence the organization's decisions. Still, in most instances, the core external environment consists of the firm's industry or market, defined by its relationship with customers, suppliers, and competitors. The internal environment refers to the organization itself, its goals and values, its resources, its organizational structure, and its systems. This means that understanding the internal environment to a large extent has to do with understanding the hierarchy and structure of all business processes.

A clear understanding of the internal environment promotes an understanding of the organization's internal *strengths* and *weaknesses*. Similarly, in-depth knowledge of the external environment provides insights into the *opportunities* and *threats* the organization is facing. Ideally, the business strategy is designed to take advantage of the opportunities by leveraging the internal strengths while avoiding the threats and protecting its weaknesses. This basic approach is referred to as a *SWOT* analysis.

The link between business process design and the business strategy is obvious when it comes to the internal environment; the internal strengths and weaknesses are mostly embodied in the form of well-designed and poorly designed business processes. By carefully redesigning processes, weaknesses can be turned into strengths, and strengths can be further reinforced to ensure a competitive advantage. The link to the external environment might be less obvious, but remember that a prerequisite for achieving an effective process design is to understand the customer requirements. Furthermore, for it to be an efficient design, the process input must be considered, which implies an understanding of what the suppliers can offer. For the business strategy to be successful, the organization must be able to satisfy the requirements of the targeted customers in an efficient, competitive way.

Our discussion so far can be synthesized through the concept of *strategic fit*. Strategic fit refers to a match between the strategic (or competitive) position the organization wants to achieve in the external marketplace and the internal capabilities necessary to take it there. Another way of looking at this is to realize that strategic fit can be attained by a market-driven strategy, a process-driven strategy, or most commonly, a combination of the two. In a market-driven strategy, the starting point is the key competitive priority, meaning the strategic position the firm wants to reach. The organization then must design and implement the necessary processes to support this position. In a process-driven strategy, the starting point is the available set of process capabilities. The organization then identifies a strategic market position that is supported by these processes. This perspective illustrates that business process design is an important tool for linking the organization's internal capabilities to the external environment so that the preferred business strategy can be realized.

1.4 Why Do Inefficient and Ineffective Business Processes Exist?

Throughout most of this chapter, we have been praising business process design and advocating the importance of well-designed processes. So, if businesses acknowledge this, then why were inefficient and ineffective (i.e., broken) processes designed in the first place? Hammer (1990), one of the founding fathers of the reengineering movement, provided an insightful answer: Most of the broken processes and procedures seen in a business are not designed at all—they just happened. The following examples of this phenomenon are adapted from Hammer's article.

Consider a company whose founder one day recognizes that she doesn't have time to handle a chore, so she delegates it to Smith. Smith improvises. Time passes, the business grows, and Smith hires an entire staff to help him cope with the work volume. Most likely, they all improvise. Each day brings new challenges and special cases, and the staff adjusts its work accordingly. The potpourri of special cases and quick fixes is passed from one generation of workers to the next. This example illustrates what commonly occurs in many organizations; the ad hoc has been institutionalized, and the temporary has been made permanent.

In another company, a new employee inquires: "Why do we send foreign accounts to the corner desk?" No one knew the answer until a company veteran explained that 20 years ago, an employee named Mary spoke French, and Mary sat at the corner desk. Today, Mary is long gone, and the company does not even do business in French-speaking countries, but foreign accounts are still sent to the corner desk.

An electronics company spends $10 million per year to manage a field inventory worth $20 million. Why? Once upon a time, the inventory was worth $200 million, and the cost of managing it was $5 million. Since then, warehousing costs have escalated, components have become less expensive, and better forecasting techniques have minimized the number of units in inventory. The only thing that remains unchanged is the inventory management system.

Finally, consider the purchasing process of a company where initially the only employees were the founders of the company. Because the cofounders trusted each other's judgment with respect to purchases, the process consisted of placing an order, waiting for the order to arrive, receiving the order, and verifying its contents (see "Original Purchasing Process" in Figure 1.8). As the company grows, the cofounders believe that they cannot afford to have the same level of trust with other employees in the company. Control mechanisms are installed to guarantee the legitimacy of the purchase orders. The process expands, but the number of value-adding activities (i.e., Ordering and Receiving) remains the same. (See "Evolved Purchasing Process" in Figure 1.8.)

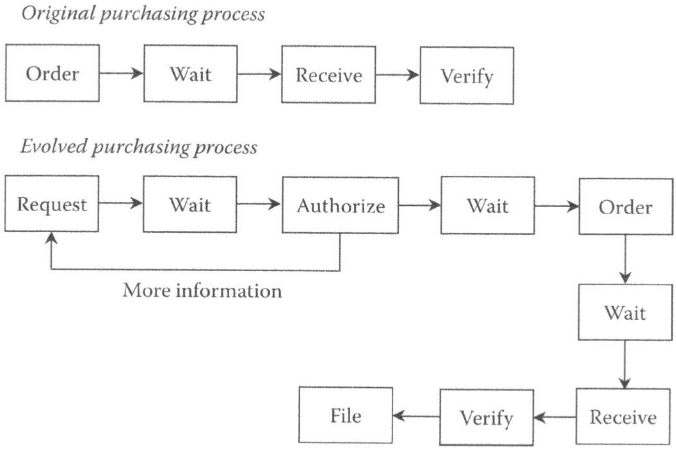

FIGURE 1.8
Process evolution.

Hammer introduced the notion of "information poverty" to explain the need for activities that make existing processes inefficient. Many business processes originated before the advent of modern computer and telecommunications technology and therefore contain mechanisms designed to compensate for information poverty. Although companies today are information affluent, old mechanisms are often embedded in their automated IT systems. There was simply not enough time or resources to carefully analyze how the new technology could enable completely new process designs when new IT systems were implemented. Large inventories are an example of a mechanism to deal with information poverty, which in this case means lack of knowledge with respect to demand. IT today enables most operations to obtain point-of-sales data in real time, which can be used to drastically reduce the need for inventories in many settings. However, leveraging its use, particularly across organizational boundaries in a supply chain, is still a challenge for many companies. Of course, over time, competition will force process changes in response to new technology and access to better information. A company that does not keep up will eventually go out of business. The question is how long it will take for process changes to occur and the extent to which the new process designs will fully leverage the potential of the new technology.

As opposed to information poverty, "information overload" may also lead to inefficient and ineffective business processes. In recent years, the amount and range of data available to organizations have exploded, and the development continues to accelerate. Clearly, this vast amount of data offers tremendous opportunities for those who can turn this into useful information. However, there is also a risk of drowning in data and information that does not add much value. Such information overload may hinder rather than enable efficient process designs. To avoid information overload, and to leverage the power harnessed in Big Data, using appropriate techniques for data analysis is key; hence, the growing interest of industry in data analytics, data science, and artificial intelligence. An example of how such new technologies can enable new process designs is the use of *demand sensing* at Amazon. By using new digital technologies and increased computing power to analyze new data sources (internal as well as external), the company can more accurately predict (or sense) ahead of time where, when, and in what quantity potential consumers will demand different products. With this information, the company can redesign its replenishment processes and proactively move inventories to be in the right location when the demand occurs. As a result, customers can get their products faster with less inventory, and fewer unsold products have to be returned to the suppliers.

To summarize, inefficient or ineffective processes can be the result of an organization's inability to take advantage of changes in the external and internal environments that have produced new design enablers such as IT.

Another reason for the existence of inefficient process structures is the local adjustments made to deal with changes in the external and internal environments. When these incremental changes accumulate over time, they create inconsistent structures. Often, adjustments are made to cope with new situations, but seldom does anyone question whether an established procedure is necessary.

In other words, inefficient and ineffective processes usually are not designed; they emerge as a consequence of uncoordinated incremental change or the inability to take advantage of new design enablers.

1.5 Summary

Business processes describe how things are done in a business and encompass all activities taking place in an organization, including manufacturing processes as well as service and administrative processes. A more precise definition, useful for purposes of modeling and analysis, is that a business process is "a network of connected activities with well-defined boundaries and precedence relationships that uses resources to transform inputs to outputs with the purpose of satisfying customer requirements." Depending on the organizational scope, a business process can be categorized hierarchically as cross functional, functional, or individual; typically, cross-functional processes offer the greatest potential for improvement.

The essence of business process design is to determine how to do the right things in the right ways. More formally, business process design can be described as a configuration of the process architecture (i.e., the inputs and outputs, the flow units, the network of activities and buffers, the resources, and the information structure) so as to satisfy external and internal customer requirements in an efficient way. In this context, it is important that the requirements of the internal customers are aligned with the overall business goals, which ultimately boils down to satisfying the desires of the external customers targeted in the business strategy.

A linkage between overall business performance and business process design can be made through the fundamental need for any business, profit maximizing or not, to satisfy and attract customers while maintaining an efficient use of resources. For a nonprofit organization, this is a necessity for survival because it enables the company to continue fulfilling its purpose. For profit-maximizing organizations, the ability to reach the overarching objective of generating long-term profits and returns above the cost of capital is contingent on the firm's long-term ability to maximize revenues and minimize costs. In the long run, this requires well-designed business processes. Similarly, in developing a business strategy, business process design is an important vehicle for linking internal capabilities with external opportunities. This allows the company to achieve a strategic fit between the firm's desired competitive position in the external market and the internal capabilities necessary to reach and sustain this position.

Finally, we discussed the fact that the emergence of inefficient and ineffective processes is inherent in most organizations simply because they must adjust to changes in their internal and external environments. These incremental adjustments, which usually make good sense when they are introduced, at least locally, tend to accumulate with time and create inefficient structures. On the other hand, another common reason for inefficiencies is the inability to take advantage of new design enablers such as IT, and automating old work processes instead of using the new technology to do things in new ways. This is why we have emphasized that inefficient and ineffective processes usually are not designed; they emerge as a consequence of uncoordinated incremental change or inability to take advantage of new design enablers.

Discussion Questions and Exercises

1. Describe a process you are familiar with; for example, getting ready for work or school in the morning. Who is the customer? Identify the components in the process hierarchy and classify the activities as value adding or nonvalue adding. Are there opportunities for improvements?
2. *Requisition Process* (adapted from Harbour, 1994): A large company is having trouble processing requisition forms for supplies and materials. Just getting through the initial approval process seems to take forever. Then, the order must be placed, and the supplies and materials must be received and delivered to the correct location. These delays usually cause only minor inconveniences. However, a lack of supplies and materials sometimes stops an entire operation. After one such instance, a senior manager has had enough. The manager wants to know the reason for the excessive delays. The manager also wants to know where to place blame for the problem! A process analysis is done at the manager's request. The requisition process is broken down into three subprocesses: (1) requisition form completion

TABLE 1.3

Steps in the Requisition Process

Step	Time (minutes)
1. Requisition form initiated	10
2. Form mailed to procurement	720
3. Form sits in the IN basket	75
4. Requisition form completed	18
5. Form sits in the OUT basket	75
6. Form mailed to authorization	720
7. Form sits in the IN basket	45
8. Form reviewed and authorized	12
9. Form sits in the OUT basket	90
10. Form mailed to ordering	720

and authorization, (2) ordering, and (3) receiving and delivery. The process analysis also reveals that the first subprocess consists of the steps shown in Table 1.3.

a. Classify the activities in this process.
b. Who is the customer?
c. Translate what you think the customer wants into measures of process performance.
d. How is the process performing under the measures defined in part c?
e. Comment on the manager's attitude.

3. *Receiving and Delivery Process* (adapted from Harbour, 1994): The company in Exercise 1.2 has a large manufacturing complex that is spread over a large area. You are assigned to study the receiving-and-delivery subprocess to make recommendations to the senior manager for streamlining the process. The subprocess begins with receiving goods on the warehouse loading dock, and it ends with the goods being delivered to the correct location. Before you were assigned to this project, the senior manager had a student intern do a simple process analysis. The manager instructed the student to observe the process and describe the various steps. Also, the student was told to find average processing times for each step in the process. The information collected by the student is summarized in Table 1.4.

   a. Classify the activities in this process.
   b. Who is the customer?
   c. Translate what you think the customer wants into measures of process performance.
   d. How is the process performing under the measures defined in part (c)?
   e. Comment on the instructions given to the intern by the senior manager in reference to the process analysis.

TABLE 1.4

Steps in the Receiving and Delivery Process

No.	Step	Time (minutes)
1	Received goods temporarily sit on the loading dock	120
2	Goods are visually inspected for damage	3
3	Goods are carried to the warehouse	10
4	Goods are stored in the warehouse	1440
5	Goods are removed from the warehouse and carried to the loading dock	10
6	Goods sit on the loading dock awaiting loading	60
7	Goods are carried to a truck	10
8	Goods are trucked to a satellite storage area	20
9	Goods are carried from the truck to a satellite storage area	10
10	Goods are stored in the satellite storage area	320
11	Goods are inspected for damage	2
12	Goods are carried to a truck	10
13	Goods are trucked to the required location	15
14	Goods are carried to the required location	10

4. *Hospital Administrative Processes* (adapted from Harbour, 1994): A director of administration for a large hospital complex receives some disturbing news. A recent auditor's report states that 28 percent of all hospital costs are related to administrative costs. The director is determined to lower this figure. She has read some magazine and newspaper articles about business process design and has decided to try it to see if it works. She calls a special off-site meeting. The meeting is held at a luxury hotel, and only senior-level managers are invited. At the meeting, she presents her concerns. She then states that the purpose of the meeting is to redesign the various administrative processes. A brainstorming session is conducted to identify potential problems. The problems are then prioritized. The meeting breaks for lunch. After lunch, everyone works on developing some solutions. A number of high-level process maps are taped to the wall, and the director discusses each of the identified problems. One suggested solution is reorganization. Everyone loves the idea. Instead of 12 major divisions, it is suggested to reorganize into 10. After the meeting is over, the director spends 4 hours hammering out the details of the reorganization. She returns to work the next day and announces the reorganization plan. Sitting in her office, she reflects on her first process redesigning efforts. She is pleased.
   a. How would you rate the director's redesign project? Would you give her a pay raise?
   b. How would you conduct this process redesigning effort?

5. *Environmental Computer Models* (adapted from Harbour, 1994): An environmental company specializes in developing computer models. The models display the direction of groundwater flow, and they are used to predict the movement of pollutants. The company's major customers are state and federal agencies. The development of the computer models is a fairly complex process. First, numerous water wells are drilled in an area. Then, probes are lowered into each well at various depths. From instruments attached to the probes, a number of recordings are made. Field technicians, who record the data on paper forms, do this. The data consist mostly of long lists of numbers. Back in the office, a data entry clerk enters the numbers into a computer. The clerk typically enters hundreds of numbers at a time. The entered data are then used to develop the computer models.

   Recently, the company has experienced numerous problems with data quality. Numerous data entry errors have resulted in the generation of inaccurate models. When this happens, someone must carefully review the entered data. When incorrect numbers are identified, they are reentered, and another computer model is generated. This rework process often must be repeated more than once. Because it takes a long time to generate the models, such errors add considerable cost and time to a project. However, these additional costs cannot be passed on to the customer. Prices for the computer models are based on fixed bids, so the company must pay all rework costs. Alarmingly, rework costs have skyrocketed. On the last two jobs, such additional costs eliminated almost all profits. The company has decided to fix the problem. The company's first step has been to hire an expert in data quality.

   The consultant makes a series of random inspections. First, the consultant checks the original numbers recorded by the field technicians. They all seem correct. Next, the consultant examines the data entered by the data entry clerks. Numerous transposition errors are identified. Some of the errors are huge. For example, one error changed 1,927 to 9,127; another changed 1,898 to 8,198. The consultant proposes some process changes, including adding an inspection step after the computer data entry step. The consultant suggests that someone other than the data entry clerk should perform this inspection. Because each model has hundreds of numbers, the additional inspection step will take some time. However, the consultant can think of no other way to prevent the data entry errors. The proposed process is outlined in Table 1.5.

   The consultant was paid and left for what he described as a "well-deserved vacation" in Hawaii. The company currently is hiring people to staff the third step in the process.

   a. What measures should be used to assess the performance of this process?
   b. How would the consultant's proposed process perform according to the measures defined in part (a)?

TABLE 1.5

Proposed Process for Environmental Computer Models

Step	Description
1	Field technicians record data on paper forms
2	Data entry clerks then enter the data into a computer
3	A printout of the database is made, and a second group of employees cross-check the data and the original forms
4	Any necessary corrections are made
5	The computer model is generated

c. Do you agree with the consultant's methodology and improvement strategies?

d. Would you propose a different process?

6. Some time ago, Toronto Pearson Airport implemented a new process to connect international flights (e.g., those arriving from Europe) and the United States–bound flights. The new process was described on Connection Information Cards that were made available to arriving passengers. The card outlined the following nine steps for an "easy and worry-free connection":

   a. Complete Canadian and U.S. forms during the flight.

   b. Proceed through Canada Immigration/Customs.

   c. Follow the signage upstairs to Baggage Claim and pick up your bags.

   d. Hand the Canada customs card to the customs officer; follow the sign for connecting customers.

   e. Proceed to the connecting baggage drop-off belt and place your bags on the belt.

   f. A frequent shuttle bus to Terminal 2 leaves from the departure level (upstairs).

   g. Disembark at U.S. Departures; proceed to the carousel behind U.S. check-in counters ("Connecting to U.S.A.") and pick up your bags.

   h. Proceed through U.S. Customs and Immigration.

   i. Leave your bags on the baggage belt before going through security to your gate.

   Construct a list of activities and include all non-value-adding activities associated with waiting (e.g., waiting for luggage, waiting for an immigration officer, waiting for a shuttle). Estimate the percentage of time passengers spend on value-adding activities versus the time spent on non-value-adding activities. Identify potential sources of errors or additional delays that can result in a hassle to the passengers. Do you expect this process to deliver on the promise of an "easy and worry-free connection"?

7. Consider the process to connect international flights (e.g., those arriving from Europe) and United States–bound flights in Exercise 6. Identify the main components of this process's architecture, that is, the inputs and outputs, the flow units, the network of activities and buffers, the resources, and the information structure. Who is the customer, and what do you think are his or her most important requirements regarding the process? Propose a new process design with a potential to satisfy the customer requirements more efficiently.
8. A city council hires a consultant to help redesign the building permit process of the city. After gaining an understanding of the current process, the consultant suggests that the best solution is to create a one-size-fits-all process to handle all building-related applications. What is your opinion about the consultant's solution?
9. The fast development and widespread use of smartphones, portable computers, global positioning systems, Internet, and mobile services has drastically changed many of the processes in our everyday lives. It has also profoundly changed the business processes in many industries. Consider, for example, the processes of buying music, renting (DVD) films, planning your trip to a new location (by car or public transportation), or shopping for a particular product. How has new technology enabled new, more efficient and effective, process designs that satisfy customer requirements better at lower costs (eliminating non-value-adding activities and buffers, resources, etc.)? What has happened to companies that were not able to change their process designs and business strategies fast enough?

Notes

1. The term *suboptimization* refers to situations in which optimizing a subsystem according to its local objectives leads to an inferior solution for the overall system. For example, a production manager under budgetary pressure might decide not to allow any more overtime. As a consequence, the orders accepted by the sales staff cannot be met on time, which leads to order cancellations. For profit-maximizing firms, this implies revenue reductions due to lost sales and also long-term effects in terms of loss of goodwill and reputation.

2. Traditionally, companies have developed organizational structures based on the type of work performed. This type of organization is commonly known as a *functional organization* because work is organized on the basis of the function it serves. A restaurant that classifies its workers as chefs, dishwashers, bartenders, and so on has classified them on the basis of the function they serve or the type of work they perform. Likewise, a manufacturer of consumer goods with accounting, finance, sales, and engineering departments has organized its employees on the basis of the tasks they perform.
3. In the field of discrete event simulation, the term *transient entity* is given to jobs that enter the simulated system, where activities are performed to complete such jobs before exiting. Therefore, in that context, transient entities and flows are synonymous. In simulation, there is also the notion of resident entities, which are the resources needed to perform activities.
4. Concurrent engineering is one of the most effective strategies designed to shorten product development cycle times and increase the likelihood of designing robust products that can be manufactured efficiently. In simple terms, the strategy consists of overlapping various phases of the design and development process, allowing downstream work to begin before the prior phase of work has been completed. However, implementing concurrent engineering is no trivial matter because of the increased need for information to flow between the different phases of the process.
5. Reengineering or business process reengineering (BPR) is an improvement program, based on the ideas of Hammer, Champy, and Davenport published in the early 1990s, that advocates radical redesign and revolutionary change of core business processes as a recipe for success. The principles of BPR will be discussed further in Chapter 2.
6. Adapted from Roberts (1994).

References

Anupindi, R., Chopra, S., Deshmukh, S.D., Van Mieghem, J.A., and Zemel, E. 2012. *Managing business process flows – Principles of operations management*. 3rd ed. Upper Saddle River, NJ: Prentice Hall.

Cross, K.F., Feather, J.J., and Lynch, R.L. 1994. *Corporate renaissance: The art of reengineering*. Cambridge, MA: Blackwell Business.

Davenport, T. 1993. *Process innovation: Reengineering work through information technology*. Boston: Harvard Business School Press.

Davenport, T., and Short, J. 1990. The new industrial engineering: Information technology and business process redesign. *Sloan Management Review* 31(4): 11–27.

Dumas, M., La Rosa, M., Mendling, J., and Reijers, H.A. 2013. *Fundamentals of business process management*. Berlin/Heidelberg: Springer-Verlag.

Gabor, A. 1990. *The man who discovered quality*. New York: Penguin Books.

Goldsborough, R. 1998. PCs and the productivity paradox. *ComputorEdge* 16(43): 23–24.

Grant, R.M. 1995. *Contemporary strategy analysis*. Cambridge, MA: Blackwell Publishers.

Grover, V., and Malhotra, M.J. 1997. Business process reengineering: A tutorial on the concept, evolution, method, technology, and application. *Journal of Operations Management* 15: 193–213.

Hammer, M. 1990. Reengineering work: Don't automate, obliterate. *Harvard Business Review 71*(6): 119–131.

Hammer, M., and Champy, J. 1993. *Reengineering the corporation: A manifesto for business revolution*. New York: Harper Business.

Harbour, J.L. 1994. *The process reengineering workbook*. New York: Quality Resources.

Lowenthal, J.N. 1994. *Reengineering the organization: A step-by-step approach to corporate revitalization*. Milwaukee, WI: ASQC Quality Press.

Manganelli, R.L., and Klein, M.M. 1994. *A framework for reengineering*. Management Review (June): American Management Association, 10–16.

Petrozzo, D.P., and Stepper, J.C. 1994. *Successful reengineering*. New York: John Wiley and Sons.

Roberts, L. 1994. *Process reengineering: A key to achieving breakthrough success*. Milwaukee, WI: ASQC Quality Press.

Shafer, S.M., and Meredith, J.R. 1998. *Operations management: A process approach with spreadsheets*. New York: John Wiley and Sons.

Van Gigch, J.P. 1978. *Applied general systems theory*. New York: Harper and Row Publishers.

2

Data Analytics and Process Improvement

Although the focus of this book is on the modeling, evaluation, and design of processes rather than implementation and change management, the best design in the world is of little interest unless it can be implemented and managed successfully. Therefore, this chapter will look at the governing principles behind successful process management, which is at the core of all process-oriented improvement programs that have surfaced over the last couple of decades. We also will take a closer look at the basic ideas behind Six Sigma, one of the most influential programs of this sort. The integration and consolidation of these governing principles and process improvement programs from different disciplines are today often referred to as *business process management* (BPM) systems. BPM is a discipline that uses various tools and methods to design, model, execute, monitor, and optimize business processes. A business process coordinates the behavior of people, systems, and information to produce business outcomes in support of a business strategy. BPM focuses on putting an efficient process in place for routine transactions and human interactions. It helps to increase the overall efficiency of the company by reducing operational costs associated with waste and rework. Organizations engaged in BPM can choose to follow one of the various BPM methodologies, which include Six Sigma and Lean.

2.1 Process Management and a Process View

Process management, as defined in this chapter, deals with the issues of managing, controlling, and improving processes. Process design is an important element in successfully managing a process; however, so are the implementation of the design and the continuous improvement and control systems necessary to reach a desirable level of process performance. Another important aspect is managing the people involved in the process.

Process management originated in the field of modern quality management. The focus on processes and process control is a fundamental component of quality management today and an integral part of the ideas put forward by its founding fathers: Deming, Juran, Crosby, Feigenbaum, Ishikawa, and others. The power of adopting a focus on processes was made all too obvious by the Japanese during the 1970s and 1980s when they expanded the notions of quality and productivity to completely new levels.

DOI: 10.1201/9781032617237-2

The strength of adopting a process view and the importance of process management is that it reveals and addresses the weaknesses of functional organizations. Although the functional structure has many virtues, not least in terms of achieving economies of scale and scope, it has the inherent weakness of fostering a focus on skills and resource use rather than work output. As mentioned in Chapter 1, this tends to lead to suboptimization of resources due to insufficient coordination and transparency of the overall process. According to Melan (1993), this tendency can be further explained by the following conditions, which are usually built into the functional structure.

- Reward systems that promote the values and support the objectives of the functional department rather than the business in its entirety. For example, the performance of an employee in the manufacturing department might be based on his or her output quantity. However, the quality of the output is deemed more important for the firm, which must satisfy customer requirements.
- Group behavior, which encourages a strong loyalty within the department and an "us versus them" attitude toward other departments within the firm.
- Strong cultural and behavioral patterns within a function. These patterns can differ across departments, reinforcing the "us versus them" mentality.
- A high degree of decentralization, creating "firms within the firm," each with its own agenda.

The deficiency of the functional organization in terms of suboptimization due to lack of coordination has also been discussed extensively in the organizational literature. To quote Mintzberg (1979), "The functional structure lacks a built-in mechanism for coordinating the workflow" and has given rise to concepts such as matrix and project organizations.

A process focus remedies the issues of coordination and transparency by creating a clear emphasis on work output and the fact that this output must meet the customers' requirements. In terms of people management, adopting a process view means that the process structure needs to be made clear to everyone involved. This increased transparency tends to improve the employees' understanding of how their efforts fit into the big picture and why their contribution matters. This, in turn, is usually a good way to make people feel proud of what they do. Pride encourages involvement and participation, a cornerstone in modern quality management and a stepping-stone for successfully empowering the workforce. Another effect of adopting a process orientation is that it helps break down barriers among departments and hampers the "us versus them" attitude by creating a sense of loyalty to the process and everyone involved.

Historically, quality management has focused largely on the manufacturing sector. It was not until the broader concept of TQM got wide recognition in the mid- to late 1980s that the principles of quality management and process thinking were applied to a significant degree in the service sector. The principles of process management, therefore, have been drawn primarily from managing manufacturing processes. However, time has proved them equally valuable in managing service processes. The explanation is that the main challenges with process management lie in the focus on workflow; that is, how a job moves through an organization.

Whether the job is a product or a service is inconsequential, at least as far as the basic principles of process management are concerned.

Having acquired a general understanding about the purpose of process management and why process orientation is a powerful approach, the next step in understanding what successful process management entails is to identify the characteristics of well-managed processes. Following Melan (1993), the core principles for successful process management include:

1. Establishing process ownership.
2. Analyzing boundaries and interfaces.
3. Defining the process by documenting its workflow.
4. Identifying control points and measurements.
5. Monitoring the process for control purposes by implementing the measures.
6. Taking corrective action and providing feedback if deviations indicate that the process is no longer in control.

The following example illustrates these basic principles of process management.

2.1.1 An Illustrative Example: Managing a Document Distribution Process

Joe recently started his new position as associate manager of the business development department, with special responsibility for strategic planning. In January and June, Joe is now responsible for formulating and issuing copies of the company's strategic plan to the firm's key managers. In fact, this is one of his more important functions. The managers then have to respond to Joe within 2 weeks regarding a wide range of strategic issues before the plan can be finalized. The importance and visibility of this document mean that any kind of problems with the process will reflect badly on Joe personally and on his department.

In early January, copies of the document were delivered from the print shop. Joe's staff then sorted the documents, put them in envelopes addressed to the designated managers, and left them to be picked up by internal mailing services. The address of each manager eligible to receive a copy was easily available from a distribution list kept by the departmental secretary. No mailing problems were reported to Joe. By mid-January, only four managers had responded to Joe, who was disappointed and embarrassed to report to his manager that he would not be able to finalize the strategic plan before the agreed deadline. He then immediately began making phone calls to the 24 managers who had not responded and was astonished by their responses.

1. Three had not yet received the document.
2. Two had just received the document at their new location.
3. Seven complained about missing pages.
4. Four reported pages out of sequence.
5. Five said they could not respond in only 1 week.
6. Three had left the organization.

The following day, Joe received a message from his manager asking him to be in the manager's office the next morning with an analysis of the problem and an action plan for how to avoid this type of debacle in the future.

Clearly, this process was not well managed. The following pages will discuss how the basic principles of process management can be used to improve it.

2.1.1.1 Assign Process Ownership

It is clear that Joe is the manager responsible for formulating and distributing the strategic plan document. Although he has no managerial control over the print shop or the mailing services, he is perceived as the owner of the entire process by his managers and is held accountable for its failure. Neither the print shop nor the mailing service is aware of its role in the process and that Joe is the owner of the process; they just perform an operation without understanding its impact. This is why it is important to clarify to all parties involved who the process owner is.

2.1.1.2 Analyze Boundaries and Interfaces

With the process ownership in place, Joe needs to analyze the process boundaries and interfaces. Based on the detected problems, he decides to focus his attention on the document distribution process, starting when the print shop receives the original document to be copied and ending when the mailing services deliver the mailed copy to the managers. Joe also defines two critical internal interfaces: the first one when the job (i.e., the copies) leaves the print shop and is handed off to his staff and the second one when the addressed envelopes are handed off to the mailing service. The external interfaces coincide with the boundaries. Furthermore, it should be noted that for the document distribution process, the input to the print shop is the original master document, and the supplier is Joe's department. The customers at the other end are the managers who are to receive a copy of the document. A high-level flowchart describing the process along with its boundaries and interfaces is given in Figure 2.1.

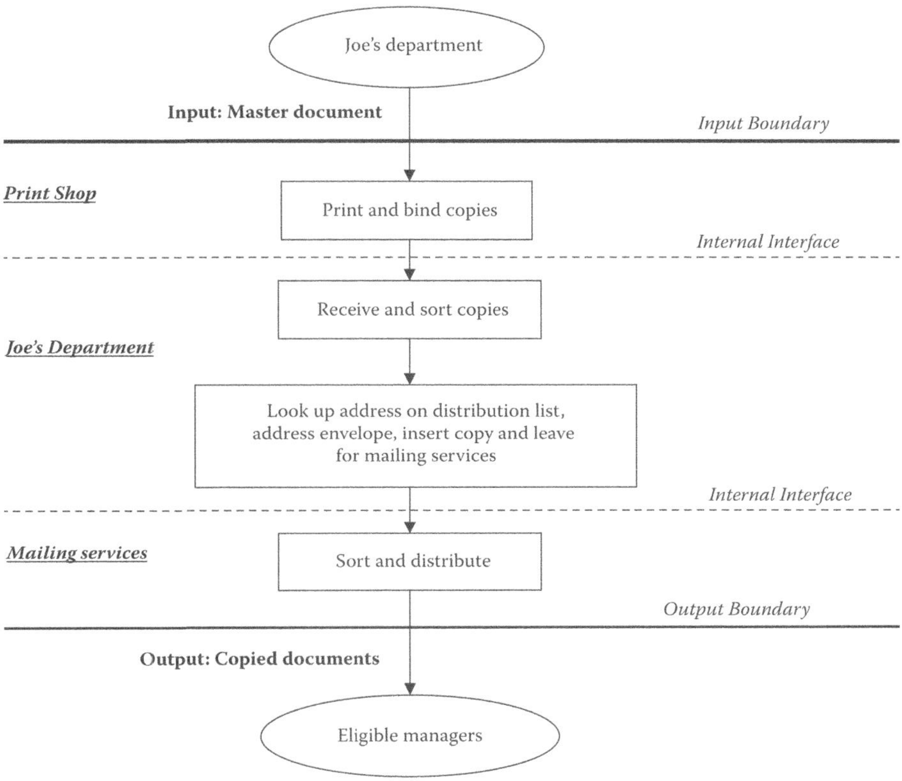

FIGURE 2.1
High-level flowchart of the document distribution process.

From the telephone conversations Joe had with the managers, he can conclude that many of the problems are related to the critical interfaces and lack of communication and coordination across them. Delivery delays and poor quality of the document copies appear to be the main issues. To approach the interface problems, Joe decides to use a customer–producer–supplier (CPS) model, starting with the requirements of the final customers and working his way back through the process.

The CPS model is a useful approach for dealing with interface-related workflow problems. This model is based on the simple premise that a producer's output should satisfy the customer's requirements. (Notice that this is closely related to the definition of a business process provided in Chapter 1.) The CPS model has three agents: the supplier who provides the inputs, the producer who transforms the input into value-added output, and the customers who are the recipients of the output. The interactions among these agents can be divided into three phases: the customer requirement phase, the process (or production) capability phase, and the producer requirement phase.

The customer (output) requirement phase is the first step in the CPS approach and involves defining and reaching an agreement on the customer requirements for the final process output. Having acquired a good understanding of the customer requirements, the next step in the CPS approach is the *process capability phase*, during which the producer assesses the process capabilities to see if it can meet the customer requirements. The final step in the CPS approach is the *producer (input) requirement phase*, during which the producer negotiates with the suppliers to agree on the input requirements necessary to satisfy the producer's needs. The input requirements usually revolve around cost, timeliness, quantity, and quality characteristics. A fundamental principle in the CPS approach is that the parties involved must mutually agree on all requirements.

The phone conversations with the managers who received the copied document (i.e., the final customers) revealed that they have two main requirements: a perfect copy of the document and 2 weeks to prepare a reply. Following the CPS model, Joe now focuses on the process (or production) capabilities of the mailing services. A manager at mailing services informs Joe that for mail to be distributed the next day, the sender must take it directly to the mail room. If it is left at the department to be picked up by mailing services, it takes an additional day before the envelopes are delivered to their recipients. Furthermore, if the mail is going to one of the company's satellite locations, which is the case for several managers on Joe's distribution list, it might take up to 4 days until the envelopes are delivered, assuming they are brought straight to the mail room. In the current process, the envelopes were left at Joe's department and the assumption was next-day delivery.

The input to the mailing services is the addressed envelopes supplied by Joe's department. Consequently, following the CPS model and assuming that Joe intends to continue using the internal mailing services, the supplier requirements indicate that Joe needs to deliver the copied documents to the mail room 4 days earlier than was done this time to satisfy the customer requirements. Another option would be to use an external courier, but Joe has no room in his budget for that. An alternative would be to change the mailing services process to shorten the delivery time, but Joe has no managerial jurisdiction over mailing services, so this is not a viable option either.

The next interface of interest is that between Joe's department and the print shop, where the former is the internal customer to the latter. The requirements posed by Joe's department are linked directly to the end-customer requirements: the copies should be of good quality and be delivered promptly. The manager of the print shop informs Joe that for a large-size document such as the strategic plan, a 2-day lead time is to be expected. A rush job increases the risk of receiving bad-quality copies. When the last job was done, Joe's secretary, not understanding the production capabilities, requested "same-day service" but received the copies 2 days later. Also, the print shop staff noticed a reversed sequence of page numbers but assumed that this was desired because that was how the original master document was arranged.

The print shop just copies the master document as is. Assuming that Joe will continue using the print shop, understanding the print shop's capabilities renders some clear requirements on the supplier of the master document, which is Joe's department: the document must be in perfect shape and be delivered to the print shop 2 days before it needs to be sorted, addressed, and put in envelopes.

To summarize, applying the CPS model has revealed some important requirements that Joe's department must consider to get the copied documents out on a certain day. Basically, the print shop needs the master document delivered on the morning of day 1, and the master document must at this point be in perfect shape because it is copied as is. The print shop delivers the copied documents to Joe's department on the morning of day 3; then Joe's staff have half a day to sort, address, and deliver the sealed envelopes to the mail room. Internal mailing services guarantees delivery to all locations during day 7. In addition to the timing issues that Joe must consider, his department also must pay special attention to the quality of the master document and the correctness of the distribution list.

2.1.1.3 Define the Process

Having determined the scope of the process, Joe is in a position to define the activities and workflow that are within the boundaries. The most common approach is to start with a high-level flowchart, such as the one in Figure 2.1, and then create a hierarchy of successively more detailed flowcharts over the different subprocesses and activities until the right level of detail is obtained in the description. The right level is found when additional details no longer appear to offer further opportunities for improvement. In this particular case, Joe decides not to proceed beyond the high-level or macro flowchart in Figure 2.1.

2.1.1.4 Establish Control Points

The current process does not contain any internal control points; as a result, Joe might not realize that something is wrong with the process until it is too late. This situation represents a high "failure cost" for Joe in terms of lost credibility within the firm. Based on the customer feedback and the analysis so far, Joe finds three instances where control points appear to make sense.

1. Inspection of the master document before it is delivered to the print shop to ensure that what goes to the print shop is flawless. This responsibility falls on Joe's secretary and his other staff members.
2. Inspection of the copies received from the print shop by Joe's staff before they are put into envelopes and mailed. This is done to ensure that pages are not missing and that pictures are not blurred due to

quality problems in the print shop. Applying the CPS model, Joe should negotiate with the print shop and require them to change their process to avoid quality problems of this sort. This would make the inspection of the final copies redundant and save Joe some resources. However, even if Joe has enough clout to make this happen, it makes sense in the short run to install the control point to make sure the print shop process performs as intended, simply because the outcome is so important to Joe.

3. Verification that the distribution list is accurate (the right people are on it, and their addresses are correct). This job is appropriate for Joe's staff. The purpose is to avoid sending copies to those who are no longer with the company or sending copies to the wrong address. The problem is magnified by the confidentiality of the strategy document.

As the quality of the process improves, these inspections will become redundant. Installing inspection points is not an efficient way to achieve quality output in the long run. However, it enables data gathering and identification of the root causes of problems occurring in dysfunctional processes.

2.1.1.5 Develop and Implement Measures

After determining the control points, the next step for Joe is to specify what to measure. Starting with the inspection of the master document at Control Point 1, it is important to record not just that a mistake occurred but what the mistake is. This assures that the data can be used to facilitate root cause analysis (RCA) and improve the process that produces the original master document. Based on the customer feedback, it appears that the frequency of missing pages and pages out of sequence should be tracked. In addition, it seems wise to look for typographical errors, the types of mistakes, and their frequency.

Control Point 2, dealing with the quality of the copies coming out of the print shop, will focus again on document quality, so the same measures should be used (i.e., frequency of missing pages; pages out of sequence; and a range of typographical errors including blurred figures, text alignment, and so on).

For Control Point 3, the verification of distribution list accuracy, interesting measures to track are the number of address changes and the turnover of managers on the list. Joe decides to use these categorized frequency measures to track the copying and address quality.

In addition to the three internal control points, Joe also decides to solicit regular feedback from the end customers regarding copy quality and delivery time. To track the performance of the last step in the distribution process, the mail delivery, Joe also decides to record the time when the envelopes are delivered to the mailroom and then ask the final recipients in their response to the strategic plan document to specify when they received the envelope. In this way, Joe can track the total distribution time because he knows when the master document was delivered to the print shop. Joe also intends to ask the managers to comment on the copying quality.

2.1.1.6 Perform Feedback and Control

An important issue for Joe is to try to improve process efficiency without compromising its effectiveness. In the proposed setup, inspection time decreases efficiency. By tracking the measurements discussed previously and using them to provide feedback to the people working in the process, he hopes to stimulate employee involvement and also put pressure on the managers of the print shop and internal mailing services by showing them data on their performance. Joe anticipates that this will stimulate them to take corrective action to improve their internal processes and eliminate the demand for inspection. To present the data in a pedagogical and convincing way, Joe has a wide range of charts and statistical tools at his disposal. To keep things simple, he has decided to start with basic histograms and bar and pie charts. The data gathering will be initiated the next time the distribution process is activated in about 6 months.

2.1.2 Summary and Final Remarks

To complete the picture, let us review what Joe, based on the basic principles of process management, has come up with in terms of an action plan to improve process performance and achieve a well-managed document distribution process.

First, Joe will make sure that the master document is finished at least 7 working days before it should be delivered to the designated managers. He will allow them at least 2 weeks to prepare a response. Furthermore, Joe's staff are instructed to inspect the master copy of the strategic plan, searching especially for any missing pages, pages that are out of sequence, or apparent typographical mistakes. These mistakes should be categorized and counted. The staff also need to make sure that the 2-day production lead time is taken into consideration.

Second, Joe's staff are instructed to inspect each copy of the strategic plan for missing pages, pages that are out of sequence, and typographical errors and record the mistakes found. If enough mistakes are found, and the timeline permits it, the whole batch should be returned to the print shop to be redone. If there is not enough time to return the whole batch, the faulty documents will be reworked at Joe's department, and the time for this should be recorded. This rework time can easily be translated into costs and will provide Joe with some convincing arguments when giving feedback to the print shop manager. An important issue in this respect is to explain the effect of the print shop quality problems on the document distribution process and the overall strategy development process.

Third, the distribution list should be updated shortly before the envelopes are addressed. The number of changes to the list should be recorded; tracking the changes over time will help Joe decide how often the list needs to be updated. For now, updates will be done prior to every distribution round.

Fourth, the date and time that the envelopes are taken to the mail room should be recorded as a means of tracking how long they spend in the internal mailing system. To obtain the arrival times, Joe will ask that the managers disclose in their response to the strategic plan exactly when the document was delivered. He also will ask them to comment on the document copying quality.

By implementing these changes, Joe takes a huge step toward proactive process management so that he can avoid situations in which the process problems are not discovered until it is too late to correct them. Joe is also in a good position to explain to his boss at their meeting the next day what caused the problems with the document distribution process and what he plans to do to prevent future mishaps of this kind. Apart from the immediate actions to assure the effectiveness of the process in the near future, Joe also can point to the necessary long-term improvements. If the print shop and mailing services improve their internal efficiency, Joe can eliminate the inspection activities and thereby improve the overall efficiency of the document distribution process. Most likely, drastic improvements could be made to the process performance in the print shop and the mailing services if their managers were willing to rethink their entire design.

2.2 Data-Driven Process Improvement

Data-driven process improvement is an approach to improving business processes by making decisions based on objective data and analysis. It involves collecting and analyzing relevant data to gain insights into how a process is currently performing and identifying areas for improvement. By using data, organizations can make informed decisions, measure the impact of changes, and continuously optimize processes for improved efficiency, quality, and productivity.

The key steps in a data-driven process improvement approach typically include:

1. *Define objectives*: Clearly define the goals and objectives of the process improvement initiative. This could be reducing costs, improving customer satisfaction, increasing productivity, or any other relevant objective.

2. *Identify metrics*: Determine the key performance indicators (KPIs) that will be used to measure the performance of the process. These metrics should align with the objectives and provide meaningful insights into the process's strengths and weaknesses.
3. *Collect data*: Gather relevant data related to the process, such as process times, error rates, customer feedback, or any other data points that are indicative of process performance. This data can be collected from various sources, including operational systems, surveys, or other measurement tools.
4. *Analyze the data*: Use various statistical, analytical, and visualization techniques to analyze the collected data. This may involve identifying trends, patterns, or correlations that provide insights into process bottlenecks, inefficiencies, or areas of improvement.
5. *Identify improvement opportunities*: Based on the data analysis, identify specific areas within the process that require improvement. This could be eliminating redundant steps, automating manual tasks, streamlining workflows, or implementing new technologies or tools.
6. *Implement changes*: Develop an action plan to implement the identified process improvements. This may involve making changes to workflows, updating standard operating procedures, providing additional training for employees, or introducing new technologies.
7. *Monitor and measure*: Continuously monitor the performance of the process after implementing the changes. Measure the impact of the improvements by comparing the new metrics with the baseline data collected earlier.
8. *Iterate and optimize*: Based on the measured results, iterate the process improvement cycle by collecting new data, analyzing it, and identifying further opportunities for optimization. This iterative approach allows organizations to continuously improve their processes over time.

By adopting a data-driven approach to process improvement, organizations can make objective decisions, minimize guesswork, and achieve more significant and sustainable improvements in their operations.

2.2.1 Data Collection

The era of data-driven process improvement and the explosion of the amount of available data have revolutionized how data is collected and analyzed. This abundance of data can be overwhelming for companies that are not sure where to start. Data-driven decisions are only as strong as the data on which they are based. Data can be qualitative (meaning contextual in nature) or quantitative (meaning numeric in nature). Many data collection methods apply to either type, but some are better suited to one over the other. The data life cycle starts with the generation of data (i.e., the data source) and it is followed by data collection. After that, data can be processed, stored, managed, analyzed, and visualized to assist the organization's decision-making.

The data to be collected and the collection method depend on the objective of the process improvement project and the metrics. Collecting data is an integral part of the success of process improvement. Some business processes generate data in digital form. For instance, e-commerce and point-of-sale platforms allow companies to store data as soon as it is generated, resulting in a seamless data collection method. However, there are processes for which data must be captured. Manual capture refers to entering data into a computer by hand. It might include typing data from a handwritten form into a document or spreadsheet, manually transcribing audio, or copy-pasting information from one location to another. Studies from the healthcare sector—where mistakes in manual data entry can be life-threatening—have found an average error rate for manual data capture of 3–4 percent, with text-based entries accounting for most of these errors. For businesses with minimal data capture needs, manual data entry is an acceptable solution. But for companies with significant data capture load, better solutions are needed to make sure employees are not taken away from core business tasks that require their knowledge and skills.

When a manual entry is not practical, organizations can explore automated data capture methods as a potential cost-effective option. Automated data capture refers to the use of technological tools like artificial intelligence and image recognition software to convert data into a computer-usable format. Automated data capture is faster than manual data capture, less error-prone, and frees up employees to focus on more important work. A study published in the Journal of Nursing Administration (Bauer et al., 2020) tested the introduction of an automated data entry process technology to the nursing unit of a large hospital. The automated data entry tool freed nurses from data entry tasks, thus creating time for patient care and other activities associated with improvements in performance and job satisfaction. Typical automated data capture includes:

- Optical Character Recognition recognizes and extracts text from physical documents and turns it into digital form.
- Digital Forms are designed to replace paper-based surveys and allow users to input information via web or mobile applications. On the backend, data flows directly into a designated database for instant access.
- Barcodes and QR Codes enable users to transmit data to others by running their code through a scanner.

Understanding the variety of data collection methods available helps analysts decide which is best, considering the timeline and budget of an improvement project. When stored together and combined, multiple data types collected through different methods provide the foundation for improving business processes.

2.2.2 Data Visualization

Data visualization plays a crucial role in process improvement by helping teams understand complex data, identify patterns, and communicate insights effectively. Visualizing data in a clear and intuitive manner helps in communicating insights and findings to stakeholders. Charts, infographics, or interactive dashboards make it easier for analysts to present complex data and analysis results to management, colleagues, or clients, facilitating better understanding and decision-making. For instance, process flow diagrams, which we introduce in Chapter 4, help analysts understand the sequence of steps in a process and identify areas where delays or redundancies may occur. This visual representation enables them to identify potential improvements, such as streamlining workflows or eliminating unnecessary steps. Visualizing process data also allows analysts to identify bottlenecks or inefficiencies in a process. Real-time data visualization dashboards provide a visual representation of key performance metrics, such as cycle times, wait times, or error rates, allowing analysts to pinpoint areas that require attention. Chapter 5 shows how some of these metrics are calculated.

2.2.2.1 Charts and Diagrams

Data visualization enables the comparison of process performance over several time periods. All processes exhibit variation, and therefore single point measurements provide limited information. Displaying data over time increases understanding of the real performance of a process, particularly about an established target or goal. Trend charts can detect performance changes, helping teams identify whether process improvements are having the desired impact. Figure 2.2 shows a trend chart for an order fulfillment process that tracks the fill rate over time. Fill rate, also called order fulfillment rate, is the percentage of orders that a company can ship from its available stock without any lost sales, backorders, or stockouts. It is a good reflection of the company's ability to meet customer demand and the overall effectiveness of its operations. It also helps the company gauge the demand for various stock-keeping units (SKUs) in its inventory and plan its procurement accordingly. In this example, the target fill rate is 95% and the performance for the first part of the year is significantly below the desired standard. The chart shows that whatever remedial actions were

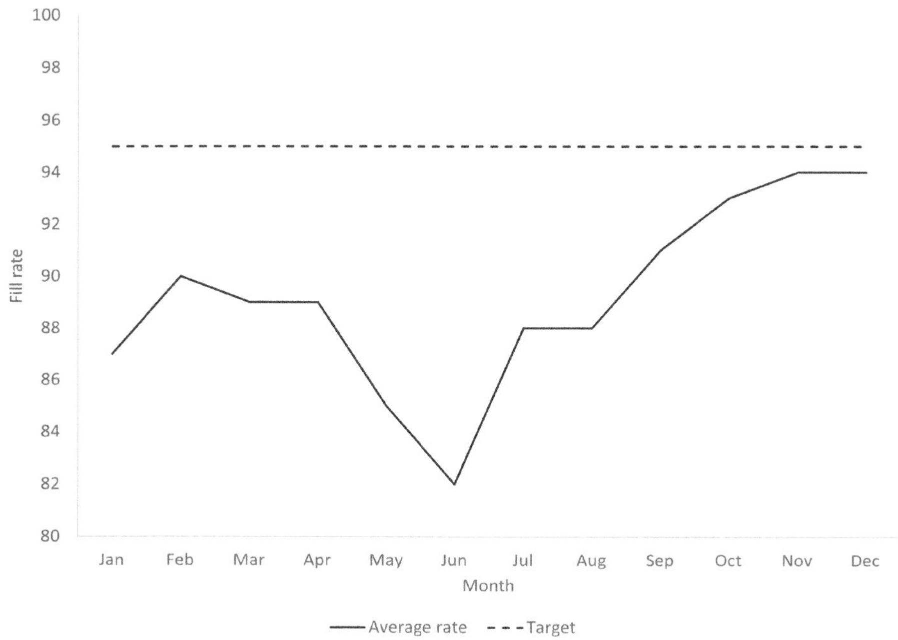

FIGURE 2.2
Trend chart for an order fulfillment process.

taken in June, they are resulting in a positive trend toward achieving the target fill rate.

Root cause analysis is another area where visualization can help. RCA is a systematic approach used in business processes to identify the underlying causes of problems or issues rather than just addressing the symptoms. It aims to uncover the fundamental reason behind a problem to prevent its recurrence and implement effective corrective actions. Fishbone diagrams (also known as Ishikawa or cause-and-effect diagrams) systematically explore potential causes and their relationships. This involves identifying major categories of potential causes, such as people, processes, equipment, materials, or external factors, and brainstorming specific causes within each category. Figure 2.3 shows an example of a fishbone diagram that identifies the major causes of missing delivery deadlines. For each main category (i.e., People, Process, Control, Equipment, Space, and Product), the diagram includes the main factors that contribute to failing to deliver orders on time.

Prior to creating a fishbone diagram, RCA engages in data collection to gather relevant information related to the problem. This may include process

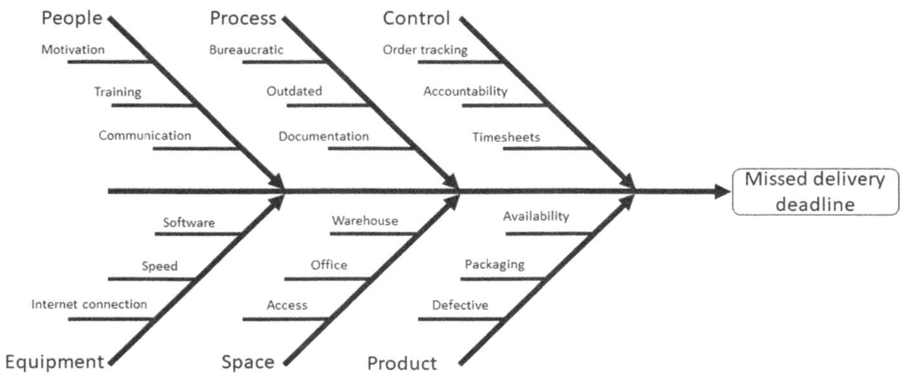

FIGURE 2.3
Fishbone diagram for missing delivery deadlines problem.

data, customer feedback, incident reports, or any other data sources that provide insights into the problem.

As we will see in Chapter 8, visualization plays a role in process simulation, where virtual models are created to simulate and analyze process scenarios. Visualization tools help in presenting simulation results, such as average waiting time or resource utilization, making it easier for analysts to understand and compare different process configurations or scenarios. As shown in Chapter 9, visualization tools enhance the outcomes of predictive analytics and forecasting models. Visualization helps to anticipate potential process issues to proactively implement improvement measures that could prevent them.

Data visualization facilitates process benchmarking, where organizations compare their process performance against industry standards or best practices. The visualization of benchmarking data can identify performance gaps to help companies understand areas where they lag competitors and establish improvement targets.

2.2.2.2 Heatmaps

Heatmaps are another popular and useful data visualization tool. A heatmap (or heat map) is a graphical representation of numerical data, where individual data points contained in a dataset are represented using different colors. The key benefit of heatmaps is that they simplify complex numerical data into visualizations that can be understood immediately. Heatmaps were first developed in the 1800s, originating in the 2D display of data in a matrix. The first known usage of heatmaps is credited to Toussaint Loua in 1873.

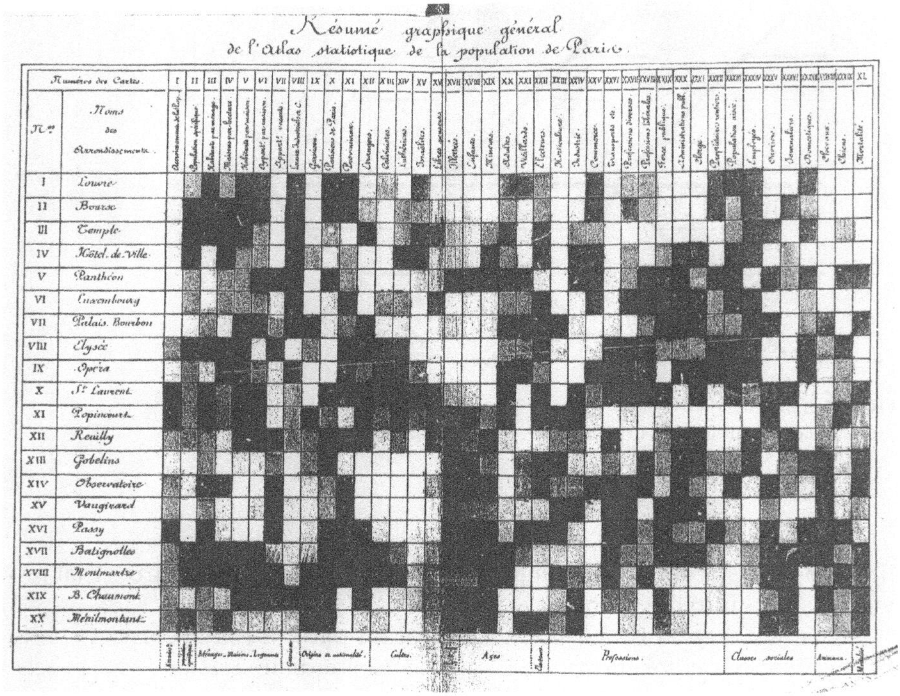

FIGURE 2.4
Loua's seminal heatmap.

Figure 2.4 shows Loua's heatmap, which summarizes 40 characteristics (columns) of 20 districts (rows) in Paris. The heatmap uses dark gray and black to denote higher value metrics and light colors like white and light gray to denote lower value metrics.

Modern applications of heatmaps include:

1. *Website heatmaps*: These heatmaps are commonly used by organizations that have an online presence or use the Internet as their main revenue channel, such as e-commerce stores, travel and hospitality websites, media services, and B2B SaaS companies. Website heatmaps depict areas of a company's web pages using a hot-to-cold color scheme with the warm-toned colors indicating the most popular sections and cool-toned colors indicating the unpopular ones. Heatmaps help measure a website's performance. They simplify numerical data and help managers understand visitors' thinking, enabling changes that positively impact conversion rates.
2. *Geographical heatmaps*: Geo heatmaps identify areas of high and low concentration of a certain value (e.g., population). Companies in the

travel and hospitality industry use data available in tourism departments (e.g., bookings and type of accommodations) to run targeted marketing campaigns in specific geographical areas. Brick and mortar companies use heatmaps to identify promising areas for new stores based on relevant demographic data.

3. *Financial services*: The market value of assets continually fluctuates. Manually logging changes into a document, drawing inferences and patterns, and revisiting numerical data can be a long process. Heatmaps simplify multiple steps in the process, enabling analysts to visually compare the performance of companies or industry sectors.
4. *Sports*: Heatmaps have multiple uses in sports. For instance, by plotting heatmaps of players' on-field performance, coaches and managers can reveal patterns within and across games. These patterns can be used to identify performance areas that need improvement or to study a rival's possible game plan and strategy.

Tracking KPIs is important for improving business process performance. Heatmaps are useful for visualizing KPI data. For instance, consider the following KPIs for a manufacturing facility:

1. *On-time delivery*: There are multiple reasons for deliveries not being made on time, such as delayed supplies, unrealistic production schedules, or machine breakdowns. Having a high delivery success rate is important to keep current customers and attract new ones. On-time Delivery = (Units delivered on time ÷ Total units delivered) × 100.
2. *Production schedule attainment*: This KPI helps manufacturers see if production was well-planned and if employees are meeting their targets. It allows manufacturers to determine performance issues that may impact deliveries. Production Schedule Attainment = (Actual output ÷ Planned output) × 100.
3. *Average cycle time*: This KPI measures the time that it takes for a customer order to be completed, from the time that the order was placed to shipment. Therefore, it includes waiting time. It averages out the cycle times of all orders. Average Cycle Time = Cycle time per order ÷ Number of orders.
4. *Capacity utilization*: This metric measures how close a machine is to producing at its theoretical, maximum capacity. With no downtime, the production on a machine would ideally reach the maximum capacity. Capacity Utilization = (Actual output ÷ Maximum output) × 100.
5. *Return rate*: This KPI tracks the fraction of goods, out of the total delivered, that customers return due to defects. Returned goods incur rework/replacement costs as well as reputational costs and customer retention. Return Rate = (Returned goods ÷ Delivered goods) × 100.

	A	B	C	D	E	F	G
1	KPI	Jan	Feb	Mar	Apr	May	Jun
2	On-time delivery	88%	94%	91%	94%	80%	89%
3	Production schedule attainment	97%	94%	92%	96%	92%	98%
4	Average cycle time	4.11	2.43	4.49	4.17	4.09	3.78
5	Capacity utilization	86%	87%	85%	88%	87%	85%
6	Return rate	1.4%	6.1%	1.9%	2.1%	4.7%	1.1%

FIGURE 2.5
Example of a KPI heatmap.

Figure 2.5 shows an example of a heatmap for the KPIs described above during the first semester of a year. A visual inspection of the heatmap indicates that March's performance should be further investigated as three out of five KPIs resulted in subpar values.

The KPI heatmap in Figure 2.5 was created using Excel's conditional formatting. For KPIs for which larger values are better, we used the first option of the Color Scales (as shown in Figure 2.5). The extremes of this option are red (dark gray in Figure 2.5) for the smallest value and green (light gray in Figure 2.5) for the largest value. For KPIs for which smaller values are better, we used the second option of the Color Scales (as shown in Figure 2.6). The extremes of this option are red (dark gray Figure 2.5) for the largest value and

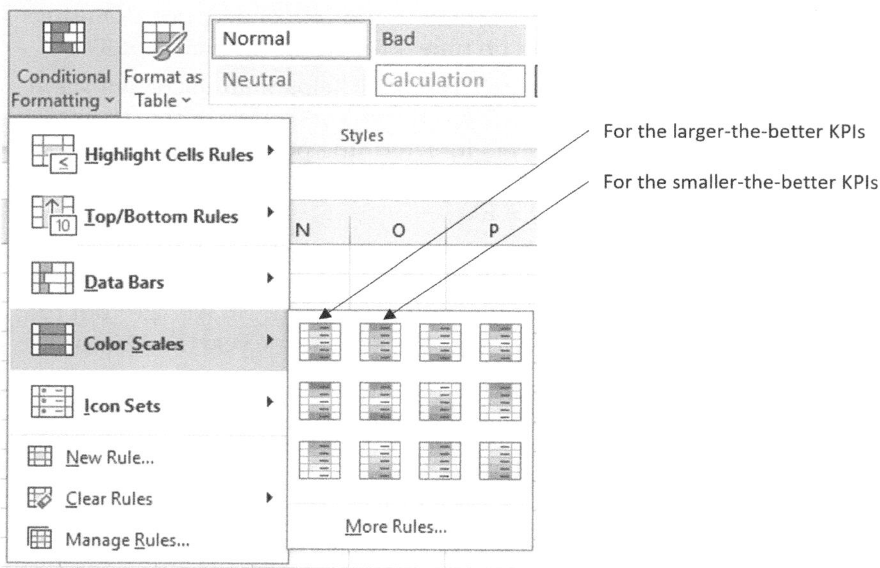

FIGURE 2.6
Excel's conditional formatting for heatmaps.

green (light gray in Figure 2.5) for the smallest value. This means that green areas (light gray) of the heatmap indicate high performance while red (dark gray) areas indicate low performance. Yellow areas (medium gray) show average (or mid-range) performance.

Data visualization of the type described above is instrumental in executive reporting. Executives typically require concise and visually appealing summaries of process performance. Visualizations such as executive dashboards or scorecards provide high-level overviews of key metrics, enabling executives to quickly grasp performance trends, make strategic decisions, and allocate resources for process improvement initiatives. Overall, data visualization in process improvement improves data comprehension, supports analysis, facilitates collaboration, and enables effective communication of insights. It empowers companies to make data-driven decisions and drive continuous improvement in processes.

2.3 Six Sigma Quality Programs

Launched in 1987, Six Sigma was the name of a companywide, process-oriented initiative at Motorola for achieving a breakthrough improvement in quality and productivity. It represented Motorola's customized approach to quality management and was heavily influenced by the principles of modern quality thinking. The novelty of the Six Sigma approach was how these basic principles were combined into an integrated program for process and quality improvement. Because of the tremendous success of the Six Sigma approach at Motorola, rendering them the first Malcolm Baldrige National Quality Award in 1988 (the most prestigious U.S. quality award), other companies around the world—including IBM, ABB, Kodak, Allied Signal, and GE—embarked on Six Sigma initiatives of their own during the early to mid-1990s. Over the years, the original Motorola concept has evolved into what today is one of the most renowned and rapidly spreading quality improvement programs. The ongoing success of these programs has led to exponential growth in the number of prestigious global firms in a wide range of industries to adopt the Six Sigma approach. Examples include Ford, American Express, Honeywell, Nokia, Ericsson, Phillips, Samsung, Johnson & Johnson, J.P. Morgan, Maytag, Sony, and Dupont. Other examples and case studies of Six Sigma implementations are concurrently reported by the American Society for Quality (www.asq.org).

2.3.1 Six Sigma Definitions

Six Sigma can be defined as an improvement program aimed at reducing variability and achieving near elimination of defects from every product, process, and transaction; see, for example, Tomkins (1997). However, this somewhat narrow definition does not capture the strategic implications of Six Sigma initiatives. In a broader sense, a Six Sigma program can be described as a companywide strategic initiative for process improvement in manufacturing and service organizations. Six Sigma has the clear objective of reducing costs and increasing revenues; that is, increasing process efficiency and process effectiveness.

The Six Sigma approach is built around a project and a result-oriented, quantitative, and disciplined improvement methodology. This methodology focuses on variance reduction but also emphasizes cycle time reduction and yield improvement. The methodology usually is divided into five steps: define, measure, analyze, improve, and control (DMAIC). (Sometimes, the "define" step is omitted and presumed to have been performed before the methodology is applied.) This methodology is embedded in a corporate framework based on top management commitment, stakeholder involvement, training programs, and measurement systems. We will look more closely at these issues, including the improvement methodology, in Section 2.3.4. However, it is worth noting the similarity between the DMAIC methodology and Phases II and III in the process management framework discussed in Section 2.1.

The goal of the Six Sigma program, as it was conceived at Motorola, is to reduce the variation of the individual processes so that they render no more than 3.4 defects per million opportunities (dpmo). To explain this further, we need to recognize that any product or process characteristic that is measured has some desired target value, T, and some upper and lower specification limits, USL and LSL, respectively. If the measured characteristic falls outside these limits, the product or service is considered defective. Furthermore, there is variation in any process output, so the measured characteristic will not be the same for every unit of output. Assuming that the distribution of the process output follows a normal distribution with a population mean, μ, and standard deviation, σ, Six Sigma Quality refers to a situation in which the distance between the target value and the closest of the specification limits is at least 6σ (see Figure 2.7). The number 3.4 dpmo is obtained as a one-sided integration under the normal distribution curve beyond 4.5σ from the process mean, μ. The explanation for this construct, illustrated in Figure 2.7, is that the process mean, μ, is allowed to shift over time and deviate from the target value by as much as 1.5σ. It is worth emphasizing that not all companies using Six Sigma programs choose to adhere to these numerical goals. An overarching Six Sigma objective is to achieve near elimination of defects, but ultimately, it is up to each organization to define what it considers "near elimination."

2.3.2 The Six Sigma Cost and Revenue Rationale

A characterizing feature of Six Sigma that distinguishes it from most other quality management initiatives is its fierce focus on bottom-line results. The

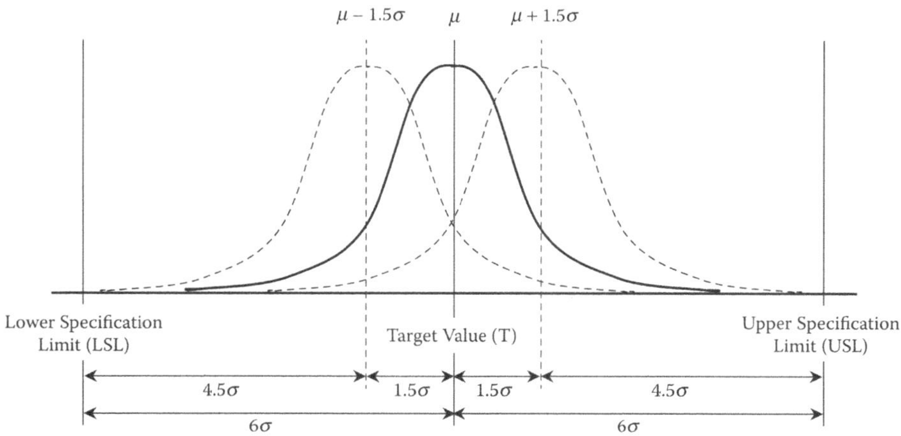

FIGURE 2.7
Technical definition of Six Sigma quality as achieved when the distance between the process target value and the closest specification limit is at least six standard deviations (sigma), and the process mean does not deviate more than 1.5 sigma from the target value, T.

top priorities and explicit objectives in every Six Sigma project are to decrease costs by improving process efficiency and to increase revenues by improving process effectiveness. The success of Six Sigma, and the reason why it is heralded in corporate boardrooms, is its ability to achieve these goals. For instance, Pulakanam (2017) concludes that over the period of implementing Six Sigma in 28 organizations, there were average savings corresponding to 1.7 percent of total revenues, and an average return of more than $2 in direct savings for every dollar invested in Six Sigma.

To acquire a better understanding of Six Sigma, we need to take a closer look at how its process improvement program is related to the underlying cost and revenue rationale.

2.3.2.1 The Cost or Efficiency Rationale

In simple terms, a firm's profit or bottom line is given by its revenues minus its costs during a specified accounting period. Consequently, by decreasing costs, a firm experiences an immediate positive effect on the bottom line. Clearly, this is the logic behind all efficiency improvement programs and cannot be attributed to Six Sigma alone. However, Six Sigma does bring an aggressive and focused approach to the table, based on attacking all types of costs, including labor costs. However, to assure cooperation and involvement from employees, Six Sigma ensures that labor cost reductions are realized through increased productivity instead of layoffs.

A fundamental principle is that every improvement project must render measurable cost savings, and every training course must include cost-cutting

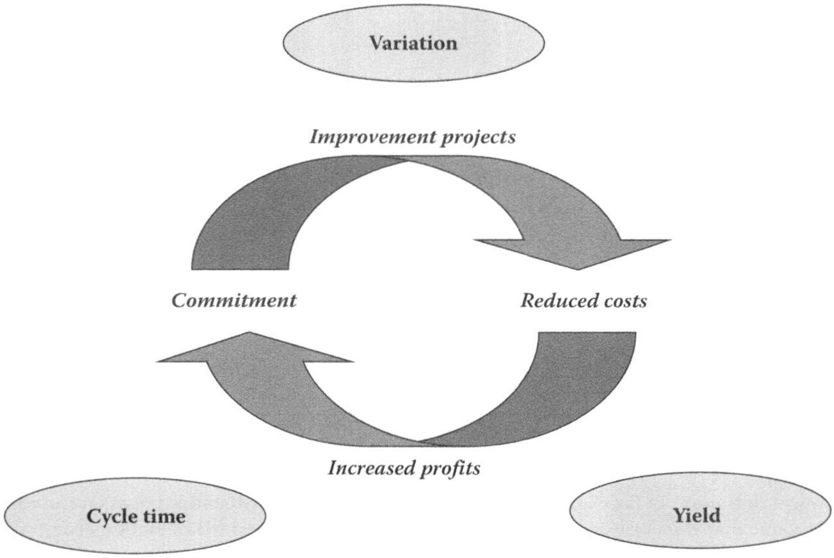

FIGURE 2.8
Six Sigma efficiency loop.

projects. To further understand the Six Sigma approach to cost reduction and improved process performance, it is helpful to consider the dimensions of variation, cycle time, and yield. This ultimately will lead to the efficiency loop (or bottom-line loop) in Figure 2.8 summarizing the cost rationale.

Variation is embedded in any process through the input used and the actual work or transformation being performed. A key operative objective in Six Sigma (see the technical definition in Section 2.3.1) is to decrease process variation and thereby increase quality and reduce costs. The first step to achieving this is to measure and gather data on important process, product, and/or service characteristics. To quantify and track the variation in these characteristics, the data need to be statistically analyzed and visualized graphically. After the measuring system is in place, the focus is placed on identifying the sources of variation.

Based on what is causing it, the variation usually is divided into two types: common cause variation and special cause variation. The latter, also referred to as *nonrandom variation*, typically is due to relatively few identifiable causes but tends to produce large, unpredictable contributions to the overall variation. Some recurrent special causes of variation are differences in the quality of input material from different suppliers, faulty equipment, and process tampering due to bad measurement systems or inadequate training of employees. The first step in reducing the overall variation is to eliminate the special cause variation by attacking its roots. The random or common cause variation, on the other hand, is typically the result of many contributing sources, such that it is considered inherent in the process and can be affected only by implementing a new design.

Three important concepts in understanding the impact of variation on process performance are dispersion, predictability, and centering. In general terms, large dispersion refers to large variation or variability in the measured process characteristics. Predictability implies that the process characteristics over time belong to the same statistical distributions. In other words, a predictable process is considered to be in statistical control, meaning that over time, the measured characteristic belongs to the same statistical distribution; that is, with the same mean and dispersion (often measured in terms of standard deviation). Given a process in statistical control, the dispersion refers to the width of the distribution; high dispersion implies that it is more likely to end up far away from the process mean. Centering refers to how well the mean of the process distribution is aligned with the target value. Ideally, we would like the process to be predictable, and the corresponding distribution should have a low dispersion and be well centered.

The standard approach for variability reduction in Six Sigma is as follows:

1. Eliminate special cause variation to reduce the overall dispersion and bring the process into statistical control (i.e., improve the predictability).
2. Reduce the dispersion of the predictable process.
3. Center the process to target.

The Six Sigma philosophy is well in line with Deming's mantra of striving for continuous improvements as well as Taguchi's perception that any deviation from the process target value will cause excess costs.

The tools used in Six Sigma approaches to reduce variation are the same as in traditional quality and process control: the 7QC tools and factorial experiments. These tools will not be explored further in this book. The interested reader can turn to any basic textbook about quality control or quality management, such as Foster (2017) or Evans and Lindsay (2011).

Cycle time and *yield* are important characteristics of any process. The cycle time is the time a job spends in the process, sometimes also referred to as *process* or *production lead time*. The process yield or productivity is the amount of output per unit of input or per unit of time. Consequently, cycle time and yield can be used to describe a variety of process performance aspects including input material, equipment use, setup times, lead times, capacity, and productivity.

Six Sigma improvement projects that focus on cycle times and yield follow the same approach as for variation; namely, to gain predictability, reduce dispersion, and center with the target. However, when it comes to centering, the target is usually more broadly defined to minimize cycle time and maximize yield improvement. For example, consider the operations of a regional distribution center (DC) where trucks arrive with incoming goods. At the DC, the trucks are off-loaded and then loaded again with outgoing goods before they depart. For scheduling purposes, it is desirable to have trucks stay the same amount of time at the depot. The cycle time—the time a truck spends at the DC—is currently 2 $\pm$ 0.5 hours. Improving the performance in terms of predictability and reduced dispersion means that the trucks would depart more consistently at 2-hour intervals. To improve the centering, we would like to reduce the average cycle time without increasing the dispersion. Tentatively, by changing the off-loading and loading processes by acquiring more or better forklift trucks, the average cycle time could be reduced to 1.5 hours. This increases the number of trucks and the tonnage handled by the DC daily. In Six Sigma companies, this type of improvement would be recorded in terms of the cost savings due to the increased capacity and in terms of the reduced variability.

The focus on cycle time and yield is not unique to Six Sigma; it has been emphasized by many business improvement programs including Lean Manufacturing, time-based management, BPR, and ABM. However, Six Sigma adds the important requirement that improvements in average cycle time or yield must not be made at the expense of increased process variability.

The Six Sigma cost rationale can be summarized by the *efficiency loop* (sometimes called the *bottom-line loop*) illustrated in Figure 2.8. Improvement projects lead to cost savings by targeting the dimensions of variation, cycle time, and yield. Moreover, lower costs lead to increased profits. To complete the loop and achieve long-term effects, the last component is a commitment to the Six Sigma initiative and to individual improvement projects. Without a commitment from top management and other stakeholders, the Six Sigma initiative will not survive. Therefore, a core concern in Six Sigma is to generate this commitment and attention. An efficient Six Sigma strategy in this respect is to report the cost savings of every individual improvement project directly to top management.

2.3.2.2 The Revenue or Effectiveness Rationale

The determinants of a company's revenues are its sales volume, closely related to its market share, together with the prices it can charge for its products and services; both are highly dependent on the level of customer satisfaction. To put it differently, the firm's ability to generate revenues is contingent on how well it can satisfy the external customers' requirements and desires. According to Motorola, GE, and other Six Sigma companies, the Six Sigma recipe of improved process performance in terms of predictability, small dispersion, and good centering of characteristics important to the customers has been successful in this respect. The key to success is the focus in every improvement project on identifying and meeting the customer requirements, internal as well as external. An important concept in this respect is the CPS model discussed in Section 2.1. A critical insight is that the measured characteristics must be important not only for the process internally but also for its external customers.

To identify customer requirements and translate them into product and service characteristics, Six Sigma advocates the use of QFD. The customers are then asked not only about their critical requirements but also about the desired target values and specification limits for these requirements. The same approach is used whether the customers are external or internal. However, as discussed in Section 2.1, the internal customer requirements must be aligned with those of the external customers, who ultimately generate the revenues. A rigorous measurement system to make sure the processes consistently match the defined characteristics—produce few defects in the eyes of the customer—is another important ingredient in the Six Sigma pursuit of increased revenues.

The revenue rationale can be summarized by the effectiveness loop (sometimes referred to as the *top line loop*): *Improved customer satisfaction leads to increased market share and larger sales volumes as well as a potential to charge higher prices, together implying increased revenues.* Combining the effectiveness loop with the efficiency loop, as shown in Figure 2.9, summarizes the Six Sigma cost and revenue rationale.

2.3.3 Six Sigma in Product and Process Design

Six Sigma has proved to be a successful strategic initiative for improving process performance when considering the dimensions of variation, cycle time, and yield by eliminating special cause variation. However, improvement projects eventually will arise for which it is not possible to reach the desired levels of performance simply by eliminating the special cause variation. In these situations, it is necessary to question the very design of the process.

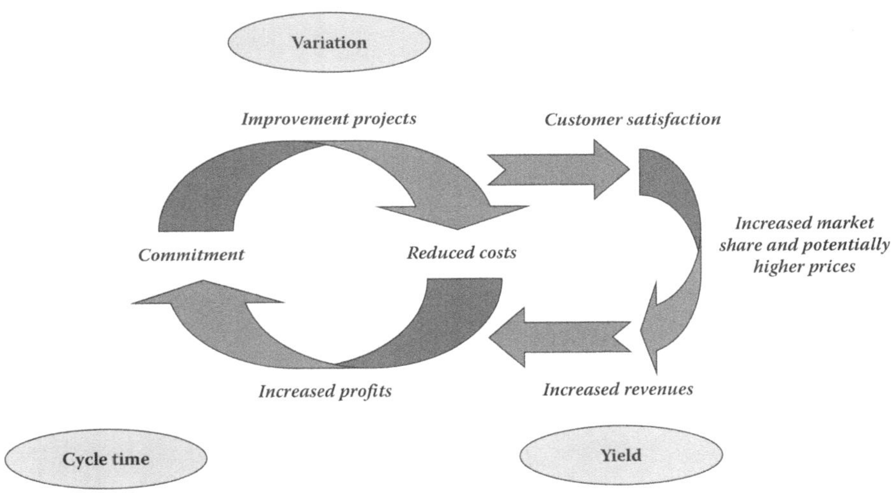

FIGURE 2.9
Six Sigma cost and revenue rationale combining the efficiency and effectiveness loops.

Six Sigma prescribes applying the same DMAIC methodology and the same statistical tools for improving the process or product design as for improving the performance of a process with a given design. In addition, prototyping and simulation are typical Six Sigma design activities. Experience shows that using the Six Sigma program for the design and implementation of new processes is considerably more complex than using it within an existing design. At the same time, it offers tremendous bottom-line potential.

The Six Sigma approach to product and process design has three complementary objectives or design steps.

1. *System design*: To design the product/process to satisfy and delight the customers
2. *Parameter design*: To make the design less sensitive to variation by determining appropriate values for the design parameters
3. *Tolerance design*: To reduce the process variability by narrowing tolerances on input material and work performed

2.3.4 The Six Sigma Framework

As mentioned earlier, the Six Sigma framework, depicted in Figure 2.10, is the vehicle used to attain the stated objectives and fulfill the Six Sigma rationale. The framework encompasses five major components: top management commitment; stakeholder involvement; training; measurement system; and at the very core, the five-step improvement methodology DMAIC. The formalized framework, with its disciplined and quantitatively oriented improvement methodology, is what sets Six Sigma apart from many other management programs. Therefore, this section is devoted to taking a closer look at each of these framework components.

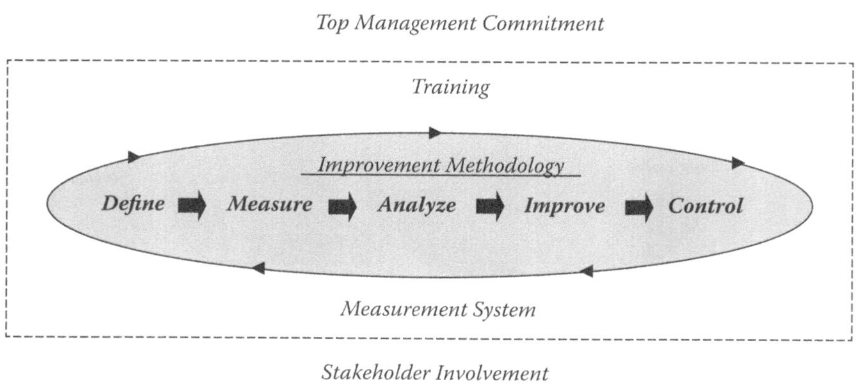

FIGURE 2.10
Six Sigma framework.

2.3.4.1 Top Management Commitment

To leverage the full potential of Six Sigma, it must be an integral part of the firm's business strategy. This requires long-term management commitment. In a sense, the entire Six Sigma framework is relying on top management commitment to materialize. Without it, the ambitious Six Sigma program will not survive. The commitment must go beyond lip service and manifest itself in pragmatic and involved management to push the Six Sigma concept out to every part of the organization. However, this does not imply that top management should be engaged in daily improvement activities; their role is as owners of the Six Sigma initiative and project sponsors. A strength of Six Sigma, compared with many other improvement programs, is its bottom-line focus, which tends to help keep top management's motivation and commitment at a high level.

2.3.4.2 Stakeholder Involvement

A successful Six Sigma program needs active support and involvement from all its important stakeholders, with particular emphasis on the employees, the suppliers, and the customers. Other stakeholders are owners and, in a broad sense, society.

The most important stakeholder group with respect to Six Sigma is the people in the organization. Without their active support and involvement, the initiative will never take off. Six Sigma tries to ensure this involvement through attractive training programs, the formalized improvement methodology, and frequent feedback on process performance and rates of improvement. Constructive union involvement is in many cases also a necessary prerequisite for workforce involvement.

The suppliers are another critical stakeholder group because wide variations in the quality and characteristics of the input immediately affect the firm's processes and their output. Therefore, key suppliers are often encouraged and supported in launching their own Six Sigma initiatives. Furthermore, a common Six Sigma strategy for involving suppliers is to share the performance data on the input they deliver.

As recognized in the Six Sigma revenue rationale, the importance of the customers and their perceptions cannot be overestimated. Six Sigma cannot reach its full potential without understanding and involving the organization's customers. Ways for involving the firm's customers include identifying their requirements and desires, having customers participate in internal Six Sigma training courses, and actively helping customers improve their processes using the Six Sigma methodology.

2.3.4.3 Training

One important characterizing feature of Six Sigma is a comprehensive and formalized training program. The training is largely focused on understanding the statistical tools used for the data analysis, which is a cornerstone of the Six Sigma improvement methodology. However, the courses also cover topics such as process performance, the DMAIC improvement methodology, project management, quality function deployment, and the Six Sigma framework. For Six Sigma to succeed, this knowledge has to be diffused throughout the entire organization.

Although the structure might vary, the Six Sigma strategy includes three fairly standardized training course levels: basic, medium, and comprehensive. They differ in scope, detail, and practical application. Comprehensive training is reserved for those employees who will devote the majority of their time to running Six Sigma improvement projects throughout the organization. Basic training is offered to all employees to prepare them for participating in improvement projects at their own workplace. As more and more projects are being performed, an increasing number of people are exposed to the Six Sigma approach, which raises the level of understanding even further. A distinguishing and appreciated feature of the Six Sigma training programs is that they go beyond classroom teaching and require the participants to apply their knowledge in hands-on improvement projects.

A common way to denote the hierarchical roles in the Six Sigma program is to use a martial arts–inspired belt rank system. Typically, the belt levels are White Belts, Green Belts, Black Belts, Master Black Belts, and Champions in increasing order of training and responsibility. Sometimes, Yellow Belts are included between the White and Green Belts. Relating the training program to these different roles, the White Belt course is typically a 1-day course that provides a basic introduction to Six Sigma. It is offered to all or most of the workforce. The Green Belt course usually targets foremen and middle management. It teaches the participants to apply the DMAIC methodology, including the necessary statistical tools, in a real project. The Black Belt course is a comprehensive course for full-time improvement experts. Although it has no formal prerequisites, it is beneficial if the participants on this level have a basic understanding of mathematics, statistics, and the firm's core processes. The Black Belt training lasts for about 6 months, and it consists of several weeks of seminars, between which the participants are required to work on different improvement projects with specified cost-savings requirements. The Black Belts represent the force that drives the operational process improvement activities. They typically are handpicked from the ranks of the most promising young leaders in the organization. The Master Black Belt is a qualified Black Belt who works full-time teaching Six Sigma courses. The Champion is a senior executive whose primary duties are to drive the improvement work forward, to be an advocate for the Six Sigma program in the top management circle, and to serve as a source of knowledge and experience. In addition, the Champion often participates in the selection of improvement projects.

Apart from the three courses mentioned, the standardized Six Sigma training program also includes a course in Six Sigma engineering and one in Six Sigma management. The former is a Black Belt course that focuses on the design of products and processes, and the latter deals with Six Sigma rollout issues and how to create the necessary momentum to keep the program moving.

2.3.4.4 Measurement System

A central issue in Six Sigma is to base decisions on factual data and to quantify the effects of implemented improvements. Therefore, a measurement system that collects the relevant data is of critical importance. As discussed previously, the preferred measurement in Six Sigma is variation, because it can be used to measure the dispersion and centering of any characteristic of interest, including cycle time or yield. In deciding which characteristics to monitor, the focus should be on what is important to the customers. The universal variation metric used in Six Sigma is dpmo, and the goal is to achieve fewer than 3.4 dpmo for all products and processes. Because all individual characteristics are measured using the same metric, it is easy to track performance over time; to compare different processes; and to consolidate individual measures for larger processes, projects, classes of products and services, and even the entire company. The measurement system helps emphasize the process performance issues throughout the entire organization. A keyword in this respect is *simplicity*—a single metric that is easy to understand and easy to remember.

2.3.4.5 The Improvement Methodology

At the core of every Six Sigma project is the same formalized improvement methodology, depicted in Figure 2.11, consisting of the five phases: define, measure, analyze, improve, and control. As mentioned earlier, the methodology rests on a quantitative philosophy, emphasizing that decisions must be based on factual data, not guesses. The Six Sigma training, therefore, focuses on applying appropriate statistical tools in each of the five phases.

The *define* phase deals with the selection of appropriate improvement projects.

Valuable sources of information in identifying candidate projects and areas of improvement include the aforementioned measurement system, customer complaints, customer satisfaction surveys, nonconformity reports, returns, warranty claims, and employee suggestions. After identifying a potential

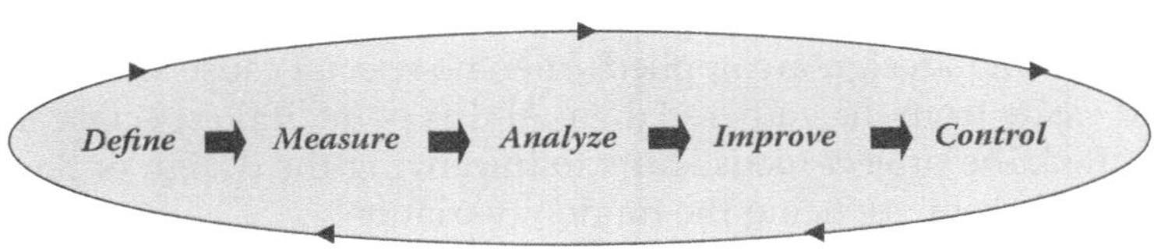

FIGURE 2.11
Six Sigma improvement methodology DMAIC.

area of improvement, the next step is to define the project scope and the processes involved. In some cases, the scope might be limited to improving the performance of just one characteristic, but other projects might involve a complete redesign of entire processes or products. Having identified a number of candidate projects, the question is which project to pursue. Commonly used selection criteria include customer and company benefits, complexity, cost savings potential, and likelihood of success.

After a project has been selected, it is assigned to an improvement team consisting of an appropriate mix of Black, Green, and White Belts. Larger projects also include a top management sponsor or champion.

The *measure* phase involves decisions on which characteristics to improve and what to measure in more detail. The general measurement system might not be detailed enough for the project in question, implying the need for additional control and measure points. Apart from identifying the characteristics or result variables to be improved, it is also important at this stage to identify and measure the critical input variables affecting the output. The input variables are classified as either control factors or noise factors. Control factors are input variables that can be affected and controlled in the short term, and noise factors are outside the direct control of the improvement team. The control factors typically have an immediate impact on the special cause variation in the output variables; this is the natural starting point for improvement. Another important issue in the measure phase is to define what constitutes unacceptable performance or a defect. Finally, enough data need to be collected to assess the current performance of the process and provide the foundation for improving it.

The *analyze* phase uses the preliminary data gathered in the *measure* phase to document the current performance and create a baseline against which to gauge the improvement efforts. The analysis typically involves calculations of means, standard deviations, and dpmos for the targeted characteristics and control factors. Moreover, control charts are often used to assess whether the process can be considered to be in statistical control. To better assess the relative performance and set appropriate improvement goals, it is also common to benchmark against other similar processes or products internally or externally.

Based on the analyzed data, the *improve* phase focuses on how to achieve the necessary improvements in predictability, dispersion, and centering. A core activity is to identify and eliminate root causes for nonrandom variability by going for the simplest improvement opportunities first. The 7QC tools are used extensively for these purposes. Advanced statistical tools including experimental design are applied when no special cause variation can be detected easily from the gathered data. At this point, if no special causes can be identified, the project focus shifts to improving the design of the process or product, thereby reducing the random variation.

After the desired improvements have been achieved, the *control* phase is initialized to verify and institutionalize the change. Important activities include process monitoring using control charts and documentation of changes that are made using, for example, flowcharts and formalized reporting of estimated cost savings. The experiences and results from the improvement project also need to be made available to the rest of the organization. An important lesson from Six Sigma companies is that projects breed projects, and sharing experiences has the twofold effect of transferring knowledge and spurring interest in the program.

2.3.5 Control Charts

A powerful approach for monitoring the performance of the process according to the measures established during the design effort is statistical process control (SPC). One of the main tools in SPC is process control charts (or Shewhart charts). Control charts are essential tools in the Six Sigma methodology, visually representing process performance over time and highlighting when a process is out of control.

A control chart is a statistical tool used in quality control to monitor and analyze process variation. No process is free from variation, and it is vital to understand and manage this variation to ensure consistent and high-quality output. The control chart is designed to help visualize this variation over time and identify when a process is out of control. The control chart achieves this by providing a visual representation of the process data over time and highlighting any special causes of variation that may be present. The primary objective of using a control chart in Six Sigma is to ensure that a process is in a state of statistical control. This means that the process is stable and predictable, and any variation is due to common causes inherent in the process.

The chart typically includes a central line, which represents the average or mean of the process data, and upper and lower control limits, which are set at a certain number of standard deviations from the mean. The control limits are usually set at three standard deviations from the mean, encompassing about 99.7 percent of the process data. If the process data falls within these control limits, the process is considered in control, and variation is deemed to be coming from common causes. If the data points fall outside these control limits, this indicates that there is a special cause of variation, and the process needs to be investigated and improved.

There are seven Six Sigma charts. Each chart has its specific use and is suitable for analyzing different data types.

1. The I-MR Chart, or Individual-Moving Range Chart, analyzes one process variable at a time. It is suitable for continuous data types and is used when the sample size is one. It consists of two charts, one for individual values (I Chart) and another for the moving range (MR Chart).
2. The X Bar R Chart is used to analyze process data when the sample size is more than one. It consists of two charts, one for the sample averages (X Bar chart) and another for the sample ranges (R Chart). It is suitable for continuous data types.

3. The X Bar S Chart is like the X Bar R Chart but uses the sample standard deviation instead of the range. It is suitable for continuous data types. It is used when the process data is normally distributed, and the sample size is more than one.
4. The P Chart, or the Proportion Chart, is used to analyze the proportion of nonconforming units in a sample. It is used when the data is binary (conforming or nonconforming), and the sample size is large.
5. The NP Chart is like the P Chart but is used when the sample size is fixed. It monitors the number of nonconforming units in a sample.
6. The C Chart, also known as the Count Chart, is used to analyze the number of defects in a sample. It is used when the data is discrete (count data), and the sample size is large.
7. The U Chart, or the Unit Chart, is used to analyze the number of defects per unit in a sample. It is used when the sample size is variable, and the data is discrete.

In the following two subsections, we use a call center example to illustrate how to create a X Bar R Chart and an I-MR Chart. Additional principles of SPC and the use of control charts are covered in any basic textbook on quality control or quality management. See, for example, Foster (2017) and Evans and Lindsay (2011). For a more advanced treatment of the technical aspects, see, for example, Duncan (1986).

2.3.5.1 Average Waiting Time in a Call Center

Waiting on-hold to speak to an agent at a call center can be frustrating. The manager of a call center wants to analyze the performance of the process of answering phone calls by means of an X Bar R Chart. The manager assumes that the individual measurements (waiting time) follow a normal distribution (waiting time is usually not normally distributed, but we will assume it for this example). The manager collects a sample of the waiting time for five calls per day for 20 days, as shown in Table 2.1.

To correctly interpret an X Bar R Chart, the manager must examine the R chart first. Therefore, the first step after collecting the data is to build the R chart. The center line of the chart is the average range (denoted by $\bar{R}$), which is calculated as the sum of the range in each sample (i.e., day) divided by the number of samples (i.e., number of days). We denote with X_{ij} the value (waiting time) of observation (call) j for sample (day) i. For a dataset with n samples and m observations per sample, the range of sample i is given by:

$$R_i = \max_{j=1,\ldots,m} X_{ij} - \min_{j=1,\ldots,m} X_{ij}$$

In our example, the range of waiting times on day 1 is:

$$R_1 = 4.5 - 1.7 = 2.8$$

TABLE 2.1

Waiting Time (in Minutes) for 100 Calls

Day	Call 1	Call 2	Call 3	Call 4	Call 5
1	3.8	1.9	1.7	1.9	4.5
2	3.3	3.5	3.1	2.7	4.1
3	2.5	3.2	2.0	4.2	4.5
4	2.5	4.0	1.8	3.8	2.5
5	1.7	2.9	4.1	3.1	3.2
6	1.3	2.0	1.8	2.6	3.2
7	1.6	3.8	3.8	3.2	1.4
8	5.0	1.4	1.2	3.4	2.2
9	3.8	2.5	2.0	5.0	8.0
10	3.4	3.3	2.9	3.4	4.9
11	2.8	3.6	3.9	2.7	4.3
12	2.6	1.2	4.3	4.1	3.9
13	1.5	1.6	3.2	3.2	4.8
14	1.3	3.2	3.8	2.4	1.6
15	1.6	2.1	2.4	1.1	3.5
16	2.7	4.3	3.6	1.3	1.5
17	1.8	1.1	2.5	3.1	4.1
18	2.3	3.3	2.9	4.6	4.2
19	2.0	4.9	1.7	3.3	4.2
20	1.5	2.6	1.6	3.8	3.8

The average range is then given by:

$$\bar{R} = \frac{\sum_{i=1}^{n} R_i}{n}$$

In our example, the manager calculates the average range as follows:

$$\bar{R} = \frac{2.8 + 1.4 + \cdots + 2.3}{20} = 2.7$$

The lower control limit (*LCL*) and the upper control limit (*UCL*) for the R chart are calculated as follows:

$$LCL_{\bar{R}} = D_3\bar{R}$$

$$UCL_{\bar{R}} = D_4\bar{R}$$

The values of the D_3 and D_4 constants depend on the number of observations in the sample (i.e., the sample size). Table 2.2 shows the values for samples of size ranging from 2 to 10.

TABLE 2.2

Constants for X Bar and R Charts

Sample Size	A_2	D_3	D_4
2	1.880	0	3.267
3	1.023	0	2.574
4	0.729	0	2.282
5	0.577	0	2.114
6	0.483	0	2.004
7	0.419	0.076	1.924
8	0.373	0.136	1.864
9	0.337	0.184	1.816
10	0.308	0.223	1.777

In the call center data, the number of observations in each sample is 5. The manager uses this information to calculate the control limits for the R chart as follows:

$$LCL_{\bar{R}} = 0 * 2.7 = 0.0$$

$$UCL_{\bar{R}} = 2.114 * 2.7 = 5.7$$

In an X Bar R Chart, the X Bar chart control limits are derived from the R Bar (average range) values. If the R chart's values are out of control, the X Bar chart control limits are inaccurate. Figure 2.12 shows the R chart associated with the waiting time data. The chart shows that the range on day 9 falls outside the upper control limit. Therefore, the R chart indicates that the process is out of control.

A point in the R chart that is out of control indicates a special cause of variation. Note that all other points in the chart vary inside the control limits. This is considered a "normal" variation caused by everyday randomness. However, the manager must investigate the root cause of the variation on day 9. Once the cause is determined and the issue is addressed, the sample can be removed, and the manager can build the X Bar chart. The process starts with the calculation of $\bar{X}_i$ for each sample (day) i.

$$\bar{X}_i = \frac{\sum_{j=1}^{m} X_{ij}}{m}$$

The manager can use this formula to calculate the average waiting time on each day. For example, for the first day the average waiting time is:

$$\bar{X}_1 = \frac{3.8 + 1.9 + 1.7 + 1.9 + 4.5}{5} = 2.8$$

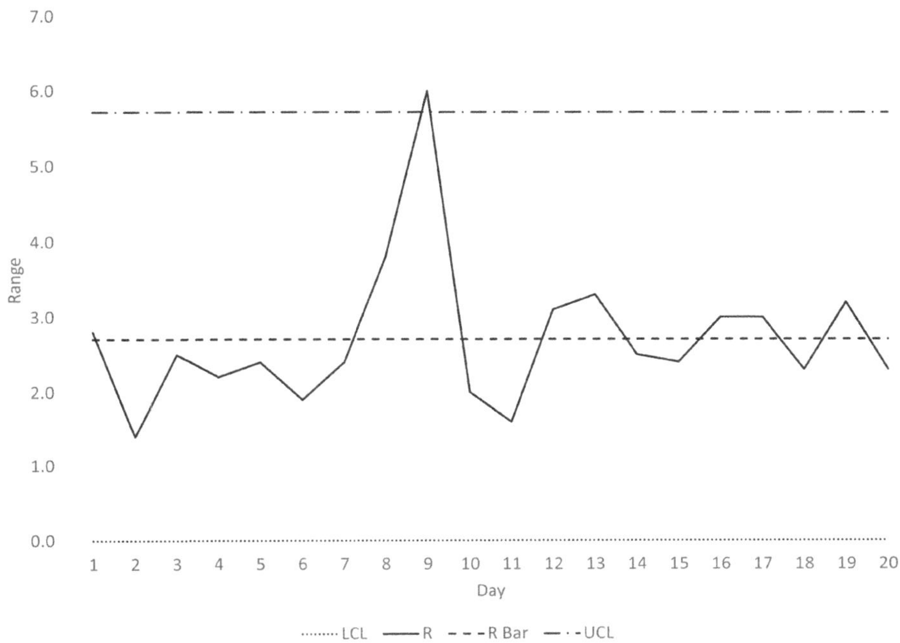

FIGURE 2.12
R chart for waiting time data.

Then, the overall average $\overline{\overline{X}}$ becomes the center line of the X Bar chart:

$$\overline{\overline{X}} = \frac{\sum_{i=1}^{n} \overline{X}_i}{n}$$

For the waiting time data, the center line is:

$$\overline{\overline{X}} = \frac{2.8 + 3.3 + \cdots + 2.7}{19} = 2.9$$

Note that we divide by 19 because we assume that the manager found the root cause of the special variation on day 9 and therefore that sample was removed from the data. The control limits for the X Bar chart are calculated with the following formulas:

$$LCL_{\overline{X}} = \overline{\overline{X}} - A_2\overline{R}$$

$$UCL_{\overline{R}} = \overline{\overline{X}} + A_2\overline{R}$$

The value of the constant A_2 depends on the sample size, as shown in Table 2.2. The manager then calculates the control limits for the X Bar chart as follows:

$$LCL_{\bar{X}} = 2.9 - 0.577 * 2.5 = 1.4$$

$$UCL_{\bar{R}} = 2.9 + 0.577 * 2.5 = 4.4$$

Note that the new value of $\bar{R}$ is 2.5, once day 9 has been removed from the data. Figure 2.13 shows the X Bar chart. The chart shows that the process is in control because all points fall within the control limits.

The manager now must determine whether the average performance of the process is acceptable. After the special cause of variation was identified and the observations on day 9 were removed, the X Bar control chart indicates that the average waiting time is 2.9 minutes. It also predicts that as long as the process remains within control, the average waiting time on any given day may be as low as 1.4 minutes and will not exceed 4.4 minutes. Note that the chart gives information on the average waiting time and not on individual waiting times. Therefore, with this analysis, the manager cannot make any claims about an individual waiting time, which can certainly fall outside the control limits even when the process is in control (e.g., call 5 on day 13 in Table 2.1 had a waiting time of 4.8 minutes). For information on the

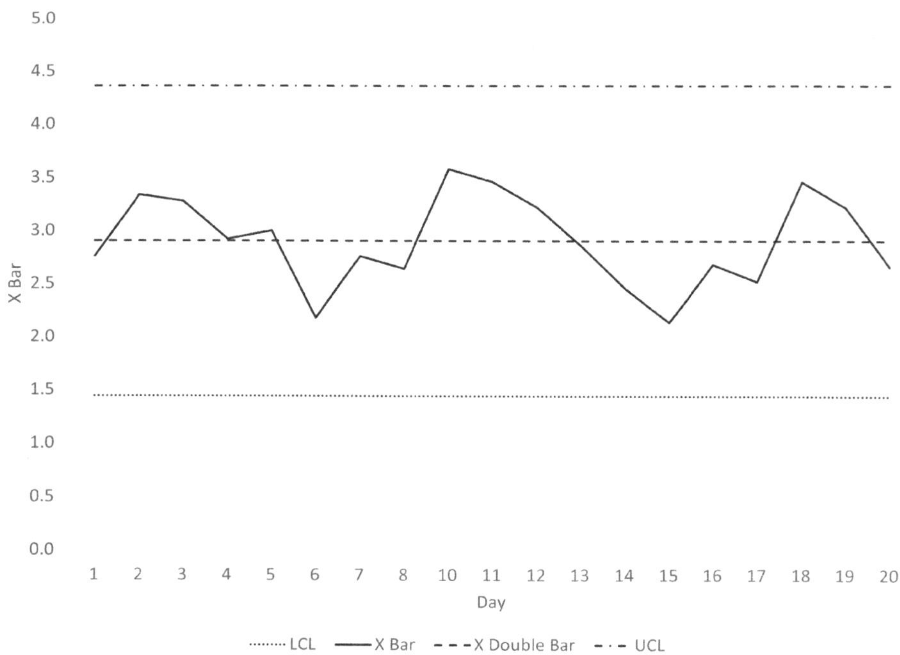

FIGURE 2.13
X Bar chart for waiting time data.

variability of individual calls, the manager would need to analyze the process with an Individual-Moving Range Chart.

2.3.5.2 Individual Waiting Time in a Call Center

The manager in the previous example would like to give customers a guarantee for the maximum waiting time. In particular, the manager would like to promise that the wait will never be longer than 5 minutes. The manager decides to use the I-MR Chart to test whether the current process meets the proposed operational standard. (For this example, we assume that the observations, that is waiting times, were made at regular intervals of time within days and across days. This assumption is necessary to be able to build an I-MR Chart.) The manager must first calculate the average moving range ($\overline{MR}$):

$$\overline{MR} = \frac{\sum_{i=2}^{n} |X_i - X_{i-1}|}{n-1}$$

The formula assumes that the data contains n individual observations. In the call center example introduced in Section 2.3.5.1, $n = 95$ calls after day 9 was removed due to the out-of-control situation identified in the previous analysis. Individual ranges are calculated using consecutive observations, starting with the second observation in the dataset. Since the formula is the absolute difference between two consecutive observations, range values cannot be negative. Because the formula starts with the second observation, the number of ranges is $n - 1$. For the call center, the manager has 95 range values, starting with $|1.9 - 3.8| = 1.9$ and ending with $|3.8 - 1.6| = 2.2$. With these values, the manager calculates:

$$\overline{MR} = \frac{1.9 + \cdots + 2.2}{94} = 1.3$$

The center line of the MR Bar chart is $\overline{MR}$ and the control limits are:

$$LCL_{\overline{MR}} = D_3 \overline{MR}$$

$$UCL_{\overline{MR}} = D_4 \overline{MR}$$

Since in our example the manager used a moving average of two consecutive observations, the sample size is two. The constant values from Table 2.2 corresponding to a sample size of two are $D_3 = 0$ and $D_4 = 3.267$, resulting in the following control limits for the MR bar chart:

$$LCL_{\overline{MR}} = 0 * 1.3 = 0$$

$$UCL_{\overline{MR}} = 3.267 * 1.3 = 4.1$$

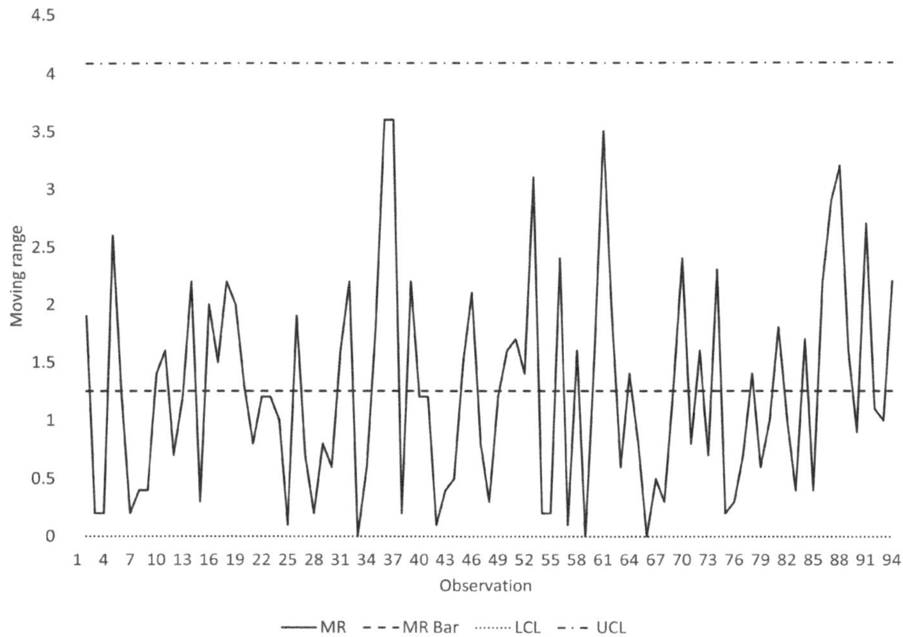

FIGURE 2.14
MR Bar chart for waiting time data.

Figure 2.14 shows the MR chart. All moving range values are less than the upper control limit and therefore the process is in control regarding variation. This means that the manager can proceed with the analysis of the individual waiting times.

To create a X Bar chart for individual observations, the center line is the average of all the individual values:

$$\bar{X} = \frac{\sum_{i=1}^{n} X_i}{n}$$

For the call center data, the manager calculates:

$$\bar{X} = \frac{3.8 + 1.9 + \cdots + 3.8}{95} = 2.9$$

The control limits are calculated as follows:

$$LCL_{\bar{X}} = \bar{X} - \left(A_2\sqrt{m}\right)\overline{MR}$$

$$UCL_{\bar{X}} = \bar{X} + \left(A_2\sqrt{m}\right)\overline{MR}$$

In these formulas, m is the sample size and A_2 is the constant in Table 2.2. For the call center, the sample size is 2 and $A_2 = 1.88$:

$$LCL_{\bar{X}} = 2.9 - \left(1.88\sqrt{2}\right)1.3 = -0.6$$

$$UCL_{\bar{X}} = 2.9 + \left(1.88\sqrt{2}\right)1.3 = 6.3$$

For the call center, the lower control limit can be adjusted to zero since the waiting time cannot be negative. Figure 2.15 shows the X Bar chart for the waiting time data.

The X Bar chart in Figure 2.15 indicates that the process is in control. This means that as long as the process stays the same, the manager can predict, within a range, how long individual waiting times will be. The control limits indicate that a customer will be put on hold for anywhere between 0 and 6.3 minutes. What does this mean for the manager's plan to commit that no call will be put on hold for more than 5 minutes? It means that the current process is not capable of meeting that guarantee. Based on the data collected by the manager, no call has been put on hold for more than 5 minutes yet. But it is just a matter of time before one will.

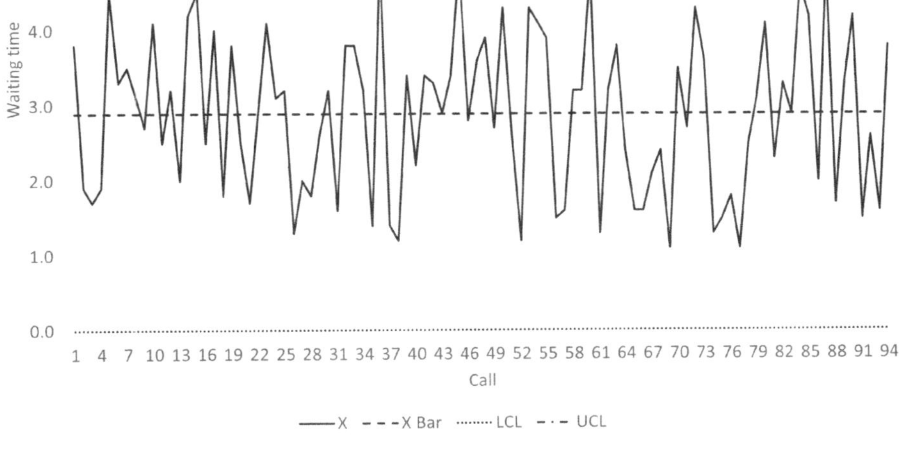

FIGURE 2.15
X Bar chart for the individual waiting time data.

2.3.6 Key Reasons for the Success of Six Sigma

To conclude our discussion on Six Sigma, we will summarize some of the acknowledged reasons for its ongoing success.

The bottom-line focus and big dollar impact encourages and maintains top management commitment and support, which is a necessity for the program to succeed in the long term.

The emphasis on and consistent use of a unified and quantitative approach to process improvement facilitates communication and real results. The disciplined application of the DMAIC methodology creates a common language through which people from different business units can share experiences and learn from each other's successes and failures with regard to process improvement efforts. It also creates awareness throughout the organization that successful process improvement must be based on factual data and not guesswork. Vague statements such as "We believe that the cycle time has been significantly reduced" are no longer acceptable. The standard answer to this is "Show me the data."

The emphasis placed on understanding and satisfying customer needs ensures that the process improvements focus not only on the efficiency aspects (to do things right) but also on the effectiveness issues (to do the right things). This ensures that the Six Sigma revenue rationale is fully leveraged. The focus on quantitative metrics forces Six Sigma companies to quantify and document customer needs and the current ability to satisfy them. Anecdotal information is replaced by reliable factual data gathered through formal customer interactions and evaluations.

The combination of the right projects, the right people, and the right tools is probably the most notable strength of the Six Sigma approach. By carefully selecting important projects and training the most talented people to apply the appropriate statistical tools on these projects, Six Sigma companies have achieved remarkable synergies. In the past, tools-oriented approaches have had mixed results because of the inherent risk of focusing on finding applications for the tools rather than finding appropriate tools for the most important problems. The tools used in Six Sigma are not new; the novelty is in the way the Six Sigma approach manages to integrate them into the improvement process. A key success factor is the rigorous Six Sigma training program and its focus on the hands-on application of the course material to real projects.

2.4 Business Process Management

Reengineering, or BPR (business process reengineering) as it is commonly known, became a buzzword in the 1990s, spurring a great interest in process design. The essence of the reengineering philosophy is to achieve drastic improvements by completely redesigning core business processes, that is, by rethinking the way business is conducted. To relate this to the distinction between process design and implementation, reengineering advocates radical design changes and fast revolutionary implementation to achieve drastic improvements. Numerous articles and books addressing this topic have been written since the publication of Michael Hammer's seminal article "Reengineering work: Don't automate, obliterate" (1990) and the contemporaneous work by Thomas Davenport and James Short, "The new industrial engineering" (1990). The popular press also actively reported on success stories as well as implementation failures of BPR. Some of the success stories are familiar, as reported in a BPR tutorial article by Grover and Malhotra (1997):

- "Ford cuts accounts payable headcount by 75 percent"
- "Mutual Benefit Life (MBL) improves insurance underwriting efficiency by 40 percent"
- "Xerox redesigns its order-fulfillment process and improves service levels by 75 percent to 97 percent and cycle times by 70 percent with inventory savings of $500 million"
- "Detroit Edison reduces payment cycles for work orders by 80 percent"

On the downside, estimates suggested that approximately 50–70 percent of reengineering projects failed to achieve the dramatic results anticipated in their objectives (Hammer and Champy, 1993). Moreover, even the so-called successful projects, such as the reengineering of Ford Motor Company's accounts payable process, sometimes took 5 years or longer to implement and yield positive results (Petrozzo and Stepper, 1994). Several issues were cited as reasons for these failures and long delays:

- Lack of top-management support and leadership (a champion) for the project.
- Poor understanding of the organization and infrastructure needed to support the new design.
- Inability to deliver the necessary advances in technology.
- Lack of expert guidance and motivation.

Hammer and Champy's reasons for failure are more detailed and include several observations gathered during business process reengineering efforts in which they have participated:

- Attempting to fix a process instead of changing it.
- Lack of focus on business processes.
- Neglecting people values and beliefs.

- Willingness to settle for minor results.
- Quitting too early.
- Placing prior constraints on the definition of the problem and the scope of the redesign effort.
- Allowing existing corporate cultures and management attitudes to prevent the reengineering effort from getting started.
- Trying to make business process design changes happen from the bottom up.
- Selecting a project leader who does not understand the key issues associated with reengineering business processes.
- Skimping on the resources devoted to the reengineering effort.
- Burying reengineering efforts in the middle of the corporate agenda.
- Dissipating energy across too many reengineering projects.
- Attempting to change core business processes when the chief executive officer (CEO) is 2 years from retirement.
- Failing to distinguish process reengineering from other business improvement programs.
- Concentrating exclusively on design (and hence ignoring everything except process redesign).
- Trying to change business processes without making anybody unhappy.
- Pulling back when people resist making design changes; dragging out the effort.

Hammer and Champy believed that the common thread that runs through all of these pitfalls could be traced to the role of senior management. It is interesting to note that for many years, W. Edwards Deming advocated the same notion in reference to failures when implementing total quality programs (Gabor, 1990). Furthermore, it falls in line with the focus on top management commitment in Six Sigma discussed in Section 2.3. Although the frequency of failure is considerable, organizations that approach the design of business processes with understanding, commitment, and strong executive leadership are likely to reap large benefits. This implies that companies must be clear about what they are trying to accomplish when designing effective business processes and how they will go about doing it. Many experts believe that a considerable number of failures associated with the reengineering of business processes can be attributed directly to misunderstanding the underlying philosophy behind BPR.

Just as the experts have identified reasons for failure, they also point to several reasons for the success of projects that change processes with the goal of radically improving their efficiency. For example, Cross et al. (1994) mention that some common themes appear in many highly successful reengineering efforts. The following similarities can be found in successful implementations:

1. Companies use process design to grow the business rather than retrench.
2. Companies place emphasis on serving the customer and on aggressively competing with the quantity and quality of products and services.
3. Companies emphasize getting more customers, more work, and more revenues, not simply concentrating on cutting back through cost reductions and downsizing.

BPR critiques and failures as well as abuses and misuses of the concept resulted in a decline in corporate interest for engaging in full-blown BPR projects. However, the focus on business process improvement has become a widely accepted approach and is standard within the change methodology portfolio. Changes in this framework are performed in a less radical way than originally proposed within BPR, leading companies toward the concept of business process management.

Business process management employs methods to discover, model, analyze, measure, improve, and optimize business strategy and processes. BPM can be thought of as a continuous process reengineering to streamline workflows, leading to increased efficiencies and cost-savings. Therefore, Six Sigma and Lean principles can be thought of as examples of BPM methodologies. Business process management, through advanced analytics and activity monitoring, coordinates people, systems, and information to achieve business outcomes. As a result, BPM has been particularly helpful in accelerating digital transformation strategies. The following subsections have been adapted from IBM's view of business process management.[1]

2.4.1 Types of BPM

There are three main types of business process management:

Integration-centric BPM: This type of BPM focuses on processes that do not require much human involvement. These processes are typically dependent on application programming interfaces and mechanisms that integrate data across systems. Human resource management and customer relationship management are examples of such systems.

Human-centric BPM: This type centers around human involvement, typically where approvals are required. Intuitive user interfaces with drag-and-drop features allow those involved in the process to assign tasks to different roles, making it easier to hold individuals accountable along the entire process.

Document-centric BPM: This type of BPM focuses on documents, for instance, a contract. When companies purchase a product or service, the purchase involves various forms and rounds of approval to arrive at an agreement between the client and the vendor.

2.4.2 BPM Lifecycle

A successful BPM system starts with the identification of the stages in a workflow. This helps management identify areas of improvement and metrics designed to track progress. By applying business process management, organizations can improve their operations, leading to better business outcomes. To achieve these results, companies must understand the BPM lifecycle. These are the five lifecycle steps:

Model: Creating a visual representation of the process is the first step in the modeling activity. The model should include details such as timelines, task descriptions, and data flows in the process. Business process management software (such as computer simulation) is a helpful tool at this stage.

Process design: The model is then used to configure a process design capable of achieving the process milestones. The individual tasks within the model and the workflow logic are instrumental in determining resource responsibilities. The resulting design should include operational rules, resource capacity, workflow logic, and all the metrics needed for future improvement and optimization.

Execute: The simulation model should be capable of predicting performance. Based on those results, management might decide to conduct a proof of concept, testing the new BPM system on a limited basis. After incorporating any feedback, making changes to the model, and estimating future performance, the team can begin to roll out the process to a broader audience.

Monitor: During this phase, the process is monitored to measure performance and identify areas of improvement.

Optimize: With the information obtained from the monitoring activity, process optimization is performed on the simulated process. Typical optimization is associated with improving efficiency (i.e., increased output with the same or fewer resources), higher resource utilization, decreased waste, and a higher level of customer service. After an optimization round, the cycle goes back to the execution step.

Successful BPM projects require careful planning, open communication, and management commitment. When done well, the benefits of BPM projects become immediately obvious to multiple stakeholders within the company.

2.4.3 BPM Potential Benefits

BPM solutions increase organizational value through process improvement, yielding several benefits. Some of these include:

Increased efficiency and cost savings: BPM systems support the optimization of existing processes and incorporate a high level of structure into the development of new processes. This is achieved by removing process redundancies and bottlenecks, resulting in improved efficiency and productivity. With increased agility, businesses can achieve their intended business outcomes at a faster rate, and they can allocate any surplus resources to other high priority work.

Enhanced employee and customer experience: BPM strategies and tools focus on eliminating wasteful work and on making relevant information accessible where it is more useful. By removing waste, employees focus on productive work and customer satisfaction, leading to achieving business objectives. Well-designed business processes shorten the learning curve of employee onboarding, enhancing productivity and engagement.

More scalable processes: Since BPM seeks to enhance process execution and enable workflow automation, this facilitates transferring and scaling to regions across the world. Business process management principles and tools can bring clarity to roles, ensuring consistency along the process, and they can also reveal opportunities to incorporate business rules for automation, encouraging innovation.

Greater transparency: Accountability and transparency are the result of process design and automation. Transparency and accountability foster communication along the process, resulting in fewer errors and rework.

Less dependency on development teams: BPM tools typically offer low-code features that remove potential dependencies on specialized-software development. BPM strives for simple onboarding of users, breaking barriers to technology adoption across the company.

2.4.4 Typical Areas of Application

BPM tools provide process structures that attempt to remove inefficiencies in workflows. Some examples where BPM has been applied successfully include:

Content distribution: Media firms can use BPM to automate the process of content preparation and delivery, from content creation to distribution. A business process management system can be designed to interact with content management, rights management, content traffic, and work order systems.

Customer service: Customer service representatives can identify frequently asked questions for chatbots to handle, increasing efficiency during a high volume of service requests. Transcript data from call centers and chatbots can also help automate processes further as well as inform more personalized answers for customers.

Procurement: Companies can improve processes to control purchase-order submissions from various parts of the organization. Workflows can be established for routine transactions as well as being effective at handling exceptions.

Human resources (HR): HR can employ BPM to streamline document and workflow management. It provides a more structured environment for processing HR forms, such as employee onboarding and off-boarding, performance evaluations, vacation requests, and timesheet approval.

Banking: When processing individuals or businesses for loans, banks need to evaluate applicants for potential credit risk. This involves collecting information from multiple sources, such as applicants, employers, and credit rating agencies. BPM attempts to expedite decisions around loan eligibility by managing the flow of information throughout the process and reducing errors in documentation.

Order fulfillment: Firms can improve the operational efficiency of their order fulfillment systems using BPM. They can use BPM to manage special offers, order capture, and order fulfillment. This has the potential of configuring processes that are customer centric, delivering greater business value through effective and efficient order management.

2.5 Revolutionary versus Evolutionary Change

The traditional definition of *reengineering* emphasized radical redesign and rapid implementation leading to revolutionary change. However, in reality, constraining factors often arise, such as union concerns, training and IT needs, strong culture, and cemented organizational boundaries that make a rapid implementation at best very costly. In fact, this is one of the main reasons that many reengineering projects took a long time to implement and, therefore, were considered failures. A revolutionary change that turns the whole organization on its head has the potential to achieve vast improvements, but it is also risky, and if it fails, it might cause more harm than good.

Consequently, many companies have ended up pursuing a successful strategy for breakthrough improvements based on redesign combined with an evolutionary change process, implementing the plans they can, given the current restrictions. See, for example, Cooper and Markus (1995), Stoddard and Jarvenpaa (1995), and Jarvenpaa and Stoddard (1998). The radical redesign constitutes the blueprint for change and provides motivation as well as a clear goal for the organization. As the implementation proceeds, the original design usually must be revised to accommodate the organizational constraints and the changing market conditions faced by the company. The process design that is finally fully implemented is therefore usually a compromise between the blueprinted "ideal" design and the original process. This implies that the improvements might not be as drastic as advocated by the reengineering movement, but at the same time, the risk of failure is considerably lower.

The *evolutionary change model* is based on the assumption that the people directly affected by and involved in a change process also must take an active part in the design and implementation of that change. See, for example, Jick (1993), Cooper and Markus (1995), Jarvenpaa and Stoddard (1998), and references therein. In other words, the evolutionary change theory advocates that change should come from within the organization itself, managed by the current leadership and carried out by the current employees. The communication should be broad and open, and the pace of change should be adapted to the available capabilities and resources, meaning that milestones and yardsticks are flexible.

The motivation for change arises from internal feelings of dissatisfaction with the current situation and a desire to do better. Moreover, new processes and procedures are piloted and implemented before they are cemented by the introduction of new IT systems. Inherently, the evolutionary model assumes that real change is best achieved through incremental improvements over time.

The advantages of evolutionary change are not only that it is less disruptive and risky than its revolutionary counterpart, but also that it tends to increase the organization's general capacity for change. The disadvantage is that it might take a long time to achieve the desired vision, and it must be kept alive and adjusted over time as market conditions change. It is worth noting that the continuous improvement and empowerment focus in quality management is closely related to the evolutionary change philosophy.

The *revolutionary change model* is based on the so-called punctuated equilibrium paradigm. See, for example, Gersick (1991), which views radical change as occurring at certain instances in time with long periods of incremental change in between. Revolutionary change unfolds quickly and alters the very structure of the organization, the business practices, and the culture. Together with this type of change come disorder, uncertainty, and identity crises.

The radical change needs to be top driven, typically by the CEO, and externally imposed on the organization. It requires external resources and an outside viewpoint. The change is driven by senior management, and these individuals need to provide the right vision, create the right culture, and build the necessary political alliances. See, for example, Nadler et al. (1995), Gersick (1991), and Jarvenpaa and Stoddard (1998).

Outsiders, in terms of consultants or new executives, are brought in to lead and participate in the change process, because they are not afraid to challenge the status quo. The change team is typically small and secluded from the rest of the organization so that they are not unduly influenced by the current way of doing things. Communication regarding what is about to happen is minimal and on a need-to-know basis.

The motivation for change is a crisis, financial or otherwise, and the milestones are firm to clearly mark that a new order is in place. As part of the change process, every member of the organization needs to be qualified for the new organization; if a person doesn't measure up, he or she has no place.

Revolutionary change is also characterized by tough decisions, including relentless cost cutting and conflict resolution. The advantage of the revolutionary approach is that drastic results can be achieved quickly. However, the enormous strain put on the organization also makes the project risky, and the probability of failure is not to be discarded lightly. The personal involvement required from top management also might divert their attention from the external marketplace, implying an increased risk of lost market opportunities and possibly misaligned strategies. Another disadvantage is that the nature of revolutionary change goes against the core values that most organizations want to instill—namely, empowerment, bottom-up involvement, and innovation. The secrecy and exclusion of employees from the change process also creates insecurity about their roles, leading them to resist rather than embrace change, not just for the moment but also in the long run. Table 2.3 summarizes some important elements of the evolutionary and revolutionary change models.

TABLE 2.3

Elements of Evolutionary and Revolutionary Change Theories

Element	Evolutionary Change Model	Revolutionary Change Model
Leadership	Insiders	Outsiders
Outside resources	Few, if any, consultants	Consultant-led initiative
Physical separation	No, part-time team members	Yes, greenfield site
Crisis	None	Poor performance
Milestones	Flexible	Firm
Reward system	Unchanged	New
IT/process change	Process first	Simultaneous process and IT change

Source: Adapted from Jarvenpaa and Stoddard (1998).

Clearly, the descriptions of the evolutionary and revolutionary change models constitute the extreme points on a continuum of different change strategies, and no definite answers clarify when to use a certain approach. However, based on empirical research, Jarvenpaa and Stoddard (1998) conclude that revolutionary implementation tends to be appropriate under special conditions, such as when the following situations are true.

1. A true performance crisis exists.
2. The change concerns a smaller, self-contained unit.
3. The organization has deep pockets, for example, in terms of a rich parent company.
4. The organization is free from the "not-invented-here" syndrome and can freely import solutions from outside the company or other parts of the company, such as using purchased software packages.

However, if not facing an immediate crisis that leaves little choice but to pursue a risky revolutionary change tactic or go broke, many companies tend to use a more cautious evolutionary implementation strategy.

The critical element is time. If management is in a reactive mode and must respond to an immediate threat, a revolutionary change tactic might be the only sound choice. On the other hand, if management is in a proactive mode and initiates the change process in anticipation of a future need, there might be time to implement the radical design through an evolutionary approach with less risk. Although the vast improvements advocated in the reengineering movement might not be attained in this case, empirical evidence shows that a well-managed organization can accomplish breakthrough results using a combination of radical redesign and evolutionary implementation. By deploying an evolutionary tactic, the organization also learns how to change. This means it will be in a better position to implement future designs faster.

The combination and consolidation of different approaches for process management and design (such as radical redesign and evolutionary implementation) are a characterizing feature of modern BPM. As described in the comprehensive handbooks edited by vom Brocke and Rosemann (2010a and 2010b) and in Dumas et al. (2013), the term BPM emerged as a collective name for process-centered approaches for improvement and management. As we mentioned before, BPM is not defined by a single framework or model; instead, it may be described as a smorgasbord of principles and approaches originating from different improvement programs and disciplines; particularly, important intellectual influences stemming from the quality management area, with Six Sigma as its most recent avatar, and the management area, with the BPR movement as a prominent representative. A third tradition that influences the BPM domain is the IT and information systems area, with its roots in workflow management systems and automation.

2.6 Summary

This chapter has looked at the governing principles of successful process management and discussed why it is important to adopt a process view. We also have described the basis for Six Sigma, one of the most influential improvement programs developed in recent years. Six Sigma has a process orientation and, to various degrees, centers on the principles of process management. The importance of distinguishing between process design and implementation, as well as between evolutionary and revolutionary change tactics as means to carry out the implementation, has also been explored. We have also noted that modern BPM is characterized by a consolidation and synthesis of all these principles, programs, and tools combined with an IT perspective.

The importance and power of adopting a process viewpoint directly to the weaknesses of the functional organization in terms of coordination and transparency across functional areas. The lack of coordination and the tendency to focus on internal functional goals, in less than perfect alignment with the overall organizational objectives, make the functional organization prone to suboptimization. As a remedy, the process view, with its focus on horizontal workflows, emphasizes coordination and customer needs. In terms of people management, adopting a process view means that the process structure needs to be made clear to everyone involved. Creating a better understanding of how everyone's contribution fits into the big picture encourages involvement and participation, which are cornerstones in modern quality management and stepping-stones for successfully empowering the workforce.

The core principles for successful process management can be grouped into the following activities:

- Establishing process ownership
- Analyzing boundaries and interfaces
- Defining the process by documenting its workflow
- Identifying control points and measurements
- Monitoring the process for control purposes by implementing measures
- Taking corrective action and providing feedback if deviations indicate that the process is no longer in control

The road to successful process management of an already designed process can be divided into three phases comprising the activities previously listed: initialization, definition, and control. In terms of attacking the difficult coordination issues across functional boundaries and interfaces, a useful concept is the CPS model. Applying this model to every critical interface in the process is an effective way of aligning internal and external customer needs, and it fosters a view of the process as a chain of customers that need to be satisfied.

In a broad sense, Six Sigma can be described as a companywide strategic initiative for process improvement in manufacturing and service organizations. Its clear objective is to improve bottom-line results by reducing costs and increasing revenues. The goal of the Six Sigma program as it was originally conceived at Motorola is to reduce the variation of the individual processes to render no more than 3.4 dpmo.

At the core of the Six Sigma approach is a project- and result-oriented, quantitative, and disciplined improvement methodology focused on not only reducing variability but also improving cycle times and yield. The methodology consists of five steps: *define*, *measure*, *analyze*, *improve*, and *control*. In many ways, these steps relate to the basic principles of process management. The methodology is embedded in a corporate framework based on top-management commitment, stakeholder involvement, training programs, and measurement systems. Although Six Sigma was originally Motorola's customized approach to quality management and process improvement, its ongoing success has brought it into the boardrooms of some of the world's most successful companies. Six Sigma's success is often attributed to a strong bottom-line focus and big dollar impact; the emphasis on and consistent use of a unified and quantitative approach to process improvement; the emphasis placed on understanding and satisfying customer needs; and the combination of the right projects, the right people, and the right tools.

The reengineering philosophy of achieving drastic performance improvements by a complete redesign of core business processes, combined with rapid implementation of the new design, created tremendous interest in the 1990s. However, quickly achieving revolutionary change involves high risks as well as potentially high returns. The large number of failed initiatives caused more harm than good and gave reengineering a mixed reputation. Several reasons for these failures have been cited, including lack of top-management support and project leadership, poor understanding of the current organization and operations, inability to obtain and adapt the necessary technology, neglecting people's values and beliefs, placing prior constraints on the scope and problem definition of the design effort, and quitting too early when people resist making changes. Most of these issues relate to the rapid implementation of the radically new design and the compressed time perspective that is inherent in the traditional reengineering definition. The soundness of the general principles for business process design advocated by the reengineering movement is unabated, including the principles for deciding when and what to redesign.

The difficulty and high risk of using a revolutionary change tactic for implementation have led some companies to embark on a successful strategy for breakthrough improvements based on radical process design combined with a more evolutionary implementation tactic. The advantages include a lower risk of failure and improved long-term organizational ability to accept change. A disadvantage is that the final implementation might not give the order-of-magnitude improvements that a revolutionary change might be capable of. The approach also requires more time; therefore, to use this strategy, management must be proactive. If a company is facing an immediate crisis, a revolutionary change tactic might be the only way forward.

Discussion Questions and Exercises

1. What are some of the main advantages and challenges of adopting a process-oriented view of the organization? How are these related to the traditional functional structure?
2. What are the purposes and challenges of each of the six main principles or activities that define successful process management? Does it make sense to group the activities into the three phases: initialization, definition, and control? Why or why not?
3. How is the Six Sigma improvement methodology related to the basic principles of process management and the three phases of initialization, definition, and control?
4. Explain the CPS model and the "chain of customers" concept. How does this help to facilitate an effective process orientation?
5. Considering the document distribution example in Section 2.1.1:
    a. What would you say are the main problems with the management of this process?
    b. Do you agree with Joe's approach for arriving at a better-managed process?
    c. Is there anything you would do differently?
    d. Do you see any potential for a radical redesign of the entire process? What do you suggest?
6. What is the technical definition of Six Sigma quality, and how is this definition related to the measure of 3.4 dpmo?
7. How would you define or describe a Six Sigma initiative to a friend who has never heard about this before? What is new about Six Sigma compared with other quality management programs?

8. What are the components of the Six Sigma framework? Are there any similarities with issues emphasized in reengineering and general process management? What is the importance of each individual component for the overall approach? What are the implications if a certain element is not in place?
9. Do the Six Sigma cost and revenue rationale and the combined effect of the efficiency and effectiveness loops make sense? How would you interpret this in a situation where Six Sigma is applied at a hospital or an educational institution such as a university?
10. What is the standard approach in Six Sigma for reducing variability? How does it relate to improvements in cycle time and yield?
11. What are the traditional roles in a Six Sigma improvement project, and how are they related to the training scheme? Does this approach make sense to you? Can you see any potential drawbacks, for example, in light of the theories for evolutionary and revolutionary change?
12. At the core of Six Sigma programs lies the improvement methodology DMAIC. Discuss how each of its five steps relates to the other components of the Six Sigma framework.
13. What are some of the reasons often mentioned for the success of Six Sigma? What are the rationales? Do you think they make sense?
14. What are some of the reasons often cited for reengineering failures? Can you see any connections to the Six Sigma framework or the general principles for process management?
15. How would you position Six Sigma in relation to reengineering and TQM?
16. What is the meaning of the process paradox? How can it be avoided?
17. Fill rate, also called order fulfillment rate, is the percentage of orders that a company can ship from its available stock without any lost sales, backorders, or stockouts. Table 2.4 shows 24 samples of four fill rate values for an e-commerce company. Each sample corresponds to a month for which four fill-rate observations are made.
    a. Use Excel to create an R chart to determine whether the variability in the process is in control.
    b. Eliminate any special causes of variation identified in part a and construct an X Bar chart with Excel to determine whether the process is in control.

TABLE 2.4

Fill Rate Samples for Exercise 17

Month	Fill Rate 1	Fill Rate 2	Fill Rate 3	Fill Rate 4
1	99%	90%	98%	89%
2	85%	92%	94%	95%
3	97%	90%	93%	85%
4	90%	86%	96%	100%
5	97%	91%	91%	87%
6	91%	93%	100%	90%
7	90%	98%	92%	92%
8	96%	98%	100%	90%
9	97%	89%	87%	97%
10	94%	88%	87%	95%
11	90%	91%	94%	90%
12	89%	86%	90%	94%
13	96%	94%	89%	85%
14	100%	99%	100%	85%
15	91%	92%	89%	89%
16	87%	92%	92%	87%
17	96%	93%	86%	93%
18	89%	88%	98%	100%
19	90%	91%	98%	86%
20	92%	85%	90%	90%
21	89%	95%	99%	98%
22	97%	95%	97%	86%
23	93%	94%	98%	98%
24	97%	94%	92%	97%

18. A delivery process would like to guarantee that all deliveries are made in no more than 30 minutes. Table 2.5 shows the time (in minutes) of 20 deliveries. These observations were taken at regular intervals and the delivery time is assumed to follow a normal distribution.

    a. Use control charts and Excel to determine whether the process is in control.

    b. If the process is in control, can the company guarantee that no delivery will ever take more than 30 minutes? Why?

TABLE 2.5

Delivery Times for Exercise 18

Day	Time
1	18
2	23
3	21
4	29
5	15
6	25
7	28
8	15
9	18
10	16
11	15
12	15
13	17
14	16
15	19
16	27
17	28
18	25
19	25
20	19

19. As mentioned in Section 2.3.5, a P chart is a type of control chart used to monitor the proportion of nonconforming observations in a sample. For each sample, the proportion p is calculated as:

$$p = \frac{\textit{Number of noncomforming observations}}{\textit{Sample size}}$$

The chart assumes that the p values follow a binomial distribution and therefore the control limits for a sample of size n are calculated as follows:

$$LCL_p = \overline{p} - 3\sqrt{\frac{\overline{p}\left(1-\overline{p}\right)}{\overline{n}}}$$

$$UCL_p = \overline{p} + 3\sqrt{\frac{\overline{p}\left(1-\overline{p}\right)}{\overline{n}}}$$

where $\bar{p}$ is the average proportion (i.e., the average of the p values) and $\bar{n}$ is the average sample size. The supervisor for a call center wants to evaluate the process for answering customer phone calls. The supervisor recorded the number of incoming calls and the number of dropped (nonconforming) calls for 21 days, as shown in Table 2.6.

The supervisor wants to create a P chart to determine whether the process is in control.

a. Use Excel to calculate the proportion of dropped calls each day. What is the average proportion of dropped calls (i.e., what is the value of $\bar{p}$)? Also, calculate the average number of calls received each day (i.e., calculate the value of $\bar{n}$).

b. Use the values calculated in part a to create a P chart. Is the process in control? That is, are all the p values within the control limits?

c. If the process is not in control, discuss possible special causes.

TABLE 2.6

Number of Daily Calls (Sample) and Number of Dropped Calls for Exercise 19

Day	Calls	Dropped
1	160	8
2	166	6
3	166	10
4	150	7
5	147	6
6	131	8
7	126	6
8	187	5
9	149	20
10	179	10
11	162	7
12	194	7
13	181	8
14	183	9
15	177	10
16	181	7
17	176	5
18	133	8
19	168	7
20	103	8
21	187	10

20. A project manager at a consulting firm would like to use SPC to monitor the resource utilization rate, which she defines as the ratio of billable hours versus total available hours. Billable hours are labor hours that are charged to a client once the work is completed. The project manager aims for a utilization rate of 0.8 or 80 percent. She thinks that this is a billable resource utilization that leaves room for employees to have a healthy life–work balance. A lower billable utilization rate would indicate that employees are spending too much time on non-critical tasks. She has collected the resource utilization data in Table 2.7. The data consists of 24 months (samples) of five resource utilization observations each.
    a. Use Excel to create R and X Bar control charts to determine whether the resource utilization rate is in control. Do the charts reveal any out-of-control situations?
    b. What course of action can the project manager take based on the information provided by the control charts?

TABLE 2.7

Resource Utilization Data for Exercise 2.20

Month	Obs 1	Obs 2	Obs 3	Obs 4	Obs 5
1	80%	80%	81%	79%	81%
2	80%	82%	80%	81%	80%
3	80%	80%	81%	80%	78%
4	80%	79%	80%	80%	81%
5	80%	80%	80%	80%	80%
6	78%	80%	81%	79%	80%
7	80%	81%	79%	79%	79%
8	82%	79%	78%	79%	82%
9	80%	80%	79%	81%	79%
10	81%	82%	80%	80%	82%
11	82%	81%	80%	80%	80%
12	80%	80%	80%	78%	80%
13	81%	81%	80%	80%	80%
14	79%	80%	81%	79%	78%
15	81%	80%	79%	79%	80%
16	67%	66%	73%	82%	72%
17	83%	79%	78%	83%	67%
18	73%	76%	76%	75%	65%
19	74%	79%	78%	84%	66%
20	77%	77%	85%	78%	76%
21	78%	79%	67%	77%	77%
22	77%	65%	75%	79%	84%
23	79%	70%	71%	74%	66%
24	73%	84%	84%	65%	75%

21. How can the general revolutionary change model help explain the many reengineering failures? What is the pragmatic strategy many companies tend to use to lower the risk of failure and the cost of change but still leverage the potential of business process design? What are its main advantages and disadvantages? What is the prerequisite for this strategy to be a viable option?
22. *PCs and the Productivity Paradox* (adapted from Goldsborough, 1998)—Despite the riotous instability of stock prices, some prognosticators are advising us not to worry. Sure, the stock market has experienced unprecedented growth without significant downslide for several years. And sure, the market has always been cyclical in the past. But we're in the midst of a revolution, say the pundits, an information revolution. Spurred by the fast development of the computer, mobile, and information technology and the widespread use of personal computers, smartphones, Internet services, and social media, the economy has morphed in a fundamental way. Information technology has ushered in a paradigm-smashing leap in productivity that might have made recession passé.

    Not so fast, say those who've studied these issues. It's not even clear that personal computers (PCs) have affected productivity appreciably, let alone leading to the kind of improvements that would allow us to sail off with our mutual fund investments into a tranquil prosperity. Bureau of Labor Statistics figures show that despite the PC "revolution" and the billions invested in technology, productivity gains measured in output per hour have remained at a feeble annual rate of 1 percent for the past 30 years, which pales in comparison to the brawny productivity growth of 3 percent annually experienced during the 1950s and 1960s.

    Common sense indicates that PCs and digitalization should increase productivity. It lets individuals plan and budget far more effectively. It makes it possible to keep track of people and things far more easily, and it helps people communicate far more efficiently. They can tap far more research sources than the largest collection of periodicals or books. Even though some studies have shown that PCs have had a positive impact on productivity, and even though some experts contend that such intangibles as convenience and service don't show up in the statistics, the fact remains that the productivity figures haven't budged. This anomaly has been called the "productivity paradox," and if you look at your own habits and those of people around you, you'll see some of the reasons why.

    - Those memos with their fancy fonts and elaborate formatting take longer to create than the simple typewritten memos of the past.
    - Likewise with those presentations adorned with graphics, sound effects, and animation.
    - E-mail makes it easy to stay in the loop, but wading through scores of nonessential messages each day is definitely a time sink.

- The Web can be an invaluable informational resource, but the temptation is great to jump from one site to another, each in turn less relevant to your work needs. Not to mention using the Web to shop, check out sports scores, and engage in chitchat.
- Then, there is the equipment maintenance. Whereas in the past only specialists got silicon under their fingernails, today everybody has to deal with software bugs, hardware conflicts, and system crashes. And when the machine is not cooperating, it lures you to tinker endlessly in pursuit of perfection.
- A few years ago, a survey by SBT Accounting Systems of San Rafael, California, showed that the typical computer user in a business setting wastes 5.1 hours each week on PCs.
- Another study by Forrester Research of Cambridge, Massachusetts, showed that 20 percent of employees' time spent on the Internet at work doesn't involve their jobs.

This is not to say that you should trade your laptop for a typewriter or prevent workers from having access to the Internet. It's not the technology that's the villain. It's how we use it. Because the machines are so dumb—all they really do is add and subtract zeros and ones—we have to be smart in managing them.

a. What policies can a company establish to remedy some of the causes of the productivity paradox?
b. Should the productivity paradox be a factor when considering the application of information technology to business process design?

Note

1. https://www.ibm.com/topics/business-process-management

References

Bauer, J., John, E., Wood, C., Plass, D., and Richardson, D. 2020. Data entry automation improves cost, quality, performance, and job satisfaction in a hospital nursing unit. *The Journal of Nursing Administration 50*(1): 34–39.

Cooper, R., and Markus, M.L. 1995. *Human reengineering. Sloan Management Review 36*: 39–50.

Cross, K.F., Feather, J.J., and Lynch, R.L. 1994. *Corporate renaissance: The art of reengineering*. Cambridge, MA: Blackwell Business.

Davenport, T. 1993. *Process innovation: Reengineering work through information technology*. Boston: Harvard Business School Press.

Davenport, T., and Short, J. 1990. The new industrial engineering: Information technology and business process redesign. *Sloan Management Review 31* (summer): 11–27.

Dumas, M.M., La Rosa, J., Mendling, M., and Reijers, H.A. 2013. *Fundamentals of business process management*. Berlin Heidelberg: Springer-Verlag.

Duncan, A.J. 1986. *Quality control and industrial statistics*. 5th ed. Homewood, IL: D. Irwin.

Evans, J.R., and Lindsay, W.M. 2011. *The management and control of quality*. 8th ed. Mason, Ohio, South-Western, Andover: Cengage Learning.

Foster, S.T. 2017. *Managing quality: Integrating the supply chain*. 6th ed. Upper Saddle River, NJ: Prentice Hall.

Gabor, A. 1990. *The man who discovered quality*. New York: Penguin Books.

Gersick, C.J.G. 1991. Revolutionary change theories: A multilevel exploration of the punctuated equilibrium paradigm. *Academy of Management Review 16*: 10–36.

Goldsborough, R. 1998. PCs and the productivity paradox. *Government Technology, December 31 1998*

Grover, V., and Malhotra, M.J. 1997. Business process reengineering: A tutorial on the concept, evolution, method, technology, and application. *Journal of Operations Management 15*: 193–213.

Hammer, M. 1990. Reengineering work: Don't automate, obliterate. *Harvard Business Review 71*(6): 119–131.

Hammer, M., and Champy, J. 1993. *Reengineering the corporation: A manifesto for business revolution*. New York: Harper Business.

Harry, M.J. 1998. *The vision of Six Sigma, 8 volumes*. Phoenix, AZ: Tri Star Publishing.

Harry, M.J., and Schroeder, R. 2000. *Six Sigma: The breakthrough management strategy revolutionizing the world's top corporations*. New York: Doubleday.

Hoerl, R.W. 1998. Six Sigma and the future of the quality profession. *Quality Progress 31*(6).

Jarvenpaa, S.L., and Stoddard, D.B. 1998. Business process redesign: Radical and revolutionary change. *Journal of Business Research 41*: 15–27.

Jick, T.D. 1993. *Managing change: Cases and concepts*. Burr Ridge, IL: Irwin.

Magnusson, K., Korslid, D., and Bergman, B. 2000. *Six sigma: The pragmatic approach*. Lund, Sweden: Studentlitteratur.

Melan, E.H. 1993. *Process management: Methods for improving products and services*. New York: McGraw-Hill.

Mintzberg, H. 1979. *The structuring of organizations*. Upper Saddle River, NJ: Prentice Hall.

Mizuno, S., and Akao, Y. 1994. *QFD: The customer-driven approach to quality planning and development*. Tokyo, Japan: Asian Productivity Organization.

Munro, R.A. 2000. Linking six sigma with QS-9000: Auto industry adds new tools for quality improvement. *Quality Progress 33*(5).

Nadler, D.A., Shaw, R.B., and Walton, A.W. 1995. *Discontinuous change*. San Francisco: Jossey-Bass.

Petrozzo, D.P., and Stepper, J.C. 1994. *Successful reengineering*. New York: John Wiley and Sons.

Pulakanam, V. 2017. Costs and savings of six sigma programs: An empirical study. *Quality Management Journal, 19*(4): 39–54.

Robbins, S.P. 1990. *Organization theory: Structure designs and applications*. 3rd ed. New Jersey: Prentice Hall.

Shafer, S.M., and Meredith, J.R. 1998. *Operations management: A process approach with spreadsheets*. New York: John Wiley and Sons.

Stoddard, D.B., and Jarvenpaa, S.L. 1995. Business process redesign: Tactics for managing radical change. *Journal of Management Information Systems 12*: 81–107.

Tomkins, R. 1997. GE beats expected 13% rise. *Financial Times*.

van Ark, B. 2016. The productivity paradox of the new digital economy. *International Productivity Monitor, 31* (Fall): 3–18.

vom Brocke, J., and Rosemann, M. (Eds). 2010a. *Handbook on business process management 1*. International Handbooks on Information Systems. Berlin Heidelberg: Springer-Verlag.

vom Brocke, J., and Rosemann, M. (Eds). 2010b. *Handbook on business process management 2*. International Handbooks on Information Systems. Berlin Heidelberg: Springer-Verlag.

3

A Framework for Business Process Design Projects

Chapters 1 and 2 defined the concepts of business processes, process design, process management, and evolutionary as opposed to revolutionary implementation tactics. We also looked closely at the basic principles of Six Sigma and business process management as representatives of well-known process-oriented improvement programs. This chapter introduces a framework for structuring business process design projects. The framework is expressed as a number of important issues or steps that typically need to be addressed in these types of projects. It also points to the usefulness of simulation. Because the general purpose of the analytical tools investigated in Chapters 4 through 11 is to support process design projects, the framework also can be viewed as a road map for the remaining chapters of this book. This means that many of the design principles and issues touched on in this chapter are revisited in subsequent chapters for a more thorough analysis.

The scope of a design project as defined in this chapter ranges from the strategic vision of what needs to be done to the final blueprint of the new design. Important intermediate issues include defining and selecting the right process to focus on, evaluating potential design enablers, and acquiring an understanding of the current process if one exists. In adherence to our distinction between process design and implementation, we do not consider *detailed* implementation issues to be part of the design project. Consequently, the framework does not deal explicitly with how to implement the design or how to manage organizational change and make it stick. At the same time, we cannot completely separate the design project from the implementation issues. First, it is pointless to design processes that will never be implemented. Therefore, in selecting which processes to design or redesign, we need to consider the expected implementation challenges and corresponding costs. Second, as discussed in Section 2.5, an evolutionary implementation tactic implies sequential adjustments of the original blueprinted design due to emerging implementation restrictions and changes in market demands. To that end, high-level implementation issues are included in our framework (see Figure 3.1).

As will be apparent, many of the ideas and principles underlying the framework have sprung from the reengineering movement with its focus on radical business process redesign. However, it is important to remember that our focus is on business process design (or synonymously, redesign) per se, which in our connotation refers to developing the blueprint

 DOI: 10.1201/9781032617237-3

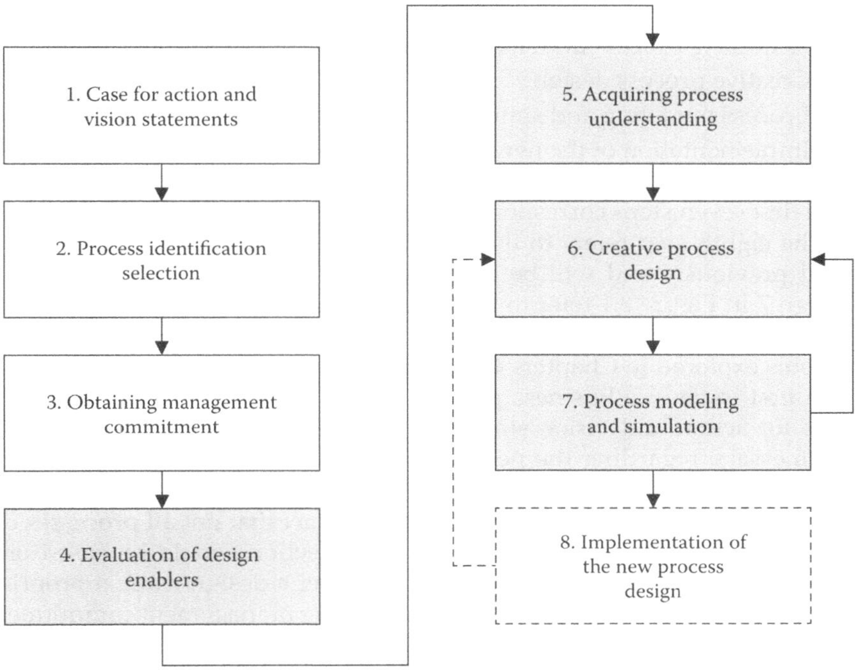

FIGURE 3.1
A framework for business process design projects. Aspects of the shaded steps are the focus of Chapters 4–11 and can be considered as the core activities relating to analytical process modeling and design.

for a new process. This does not prescribe the strategy for how the design is going to be implemented. Reengineering, on the other hand, refers to radical process design and revolutionary implementation. Consequently, in the following sections, we will draw on the sound principles for process design stemming from the process improvement and process management literature, but we do not make any assumptions regarding how the design ultimately is implemented.

The framework is influenced by an approach in Chase et al. (1998) that explicitly advocates the use of computer simulation for the modeling and testing of proposed process designs. However, there are also similarities with the frameworks suggested by proponents of the reengineering movement as well as with the Six Sigma framework covered in Section 2.3.4. Our framework consists of the following eight steps, also depicted in Figure 3.1:

1. Case for action and vision statements.
2. Process identification and selection.
3. Obtaining management commitment.

4. Evaluation of design enablers.
5. Acquiring process understanding.
6. Creative process design.
7. Process modeling and simulation.
8. Implementation of the new process design.

The first seven steps correspond directly to our definition of a design project. The eighth step refers to the high-level implementation concerns mentioned previously and will be discussed only briefly. The shaded Steps 4 through 7 in Figure 3.1 refer to the core design activities, carried out by the design team. They will be analyzed further using the modeling approaches and tools explored in Chapters 4 through 11.

The first issue in a business process design project is the formulation of a case for action and vision statements. The purpose is to communicate a clear message regarding the need for change and a vision of where to go. This is followed by the identification and selection of the process to be designed or redesigned. This step stresses the fact that not all processes can (or should) be redesigned at once, and the selection should be based on a cost/benefit analysis. It also includes appointing a design team appropriate for the selected process. The third step, obtaining management commitment, is crucial not just for the design project itself but also, even more so, when considering the likelihood of successful implementation. The importance of top-management involvement increases with the scope of the process design and implementation.

The fourth step initiates the actual process design activities by encouraging the design team to evaluate possible enablers of a new process design. A key design enabler in most cases is new technology. After examining relevant design enablers, the design team members in step 5 should clarify the process to be designed and make sure they thoroughly understand it. Important means for acquiring this understanding based on existing processes include process charts and flowcharts, described in Chapter 4. However, it also involves an understanding of customer requirements and the overall purpose of the process and its deliverables.

Step 6 refers to the creative activity of coming up with a new conceptual design for the process. This is where team members use their imagination and general design principles to create a process capable of achieving the level of performance demanded by the marketplace. Business process benchmarking is often used to obtain ideas and stimulate creativity as well as for gaining process understanding. The seventh step in the framework emphasizes the use of modeling and simulation as a means for testing and predicting the performance of the proposed design. The feedback loop from Step 7 back to Step 6 (see Figure 3.1) indicates that process modeling and simulation should be used interactively by the design team to test ideas and to stimulate creativity. The advantage of simulation compared with pilot tests performed on the real system is that it is a cheaper and faster alternative. The drawback is that issues related to human behavior that affect the implementation of the design might go undetected. Consequently, for larger processes, a pilot test is still recommended as a first step toward full-blown implementation.

When the team is satisfied with the predicted performance of the proposed process, the implementation phase is initiated, possibly with a pilot test. The feedback loop from Step 8 to Step 6 indicates that the design might have to be modified due to unexpected implementation issues detected in a pilot test or as an integral part of an evolutionary change tactic.

The remainder of this chapter describes the methodological steps in more detail, thereby completing our suggested framework for business process design projects.

3.1 Step 1: Case for Action and Vision Statements

It has been documented that the companies that are the most successful in selling change to their employees are those that develop the clearest message about the need for change. Hammer and Champy (1993), among others, emphasize that this communication should include two key messages:

1. *A case for action statement:* This shows where the company is as an organization and why it can't stay there.
2. *A vision statement:* This is what the organization needs to become and the objectives that need to be fulfilled to get there.

Clearly, the focus and importance of the case for action and the vision statements depend on the scope of the business process design project in question. If the project has a large scope and focuses on processes that are core to the organization, the action and vision statements involve the strategic position of the entire company as indicated in the aforementioned Steps 1 and 2. When the design project has a more modest scope, the need for action and vision statements is still important for determining the direction of the design project, but they typically refer to a business unit or department rather than the entire company.

The *case for action* should be brief but meaningful. Figure 3.2 shows an example of an effective case for action formulation taken from Hammer and Champy (1993). It was prepared by senior management of a pharmaceutical company to convince the employees that the research and

We are disappointed by the length of time we require to develop and register drugs in the United States and in major international markets.

Our leading competitors achieve significantly shorter development cycles because they have established larger-scale, high-flexible, globally integrated R and D organizations that operate with a uniform set of work practices and information systems.

The competitive trend goes against our family of smaller, independent R and D organizations, which are housed in several decentralized operating companies around the world.

We have strong competitive and economic incentives to move as quickly as possible toward a globally integrated model of operation. Each week we save in the development and registration process extends the commercial life of our patent protection and represents, at minimum, an additional $1 million in annual pretax profit — for each drug in our portfolio.

FIGURE 3.2
Case for action (pharmaceutical company).

development (R&D) process was in dire need of change. This case for action contains five major elements that make it effective: (1) business context, (2) business problem, (3) marketplace demands, (4) diagnostics, and (5) cost of inaction.

The *business context* means describing what is currently happening, what is changing, and what is important in the environment in which the company operates. The leading competitors, according to the case for action in Figure 3.2, are operating with much shorter development cycles. The *business problem* is the source of the company's concerns. The company admits that it is taking too long to develop and register new drugs. The *marketplace demands* are the conditions that have established performance standards that the company cannot meet. The *diagnostics* section establishes why the company is unable to meet the marketplace demands. In this case, the company realizes that having globally integrated R&D organizations is a competitive advantage. Finally, the cost of inaction spells out the consequences of not changing. The pharmaceutical company estimates the cost of every week of delay in the development and registration process.

The *vision statement* should state the objectives for the new process in qualitative and quantitative terms. These objectives can include goals for cost reduction, shorter time-to-market, improved quality and customer satisfaction levels, and specific values for financial indicators. With these objectives in place, a meaningful reference is established against which to measure progress. It also defines what the new process design should be able to achieve and thereby provides guidance to the design team regarding where to aim. Moreover, having a clear set of goals in place helps spur ongoing action during implementation. Remember that this is one of the cornerstones in the Six Sigma motivation for focusing on one unifying measure and the goal of 3.4 defects per million opportunities.

We are a worldwide leader in drug development.

1. We have shortened drug development and registration by an average of six months.
2. We are acknowledged leaders in the quality of registration submissions.
3. We have maximized the profit potential of our development portfolio.

We have created, across our operating companies, a worldwide R and D organization with management structures and systems that let us mobilize our collective development resources responsibly and flexibly.

1. We have established uniform and more disciplined drug development, planning, decision-making, and operational processes across all sites.
2. We employ innovative technology-based tools to support our work and management practices at all levels and between all R and D sites.
3. We have developed and implemented common information technology architecture worldwide.

FIGURE 3.3
Vision statement (pharmaceutical company).

Vision statements do not need to be long, but they should have solid content and not be too simplistic. For example, the statement "We want to be number one in our industry" provides no clue as to what the company needs to do to achieve this vision. Compare this statement with the one expressed by Federal Express in its early years: "We will deliver the package by 10:30 the next morning." This statement is about operations (we will get the package delivered); it has measurable objectives (we will deliver it by 10:30 a.m.); and it changed the basis for competition in an entire industry (from long, unpredictable delivery times to guaranteed overnight delivery). Figure 3.3 shows a longer vision statement, corresponding to the case for action of the pharmaceutical company in Figure 3.2.

The Federal Express vision statement and the one in Figure 3.3 are effective because they contain three key elements: (1) they focus on operations, (2) they include measurable objectives and metrics, and (3) they define a desired situation that will provide a competitive advantage for the organization when it is reached.

3.2 Step 2: Process Identification and Selection

Process selection is critical to the success of a design project and the subsequent implementation of the designed process. In Section 2.3.4, we addressed the question of how to prioritize improvement projects in Six Sigma programs.

Initially, all business processes of an organization should be considered candidates for redesign. Of particular interest are those processes that are core to the organization because changing these offers the highest potential to impact the organization's overall performance. On the other hand, they also represent the largest commitment financially and the greatest challenge to change successfully (i.e., successful design *and* implementation). As the reengineering movement has proved, the risk of failure is high, but so are the rewards of success.

Regardless of the scope, prioritizing is an important activity because restrictions in budget and resources prevent engagement in too many design projects simultaneously. The criteria of *dysfunction, importance,* and *feasibility* can be used to screen the candidate processes and determine which process to redesign first. These criteria were discussed in Section 2.4.4 and can be summarized by the following questions:

- Which process is currently in most trouble?
- Which process is most critical to accomplishing the firm's business strategy and has the greatest impact on the company's customers?
- Which process is most likely to be designed and implemented successfully?

Other relevant questions that can help management narrow down the choices include the following:

- What is the project scope and what are the costs involved in the design project and in the subsequent implementation?
- Can a strong team be formed to redesign and possibly also implement the chosen process effectively?
- Is it likely that management will have a strong commitment to change the process?
- Does the process need a new design, or could a continuous incremental improvement approach deliver the desired results?
- Is the process obsolete or the technology used outdated?

When selecting the appropriate design project, implementation must be considered. The cost associated with the change tactic must be recognized and compared with the limits of the available budget. As discussed in Section 2.5, a revolutionary implementation tactic is usually costly but much faster than an evolutionary strategy. It is pointless to design a new process that can never be implemented due to resource limitations. Another important aspect to consider is the likelihood of configuring a design team with the right attributes to handle the task of redesigning the chosen process and, in some cases, to play a major part in its implementation. Another consideration is how likely it is that management commitment will be maintained throughout the design and implementation effort for a certain process.

When selecting processes to be redesigned, it is helpful to recognize that not all processes that perform below expectations do so because of a bad design. As discussed in Section 2.3 on Six Sigma, if the improvement goals can be reached by eliminating the sources of special cause variation (i.e., by applying a continuous improvement methodology), the process under consideration might not be the best candidate for redesign. In general, this means that the process is currently operating below its potential as defined by its design, and, therefore, continuous improvement techniques can be effective in closing the gap between actual performance and the theoretical capability. Other important issues to consider before selecting a certain process for a design project are whether the process is obsolete or the technology it uses is outdated. If replacing the technology is sufficient to achieve the desired performance results, then the process does not need to be redesigned at this time.

After a process is selected, a design team is configured and assigned to the task of conceiving a new process design. The team usually consists of a mix of insiders and outsiders with respect to the process in question. The insiders can provide profound insight into current operations. The outsiders have no stake in the current way of doing things, and they bring a fresh perspective and possibly some unique new expertise to the table.

3.3 Step 3: Obtaining Management Commitment

The organization's top management must set the stage for the design project itself and for the subsequent implementation process. Evidence shows that if top management does not buy into the proposed change, the improvement effort is bound to fail. This is reflected by the fact that top-management support is emphasized in all significant improvement programs, including Six Sigma. In fact, the more profound and strategic the change is, the more crucial top-management involvement becomes. If the scope and importance of the process to be designed and implemented are more modest and primarily tactical in nature, involving the top management of the entire corporation might be less crucial. It might suffice to engage only the top executives of the business unit or functional units directly involved in the process.

Securing management commitment, although important, is not simple. Some argue that commitment cannot be achieved without education; that is, managers will not commit to something they do not fully understand. True commitment can occur only after managers have gained enough understanding of the design and the implementation processes and have recognized the need for change.

Recall that many failures associated with reengineering core business processes have been attributed to the change-resistant attitude of the organization's middle managers and the lack of top-management commitment. These two issues are closely related because people, including middle managers, are more likely to be fearful of change when direction is lacking. If top management commits and establishes a sense of where the whole organization is heading, people can get excited about these changes and the meaning of their work within the new process design. Furthermore, as discussed in Section 2.5, revolutionary change tactics put a tremendous strain on the organization by forcing it through a rapid change. Consequently, clear direction and commitment become particularly important when using rapid implementation approaches. A more evolutionary change tactic with its slower progression has in general fewer problems with change-resistant attitudes.

3.4 Step 4: Evaluation of Design Enablers

New technology in general and information technology (IT) in particular are considered essential enablers for new business process designs. However, inappropriate use of technology can block attempts to do things in new ways by reinforcing old ways of thinking and old behavioral patterns. Other enablers could be changes in the legislature or changes in the market structure on the customer side, the supply side, or among the competition. However, because it is the most prominent process design enabler, we will focus our attention on new technology.

A design team should follow two important principles when evaluating the potential of new technology for enabling new process designs.

1. Do not equate technology with automation. It will prevent creative process design.
2. Do not look for problems first and then seek technology solutions to fix them.

In other words, embedding an existing business process in a new information system does not qualify as a new process design. Consider, for example, what automation might have accomplished at the insurance company in Section 1.2.2. The company could have tried to implement an electronic transfer of claims from the local independent agent to the claims-processing center and within the processing center. Such a system would have simplified the old process by eliminating the data entry step at the processing center and possibly replacing office mail with electronic mail. The automation, however, would have done nothing to eliminate the requests for additional information from the processing center to the customer, which not only adds time to the process but also tends to annoy the customer.

Breakthrough improvements are usually not possible with automation alone. In these situations, the technology is used to accelerate the existing process but not to do the necessary work in new ways. Doing things the wrong way faster does not lead to radical improvements.

How can a redesigning team avoid getting caught in the automation trap? A key rule is that the team should not make the mistake of evaluating technology using an existing process as the only point of reference. Consider the following two questions that a team might ask (Hammer and Champy, 1993):

- How can new technology be used to enhance, streamline, or improve what we are currently doing?
- How can new technology allow us to do new things that we are currently not doing?

Note that the first question does involve an element of automation, but the second question has a clear focus on innovation. That is, the second question focuses on exploiting state-of-the-art technology to achieve entirely new goals. The team must keep in mind that the true potential of the new technology lies in breaking compromises that were made to accommodate limitations in the old technology. Table 3.1 summarizes the ability of some disruptive technologies to break old rules and compromises.

TABLE 3.1

Technology as a Mechanism to Break Rules

Old Rule	Disruptive Technology	New Rule
Information can appear in only one place at a time	Shared databases	Information can appear simultaneously in as many places as needed
Only experts can perform complex work	Expert systems and artificial intelligence (AI)	A generalist may be able to do the work of an expert
Businesses must choose between centralization and decentralization	Telecommunication networks (including intranets)	Businesses can simultaneously reap the benefits of centralization and decentralization
Managers make all decisions	Decision support tools (database access and modeling tools)	Decision-making is part of everyone's job
Field personnel need offices where they can receive, store, retrieve, and transmit information	Wireless data communication and portable computers	Field personnel can send and receive information wherever they are
The best contact with a potential buyer is personal contact	Interactive web pages, Skype, and video link meetings technology	The best contact with a potential buyer is effective contact
People must find where things are	Automatic identification and tracking technology	Things tell you where they are
Plans get revised periodically	High performance computers	Plans get revised instantaneously

Source: Adapted from Hammer and Champy (1993).

The successful application of modern technology within the context of process design requires *inductive* rather than *deductive* thinking. In general terms, inductive reasoning means working with observed data or facts to identify structures or patterns and using these observations to reach a general conclusion. Deductive reasoning, on the other hand, starts with general conclusions in terms of hypotheses, theories, or accepted beliefs and uses these to explain or solve a specific problem through logical arguments.

Many executives and senior managers tend to feel more comfortable with deductive thinking. This means they are good at defining a particular problem associated with a managerial situation. They then seek alternative solutions to the problem based on accepted beliefs and theories, and they evaluate the impact of adopting a particular solution. A major challenge for the design team is to adopt an inductive approach and develop the ability to evaluate current and emerging technologies without getting trapped in conventional beliefs about how things should be done. They are then in a position to identify creative applications that enable breakthrough process designs. In essence, inductive thinking requires the ability to recognize that modern technology could be an effective solution to a problem that the company is not yet aware that it has.

To make our discussion on new technology as an enabler for change more concrete, the following two sections will look at how (information) technology has enabled new process designs in the banking and grocery store industries.

3.4.1 Example: The Internet-Enabling Change at Chase Manhattan Bank

The tremendous advances in IT in recent years, in particular those related to the Internet, have triggered numerous initiatives in e-commerce-related areas. Some businesses have expanded their electronic linkages with their partners and suppliers, and others have added electronic services to their customers. This section will look at how the Internet, e-commerce, and electronic banking technology enabled change at Chase Manhattan Bank.[1]

Traditionally, the financial engineers at Chase compared the efficiency of delivering services using alternative channels and their impact on the bottom line. With the availability of new IT, it was time for them to understand the effectiveness of e-business compared with other existing channels such as physical branches and ATMs. From a unit cost perspective, e-commerce allows Chase to deliver services more economically and in a timelier fashion.

The move to electronic banking cannot be achieved without redesigning processes that were historically paper based. The business process unit at Chase managed a project with the operations and legal departments to redesign the processing of money judgment documents, such as restraining notices and account levies. This was a purely paper-based process. So, Chase worked with other banks, government agencies, and creditor rights attorneys to transform the paper-based process into an e-process. The change involved modifying the laws that govern levies and subpoenas to allow electronic service of these legal documents. Other important projects at Chase included e-business changes in retail banking and e-commerce work in the global and institutional banking areas. These projects eventually allowed large corporate customers to have Web access to their account data and to perform transactions online.

For Chase and other companies, Internet services and e-commerce offer tremendous opportunities to break compromises and improve efficiency and effectiveness in the service offering. However, to realize this potential, many of their business processes need to be completely redesigned to support the new business environment. When business managers decide they want to increase their e-business, it usually means a dramatic change in several business processes. Redesign projects in these companies focus on establishing high-level process strategies that increase efficiency, productivity, and coordination among groups as opposed to concentrating on small pieces of key business processes.

3.4.2 Example: New Technology as a Change Enabler in the Grocery Industry

After a brief look at how new technology has impacted the banking industry, we now turn our attention to the traditional and well-known retail and grocery store industry. Steven L. Cohen (2000) predicted that this industry would undergo significant changes over the following 25 years. He argued that the supermarket of the future would rely heavily on technology as well as process engineering tools and techniques to compete. The following is a description of the changes that Cohen predicted in the supermarket industry and that are enabled by new technology. Many of these technologies and changes are now a reality, depending on where in the world one lives, while others are still lurking in the future.

The typical supermarket as described in Cohen (2000) has scanners at the front checkout area that read the price and description of items from their universal product code (UPC). About 85% of all items can be scanned from either manufacturer or in-store UPC labels applied to packaging. On-hand shelf inventory is set, ordered, and maintained by personnel with the experience and knowledge to know how fast a particular item sells in that store location. Items are replenished periodically.

Orders are placed by handwritten communication, telephone, or fax or by using a handheld ordering machine recording the warehouse code number. This information is transmitted to each supplier as appropriate. The orders are printed out at the warehouse, and each item is picked from the warehouse's inventory. Typically, as each case of product is pulled from the storage area at the warehouse, it is brought to a central staging area. The cases are built as a block onto a pallet, and the pallet is wrapped in plastic film for stability and marked for delivery to a particular store location.

After the order is filled, the pallets are loaded onto trucks and shipped to the store. The pallets are off-loaded at the store into the receiving area; broken down; and separated onto large, flat-wheeled carts by the grocery store workers according to the aisle or area of the store to be restocked. These wheeled carts are brought onto the sales floor. The grocery store worker checks each case against the shelf and restocks the item. If a case is damaged or contains the wrong products, these items are set aside. The grocery store manager then calls in a report to the warehouse's customer service department at the next opportunity. The overstock of items is repacked, condensed onto as few pallets as possible, and placed in store inventory to be put on the sales floor at a later time.

The billing information is printed on paper and is sent with the order to the store. In most cases, but not always, the UPC information is on the bill of sale from the suppliers. The personnel working in scanning check each bill line by line and item by item to compare it with existing information in the store's computer system. If an adjustment to the item description, a change in UPC, or a change in price information occurs, the scanning operator makes the change in the computer, prints a temporary tag, and later walks to the store to find these changed items and put up the new shelf tag.

Many of the problems and challenges with the processes described are related to a lack of communication and the application of new methods and technologies. Technology can play an important role in driving the change in how manufacturers, warehouses, and retailers communicate with each other. Consider, for example, the following:

- Communication networks can be established to allow lower inventory levels and just-in-time delivery from the manufacturer to the retailer.
- On-hand shelf inventory in the retail store can be linked to the store's main computer.
- Automatic computer reordering can be established to maintain stock on hand.
- Increases in demand due to seasonal or market changes can be flagged for store management.
- Manufacturers can develop more efficient packaging methods, allowing more flexible unit sizes, which would minimize the number of backorders at the warehouse and retail levels.
- Aisles in the store can be divided into aisle, aisle side, and section (e.g., Aisle 1, Side A, Section 14). This would facilitate the restocking of products and mapping of the store and its products and allow the manufacturer, warehouse, and retailer to apply simulation techniques to plan store and product layouts.
- Orders at the warehouse can be picked and palletized according to aisle and sometimes aisle side.
- When the truck arrives from the warehouse to the store, the pallets can be brought directly onto the sales floor for restocking without the time-consuming breakdown of pallets and sorting of stock onto carts for the appropriate aisle.

To implement process changes enabled by the appropriate application of new technology, employees would have to be better trained and computer literate. Some of the changes in employees' roles and responsibilities might include the following:

- Employees who are restocking the shelves in the store would carry small, handheld computers/printers. To find out what section of the aisle a product is in, the employee would scan the UPC tag on a case of product, which would eliminate time wasted searching for shelf location.
- If an item is damaged, or if it is simply the wrong product, employees would be able to enter that information into the handheld computer for credit or return to the warehouse.
- Updated pricing information would be forwarded to the store mainframe via the billing department so that all appropriate price changes and percentage calculations could be completed before the order arrives at the store.
- The people putting up the order would check the shelf tag and compare the item they are stocking with the posted retail information. If necessary, they would print a new tag to update the price change, thus eliminating discrepancies between shelf and register pricing.

On the customer side, IT can also produce some changes; for example, in shopping habits. Paper coupons can be phased out because retailers and manufacturers can offer electronic coupons associated with customer accounts and loyalty programs. Customers can also create electronic shopping lists. Information tracking and customer shopping cards supply manufacturers and store owners with useful data. The store may tailor its on-shelf inventory and perform better shelf management by tracking consumers' purchasing habits and, thus, increase the rate of inventory turnover.

To summarize, information and increased computer capability enable manufacturers, warehouses, retailers, and customers to communicate more effectively. As a result, members of the supply chain are able to voice their desires and needs more effectively.

3.5 Step 5: Acquiring Process Understanding

Understanding the process to be designed is a key element of any design effort. In terms of process understanding, only a subtle difference exists between redesigning an existing process in an organization and creating a design for a currently nonexistent process. In both cases, we must understand what the new process is supposed to do, and particularly what customers desire from it. In the former case, it is also important to understand what the existing process is doing and why it is currently unsatisfactory. However, even if the process to be designed is the only one of its kind within the organization, similar processes usually can be found elsewhere. Business process benchmarking, further discussed in Section 3.6.1, is a tool that is often used to gain process understanding and inspire creative new designs by studying related processes with recognized excellent performance. In the following section, we will look first at some important issues related to gaining an understanding of an existing process. Then, we will consider aspects of process understanding related to customers and their requirements.

3.5.1 Understanding the Existing Process

To acquire the necessary understanding of an existing process, whether it is internal to the organization or an external benchmark, the design team needs to seek the answers to the following general questions:

- What does the existing process do?
- How well (or poorly) does the existing process perform?
- What are the critical issues that govern the process performance?

These questions can be answered at various levels of detail. Hammer and Champy (1993) argue that the goal of a design team must be to understand the process and not to analyze it. The difference is that understanding can be achieved at a high level, but analysis implies documenting every detail of the existing process. In other words, the team can gain enough understanding of the process to have the intuition and insight to redesign it without analyzing the process at a very high level of detail.

Analyzing to an extreme generally is considered a mistake. It is easy to fall into this trap because the analysis activity tends to give a false impression of progress. However, process analysis should not be ignored or considered a negative activity either because in many situations, it helps persuade others that a new design is necessary. Conventional process analysis is also a useful tool in the implementation or post-implementation steps. At this stage, a continuous improvement approach is used to close the gap between the theoretical capabilities of the new process and its actual performance. Further investigation of several tools associated with conventional process analysis can be found in Chapters 4 and 5.

The rationale for emphasizing understanding over analyzing is fairly straightforward: the team must avoid what is referred to as "analysis paralysis." This phrase is used to describe the phenomenon of becoming so familiar with a process that it becomes virtually impossible to think of new ways in which to produce the same output. Analysis is then considered an inhibitor of creativity, which is one of the main assets that a successful design team must possess. Note that understanding a process is not less difficult than analyzing it; on the contrary, in many ways, understanding can be considered harder. It is also hard to know when the team has gained enough understanding to be able to move to the creative design and modeling phase without neglecting important pieces of information regarding the process in question. This issue has no simple answers.

We conclude this section with a discussion regarding concrete issues that are important to consider when gaining understanding of an existing process. The essential issues and activities we consider are the configuration of the design team, building a high-level process map or flowchart, testing the original scope and scale, and identifying the process owner (Lester, 1994).

Configuration of the design team

The design team performs most of the work in the process-understanding and new-process-design phases. As mentioned in Section 3.2, the team is made up of business insiders (those performing the current processes, including managers and workers) and outsiders (those not directly involved in the current process, such as consultants, human resource experts, and customers). A ratio sometimes recommended is three insiders for each outsider.

Building a high-level process map

The map should focus on the customer and be business oriented; that is, instead of consisting of names of organizations (such as marketing and distribution), the map should depict the interactions between connected business processes and how they support the customer's processes. The goal of this map is as follows:

1. To build a common understanding among the team members and other stakeholders
2. To encourage a common vocabulary that cuts across functional boundaries
3. To highlight the subprocesses that are critical to achieving customer demands
4. To test the boundaries established by the initial scope and scale
5. To identify key interface points
6. To pinpoint redundancies and other forms of wasted effort

The map should have an appropriate level of detail for the scope of the design project, neither too detailed nor too simplistic. A suggested rule of

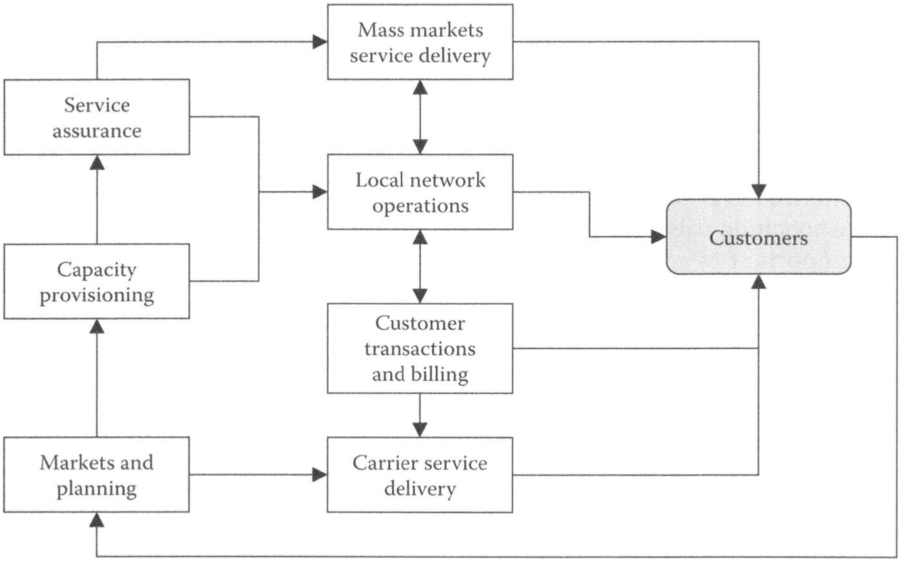

FIGURE 3.4
High-level process map for a telecommunications company.

thumb is no more than 15 subprocesses from start to finish. As an example, Figure 3.4 illustrates a high-level process map for a telecommunications company. Note that the map is neither a detailed flowchart[2] nor an organizational chart. It shows the interactions between subprocesses and not the flow of work or data. (Later, in Chapter 4, several charts will be introduced including service systems maps, which contain a higher level of detail because they focus on activities and flows.)

Testing the original scope and scale

The design team should reassess the initial scope and scale based on the new information available through self-examination, benchmarking, and customer visits. This activity is iterative in the sense that the team might have to revisit the scope and scale issue more than once.

Identifying the process owner

The process owner, as discussed in Section 2.1, is the person who will take responsibility for the new business process and will be held accountable for the results of the process. The more the process owner is involved in the design of the new process, the smoother the transition to the subsequent implementation phase will be. Therefore, identifying this individual and involving him or her in the process design project will enable valuable early interaction and input throughout the design phase.

3.5.2 Understanding the Customer

A crucial element in understanding a process is to understand its customers and their current and future needs. Customers demand basically two things from any provider of products and services: (1) they want products and services that meet their needs and requirements, and (2) they want to obtain those products and services efficiently and effectively. Understanding these requirements is fundamental to arriving at a good process design. This is why most experts agree that the best place for a team to begin to understand a process is at the customer end (see, for example, the customer–producer–supplier [CPS] model discussed in Section 2.1.1). To gain an understanding of the customers' needs, the following questions would be useful for the design team to consider (Hammer and Champy, 1993):

- What are the customers' real requirements?
- What do they say they want, and what do they really need, if the two are different?
- What problems do they have?
- What processes do they perform with the output?

Because the ultimate goal of designing a new process is to create one that satisfies customer demands, it is critical for the design team to truly understand these needs and desires. Also note that to gain this understanding, it is not sufficient to simply ask customers what they need because, in most cases, the answer is based on what they *think* they need. There is an important difference between stated and real (or hidden) needs. For example, a customer might state the need for a clothes dryer, but the real need is to remove moisture from clothes. Identifying the real or hidden need opens the door for new, creative ways to satisfy this need other than via the conventional dryer. Another example would be customers of a residential construction contractor who state that they need an efficient heating system. Their real need is a warm home and low heating costs. This real need is best fulfilled if the contractor pays attention to the insulation of the house and uses energy-conserving, three-pane windows instead of the standard single-pane windows. Focusing on the stated need would mean installing the best heating system available at great cost. However, in a poorly insulated house, it would most likely not make much of a difference. As a result, although the stated need was addressed, the customers will be unhappy because their real need was not satisfied by the building process.

3.6 Step 6: Creative Process Design

Process design is just as much a creative art as it is a science, and maybe even more so. Consequently, there exists no simple textbook solution or uniform stepwise method for arriving at a good process design. Process designers need to use creative thinking and put aside current rules, procedures, and values to find new ways of doing the necessary work. Existing processes tend to be complicated and inefficient because they have evolved over time. As discussed in Section 1.4, most processes are not designed; they just emerge as new parts of the process are added to satisfy some immediate need. Because the process expands incrementally, the intermediate decisions on how to incorporate new elements into the process affect the final design and its efficiency.

Suboptimal solutions and inefficient systems are created when local incremental methods are used to find answers to decision or design problems. At each intermediate stage, the arrangement of information might be right, but this does not ensure that the final arrangement will be entirely correct (or optimal). Consider an experiment designed by Dr. Edward de Bono, founder of the Cognitive Research Trust in Cambridge, England. When two pieces of plastic are given to someone with instructions to arrange them in a shape that can be described easily, the pieces are always arranged into a rectangle, as shown in Figure 3.5.

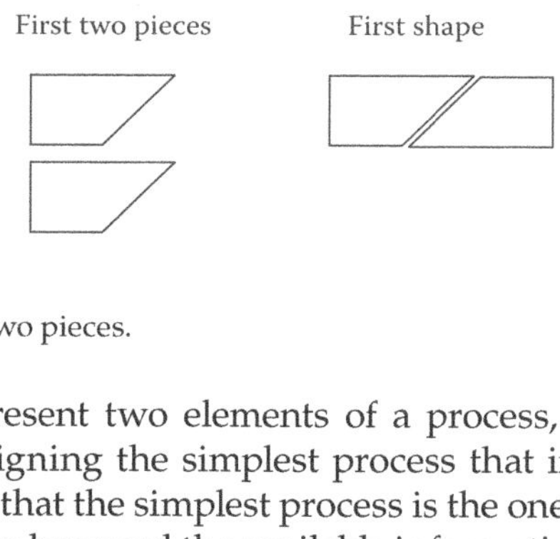

FIGURE 3.5
Arrangement of the first two pieces.

If these pieces represent two elements of a process, then the person was given the task of designing the simplest process that incorporates the given elements. We assume that the simplest process is the one that can be described most easily. The person has used the available information to find a design that maximizes simplicity. Then, a square piece is added, as shown in Figure 3.6. The task is still to arrange the pieces in a shape that is easy to describe, so the result is usually another rectangle, as shown in Figure 3.6.

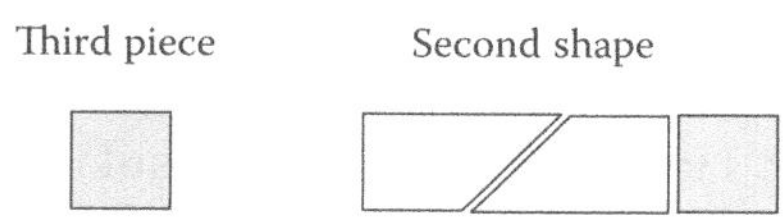

FIGURE 3.6
Arrangement after the third piece is added.

Then, two additional pieces are added, as shown in Figure 3.7. Few people are able to incorporate these two final pieces effectively; that is, in such a way that the final shape is easy to describe. For example, the configuration in Figure 3.7 might be obtained.

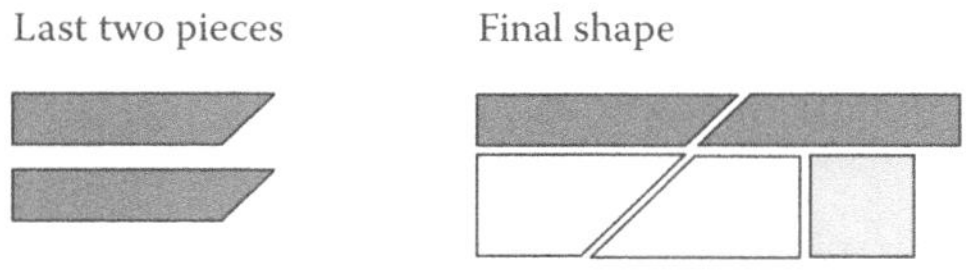

FIGURE 3.7
Arrangement after the last two pieces are added.

The difficulty stems from the tendency to add the new pieces without redesigning the existing structure. This makes the construction of the second rectangle almost inevitable after the first one has been made. Yet, considering the pieces independently of the sequence in which they appeared, the square pattern is just as good an arrangement for the second step. From the square pattern, the final arrangement, shown in Figure 3.8, is obvious, but if one starts at the rectangle, it is nearly impossible to conceive.

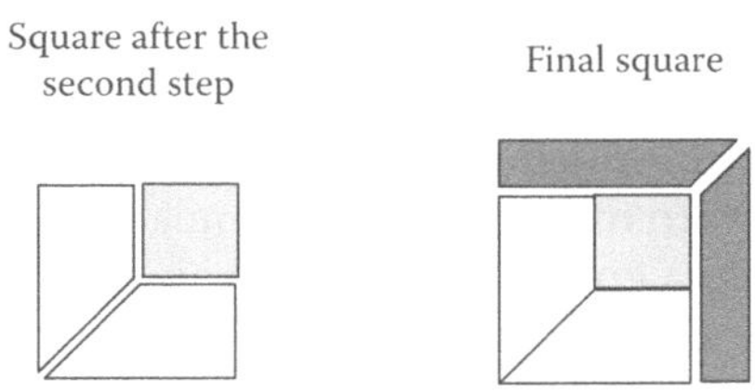

FIGURE 3.8
A more effective design.

This example illustrates how a particular arrangement of information might make the best possible sense at the time and in the sequence in which it is presented, yet it can block further development and lead to suboptimal solutions. Extending the analogy, when a design team encounters an inefficient and ineffective design based on a "rectangle," the team must break away from this configuration to introduce the "square" as an entirely different way to think about the process.

3.6.1 Benchmarking

Benchmarking essentially refers to the efforts of comparing the organization's activities and performance in certain areas with what others in the same industry or in other disciplines are doing. Every benchmarking relationship involves two parties: the target firm (or benchmark) that is being observed and the initiator firm that requests contact and observes. It is important to recognize that these roles are usually not static over time. The target firm often enters into a reciprocal agreement to study some aspects of the initiator firm's operations. Without offering something in return, it is often difficult for the initiator firm to gain access to all the desired information.

Embarking on a benchmarking effort has two basic purposes:

1. To assess the firm's or the process's current performance relative to the competition and thereby identify performance gaps and set performance goals
2. To stimulate creativity and inspire ideas for how to improve on current process performance

Based on which objective is the more important and on what is being benchmarked, it is possible to identify a number of different types of benchmarking and an infinite number of ways to execute the benchmarking activities. However, the second objective of learning how to improve process performance typically requires a more involved approach with close interaction between the initiator firm and the target firm, often including onsite visits. This type of benchmarking often is referred to as *business process benchmarking* and is what will be examined in this section. The type of benchmarking that focuses on identifying performance gaps and goals in certain metrics, whether measures of productivity, time, quality, or finance. Data envelopment analysis (DEA) can be a useful tool to analyze this type of benchmarking data and define relevant performance goals.

In the context of a process design project, both objectives are relevant. The design team must be able to determine relevant performance goals to know what to aim for in creating the new design. Even more importantly, business process benchmarking can be used to stimulate the design team's creativity and to generate ideas for how to do things in new ways. For this to happen, it is important that the target firms are chosen carefully. The idea is to learn and be inspired by the best. This could refer to the best in an industry (best-in-class benchmarks) or the best across all industries (often called *best-of-the-best* or *best-in-the world* benchmarks). Generally, the further away from its own industry a design team goes, the greater is the potential of getting breakthrough design ideas. However, at the same time, it is more challenging to identify and translate similarities between processes. A famous example of successful out-of-industry or best-of-the-best benchmarking is Xerox, which to improve its warehousing operations, turned to the mail order company L.L. Bean.

After identifying an appropriate target firm, a good starting point for a business process benchmarking effort is the 5w2h framework developed by

TABLE 3.2

The 5w2h Framework

Classification	5w2h Questions	Description
People	Who?	Who is performing the activity? Why is this person doing it? Could/should someone else perform the activity?
Subject matter	What?	What is being done in this activity? Can the activity in question be eliminated?
Sequence	When?	When is the best time to perform this activity? Does it have to be done at a certain time?
Location	Where?	Where is this activity carried out? Does it have to be done at this location?
Purpose	Why?	Why is this activity needed? Clarify its purpose.
Method	How?	How is the activity carried out? Is this the best way or are there alternatives?
Cost	How much?	How much does it currently cost? What would be the tentative cost after improvement?

Source: Adapted from Robinson (1991).

Robinson (1991), summarized in Table 3.2. This framework specifies seven questions that should be answered through the business process benchmarking effort. Five of these questions begin with the letter *w* (*who, what, when, where,* and *why*), and the remaining two start with the letter *h* (*how* and *how much*). If the initiator's design team can answer these questions after the benchmarking effort is completed, they have acquired a good understanding of the investigated process. When considering the 5w2h questions, it is important to do so in a process context.

The 5w2h framework can also be used to acquire an in-depth understanding of an existing process that is about to be redesigned. Therefore, it complements the discussion in Section 3.5.1 about how to gain understanding of an existing process.

Even though the 5w2h framework seems simple enough, finding answers to these questions for a complex cross-functional process is not easy. Effectively managing a large-scale business process benchmarking effort is a difficult matter associated with high costs. In the context of our process design project, the benefits must be weighed against these costs. As indicated before, there exists no simple uniform textbook approach to process design. Process benchmarking is sometimes useful to generate ideas for new designs, but in other cases, it can cost more than it is worth. This is why the scope of the benchmarking effort must be controlled carefully.

A more in-depth discussion of issues related to the management of benchmarking projects falls beyond the scope of this book but can be found in Camp (1995).

3.6.2 Design Principles

The creative nature of the design phase makes it impossible to specify any definite rules for how to proceed and how to succeed. However, it is possible to point to a number of general design principles that can help guide the design team and inspire their creativity. See, for example, Hammer (1990), Chase et al. (1998), and Hyer and Wemmerlöv (2002). In broad terms, these design principles relate to who does the work as well as where, when, and why it is done. However, they also involve the information and the level of integration necessary to coordinate the process. In this section, we will take a closer look at 10 general design principles. We will refer to them as *people-oriented* and *conceptual* because many of them are related directly to horizontal and vertical work integration contingent on the workforce's capabilities. Furthermore, their focus is to inspire the creativity of the design team in their efforts to come up with a new conceptual process design.

At the end of this section, we also consider seven somewhat more technical and flow-oriented workflow design principles. These principles have a long tradition in industrial engineering, where they have been used successfully for improving the efficiency of manufacturing processes. Chapter 4 provides a detailed investigation of these principles together with modeling approaches and analytical tools. It is also worth noting that all the quantitative tools and modeling approaches explored in Chapters 5 through 11 have the purpose of facilitating the analysis and evaluation of several aspects of a given process design. They represent ways to quantitatively investigate the effects of applying the design principles discussed in this section to specific real-world situations.

The people-oriented and conceptual design principles that we consider are summarized in Figure 3.9. In reviewing Figure 3.9, and for our discussion of these design principles, it is important to keep in mind that they are guiding principles and not absolute rules. Every design instance is unique and needs to be treated as such. Trying to apply these design principles without discretion is more likely a road to disaster than one to successful business process designs.

1. Organize work around outcomes, not individual tasks

The idea behind this principle is to move away from highly specialized and fragmented task assignments toward more horizontally integrated activities. Process complexity is reduced, and activity complexity is increased, as illustrated in Figure 1.3. The rationale for moving in this direction is that it leads to

General people-oriented and conceptual process design principles

1. Organize work around outcomes, not tasks	6. Treat geographically dispersed resources as though they were centralized
2. Let those who use the process output perform the process	7. Link parallel activities instead of just integrating their output
3. Merge information processing and data gathering activities	8. Design the process for the dominant flow not the exceptions
4. Capture the information once – at the source	9. Look for ways to mistake-proof the process
5. Put the decision point where the work is performed and build control into the process	10. Examine process interactions to avoid suboptimization

FIGURE 3.9
People-oriented and conceptual design principles.

fewer activities in the process that need formal coordination. It also reduces the number of handoffs and eliminates the related control steps, all of which represent some of the major causes of process inefficiencies. It also increases the work content for individual workers, which encourages them to take responsibility for their work and to take pride in what they do. These are important people issues that attempt to empower the workforce and lead to high-quality output.

Applying the principle implies that several specialized tasks previously performed by different people should be combined into a single activity. This activity can then be performed by an individual "case worker" or by a "case team" consisting of a limited number of individuals with complementary skills. The choice between the two includes considerations regarding the scope of work, the capacity needs, how critical it is to avoid handoffs, the importance of a single point of contact for customers, the required ability to handle exceptions, cultural issues, training costs, and so on. It should be mentioned that an important enabler for horizontal integration is cross-functional training, which means that each individual is trained to perform several different tasks.

The principle has spawned the integration of labor approach known as *case management* (Davenport and Nohria, 1994). The case manager's role represents a break with the conventional approach to the division of labor. Davenport and Nohria have observed four common components of a successful case manager's role. The case manager

- Completes or manages a "closed-loop" work process to deliver an entire product or service to the customer.
- Is located where the customer and various other functions or matrix dimensions intersect.
- Has the authority to make decisions and address customer issues.
- Easily accesses information from around the organization and uses information technology in decision-making.

Case management is an appropriate design for work that is performed along process lines rather than in business functions. Unlike business functions, processes are designed around the completion of a significant product or service. Furthermore, case management is particularly useful in processes that deal with customers and, therefore, involve managing the entire cycle of activities from customer order to product or service delivery, billing, and payment. In many ways, case management parallels the governing ideas for process management discussed in Section 2.1 with the case manager playing the role of the process owner.

In Section 1.1.2, we described the process at IBM Credit Corporation, which was documented in an attempt to improve performance. The entire process consumed 6 days on average, and sometimes, it took as long as 2 weeks, although the actual work could be completed in an average of 90 minutes. IBM tried several fixes with the goal of improving this process. For example, they installed a control desk that could answer questions from field salespersons. The control desk helped find the status of each request but also added more turnaround time.

After some brainstorming, IBM Credit realized that the problem did not lie in the tasks and the people performing them but in the structure of the process, which forced work to be handed from one office to another. The existing process was designed under the assumption that each request was unique and difficult to process and, therefore, required the intervention of four highly trained specialists. The reality was that most requests were simple and straightforward. So in the end, IBM Credit replaced its specialists with generalists supported by an easy-to-use computer system that provided access to the databases and tools that specialists would use. The generalists became case workers in charge of the transaction from the beginning to the end. They would get help from a small pool of specialists when requests were indeed unique and difficult to handle.

2. Let those who use the process output perform the process

Another way of expressing this is that work should be carried out where it makes most sense to do it. By avoiding excessive delegation of responsibilities, the risk of coordination inefficiencies and a "laissez faire" mentality decreases. If you are using the output you produced yourself, there is no one to blame or point a finger at if that output does not meet your expectations. This idea is closely related to the principle of horizontal integration of specialized tasks discussed previously. The objective is again to avoid handoffs and coordination inefficiencies.

A typical example of the application of this principle is allowing employees to make some of their own purchases without going through a purchasing department. The purchases can be limited to a given dollar amount to maintain a certain amount of control over expenses. Another example is the various sorts of so-called vendor managed inventory (VMI) initiatives whereby a company asks its suppliers to manage its incoming goods or parts inventory. Wal-Mart has successfully implemented this principle, using IT to transfer point-of-sale data to suppliers, who use the data to monitor inventory levels and assume the responsibility of telling Wal-Mart when to reorder.

3. Merge information processing and data-gathering activities

This principle is based on having the people who collect the data also analyze it and turn it into useful information. This idea is closely related to the previous principle of having those who use the output of the process perform the process. The difference is the particular focus on information processing, which is often a major issue in administrative business processes. Having the same people collect the data and process it into useful information eliminates the need for an additional group to process and reconcile data that they did not collect. This reduces the risk of introducing errors and presenting inaccurate information. An example is the traditional accounts payable department that receives and reconciles purchase orders and supplier invoices. Through electronic order and information processing, the need for invoices is eliminated. This increases the accuracy of the submitted information and renders much of the traditional reconciling work obsolete.

4. Capture the information once—at the source

Information should not be reentered several times into different computerized systems or have to move back and forth between electronic and paper formats. To avoid errors and costly reentries, it should be collected and captured by the company's computerized information system only once—at the source (i.e., where it was created). This idea is closely connected to Principle 3 of merging information processing into the activity that gathers the data. The objectives in both cases are to speed up the process, avoid mistakes, and assure high information quality at a lower cost. A typical example that violates this principle of capturing the data at the source is given by the original claims-processing process of the insurance company in Section 1.2.2. Information gathered by the independent agent was later rekeyed at the claims-processing center. Furthermore, the information was typically incomplete, forcing additional contact with the customer.

5. *Put the decision point where the work is performed and build control into the process*

Case management compresses processes horizontally in the same way as employee empowerment compresses the organization vertically. Vertical compression occurs when workers make decisions in instances where they used to go up the managerial hierarchy for an answer. This principle encourages decision-making to become part of the process instead of keeping it separate. An effective way to achieve this is to let workers assume some of the responsibilities previously assigned to management. Under the mass-production division of labor paradigm, the assumption is that workers have neither the time nor the inclination to monitor and control the work they are doing and, furthermore, that they lack the knowledge to make decisions about it. Today, this assumption can be discarded due to a more educated and knowledgeable workforce and the advent of decision support systems. Controls are made part of the process, reinforcing a vertical integration that results in flatter, more responsive organizations.

6. *Treat geographically dispersed resources as though they were centralized*

Modern IT makes it possible to break spatial compromises through virtual colocation of individuals and work teams. This means that although they are sitting at different locations, employees can have online access to the same information and multiple media to support instantaneous communication. As a result, geographically disbursed resources should not constrain the design team to consider only decentralized approaches. This does not imply that a centralized structure is always the best alternative, but it leaves that door open for the design team despite potential spatial restrictions. Concrete examples of enabling technologies include GroupWare software, which facilitates parallel processing of jobs performed by geographically dispersed organizational units, intranets, and videoconferencing. The last technology enables people at different locations to see the reactions and the body language of those with whom they communicate.

7. *Link parallel activities instead of just integrating their output*

An important cause of costly rework and delays in processes is related to situations where outcomes of parallel sequences of activities are being integrated or assembled. If any discrepancies or errors are detected in either of the outputs to be combined, the job is delayed and requires at least partial rework. The problem is compounded if the detection of the problem is delayed.

The heart of the problem is that most processes that perform activities in parallel let these activities operate independently. Therefore, operational errors are not found before the integration activity is performed, resulting in the need for an excessive amount of additional resources for rework. Parallel activities should be linked and coordinated frequently to minimize the risk of realizing that major rework must be done at the integration step. To illustrate, consider the simultaneous construction of a tunnel through both sides of a mountain.

Without continuous linking of the two parallel approaches, the risk is high that they will not meet in the middle as intended. The result is a costly rework project or two parallel tunnels through the mountain at twice the cost.

8. Design the process for the dominant flow, not for the exceptions

By focusing too much on the exceptions and special cases that could arise, the process design tends to be overly complicated. For example, some jobs might be so expensive or performed for such important customers that higher-level management believes it is necessary to inspect them. However, these exceptional cases do not imply that all jobs should be subject to management approval. An approval step should not be designed into the process unless it is needed for all jobs. An excellent example of a process suffering from too much focus on exceptional cases is the original process at IBM Credit Corporation discussed previously and in Section 1.1.2.

9. Look for ways to mistake-proof the process

To mistake-proof or fail-safe the process means designing it so that it becomes virtually impossible for certain mistakes or errors to occur. Due to its Japanese origins, mistake-proofing is also referred to as *Poka-Yoke*, a well-known strategy for waste reduction and improvement in product and process design. Countless examples of this conceptually simple but somewhat elusive design principle can be cited: ATMs that start buzzing if you do not remove your card, self-flushing public toilets, templates that guide correct insertions of components, order-entry systems that will not accept incorrect part numbers, color-coded fields for data entry, online process documentation guiding work execution in real time, and personal digital organizers with sound reminders for important appointments.

A 10th principle, derived from the systems design theory (Van Gigch, 1978), can be added to the nine aforementioned principles.

10. Examine process interactions to avoid suboptimization

A process improvement developed in isolation for a distinct part of the overall process might seem effective in light of the particular subprocess in question. However, by neglecting interactions with other subprocesses and the effects on the overall process, an isolated improvement could easily result in a suboptimal design. Looking for improvements by analyzing only portions of the total process leads to what is known in general systems theory as *disjointed incrementalism*. To illustrate, consider the following example of suboptimization adapted from Hammer and Champy (1993).

> A plane belonging to a major airline is grounded one afternoon for repairs at Springfield Airport. The nearest mechanic qualified to perform the repair works is at Jamestown Airport. Joe, the maintenance manager at Jamestown, refuses to send the mechanic to Springfield that afternoon. The reason is that after completing the repairs, the mechanic would have

to stay overnight at a hotel and the hotel bill would come out of Joe's budget. Therefore, the mechanic is not dispatched to the Springfield site until the following morning, enabling him to fix the plane and return home the same day. This means that Joe's budget is not burdened with the $100 hotel bill. Instead, a multi-million dollar aircraft sits idle, and the airline stands to lose hundreds of thousands of dollars in revenue. Joe is not foolish or careless; he is doing what he is supposed to do—namely, to control and minimize his expenses. Minimizing labor costs might be a worthwhile goal for the maintenance subsystem viewed in isolation, but it disregards the goals of the larger system—the airline, for which earning revenue is the overriding objective.

To conclude our discussion on the people-oriented and conceptual design principles, let us look at Figure 3.10, which summarizes the principles and some of the recurring themes and objectives that characterize them. Principles 1 through 4 have a common focus on horizontal integration of tasks into augmented activities, thereby eliminating handoffs, control points, and sources of errors. Horizontal integration also implies fewer activities

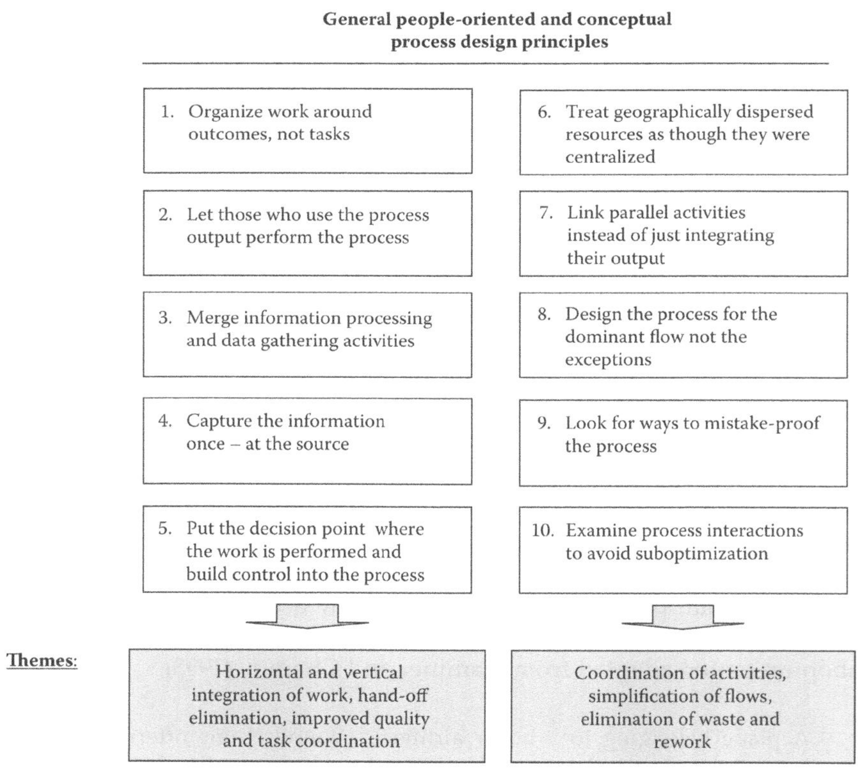

FIGURE 3.10
Summary of the considered people-oriented and conceptual design principles.

to coordinate within the process at the price of increased task complexity within each activity. A key issue is then to have cross-trained personnel who can perform the tasks in an activity without formal coordination.

Principle 5 complements the horizontal integration by focusing on the importance of vertical work integration and delegation of responsibility. Empowering workers eliminates formal control points and facilitates easier coordination through decentralization. Finally, Principles 6 through 10 focus on aspects of coordinating activities so that waste, rework, and other inefficiencies are avoided.

The people-oriented and conceptual design principles may be complemented by the seven more technical flow-oriented design principles summarized in Figure 3.11. These stem from the field of industrial engineering and are often referred to as *work flow design principles*. The principles are to (1) establish a product orientation in the process, (2) eliminate buffers, (3) establish one-at-a-time processing, (4) balance the flow to the bottleneck,[3] (5) minimize sequential processing and handoffs, (6) schedule work based on its critical characteristics, and (7) minimize multiple paths due to specialized operations for exception handling (see Figure 3.11). Industrial engineers have applied these principles successfully to the design of manufacturing systems for decades. In broad terms, they focus on achieving efficient process flows, managing resource capacities, maximizing throughput, and reducing cycle times (see Figure 3.11).

Traditional, technically oriented workflow design principles

Focus: Efficient process flows, managing resource capacity, throughput, and cycle times

- ✓ Establish product orientation in the process
- ✓ Eliminate buffers
- ✓ Establish one-at-a-time processing
- ✓ Balance the flow to the bottleneck
- ✓ Minimize sequential processing and handoffs
- ✓ Schedule work based on its critical characteristics
- ✓ Minimize multiple paths due to specialized operations for exception handling

FIGURE 3.11
Summary of the traditional workflow design principles.

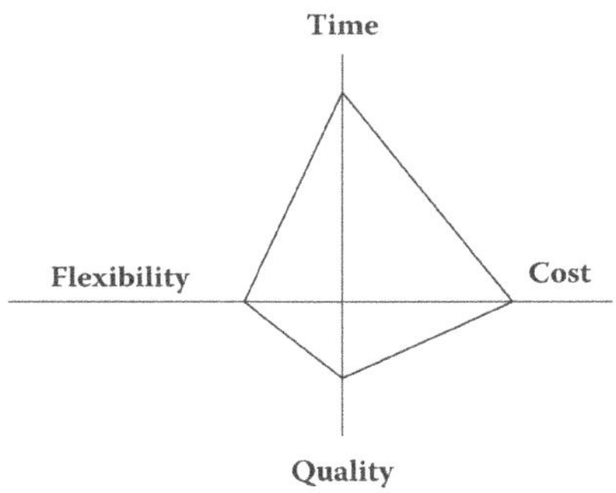

FIGURE 3.12
Devil's quadrangle framework. (Adapted from Dumas et al., 2013)

Detailed discussions of these important design principles are included in Chapter 4, where each is investigated thoroughly together with relevant tools for implementation, quantitative modeling, and evaluation.

3.6.3 The Devil's Quadrangle

When designing a new process, it is vital to know exactly what the new design should achieve. The overall goals should be clear from the case for action and vision statements discussed in Section 3.1. However, when considering concrete design alternatives, there are often trade-offs between the four performance dimensions: time, cost, flexibility, and quality. Ideally, a new process design leads to significant improvements in all performance dimensions, but quite often, improving in one dimension may negatively affect another. For example, if the main objective for the new design is to reduce the process time, there may be different design alternatives with similar time performance that impact the quality, cost, and flexibility performance in very different ways. Awareness and consideration of these trade-offs is very important for arriving at an effective design. A simple conceptual framework that can help the design team to identify and think about these issues is the "devil's quadrangle" discussed in Dumas et al. (2013) and depicted in Figure 3.12. The ominous name refers to the difficult trade-offs and compromises that sometimes are required.

3.7 Step 7: Process Modeling and Simulation

Conceptual process designs must be tested before they are implemented. It is possible for the design to be flawed or to simply not deliver the intended outcomes. The design team uses test results to decide whether to proceed with the implementation phase or go back to the drawing board to revise the design. The iteration between design modifications and testing continues until the team is satisfied with the predicted performance of the new process.

The testing can be done through implementation of pilot projects (see also Roberts, 1994). However, this is generally an expensive way to perform an initial assessment of the process performance. It also takes quite some time before enough data are available from the pilot to make a sound judgment of the effects of the new design. As a result, the design team is fairly restricted regarding how much testing can be done. Therefore, an attractive alternative is to use process modeling and quantitative tools for the initial testing. Among the available modeling tools, the most flexible and in many respects most powerful (although not always the most appropriate) is simulation. The advantage of using quantitative models to test the design is that it is much cheaper and faster than the pilot implementation approach. This means that the design team has much greater freedom in testing new ideas, which stimulates creativity and allows the team to arrive at better process designs. The drawback is that any model, by definition, is an approximation of reality, and it can never capture all its aspects. Particularly hard to model are behavioral issues, which have to do with attitudes, resistance to change, and worker behavior—all factors that to a large extent are unknown before the implementation. Therefore, process modeling and simulation can never completely replace pilot testing, and vice versa. Rather, the two approaches complement each other. In terms of supporting the design team in developing a good blueprint for a process design, process modeling and simulation is a powerful approach. To finalize the design and explore behavioral issues, pilot testing is often a worthwhile step before moving on to a full-blown implementation.

The remainder of this section is devoted to a conceptual discussion regarding the usefulness of simulation, or more specifically, discrete event simulation for process modeling and evaluation. The technical aspects of discrete event process simulation and simulation modeling are thoroughly treated in Chapters 7 through 10. Although simulation modeling has a central role in this book, it is important to bear in mind that it is not always the best method to use. Chapter 7 provides a discussion of the pros and cons of simulation, including when it is appropriate to use and when it is not.

In general terms, *simulation* means "to mimic reality in some way." This can be done through physical models such as wind tunnels or through computer-based models. Flight simulation is a well-known example of a computer-based model. Simulation in the context of process modeling actually refers to discrete event computer simulation. This is a technique that allows the representation of processes, people, and technology in a dynamic computer model. Process modeling and simulation as a concept consist of the following basic steps:

1. Building a simulation model of the process.
2. Running the simulation model.
3. Analyzing the performance measures.
4. Evaluating alternative scenarios.

The simulation model mimics the operations of a business process, including customer arrivals, truck deliveries, people missing work, and machines breaking down. This is accomplished by stepping through the events in compressed time, often while displaying an animated picture of the flow. While stepping through the events, the simulation software accumulates data related to the model elements, including capacity utilization of resources, number of jobs, and waiting times, in buffers. The performance of the process can then be evaluated through statistical analysis of the collected output data. Although it is conceptually straightforward, successfully using simulation for process modeling and design requires some technical skills. We will examine these technical issues regarding process modeling and simulation in detail in Chapters 7 through 10.

One of the major strengths of simulation modeling as a tool for business process design is that it can help reduce the risks inherent in any type of change. The use of scenario-based what-if analyses enables the design team to test various alternatives and choose the best one. The result is that goals can be met in an efficient manner (Avni, 1999). With regard to what-if analyses, the main advantages of simulation modeling over tests performed on the real system are that (1) it is done "off-line" without disturbing current operations, and (2) time is compressed in a simulation run. In a redesign effort, an existing process is observed, and strategies are developed to change it to enhance performance. Testing such strategies on the real system would disturb daily operations and negatively impact results. In a simulation model, any strategy can be tested safely without upsetting the environment. Time compression can be used because in a simulation model, events are accelerated. By only considering the *discrete events* when something happens in the system, simulation can compress time and "advance the clock" much faster than it would occur in real time, enabling the analyst to examine longer time horizons. Depending on the size of the model and the computer's capabilities, several years can be simulated in minutes or even seconds.

Another risk that simulation can help mitigate is that of suboptimization. As the simulation model encompasses various processes, analysts can study how these processes interact and how changing one process impacts the others. For example, in a manufacturing facility, maximization of the output of a group of machines can increase inventory in the assembly area. However, a simulation model containing machining and assembly information can be used to optimize the overall output and inventory.

As mentioned, simulation modeling also promotes creativity and thereby leads to better designs. With the low cost of testing a new design at virtually no risk, people are less reluctant to participate in the design process and to pitch in new ideas. The flow of ideas gains momentum, and soon a brainstorming session produces a much-needed shower of creativity.

In relation to other quantitative tools, a major strength of simulation modeling is its ability to *capture system dynamics*. Random events such as equipment breakdowns and customer arrivals are modeled by using probability distributions. As the simulation advances in time, the embedded distributions are sampled through the use of random number generators. Hence, the dynamic interaction among system elements is captured. For example, two workstations in a sequential flow might break down at the same time and cause no inventory accumulation between them; however, the occurrence of stoppages at different times might cause significant blockage or idle time and reduce output. In other words, *when* something happens can be as important as *what* happens.

Another interesting aspect of simulation that is directly related to the simulation software available today is its ability to provide animation to help visualize the process operations. Through the use of animation, ideas come alive. Equally important are the multitude of other graphical tools enabling dynamic reporting of results and trends. Time series, histograms, and pie charts are some of the dynamic reporting features of most simulation software packages today. These features help simulation modeling enhance communication. At the same time, the quantitative nature of simulation reporting brings objectivity into the picture. Improvement initiatives can be compared and prioritized. Subjective statements such as "That idea will never work" or "We don't know how much improvement we will obtain" can be put to rest.

To conclude this discussion on the usefulness of simulation modeling, consider as an example a team in charge of redesigning the customer service process of a call center. A simulation model could help the design team capture the dynamics of the system. This is done by incorporating into the model the random nature of calls arriving at the center and the time it takes to serve each of these calls. The model also can capture the process structure, including the interdependencies among customer representatives and the alternative routing schemes. With a simulation model in place, it is easy for the design team to test different scenarios and new ideas for improving the process performance. For example, it can assist in determining staffing levels, telecommunications technology requirements, and operational policies.

3.8 Step 8: Implementation of the New Process Design

As indicated in the introduction to this chapter, a detailed discussion of implementation issues and challenges is beyond the scope of our framework for business process design projects. However, the design project cannot be considered in complete isolation from the subsequent implementation step. When selecting the process to be designed (or redesigned), the implementation strategy is an important factor because it is pointless to design a process that will never be implemented. Crucial criteria in this respect are time, cost, improvement potential, and likelihood of success. In this section, we discuss some high-level implementation issues that can have a direct bearing on these issues and, therefore, on the process design project as a whole.

As discussed in Chapter 2, the implementation strategy or change tactic can be identified conceptually as revolutionary, evolutionary, or on a continuum in between. For a more detailed discussion of the characteristics of revolutionary and evolutionary change, see Section 2.5. However, in broad terms, a revolutionary implementation approach means rapid change with high costs, primarily caused by the need for external resources. It also implies a high potential for quick, dramatic improvements as well as a high potential for failure. It requires a committed and decisive chief executive officer and management team that can make harsh decisions to force the change through the organization.

An evolutionary change tactic typically requires a longer time horizon and does not offer the same potential for dramatic improvements in the short run. On the other hand, the costs are lower because the change is achieved primarily by using internal resources. An evolutionary approach also gives the organization time to adapt and embrace the change gradually. This reduces the risk of organizational collapse and destructive behavior, making it more likely that the introduced changes will last. The extended time horizon requires long-term management commitment to keep the vision alive. It also implies that the original blueprinted design probably will need to be revised as market conditions change. In other words, the original design is viewed as an ideal that typically needs to be modified in accordance with restrictions identified during implementation and considered too expensive to remove.

Regardless of the chosen implementation strategy, leadership is critical for success. Any form of significant change necessitates continued engagement on the part of senior executives and senior management (see Sections 2.3 and 2.4). Often, the design team, or part of it, also has the responsibility for implementing the new designs. However, support and buy-in from line managers accountable for delivering the expected improved performance are crucial for successful implementation. Training employees in skills needed in the new process is also important for successful implementation. Recall that many of the conceptual design principles discussed in Section 3.6.2 involve horizontal and vertical compression of work and responsibilities, which are contingent on workforce capabilities.

Another important implementation issue (discussed in Sections 2.1 and 2.2) is to measure the effects of the process change and assess the impact. When measuring the overall effects, the performance of the new process can be compared with that of the old process before the change. However, the new design can also be compared with the objectives specified at the beginning of the design project, many of which should be included in the mission statement (see Section 3.1). A comparison with the performance of the old process shows the effects compared with the status quo, and a comparison against the stated objectives gauges the performance against the specified goals.

It is also important to reflect on what can be learned from the process design and the implementation of projects. What worked, what did not, and why? What were the main challenges? What design ideas did not work out in practice, and why not? If the implemented design is different from the original blueprinted design, simulation can be used to assess the gap in specified performance measures. Comparisons can also be made with the specified performance objectives of the ideal design. It is also important to investigate why the blueprinted process design was not fully implemented. Was it because of flaws in the blueprinted design, which made it impossible to implement, or because of mistakes made in the implementation phase? This feedback is invaluable information for improving the activities related to designing and implementing new processes in the future. Recall from Section 2.3 that an important activity in Six Sigma is for the improvement teams to share their experiences and success with the rest of the company, thereby transferring knowledge and maintaining momentum for change.

3.9 Summary

This chapter has introduced a framework for structuring business process design projects, particularly emphasizing the usefulness of process modeling and simulation for the analysis and evaluation of new process designs. The framework summarized in Figure 3.13 consists of eight major steps or issues that need to be addressed in process design projects. They range from the strategic vision of what needs to be done to the final blueprint of the new process design (see Figure 3.13). The framework also includes high-level implementation concerns. However, the detailed implementation issues are outside the scope of the framework and the process design project as defined here.

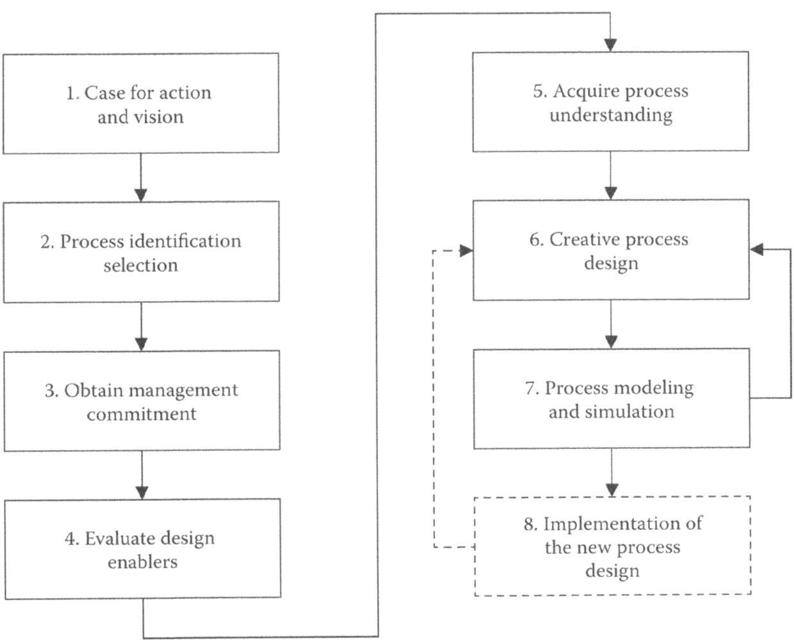

FIGURE 3.13
Summary of the framework for business process design projects.

The framework starts at a strategic level with the case for action and vision statements. The former explains why it is necessary to change, and the latter specifies where to go and includes concrete, measurable objectives. These objectives provide a direction for the entire design project and, in many cases, also for the subsequent implementation project. After the overall objectives are in place, the next steps are to identify and select the process that needs a new design and to obtain management commitment for the design project as well as for the subsequent implementation. Once the process is chosen and management support secured, the actual design activities can begin with investigating potential design enablers and acquiring an understanding of the process to be designed. The most common enabler for new process designs is new (information) technology. Process understanding has two main elements: understanding what the process is supposed to do, as defined by customer needs and desires, and understanding what the process is currently doing, if a process exists. A potential source of information for gaining process understanding as well as for generating design ideas and stimulating creativity is business process benchmarking.

With a profound understanding of available design enablers and the process to be designed, the design team must leverage their creativity to develop a new conceptual design. Ten general design principles were outlined that can help stimulate the team's creativity and provide some guidance. However, there are no set rules or simple recipes for success. Common themes through the conceptual design principles are horizontal and vertical work integration and the use of IT to improve coordination and avoid suboptimization of the process. This chapter also looked briefly at seven more technical and flow-oriented design principles to be investigated further in Chapter 4.

Having conceived a new conceptual design, the design team needs to test it to see whether it performs as intended and meets the stated objectives. Process modeling and simulation represent powerful methods for the numerical evaluation of a given design. A core strength with these methods is that it is easy to perform what-if analyses and test new ideas. Advantages compared with testing on the real-world system include low cost, no disturbance of current operations, and speed. It is important to recognize that to leverage the full potential of process modeling and simulation, they need to be used interactively by the design team to test new ideas and to stimulate creativity.

Discussion Questions and Exercises

1. Telson, a European telecommunications and mobile operator company, is facing fierce competition in its home market after a recent deregulation. Executives at Telson believe that the quality of service must increase drastically to remain competitive. Customers have identified two critical service areas: (1) technical support and

(2) general customer service. The company offers technical support online, over the phone, and to some extent at its stores. They also have service technicians who make home visits if the problems so require. Currently, a large portion (about 50%) of the technical support issues involve a home visit from a service technician, and it then takes an average of 7 days from the time the call is received until the issue is resolved. Technical support issues that do not involve a home visit from a service technician take on average 60 minutes to resolve. The waiting time on hold is then 45 minutes on average. When a customer calls the general customer service desk, which is open 24/7, it takes on average 15 minutes from when the call is received until it ends. About 60% of this is waiting time on hold. When they contact general customer services via e-mail, the average response time is 6 hours and via chat, 30 minutes.

Telson would like to redesign its processes and dramatically improve its service offering in the critical service areas. Write a vision statement and a case for action that Telson could use to launch a redesign project.

2. Customer relationship management (CRM) has emerged as an important application in business software. It has been defined as a *horizontal business strategy that focuses on customer-facing portions of business processes*. CRM software is designed to integrate a company's sales, marketing, and customer support functions to serve the customer better, while at the same time making the information available throughout the organization. Search the Web for an application of CRM software that enabled a company to achieve higher levels of customer service and improve profitability while redesigning key business processes.
3. What are some of the important considerations when selecting a process for a design project? How might the scope of the process influence the relative importance of the criteria used? Are all processes with poor performance top candidates for redesign? How does this relate to the discussions regarding differences between incremental continuous improvement and process design in Chapters 1 and 2?
4. In most improvement programs, top-management commitment is emphasized as a key success factor. Why do you think this is? Is the role of top management different in a situation with a revolutionary implementation strategy than it would be under an evolutionary change tactic? How?
5. What is the difference between process design or redesign and automation?
6. Explain in your own words the difference between inductive and deductive thinking.

7. Give an example of the application of the following disruptive technologies in a setting familiar to you:
   a. Shared databases
   b. Wireless communications
   c. Handheld computing
   d. Artificial intelligence
8. What are the main issues in understanding a process? Are there any differences between a design project and a redesign project? How does the concept of "analysis paralysis" fit in?
9. Discuss how benchmarking can be a useful tool in business process design projects.
10. Discuss the difference between stated and real (or hidden) customer needs. Why is this distinction important in the context of business process design? Give an example.
11. Identify similarities and differences among the 10 conceptual design principles discussed in Section 3.6.2. How would you summarize the underlying ideas or themes? Do the principles make sense? Why or why not?
12. Describe a potential application of case management in the following organizations:
    a. Dell Computers
    b. A telecommunications company
    c. A community hospital
13. Discuss the advantages and disadvantages of using quantitative process modeling and simulation for testing new designs. How would you relate process modeling and simulation to pilot tests on the real system?
14. Use the devil's quadrangle framework to identify and discuss which performance measures may be affected (positively or negatively) by the following process changes:
    a. A new computer application is introduced that speeds up the handling of loan applications.
    b. An e-tailer hires temporary warehouse workers in December to handle the increased order picking volume caused by Christmas orders.
    c. A video rental company with brick and mortar stores experiences a rapidly decreasing customer base. They, therefore, decide to invest in new technology and offer customers electronic streaming services to rent movies online and on demand.
15. Consider the original IBM credit process described in Section 1.1.2 and compare it with the redesigned process described in Section 3.6.2. Discuss which of the design principles summarized in Figures 3.10 and 3.11 have been used, and how they are achieved.

16. Consider the example from the insurance industry described in Section 1.2.2. Analyze which of the design principles summarized in Figures 3.10 and 3.11 are used in redesigning this process and how.
17. Based on your own experience, identify business processes that used to be competitive but have become outdated because of new (information) technology or because they are unnecessarily complex in light of what competitors offer. For instance, consider the effect the Internet and e-commerce have had on the media industry regarding the production, consumption, and distribution of music, film, computer games, newspapers, literature, and so on.

Notes

1. "Chase-ing e-business," *IIE Transactions* 32(6) (June 2000): 25.
2. A flowchart is a graphical description of the process flow. This tool is thoroughly explained in Chapter 4.
3. A bottleneck is the resource that through its capacity restriction limits the output of the entire process (see Chapter 4).

References

Avni, T. 1999. Simulation modeling primer. *IIE Solutions 31*(9): 39–41.

Camp, R. 1995. *Business process benchmarking.* Milwaukee, WI: ASQ Quality Press.

Champy, J. 1996. *Reengineering management—The mandate for new leadership.* New York: Harper Business.

Chase, R.B., Aquilano, N.J., and Jacobs, F.J. 1998. *Production and operations management: Manufacturing and services.* 8th ed. New York: McGraw-Hill/Irwin.

Cohen, S.L. 2000. The supermarket in 2010. *IIE Solutions 32*(4): 38–41.

Cross, K.F., Feather, J.J., and Lynch, R.L. 1994. *Corporate renaissance: The art of reengineering.* Cambridge, MA: Blackwell Business.

Davenport, T., and Nohria, N. 1994. Case management and the integration of labor. *Sloan Management Review* (Winter): 11–11.

Dumas, M., La Rosa, M., Mendling, J., and Reijers, H.A. 2013 *Fundamentals of business process management.* Berlin Heidelberg: Springer-Verlag.

Hammer, M. 1990. Reengineering work: Don't automate, obliterate. *Harvard Business Review 71*(6): 119–131.

Hammer, M., and Champy, J. 1993. *Reengineering the corporation: A manifesto for business revolution.* New York: Harper Business.

Hyer, N., and Wemmerlöv, U. 2002. *Reorganizing the factory: Competing through cellular manufacturing.* Portland, OR: Productivity Press.

Lester, D. 1994. Reengineering methodology and case study. In E. Snodgrass (Ed.), *Beyond the basics of reengineering: Survival tactics for the 90's*. Norcross, GA: Industrial Engineering and Management Press.
Lowenthal, J.N. 1994. *Reengineering the organization: A step-by-step approach to corporate revitalization*. Milwaukee, WI: ASQC Quality Press.
Roberts, L. 1994. *Process reengineering: A key to achieving breakthrough success*. Milwaukee, WI: ASQC Quality Press.
Robinson, A. 1991. *Continuous improvement in operations: A systematic approach to waste reduction*. Cambridge, MA: Productivity Press.
Tumay, K. 1995. Business process simulation. In Proceedings of the 1995 Winter Simulation Conference, editors C. Alexopoulos, K. Kang, W. R. Lilegdon, and D. Goldsman, 55–60. https://informs-sim.org/wsc95papers/prog95sim.html
Van Gigch, J.P. 1978. *Applied general systems theory*. 2nd ed. New York: Harper and Row.

4

Basic Tools for Process Design

Modeling business processes with the goal of creating effective designs requires more than word processors and spreadsheets. Electronic spreadsheets are useful for mathematical models, but they are less practical in the context of modeling business processes. The tools that are introduced in this chapter help designers check for feasibility, completeness, and effectiveness of a proposed design. These tools facilitate a better understanding of the processes under consideration and the design issues associated with them. The tools are associated with specific design principles, as shown in Table 4.1.

Section 4.2 discusses additional design principles not shown in Table 4.1, such as the elimination of buffers, the establishment of a product orientation, one-at-a-time processing, and the elimination of multiple paths through operations. The specific tools for these design principles are not identified, but they are discussed because it is important to keep them in mind when designing business processes.

Most of the tools presented here assume the availability of key data, such as processing times and the steps required to complete a process. Therefore, we will first address the issue of collecting relevant data that are needed for the application of the basic tools described here. In manufacturing, tagging of parts is a method that often is used to collect data associated with

TABLE 4.1

Tools and Design Principles

Design Principle	Tool
Eliminating non-value-added activities and measuring process improvement	General process chart
Understanding process flow and its relationship to infrastructure configuration	Process flow diagram and LD analysis
Understanding workflow	Flowcharting, data-flow diagrams, and IDEF modeling
Balancing flow and capacity	Line balancing
Minimizing sequential processing and handoffs	Case management
Processing work based on its characteristics	Scheduling

DOI: 10.1201/9781032617237-4

Part Tagging Sheet

Shop Order Number: [ ] Shop Order Qty: [ ] **Part Number:** [ ]

Shop Order Issue Date: [ ]

Step #	Process Description	Op #	Machine ID	Shift	Lot Size In	Lot Size Out	Activity Start Date	Activity Start Time	Activity Stop Date	Activity Stop Time	Comments ex. 3 Scrap, Split Qty
1							__/__/01		__/__/01		
2							__/__/01		__/__/01		
3							__/__/01		__/__/01		
4							__/__/01		__/__/01		
5							__/__/01		__/__/01		
6							__/__/01		__/__/01		
7							__/__/01		__/__/01		
8							__/__/01		__/__/01		
9							__/__/01		__/__/01		
10							__/__/01		__/__/01		
11							__/__/01		__/__/01		
12							__/__/01		__/__/01		
13							__/__/01		__/__/01		
14							__/__/01		__/__/01		

Purpose To track exactly where parts/components travel throughout its manufacturing process and to find MCT (Manufacturing Cycle Time).

Directions
A Person to give shop/job order to floor attaches tag sheet to paperwork and enters all fields with a double lined border
B The first line should be filled out by SET-UP PERSON if set-up is required. No Qty's are entered unless scrap is produced.
C The next line is filled out by OPERATOR. Start time is first complete piece-stop is last piece before the qty is moved to the The Sheet should be attached to parts and follow throughout. Repeat process until last operation. Pull tag off before ready to ship.

FIGURE 4.1
Example of a job tagging sheet.

a process. Tagging is the process of documenting every activity that occurs during a process. The main data collected by tagging methods are processing times. Tagging is done with two data collection instruments, known in the context of business processes as a job tagging sheet and a workstation log sheet.

A job tagging sheet is a document that is used to gather activity information. If performed correctly, tagging gives an accurate snapshot of how a process is completed within an organization. The information that is gathered can give the company many insights into what is happening in a process, including the steps necessary to complete a job as well as the processing and delay times associated with each activity. The data collected in the job tagging sheets can be used as inputs to several design tools that are discussed in this and subsequent chapters. An example of a job tagging sheet is shown in Figure 4.1. The job tagging sheet accompanies a job from the beginning to the end.

Workstation log sheets are used to document the activities that occur at each workstation. This information can be used to determine the utilization and potential capacity of a workstation. However, management must assure

Machine Log

Machine ID: [ ]

Activity										Activity Start		Activity Stop		
Set-UP	Repair	PrevMai	Run	Other	Part #	Shop / Job Order Number	Lot Size In	Lot Size Out	Shift	Date	Time	Date	Time	Comments
										___/___/01		___/___/01		
										___/___/01		___/___/01		
										___/___/01		___/___/01		
										___/___/01		___/___/01		
										___/___/01		___/___/01		
										___/___/01		___/___/01		
										___/___/01		___/___/01		
										___/___/01		___/___/01		
										___/___/01		___/___/01		
										___/___/01		___/___/01		
										___/___/01		___/___/01		
										___/___/01		___/___/01		
										___/___/01		___/___/01		
										___/___/01		___/___/01		
										___/___/01		___/___/01		

Purpose: To determine machine availability.

Directions: A All necessary Equip needs to have a Machine Log attached to Machine. Person who attaches Machine Log enters Machine ID at the top of the sheet.

B Lines should be filled in whenever an activity is performed on the machine for longer than 1/2 h. If several 20 min activities happen throughout the day--add up the total time and enter as one line Person who is performing activity is responsible for entering the information.

C Full sheets should be placed in the back of the plastic sleeve and a new log sheet should be brought to the front with the correct Machine ID filled in.

FIGURE 4.2
Example of a workstation log sheet.

employees that the information collected is part of a continuous improvement or process-design effort and not a "witch hunt." A commitment must be made to improving processes and not individual efficiencies. Figure 4.2 shows an example of a workstation log sheet.

Appendix L of *Reorganizing the Factory: Competing Through Cellular Manufacturing* by Nancy Hyer and Urban Wemmerlöv (2002) contains a detailed discussion on tagging in the context of manufacturing processes.

4.1 Process Flow Analysis

Four charts typically used for process flow analysis are the general process chart, the process flow diagram, the process activity chart, and the flowchart (Melnyk and Swink, 2002). In general, these charts divide activities into five different categories: operation, transportation (physical and information), inspection, storage, and delay. Figure 4.3 shows a common set of symbols

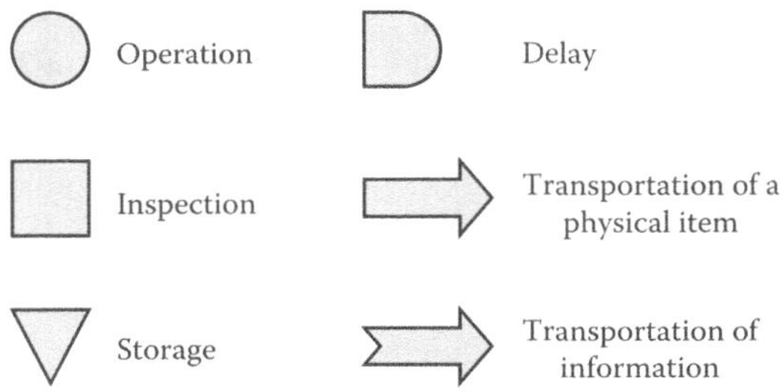

FIGURE 4.3
Activity symbols.

used to represent activities in charts for process flow analysis. The list of symbols is not exhaustive or universal. While using basic and universally known symbols (such as those in Figure 4.3) is generally preferred, charts prepared for peer groups with specialized knowledge may include special symbols. Tools such as DIAGRAM by ConceptDraw (https://www.conceptdraw.com/) include libraries of special symbols to create highly expressive charts for business process modeling. Flowcharting is discussed further in Section 4.1.4.

In a variety of settings, operations are the only activities that truly add value to the process. Transportation generally is viewed as a handoff activity, and inspection is a control activity. Note, however, that in some situations, transportation might be considered the main value-adding activity (or operation) in the process. Think, for example, about airlines, freight companies, postal services, and cab and bus companies.

The difference between storage and delay is that storage is a scheduled (or planned) delay. This occurs, for example, when jobs are processed in batches. If a process uses a batch size of two jobs, then the first job that arrives must wait for a second job before processing can continue. The time that the first job has to wait for the second to complete a batch is a planned delay (or storage).

A delay, on the other hand, is unplanned waiting time. Delays normally occur when a job (or batch of jobs) must wait for resources to be freed before processing can continue. A delay is always a non-value-added activity, but under certain circumstances, storage might add value to the customer. For example, consider an order fulfillment process that carries an inventory of finished goods to shorten the time it takes to fulfill an order. If customers are interested in the rapid delivery of their orders, then the storage activity might be valuable to them.

4.1.1 General Process Charts

The general process chart summarizes the current process, the redesigned process, and the expected improvements from the proposed changes

TABLE 4.2
Example of a General Process Chart

	Current Process			Redesigned Process			Difference	
Activity	**No.**	**Time**	**%**	**No.**	**Time**	**%**	**No.**	**Time**
Operation	5	30	10	5	30	37.5	0	0
Inspection	3	60	20	1	20	25.0	–2	–40
Transportation	10	120	40	2	20	25.0	–8	–100
Storage	0	0	0	0	0	0	0	0
Delay	7	90	30	1	10	12.5	–6	–80
Total	25	300	100	9	80	100	–16	–220

(Table 4.2). This chart characterizes the process by describing the number of activities by category, the amount of time activities in each category takes overall, and the corresponding percentage of the process total.

The information that is summarized in the general process chart indicates with single-glance major problems with the existing process and how the proposed (redesigned) process will remedy some (or all) of these problems. These problems are measured by the time (and corresponding percentage of time) spent on value-added and non-value-added activities. A redesigned process should increase the percentage of time spent on value-added activities by reducing the number of non-value-added activities or the time spent on such activities. Note that the summary focuses on the time to perform the activities (labeled *Time*) and the frequency of occurrence (labeled *No.*) in each category.

The example in Table 4.2 shows that although the number of value-added activities (i.e., the operations) did not change in the redesigned process, the percentage of time that the process performs operations increases from 10% to 37.5%. This is achieved by reducing the amount of time spent on non-value-added and control activities. The negative numbers in the columns labeled Difference indicate the reduction in the frequency of occurrence (No. column) and total time (Time column) corresponding to each category.

4.1.2 Process Flow Diagrams

The most typical application of process flow diagrams has been the development of layouts for production facilities. These diagrams provide a view of the spatial relationships in a process. This is important because most processes move something from activity to activity, so physical layouts determine the distance traveled and the handling requirements.

The process flow diagram is a tool that allows the analyst to draw movements of items from one activity or area to another on a picture of the facility. The resulting diagram measures process performance in units of time and distance. This fairly straightforward analysis must include all distances over which activities move work, including horizontal and vertical distances when the process occupies different floors or levels of a facility. The diagram assumes that moving items require time in proportion to the distance and that this time affects overall performance. This is why accurate process flow diagrams are particularly helpful in the design of the order-picking process described at the end of this chapter.

The analyst can add labels to the different areas in the process flow diagrams. These labels are then used to indicate the area in which an activity is performed by adding a column in the process activity chart. Alternatively, the process flow diagram may include labels corresponding to an activity number in the process activity chart. This labeling system creates a strong, complementary relationship between the two tools. The process activity chart details the nature of process activities, and the process flow diagram maps out their physical flows. Together, they help the operations analyst better understand how a process operates.

A process flow diagram is a valuable tool for process design. Consider, for example, a process with six work teams, labeled A to F, and physically organized as depicted in Figure 4.4. Clearly, unnecessary transportation occurs due to the current organization of the work groups within the facility. The sequence of activities is such that work is performed in the following order: C, D, F, A, B, and E. Rearranging the work groups to modify the

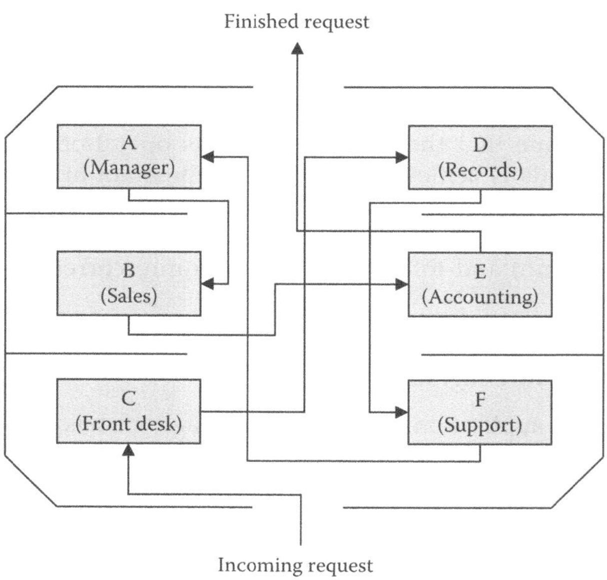

FIGURE 4.4
Example of a process flow diagram.

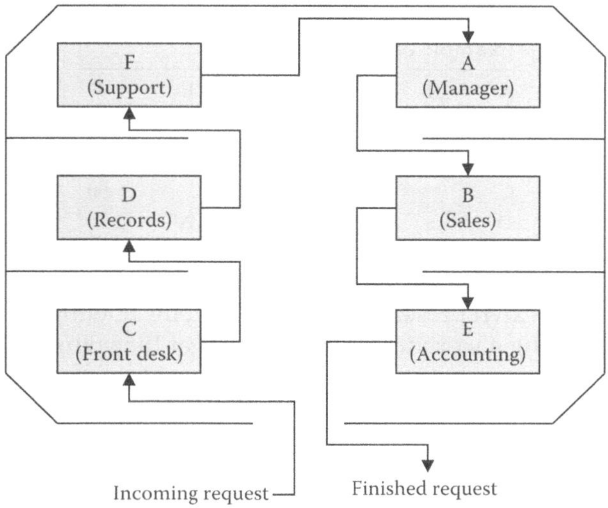

FIGURE 4.5
Redesigned process flow diagram.

flow of work through the facility, as depicted in Figure 4.5, results in a more efficient layout.

The focus of an analysis of a process flow diagram is on excessive and unnecessary movements. These may be evident by long moves between activities, crisscrosses in paths, repeated movements between two activities, or other illogical or convoluted flows. An effective design eliminates crisscrosses and locates sequential, high-volume activities close together to minimize move times.

The physical arrangement of people, equipment, and space in a process raises four important questions:

1. What centers should the layout include?
2. How much space does each workstation (or center) need?
3. How should each center's space be configured?
4. Where should each center be located?

The process flow diagram is most helpful in answering question 4 regarding the location of each center. The diagram also can be complemented with a simple quantitative method to compare alternative designs. The method is based on calculating a load–distance score (LD score) for each pair of work centers. The LD score between work centers i and j is found as follows:

$$\text{LD score}(i,j) = \text{Load}(i,j) \times \text{Distance}(i,j)$$

TABLE 4.3

Example of a Load Matrix

A	B	C	D	E	F
A	20		20		80
B		10		75	
C			15		90
D				70	

The *load* value is a measure of "attraction" between two work centers. For example, the load can represent the volume of items flowing between the two work centers during a workday. Hence, the larger the volume of traffic is, the more attraction there is between the two centers. The goal is to find a design that will minimize the total LD score (i.e., the sum over all pairs of work centers). Consider the load matrix in Table 4.3 associated with two alternative designs depicted in Figure 4.6.

The calculations in Table 4.4 compare the two designs in Figure 4.6. The comparison is in terms of the LD score. The distances in Table 4.4 are rectilinear. This means that the number of line segments (in the grid) between two centers determines the distance. For example, one needs to go through three lines in the grid to connect work centers A and F in both designs in Figure 4.6.

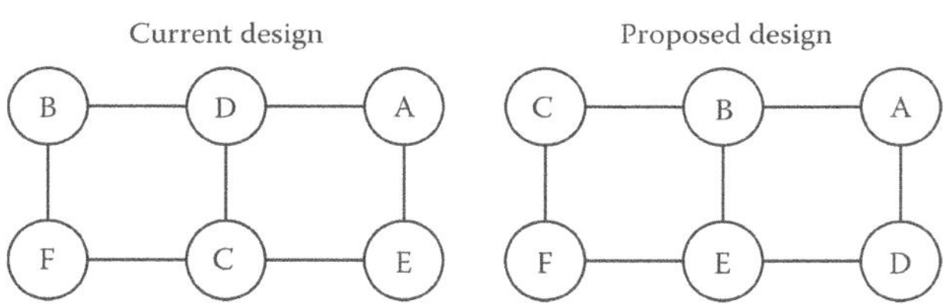

FIGURE 4.6
Location of work centers in alternative designs.

TABLE 4.4

LD Calculation for Two Designs

		Current Design		Proposed Design	
Centers	**Load**	**Distance**	**LD Score**	**Distance**	**LD Score**
(A, B)	20	2	40	1	20
(A, D)	20	1	20	1	20
(A, F)	80	3	240	3	240
(B, C)	10	2	20	1	10
(B, E)	75	3	225	1	75
(C, D)	15	1	15	3	45
(C, F)	90	1	90	1	90
(D, E)	70	2	140	1	70
Total			790		570

The total LD score for the proposed design is clearly better than the current design (570 vs. 790). However, the savings in operational costs derived from the new design must be compared with the corresponding relocation costs. It might be possible to find a solution where only a few work centers need to be moved and still achieve a significant improvement from the current design.

Distances typically are measured either as rectilinear or Euclidian for the purpose of calculating an LD score. In general, if center 1 is located in position (x_1, y_1) and center 2 is in position (x_2, y_2), the distance between the centers can be calculated as follows:

$$\text{Rectilinear distance} = |x_1 - x_2| + |y_1 - y_2|$$

$$\text{Euclidian distance} = \sqrt{(x_1 - x_2)^2 + (y_1 - y_2)^2}$$

For instance, the rectilinear distance between a center located in position (3, 7) and a center located in position (4, 2) is $|3 - 4| + |7 - 2| = 1 + 5 = 6$. The corresponding Euclidian distance is $\sqrt{(3-4)^2 + (7-2)^2} = 5.1$. The Euclidian distance is the length of a straight line between the two points.

4.1.3 Process Activity Charts

The general process chart does not place activities in sequence, which is important and necessary to gain a high-level understanding of the differences between two alternative processes. For example, two alternative processes with the same mix of activities would have identical summarized information in the general process chart, although one, for instance, may place most of the control activities at the beginning of the process and the other one at the end.

The process activity chart complements the general process chart by providing details to gain an understanding of the sequence of activities in the process (see Figure 4.7). The analyst fills in the required information and identifies the appropriate symbol for each activity in one line of the chart and then connects the symbols to show the flow of the process. The completed chart describes the exact sequence of activities along with the related information. The arrow (→) in Figure 4.7 is used to represent the transportation of physical items or information.

Although a basic chart can be built with the activity category, the time requirement, and the activity description, more elaborate charts can be developed to include other pieces of information by adding more columns to the chart. For example, information about the number of people involved in each activity can give an indication of staffing needs and overall cost.

Note that the process activity chart considers average activity times only. An estimate of the total time to process a job can then be obtained by adding the activity times. However, this estimate ignores the variation associated with the time to perform each activity. Also note that the chart is not useful when the process includes several variants because each variant needs its own chart. For example, a process where some requests might need the work of a group of

Process Activity Chart

Page: 1 of 1

Process: X-Ray

Date: 5/1/03

Current Process ☑

Proposed Process ☐

Developed by: Boulder Community Hospital

No.	Description	Time	Value Code (V/N/C)	Symbol
1	Walk to Lab	7	N	
2	Wait	10	N	
3	Fill Insurance Form	6	C	
4	Fill Lab Form	5	C	
5	Wait	7	N	
6	Undressing	3	V	
7	Take X-rays	5	V	
8	Develop X-ray	12	V	
9	Check X-ray	3	C	
10	Transfer X-ray	10	N	
11	Walk back	7	N	

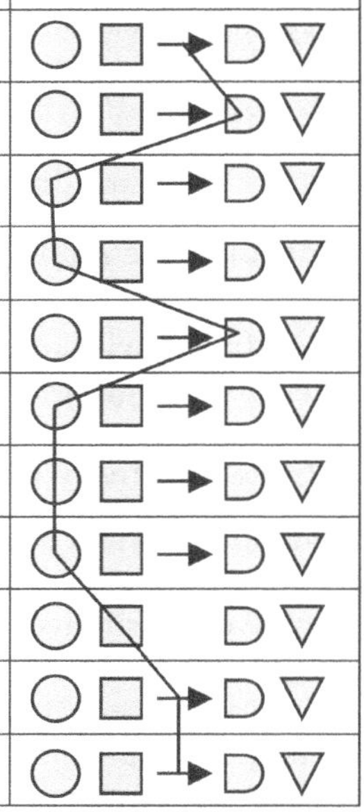

For each activity, fill in the required information. Also, connect the symbols to show the flow through the process.

The value code indicates whether the activity adds value (V), does not add value (N), or controls (C).

FIGURE 4.7
Example of a process activity chart.

specialists would require two charts: one with the flow of requests without the specialists and one with the flow of requests through the specialists. Finally, the process activity chart cannot show two or more simultaneous activities like those in processes organized in parallel. Flowcharts, described in the following section, overcome this limitation. Unlike the process activity chart, flowcharts are capable of depicting processes with multiple paths and parallel activities.

4.1.4 Flowcharts

Flowcharts are a fundamental tool for designing and redesigning processes. They graphically depict activities, typically in a sidelong arrangement such that they follow the movement of a job from left to right through the process. A flowchart can help identify loops in a process (i.e., a series of activities that must be repeated because of problems such as lack of information or errors). A flowchart also can be used to show alternative paths in the process, decision points, and parallel activities. In addition to the symbols in Figure 4.3, a symbol for a decision point is required when creating flowcharts. The decision symbol is a diamond (◆), as illustrated in Figure 4.8.

To illustrate the use of flowcharts, consider the ordering process in Figure 4.8, which begins with a telephone operator taking information over the phone. The operator then passes order information once a day to a supervisor who reviews it for accuracy and completeness. Accurate orders are fulfilled by customer service, and incomplete orders are set aside for a sales representative, who will contact the customer.

Due to the decision point after the inspection activity, the process in Figure 4.8 has two possible paths. Therefore, this process would generate two activity charts (one for each path) but only one flowchart.

Flowcharts can be used to convey additional information, going beyond the sequence of activities and the logical steps in the process. For instance, flowcharts also might include a time estimate associated with the execution of each activity. An estimate of the activity time can be obtained by using the following equation:

$$\text{Activity time} = \frac{\text{Unit processing time} \times \text{Batch size} + \text{Setup time}}{\text{Efficiency}}$$

where

Unit processing time is the time to process a single job.

Batch size is the number of jobs in a batch.

Setup time is the time necessary to get ready to process the next batch of jobs.

Efficiency is a measure that indicates the speed of processing with respect to a specified standard time.

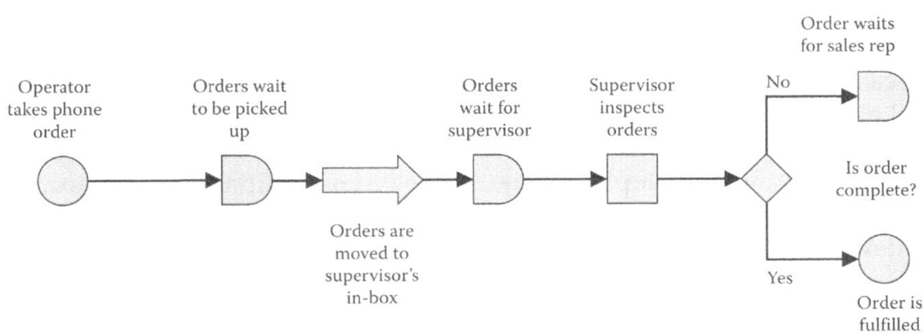

FIGURE 4.8
Example of a flowchart.

This relationship assumes that the standard time to process one unit is known (or can be estimated). It also assumes that the standard time to set up the processing of an entire batch is known. Because the standard times correspond to a worker with 100% efficiency, the activity time will increase if the efficiency of the worker decreases.

Suppose it takes 2 min to inspect one order for completeness in an order fulfillment process. Also, suppose that the supervisor inspects orders in batches of 20 and that it takes 10 min to prepare the inspection of the batch. If the supervisor was recently hired and therefore his or her efficiency is currently 80%, the inspection time for 20 orders is given by the following equation:

$$\text{Activity time} = \frac{2 \times 20 + 10}{0.8} = 62.5\,\text{min}$$

If orders are not processed in batches and no preparation time (setup) is needed before the actual inspection, then the inspection time per order is

$$\text{Activity time} = \frac{2 \times 1 + 0}{0.8} = 2.5\,\text{min}$$

Therefore, 20 orders could be processed in 50 min. These calculations assume that the mean or average values associated with unit processing and setup times are accurate estimates of the actual values. However, this assumption is not always valid. In many situations, the time values vary significantly, so a more accurate representation is a frequency (or probability) distribution. When probability distributions govern the activity times as well as the arrival time of the items to the process, a simple flowchart might not be sufficient to fully understand the process dynamics. Chapters 7 through 10 explore the use of an advanced tool for process design known as discrete event simulation, which follows the principles of a flowchart to create a dynamic representation of the process.

Other information that can be included in a flowchart includes frequency values for paths originating from a decision point. For example, in the processes shown in Figure 4.8, frequency information associated with the probability of a complete order could be displayed. If an order is complete approximately 90% of the time, then 90% would be placed next to the Yes path and 10% would be placed next to the No path. Chapter 5 will examine how to use a flowchart with activity times and frequency information in the context of managing process flows.

Flowcharting is often done with the aid of specialized software. For instance, Figure 4.9 shows a flowchart for a credit card order processing created with a flowcharting software tool called SmartDraw. The symbols used in the flowchart in Figure 4.9 are specific to SmartDraw and vary from those depicted in Figure 4.3.

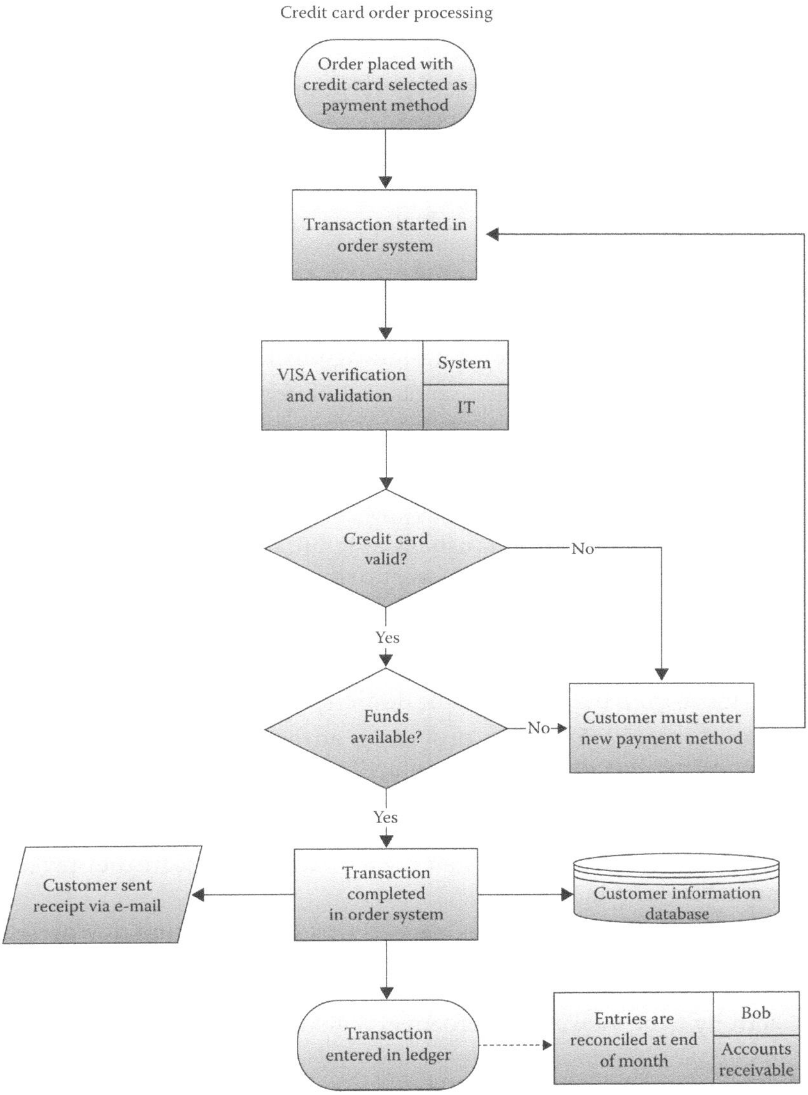

FIGURE 4.9
Flowchart of a credit card order processing created with a flowcharting software tool called SmartDraw.

4.1.5 Service System Maps

Service system mapping (SSM) is an extension of traditional flowcharting. It documents how a business process interacts with its customers and the role they play in the delivery of service. The technique was created by Gray Judson Howard in Cambridge, Massachusetts, with the purpose of viewing activities in the context of how they add value to the customers. SSM has the following goals:

- To build shared and consistent perceptions of the customers' experience with the entire core process.
- To identify all points of contact between the business process and its customers.
- To provide the basis for developing an economic model of the business.
- To identify opportunities for improving the effectiveness of the business process.
- To provide a framework for designing business processes.
- To aid in pinpointing control points and strategic performance measures.

According to Cross et al. (1994), the benefits of using SSM for process design include the following:

- *Improved communication*: By providing a graphic representation of the day-to-day processing, a map is usually a significant improvement over written descriptions (one picture is worth a thousand words). A map reduces the risk of varied and inaccurate representations of current operations and establishes a foundation for effective company-wide collaboration on business process improvements. In other words, SSM helps employees move from the current to the desired state of operations.
- *Market research focus*: By providing an accurate, step-by-step portrayal of points of customer contact from the customer's perspective, a map enhances management's ability to target and focus on areas critical to customer satisfaction. The need to improve understanding of certain aspects of customer contact may suggest further targeted market research.
- *Operations design and process management*: Upstream and downstream effects are easier to identify when additions, deletions, or alterations are made up on a map. A map also allows operations or backroom activity to be viewed in a broader context in relation to the customer.
- *Application of information technology*: Information systems (IS) can be designed from a more strategic perspective using SSM. Not only are the customer contact points, materials flows, and specific activities depicted, but the information flows required to perform key activities also are included. New IS (or enhancements to existing systems) can be designed to link value-added activities in the core process.
- *Critical performance measures focus*: Mapping facilitates the collection of key measures necessary in process redesign. SSM also fosters the development of effective policies, procedures, guidelines, and training programs that help set the context for performance management systems. With a map, a design team can pinpoint the critical measurement points. A broad pictorial perspective facilitates the design of measures to meet

> the best strategic balance between service quality, product quality, and profitability. Performance measures are most effective when controlling the horizontal flow of work as depicted by the map and when unconstrained by the vertical structure of the organization. The map provides this perspective by depicting all activities in a continuous flow rather than compartmentalizing them by their organizational function.

Although the particular characteristics of a map, including the level of detail, are specific to each application, a general template for SSM exists. This template consists of two main elements: horizontal bands and process segments.

The purpose of the horizontal bands is to organize activities in terms of the different "players" in a process (see Figure 4.10). An SSM typically contains the following five bands:

1. End user or customer band.
2. Frontline or distribution channel band.
3. Backroom activity band.
4. Centralized support or IS band.
5. Vendor or supplier band.

In Figure 4.10, the customer triggers the beginning of the process by performing activity A. For instance, activity A could be a phone call. A frontline employee performs activity B (e.g., answering the phone call and checking the status of an order). The backroom activity band could include warehouse activities in an order fulfillment process or underwriting activities in a policy-issuing process of an insurance company. Activities E, F, and I are assigned to the IS band, which could mean, for example, database queries. Activity H is performed 30% of the time; the other 70% of the time, this activity is bypassed. The process finishes with a frontline employee (e.g., a customer or sales representative) contacting the customer (see activities J and K). The bands in an SSM can be customized to each situation; however, it is generally recommended not to have more than seven bands in one map.

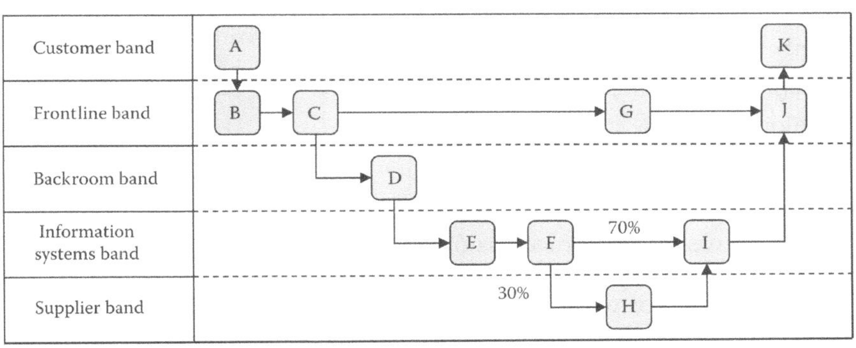

FIGURE 4.10
Example of an SSM with horizontal bands.

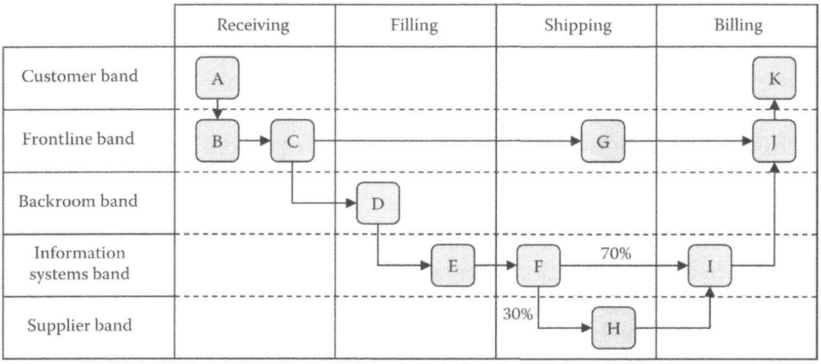

FIGURE 4.11
Example of a SSM with horizontal bands and process segments.

Process segments are sets of activities that represent a subprocess; that is, a segment produces a well-defined output given some input. For example, an order fulfillment process can be divided into receiving, filling, shipping, and billing (see Figure 4.11). Note that these segments (or subprocesses) are chronologically ordered from left to right on the map.

The process segments do not have to be similar in size as measured by the number of activities. It is more important that the process is partitioned into segments that produce meaningful output.

As originally proposed, service system maps use additional icons and conventions to convey information. For example, a computer icon can be placed next to an arrow to indicate that the communication occurs via e-mail or a file transfer. Similarly, a telephone icon indicates a phone call, or a fax machine icon indicates that a fax is sent. In addition to these icons, software companies have developed flowcharting packages with increased flexibility and functionality. The following software products are examples of flowcharting tools with service system map features:

DIAGRAM (conceptdraw.com).
Igrafx (igrafx.com).
PaceStar (pacestar.com).
RFFlow (rff.com).
SmartDraw (smartdraw.com).
TeamFlow (teamflow.com).
Visio (microsoft.com/office/visio).

Most of these software tools follow the traditional flowchart notation represented by activity symbols similar to those depicted in Figure 4.3. An example of an SSM developed with SmartDraw is shown in Figure 4.12. The

SSM depicted in Figure 4.12 is rotated 90° with respect to Figure 4.10. In other words, instead of representing process participants as horizontal bands, they are represented as vertical bands. Because the process is small, no process segments are shown.

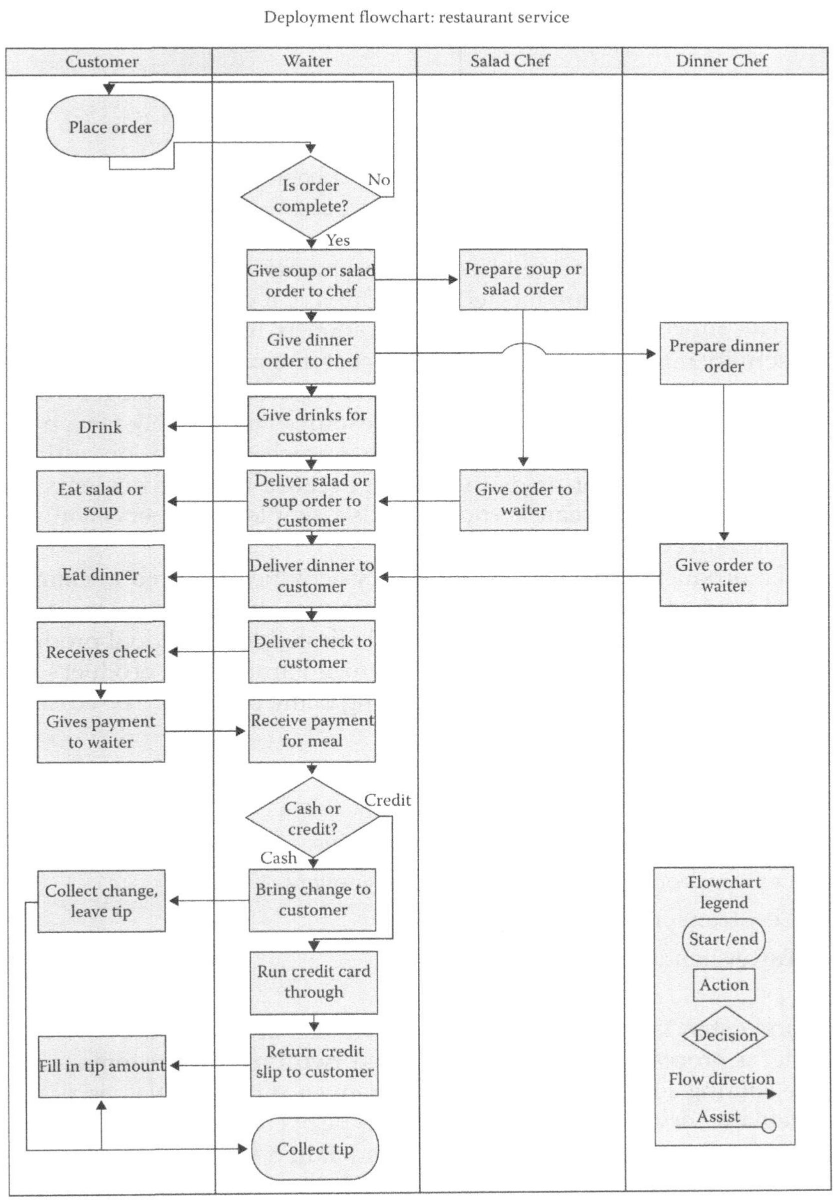

FIGURE 4.12
Example of a SSM developed with SmartDraw.

4.2 Workflow Design Principles and Tools

A set of workflow design principles was briefly introduced in Chapter 3. This section will expand the discussion of these principles and present appropriate tools for their implementation.

4.2.1 Establish a Product Orientation in the Process

A set of activities can be organized in two basic ways: by function or by product or service. In a functional (or process) orientation, workstations or departments are grouped according to their function. For example, all drilling equipment is located in one area of a machine shop, or all technical support personnel are located in one area of a computer software company. The functional (or process) orientation is most common when the same activity must serve many different customers. Because demand levels for these activities are generally low or unpredictable, management does not like to set aside human and capital resources exclusively for a particular product, service, or customer type. In a process orientation, utilization of equipment and labor tends to be high because the requirements for all products and services can be pooled. Also, employee supervision can be more specialized.

With a product orientation, all necessary activities required to complete a finished product or service are organized into an integrated sequence of work modules. This means that resources are dedicated to individual products or services. Consequently, activities are organized around the product's or service's route, rather than being shared among many products. The advantages of a product orientation are as follows:

- Faster processing rates.
- Lower work-in-process (WIP) inventories.
- Less unproductive time lost to changeovers.
- Less transportation time.
- Fewer handoffs.

To understand the difference between functional and product orientation, consider a process with five activities and two types of customers, A and B. The routing for customer A (i.e., the sequence of activities necessary to complete the service associated with this type of customer) is 1, 2, 4, 5, and 3. The routing for customer B is 3, 1, 4, 2, and 5. In a functionoriented system, each resource that performs the corresponding activity is located in one area (as illustrated in Figure 4.13a), and customers must travel to each area to be

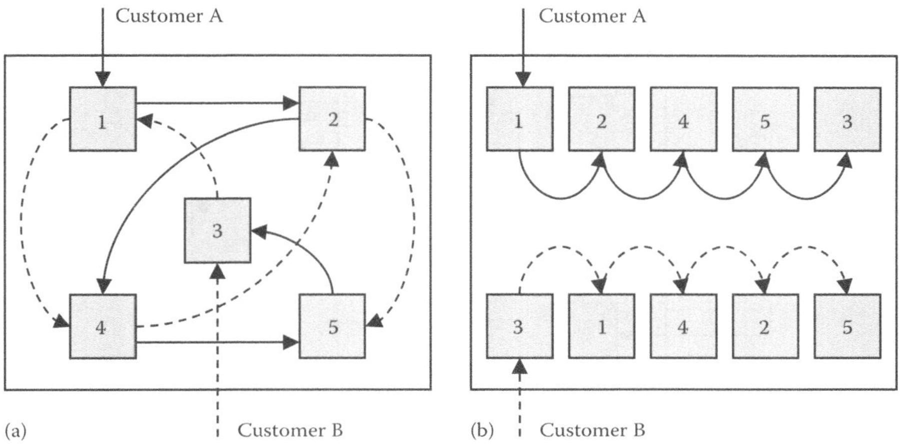

FIGURE 4.13
(a) Process versus (b) product or orientation.

served. A product-oriented system, on the other hand, has dedicated servers for each activity and customer type in order to streamline the processing of each customer (Figure 4.13b).

Note that a product orientation is a capital-intensive way of organizing activities because the resources are not pooled. Therefore, the volume of customers A and B must be sufficient to justify a product-oriented organization and keep resource utilization at an acceptable level. Unfortunately, volumes are not always high enough to justify dedicating resources to a single product or service. In such situations, it might be possible to create a hybrid orientation in the process. One popular hybrid orientation in manufacturing is known as *group technology*. This technique groups parts or products with similar characteristics into families and sets aside groups of machines for their production. The equivalent in a business process is to group jobs (e.g., requests or customers) into families. Families may be based on demand or routing requirements. The next step is to organize the resources (e.g., equipment and people) needed to perform the basic activities into separate areas called *cells*. Each cell must be able, with minor adjustments, to deal with all the jobs in the same family.

Suppose two more customers are added to the previous example: customer C with routing 3, 2, 4, 5, and 1 and customer D with routing 1, 3, 4, 2, and 5. Because customer C shares the sequence of activities 2, 4, and 5 with customer A, the family AC can be formed within a group technology orientation. Similarly, a family BD can be created with customers B and D because they share the sequence of activities 4, 2, and 5. A cell AC would

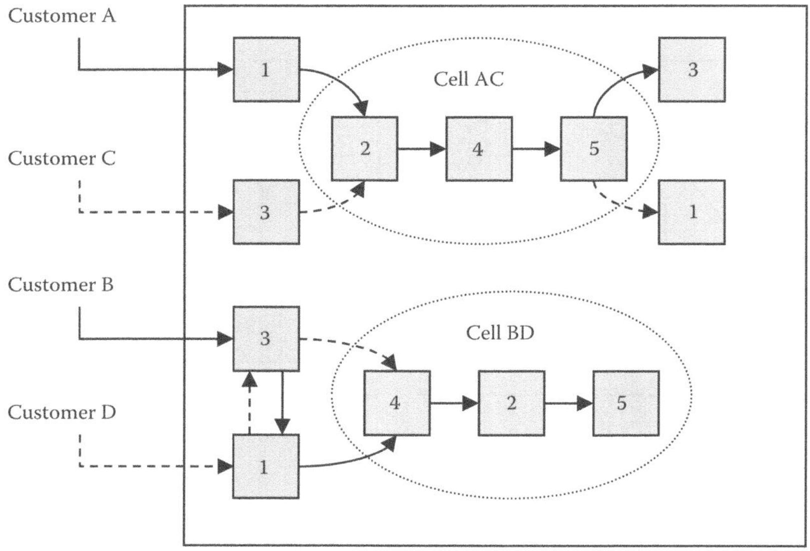

FIGURE 4.14
Hybrid orientation.

handle customers A and C, and a cell BD would handle customers B and D (see Figure 4.14).

The hybrid orientation based on group technology simplifies customer routings and reduces the time a job is in the process.

4.2.2 Eliminate Buffers

In the context of business processes, a buffer consists of jobs that are part of the WIP inventory. The term WIP was coined in manufacturing environments to denote the inventory within the production system that is no longer raw material, but not a finished product either. In more general terms, WIP consists of all the jobs that are in queues, moving from one operation to the next, being delayed for any reason, being processed, or residing in buffers. Whether intentional or not, buffers cause logistical nightmares and create unnecessary work and communication problems. When buffers exist, tracking systems are needed to identify the jobs that are in each buffer. Effort is spent in finding orders, expediting work, and communicating status to customers.

Consider the following example from Cross et al. (1994). At a computer company, the processing of changes to maintenance contracts was taking 42 days as a result of a series of paperwork buffers between functional areas in the process. Some areas were hiding many of the errors that appeared on customer bills. After establishing a product orientation and reorganizing into teams, the buffers disappeared. Processing errors were caught immediately and eliminated. The impact was an 80% reduction in backlog and a 60% reduction in cycle time. Cash flow also improved, reducing the need to borrow capital to finance operations.

Note that in this example, changing the system from a process to a product orientation eliminated buffers. However, a product orientation does not always guarantee a buffer-free operation. A product-oriented process must be well balanced in order to minimize buffers.

4.2.3 Establish One-at-a-Time Processing

The goal of this principle is to eliminate batch processing or, in other words, to make the batch size equal to one. By doing so, the time that a job has to wait for other jobs in the same batch is minimized, before or after any given activity. Every process has two kinds of batches: *process batches* and *transfer batches*. The process batch consists of all the jobs of the same kind that a resource will process until the resource changes to process jobs of a different kind. For example, a laboratory might create batches of all the blood samples requiring the same tests. An online retailer might process several orders by the same customer as a batch with the goal of reducing shipping costs. A manufacturer of auto parts requiring different colors might use batch processing to minimize the number of times painting stations have to be switched from one color to another.

Transfer batches refer to the movement of jobs between activities or workstations. Rather than waiting for the entire process batch to be finished, work that has been completed can be transferred to the next workstation. Consequently, transfer batches are typically equal to or smaller than process batches. The advantage of using transfer batches that are smaller than the process batch is that the total processing time is reduced; as a result, the amount of WIP in the system is decreased.

Figure 4.15 compares the total processing time of a process with three activities in sequence when the batch sizes are changed. In this example, it is assumed that each item requires one unit of processing time in activities 1 and 3 and 0.5 of a time unit in activity 2.

When the transfer batch is the same as the process batch, the total processing time for 100 items is 250 units of time. When the transfer batch is changed to a size of 20 items, then the total processing time for 100 items is 130 units of time. In this case, the first 20 units are transferred from activity 1 to 2 at time 20. Then at time 30, this batch is transferred from activity 2 to 3. Note that between times 30 and 40, the resources associated with activity 2 are idle (or working on activities of a different process).

From the point of view of the utilization of resources, two values are important: setup time and processing run time. Larger batch sizes require fewer setups and, therefore, can generate more processing time and more output. This is why larger batch sizes are desirable for bottleneck resources[1] (i.e., those with the largest utilization). For non-bottleneck resources, smaller process batch sizes are desirable because the use of existing idle times reduces WIP inventory.

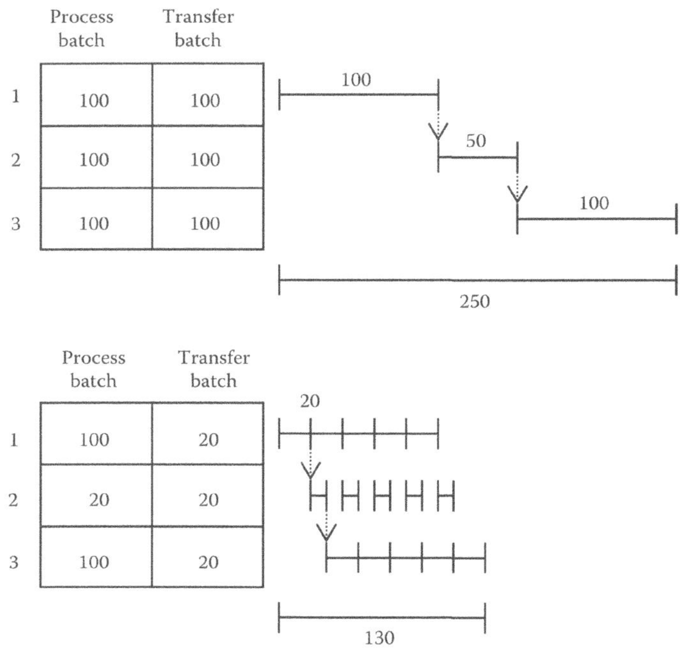

FIGURE 4.15
Effect of changing batch sizes.

4.2.4 Balance the Flow to the Bottleneck

The principle of balancing the flow to the bottleneck is linked to an operation management philosophy known as theory of constraints (TOC). Eliyahu M. Goldratt coined this term and popularized the associated concepts in his book *The Goal* (Goldratt and Cox, 1992). This book uses the Socratic method to challenge a number of assumptions that organizations make to manage operations. The novel narrates Alex Rogo's struggle to keep schedules, reduce inventory, improve quality, and cut costs in the manufacturing facility that he manages. Jonah, a physicist with a background that is "coincidentally" similar to Goldratt's, guides Alex through a successful turnaround of his plant by applying simple rules to the plant's operations. Jonah reveals the first rule in the following exchange (page 139 of *The Goal*):

> "There has to be some relationship between demand and capacity." Jonah says, "Yes, but as you already know, you should not balance capacity with demand. What you need to do instead is balance the flow of product through the plant with demand from the market. This, in fact, is the first of nine rules that express the relationships between bottlenecks and non-bottlenecks and how you should manage your plant. So let me repeat it for you: Balance flow, not capacity."

Balancing the flow and not the capacity challenges a long-standing assumption. Historically, manufacturers have tried to balance capacity across a sequence of processes in an attempt to match capacity with market demand. However, making all capacities the same might not have the desired effect. Such a balance would be possible (and desirable) only if the processing times of all stations were constant or had a small variance. Variation in processing times causes downstream workstations to have idle times when upstream workstations take longer to complete their assigned activities. Conversely, when upstream workstations' processing time is shorter, inventory builds up between stations. The effect of statistical variation is cumulative. The only way that this variation can be smoothed is by increasing WIP to absorb the variation (a bad choice because process managers should be trying to reduce WIP as explained in Section 4.2.2) or increasing capacities downstream to be able to compensate for the longer upstream times. The rule here is that capacities within the process sequence do not need to be balanced to the same levels. Rather, attempts should be made to balance the flow of product through the system (Chase et al., 1998).

When processing times do not show significant variability, the line-balancing approach of manufacturing can be applied to create workstations in a process. The goal of line balancing is to balance capacity (instead of flow, as proposed by Goldratt); therefore, it should not be used when processing times vary significantly. The line-balancing method is straightforward, as indicated in the following steps:

1. Specify the sequential relationships among activities using a precedence diagram. The diagram is a simplified flowchart with circles representing activities and arrows indicating immediate precedence requirements.
2. Use the market demand (in terms of an output rate) to determine the line's cycle time (C) (i.e., the maximum time allowed for work on a unit at each station).

$$C = \frac{\text{Process time per day}}{\text{Market demand per day (in units)}}$$

3. Determine the theoretical minimum number of stations (TM) required to satisfy the line's cycle time constraint using the following formula:

$$TM = \frac{\text{Sum of activity times}}{C}$$

4. Select a primary rule by which activities are to be assigned to workstations and a secondary rule to break ties.

5. Assign activities, one at a time, to the first workstation until the sum of the activity times is equal to the line's cycle time or until no other activity is feasible because of time or sequence restrictions. Repeat the process for the rest of the workstations until all the activities are assigned.
6. Evaluate the efficiency of the line as follows:

$$\text{Efficiency} = \frac{\text{Sum of activity times}}{\text{C} \times \text{Actual number of stations}}$$

7. If efficiency is unsatisfactory, rebalance using a different rule.

Example 4.1

Consider the activities in Table 4.5, a market demand of 25 requests per day, and a 420 min working day. Find the balance that minimizes the number of workstations.

1. Draw a precedence diagram. Figure 4.16 illustrates the sequential relationships in Table 4.5.
2. The next step is to determine the cycle time (i.e., the value of C). Because the activity times and the working day are both given in minutes, no time conversion is necessary:

$$C = \frac{420 \text{ min/day}}{25 \text{ requests/day}} = 16.8 \text{ min/request}$$

3. Calculate the TM required (i.e., the value of TM):

$$\text{TM} = \frac{63 \text{ min}}{16.8 \text{ min}} = 3.75 \text{ stations} = 4 \text{ (rounded up)}$$

TABLE 4.5

Activity Times and Immediate Predecessors

Activity	Time (min)	Immediate Predecessor
A	2	–
B	11	A
C	4	B
D	5	–
E	7	D
F	6	C
G	2	C
H	10	E
I	2	E
J	8	F, G, H, I
K	6	J

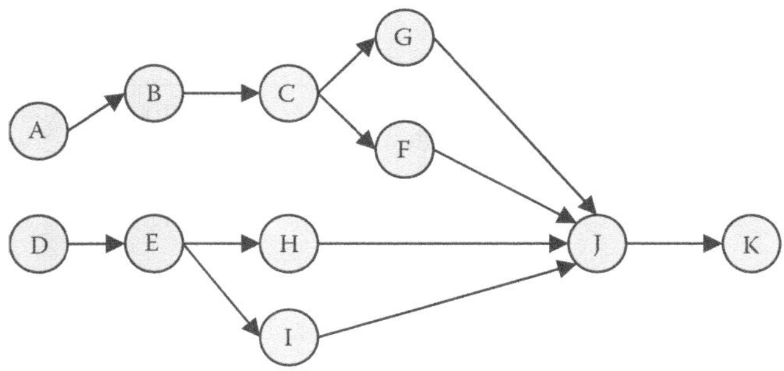

FIGURE 4.16
Precedence diagram.

4. Select a rule to assign activities to stations. Research has demonstrated that some rules are better than others for certain problem structures (Chase et al., 1998). In general, the strategy is to use a rule that gives priority to activities that either have many followers or are of long duration because they effectively constrain the final balance. In this example, our primary rule assigns activities in order of the largest number of followers, as shown in Table 4.6. The secondary rule, used for tie-breaking purposes, will be to assign activities in order of longest activity time. If two or more activities have the same activity time, then one is chosen arbitrarily.

TABLE 4.6

Number of Followers

Activity	Number of Followers
A	6
B and D	5
C and E	4
F, G, H, and I	2
J	1
K	0

5. The assignment process starts with activity A, which is assigned to workstation 1. Because activity A has a time of 2 min, there are 16.8 – 2 = 14.8 min of idle time remaining in workstation 1. We also know that after the assignment of activity A, the set of feasible activities consists of activities B and D. Because either one of these activities could be assigned to workstation 1 and both have the same number of followers, activity B is chosen because it has the longer time of the two. Workstation 1 now consists of activities A and B with an idle time of 14.8 – 11 = 3.8 min. Because activities C and D cannot feasibly be assigned to workstation 1, a new workstation must be created. The process continues until all activities are assigned to a workstation, as illustrated in Table 4.7.

TABLE 4.7

Results of Balancing with Largest Number of Followers Rule

Station	Activity	Station Idle Time (min)	Feasible Activities	Activities with Most Followers	Activities with the Longest Time
1	A	14.8	B, D	B, D	B
	B	3.8	None		
2	D	11.8	C, E	C, E	E
	E	4.8	C, I	C	
	C	0.8	None		
3	H	6.8	F, G, I	F, G, I	F
	F	0.8	None		
4	G[a]	14.8	I		
	I	12.8	J		
	J	4.8	None		
5	K	10.8			

[a] Denotes an activity arbitrarily assigned when a tie remains after applying the primary and secondary selection rules.

6. Calculate the efficiency of the line:

$$\text{Efficiency} = \frac{63\,\text{min}}{16.8\ \text{min} \times 5} = 75\%$$

7. Evaluate the solution. An efficiency of 75% indicates an imbalance or idle time of 25% across the entire process. Is a better balance possible? The answer is yes. The balance depicted in Figure 4.17 is the result of using the longest activity time rule as the primary rule and the largest number of followers as the secondary rule.

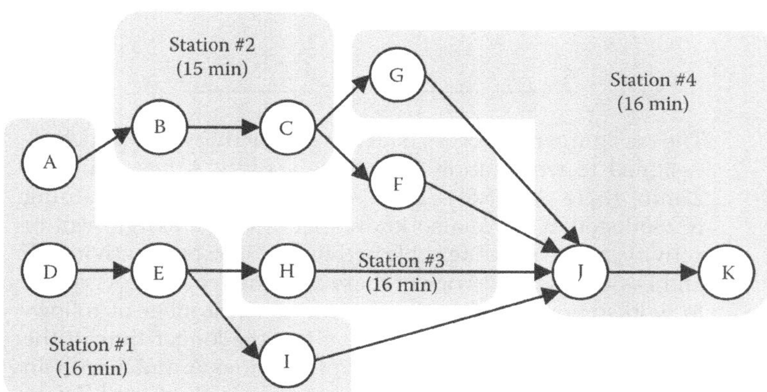

FIGURE 4.17
A more efficient solution. ■

Ragsdale and Brown (2004) developed a spreadsheet model for solving the classical line-balancing problem described earlier. The optimization model employs the evolutionary solver within Excel to conduct a heuristic search for an assignment of activities to workstations that minimizes the total number of workstations in the design. Note that the minimization of the number of workstations is equivalent to maximizing the efficiency of the design.

Constraints to enforce the precedence relationships among the activities and the cycle time are also included in the model.

In line balancing, it is possible for the longest activity time to be greater than the cycle time required for meeting the market demand. If activity H in the previous example had required 20 min instead of 10 min, then the maximum possible process output would have been 420/20 = 21 requests/day. However, there are a few ways in which the cycle time can be reduced to satisfy the market demand:

- *Split an activity*: Is there a way of splitting activity H so that the work can be shared by two workstations?
- *Use parallel workstations*: The workstations are identical and requests would be sent to either one, depending on the current load.
- *Train workers or apply technology*: Is it possible to train a worker or apply technology so that activity H can be performed faster? Note that in this case the effect of the training or application of technology should be such that activity H can be performed at least 19% faster because the time for this activity exceeds the required cycle time by $(1 - 20/16.8) \times 100\%$, or approximately 19%.
- *Work overtime*: The output of activity H is 420/20 = 21 requests/day, 4 short of the required 25. Therefore, 80 min of overtime is needed per day.
- *Redesign*: It might be possible to redesign the process in order to reduce the activity time and meet market demand.

4.2.5 Minimize Sequential Processing and Handoffs

Sequential processing can create two problems that lengthen total process time:

(1) Operations are dependent on one another and therefore constrained or gated by the slowest step, and (2) no one person is responsible for the whole service encounter (Cross et al., 1994).

Example 4.2

This example illustrates the benefits of the design principle of minimizing sequential processing. Assume that a service requires 30 min to complete. Suppose that four activities of 10, 7, 8, and 5 min are assigned each to one of four individuals. In other words, the process is a line with four stations, as depicted in Figure 4.18a. According to the concepts in the previous section, the process has an output rate of 6 jobs/h and an efficiency of $30/(10 \times 4) = 75\%$.

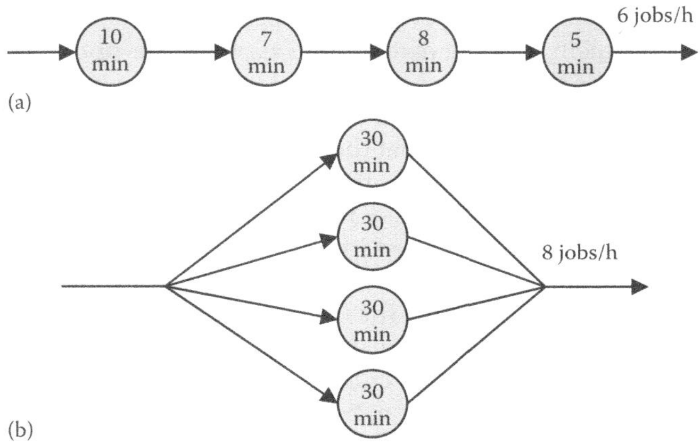

FIGURE 4.18
(a) Sequential versus (b) parallel processing.

Significant improvements can be achieved by replacing sequential processing with parallel processing (Figure 4.18b). For example, if each of the four workers performs the entire process (i.e., each performs the four steps in the process), output per hour theoretically could be increased by 33.3% to an output of 8 jobs/h. However, it might not be possible to achieve the theoretical output of the parallel process due to the differences in performance of each individual. Nevertheless, the parallel process will always produce more output than its sequential counterpart, as long as the total parallel time (in this case, $30 \times 4 = 120$) is less than the slowest station time multiplied by the square of the number of stations (in this case, $10 \times 42 = 160$).

If each worker in this example were to take 40 min to complete each job in the parallel process, then the total time in the parallel system would be 160 min, and the output would be the same as in the sequential process. On the other hand, suppose that four workers require 30, 36, 40, and 32 min to complete a single job. Then the output of the parallel process would be $(60/30) + (60/36) + (60/40) + (60/32) = 7.04$ jobs/h, which represents a 17.4% improvement over the sequential process. Creating parallel servers capable of completing a set of activities that represent either an entire transaction or a segment of a larger process is related to the concept of case management introduced in Chapter 3. In Example 4.2, the process is transformed from employing specialized workers to employing case workers. ■

4.2.6 Establish an Efficient System for Processing of Work

Jobs typically are processed in the order in which they arrive at a workstation. This mode of operation is known as the first-in-first-out (FIFO) or first-come-first-served (FCFS) processing policy. As part of the design of a process, the analyst may consider alternative ways in which work is processed. The development and implementation of alternative processing policies belong to the general area of scheduling. Scheduling involves determining a processing order and assigning starting times to a number of jobs that tie up various resources for a period of time. Typically, the resources are in limited supply. Jobs consist of activities, and each activity requires a certain amount of specified resources for a given amount of time.

The importance of considering alternative scheduling schemes increases with the diversity of the jobs in the process. Consider, for example, a process that provides technical support for a popular software package. In such a process, e-mail, chats, and telephone calls are received and tie up resources (i.e., support personnel). Some requests for technical assistance might take longer to complete due to their complexity. As a result, requests that can be processed faster might be delayed unnecessarily if the requests are simply answered in the order they are received and routed to the next available support person.

A simple scheduling approach is to classify work into "fast" and "slow" tracks, based on an estimation of the required processing time. At a major university, approximately 80% of the applications for admission to undergraduate programs are considered for a fast-track process because the applicants are either an "easy accept" or an "easy reject." The remaining 20% are applicants whose files must be examined more thoroughly before a decision can be made. The same principle applies to supermarkets, where customers are routed (and implicitly scheduled for service) to regular and express cashiers.

Other characteristics that can be used to schedule jobs through a process include the following:

- Arrival time.
- Estimated processing time.
- Due date (i.e., when the customer expects the job to be completed).
- Importance (e.g., as given by monetary value).

A key element of a scheduling system is its ability to identify "good" schedules. Ideally, a numerical measure could be used to differentiate the good schedules from the bad ones. This numerical measure is generally known as the *objective function*, and its value is either minimized or maximized. Finding the "right" objective function for a scheduling situation, however, can be difficult. First, such important objectives as customer satisfaction with quality or promptness are difficult to quantify and are not immediately available from accounting information. Second, a process is likely to deal with three different objectives:

- Maximize the process output over some period of time.
- Satisfy customer desires for quality and promptness.
- Minimize current out-of-pocket costs.

In practice, surrogate objectives are used to quantify the merit of a proposed schedule. For example, if the objective is to maximize utilization of resources, a surrogate objective might be to minimize the time it takes to complete the last job in a given set of jobs. (This time is also known as the makespan.) On the other hand, if customers tolerate some tardiness (i.e., the time elapsed after a promised delivery time) but become rapidly and progressively more upset when the tardiness time increases, minimizing maximum tardiness might be a good objective function. Some of the most common surrogate objective functions are

- Minimize the makespan.
- Minimize the total (or average) tardiness.
- Minimize the total (or average) weighted tardiness.
- Minimize the maximum tardiness.
- Minimize the number of tardy jobs.
- Minimize the amount of time spent in the system (i.e., the processing plus the waiting time).

As mentioned previously, the makespan is the total time required to complete a set of jobs. The total tardiness is the sum of the tardiness associated with each job. Tardiness is when the completion time lasts beyond the due date. The weighted tardiness is calculated as the product of the tardiness value and the importance (or weight) of a job.

Example 4.3

In order to understand the effect that scheduling decisions have on these objective functions, consider a set of five jobs, characterized by the values given in Table 4.8, to be processed by a single server.

TABLE 4.8

Scheduling Example

Job	Arrival Time (min)	Estimated Processing Time (min)	Due Date	Importance (Weight)
A	15	30	100	10
B	17	15	220	5
C	22	45	120	12
D	30	60	110	8
E	35	20	150	11

■

In this example, assume that the single server is allowed to process the jobs in any order he or she wishes. Note that if the objective is to minimize the makespan, all processing orderings give the same solutions—170 min

TABLE 4.9

FIFO Schedule

Job	Starting Time	Ending Time	Tardiness	Weighted Tardiness	Tardy Jobs
A	35	65	0	0	0
B	65	80	0	0	0
C	80	125	5	60	1
D	125	185	75	600	1
E	185	205	55	605	1
Total			**135**	**1265**	**3**

(the sum of all the processing times). In other words, the server can process the jobs in any order, and the makespan will remain the same. Also note that this is not the case when the service facility consists of more than one server.

Now suppose the server decides to process the jobs in the order they arrived; that is, the server uses the FIFO scheduling rule. In this case, the processing order is A, B, C, D, and finally E. If no idle times between jobs are considered, then the starting and finishing times for this sequence are shown in Table 4.9. (In this table, it is assumed that the server starts processing after the arrival of the last job, that is, at time 35.) The calculations in Table 4.9 show that a large weighted tardiness penalty is incurred when the jobs are processed using the FIFO rule.

A rule that is particularly useful when jobs have specified due dates is known as *earliest due date first* (EDD). Using EDD, the jobs are sequenced in increasing order of their due date values. The schedule that results from applying this rule to the example is A, D, C, E, and B. The corresponding calculations are shown in Table 4.10.

The EDD schedule is better than the FIFO schedule in terms of tardiness and weighted tardiness. However, the number of tardy jobs is the same for both schedules. Although the EDD schedule is better than the FIFO rule in

TABLE 4.10

EDD Schedule

Job	Starting Time	Ending Time	Tardiness	Weighted Tardiness	Tardy Jobs
A	35	65	0	0	0
D	65	125	15	120	1
C	125	170	50	600	1
E	170	190	40	440	1
B	190	205	0	0	0
Total			105	1160	3

terms of the tardiness measures that have been considered, the EDD schedule is not optimal. In other words, it is possible to find schedules that are better than the one produced by the EDD rule in terms of tardiness and weighted tardiness.

An important characteristic of the EDD schedule is that it gives the optimal solution to the problem of minimizing the maximum tardiness. The maximum tardiness in the FIFO schedule is 75, due to job D. The maximum tardiness in the EDD schedule is 50, due to job C. Therefore, it can be stated with certainty that no other job sequence will result in a maximum tardiness of less than 50.

In addition to FIFO and EDD, researchers and practitioners have developed many other rules and algorithms (sets of steps) for sequencing and scheduling problems. For example, consider two related rules known as (1) *shortest processing time first* (SPT) and (2) *weighted SPT* (WSPT).

The SPT rule simply orders the jobs by their processing time, whereby the job with the shortest time is processed first, followed by the job with the next shortest time, and so on. The result of applying this rule to the example is given in Table 4.11. In terms of weighted tardiness, the SPT sequence is better than FIFO and EDD in this example.

The WSPT schedule can be obtained by first calculating the weight to processing time (WP) ratio for each job. The value of this ratio for job A is 10/30 = 0.333; that is, the weight of 10 is divided by the processing time of 30. After calculating all the ratio values, the WSPT sequence is obtained by ordering the jobs from the largest WP ratio to the smallest. It is easy to verify that the WSPT sequence for this example is E, A, B, C, and D. As a matter of coincidence, this sequence results in the same tardiness and weighted tardiness values as the ones corresponding to the SPT sequence. This, however, is not necessarily always the case.

Finally, an algorithm is described that will minimize the number of tardy jobs when the importance (or weight) of all the jobs is considered equal. The

TABLE 4.11

SPT Schedule

Job	Starting Time	Ending Time	Tardiness	Weighted Tardiness	Tardy Jobs
B	35	50	0	0	0
E	50	70	0	0	0
A	70	100	0	0	0
C	100	145	25	300	1
D	145	205	95	760	1
Total			120	1060	2

procedure is known as Moore's algorithm and can be summarized by the following sequence of steps:

1. Order the jobs using the EDD rule.
2. If there are no tardy jobs, stop. The optimal solution has been found.
3. Find the first tardy job.
4. Suppose that the tardy job is the *k*th job in the sequence. Remove the job with the longest processing time in the set of jobs from the first one to the *k*th one. (The jobs that are removed in this step are inserted at the end of the sequence after the algorithm stops.)
5. Revise the completion times and return to step 2.

Example 4.4

This algorithm is applied to the data in Example 4.3:

1. See Table 4.12.

TABLE 4.12

Step 1 of Moore's Algorithm to Minimize Number of Tardy Jobs

Job	Processing Time	Due Date	Completion	Tardy Jobs
A	30	100	65	0
D	60	110	125	1
C	45	120	170	1
E	20	150	190	1
B	15	220	205	0

2. There are three tardy jobs in the sequence.
3. Job D is the first tardy job.
4. Job D must be removed from the sequence because its processing time of 60 is larger than the processing time of 30 corresponding to job A.
5. See Table 4.13.

TABLE 4.13

Step 5 of Moore's Algorithm to Minimize Number of Tardy Jobs

Job	Processing Time	Due Date	Completion	Tardy Jobs
A	30	100	65	0
C	45	120	110	0
E	20	150	130	0
B	15	220	145	0
A	30	100	65	0

■

Since no more tardy jobs remain in the sequence, the optimal solution has been found. The optimal schedule, therefore, is A, C, E, B, and D. Job D is the only tardy job with a tardiness of 205 – 11 0 = 95 and a weighted tardiness of 760. This schedule turns out to be superior to all the other ones examined so far in terms of multiple measures of performance (i.e., average tardiness, weighted tardiness, and number of tardy jobs). However, the application of Moore's algorithm guarantees an optimal sequence only with respect to the number of tardy jobs when all the jobs have the same weight. In other words, Moore's algorithm does not minimize values based on tardiness (such as total tardiness or weighted tardiness), so it is possible to find a different sequence with better tardiness measures.

4.2.7 Minimize Multiple Paths through Operations

A process with multiple paths is confusing and, most likely, unnecessarily complex. Also, multiple paths result in a process in which resources are hard to manage and work is difficult to schedule. Paths originate from decision points that route jobs to departments or individuals. For example, a telephone call to a software company could be routed to customer service, sales, or technical support. If the call is classified correctly, then the agent in the appropriate department is able to assist the customer and complete the transaction. In this case, there are multiple paths for a job, but the paths are clearly defined and do not intersect.

Suppose now that a customer would like to order one of the software packages that the company offers, but before placing the order, he or she would like to ask some technical questions. His or her call is initially routed to a sales agent, who is not able to answer technical questions. The customer is then sent to a technical support agent and back to sales to complete the transaction.

These multiple paths can be avoided in a couple of ways. The obvious solution is to have sales personnel who also understand the technical aspects of the software they are selling. Because this might not be feasible, information technology (such as a database of frequently asked questions) can be made available to sales agents. If the question is not in the database, the sales agent could contact a technical specialist to obtain an answer and remain as the single point of contact with the customer.

Alternatively, the process can be organized in case teams (see Section 3.6.2 about design principles) that are capable of handling the three aspects of the operation: sales, customer service, and technical support. In this way, calls are simply sent to the next available team. A team can consist of one (a case manager) or more people (case team) capable of completing the entire transaction.

4.3 Additional Diagramming Tools

An important aspect of designing a business process is the supporting information infrastructure. The tools reviewed so far in this chapter focus on the flow of work in a business process. In addition to the flow of work (e.g., customers, requests, or applications), the process designer must address the issues associated with the flow of data. Although this book does not address topics related to the design and management of information systems (IS), it is important to mention that data-flow diagrams are one of the main tools used for the representation of IS. Specifically, data-flow diagramming is a means of representing an IS at any level of detail with a graphic network of symbols showing data flows, data stores, data processes, and data sources/destinations. The purpose of data-flow diagrams is to provide a semantic bridge between users and IS developers. The diagrams are graphical and logical representations of what a system does, rather than physical models showing how it does it. They are hierarchical, showing systems at any level of detail, and they are jargonless, allowing user understanding and reviewing. The goal of data-flow diagramming is to have a commonly understood model of an information system. The diagrams are the basis of structured systems analysis. Figure 4.19 shows a data-flow diagram for insurance claim software developed with SmartDraw.

Another flowcharting tool that was not mentioned in the previous section is the so-called integrated definition (IDEF) methodology. IDEF is a structured approach to enterprise modeling and analysis. The IDEF methodology consists of a family of methods that serve different purposes within the framework of enterprise modeling. For instance, IDEF0 is a standard for function modeling, IDEF1 is a method of information modeling, and IDEF1x is a data-modeling method.

IDEF0 is a method designed to model the decisions, actions, and activities of an organization or system; therefore, it is the most directly applicable in business process design. IDEF0 was derived from a well-established graphical

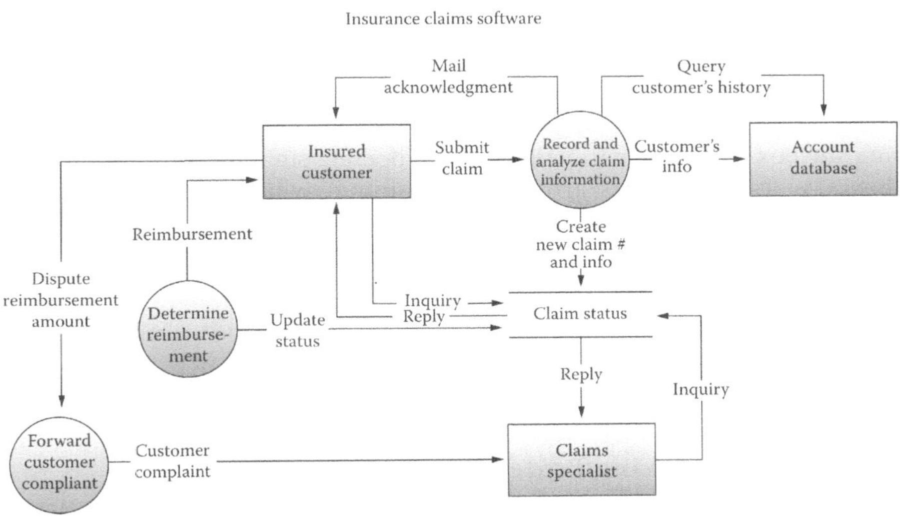

FIGURE 4.19
Example of a data-flow diagram developed with SmartDraw.

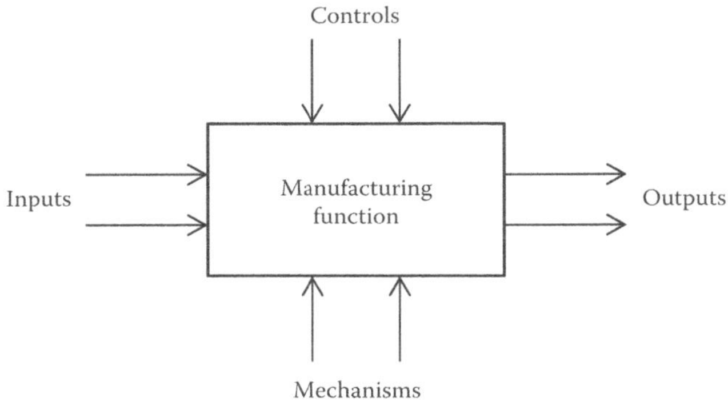

FIGURE 4.20
Basic syntax for an IDEF0 model.

language, the structured analysis and design technique (SADT). The U.S. Air Force commissioned the developers of SADT to develop a function modeling method for analyzing and communicating the functional perspective of a system. Effective IDEF0 models help organize the analysis of a system and promote good communication between the analyst and the customer. IDEF0 is useful in establishing the scope of an analysis, especially for a functional analysis. As a communication tool, IDEF0 enhances domain expert involvement and consensus decision-making through simplified graphical devices. As an analysis tool, IDEF0 assists the modeler in identifying what functions are performed, what is needed to perform those functions, what the current system does right, and what the current system does wrong. Thus, IDEF0 models are often created as one of the first tasks of a system development effort.

The "box and arrow" graphics of an IDEF0 diagram show the function as a box and the interfaces to or from the function as arrows entering or leaving the box. To express functions, boxes operate simultaneously with other boxes, with the interface arrows "constraining" when and how operations are triggered and controlled. The basic syntax for an IDEF0 model is shown in Figure 4.20. A complete description of the family of IDEF methods can be found in www.idef.com.

4.4 From Theory to Practice: Designing an Order-Picking Process

This chapter will conclude with a discussion of the use of some of the basic process-design tools in the context of the operations at a warehouse (Saenz, 2000). The goal is to point out where some basic tools may be applied in the design of a real-world process. The situation deals with the redesign of the order-picking process at a warehouse of a company that has been affected by the advent of e-commerce and increased customer demands. It is assumed that this change in the environment has affected all functions of a traditional warehouse.

To keep the company competitive, management decided to improve the performance of their warehouse operations, starting with the order-picking process. Traditional warehouse functions include receiving products, storing products, replenishing products, order picking, packing, and shipping. Order picking is the heart of most warehouse operations and impacts inbound and outbound processes. Designing an effective picking process can lead to the overall success of a warehouse operation and business. Several key issues need to be considered when designing a picking process: order profiling, picking equipment, slotting strategy, replenishing, layout, and picking methods.

Customers are tending toward making smaller, more frequent orders, which makes order profiling for each product an essential ingredient in the design of a picking process. *Order profiling* refers to defining the product activity in terms of the number of "lines" ordered per product over some period of time—in other words, the number of times one travels to a location to pick an item. Based on this criterion, products are classified as fast moving, medium moving, slow moving, or dead items. The cubic velocity of a product plays an equally important role in classifying activity. It helps determine if a product requires broken-case (each), case, or pallet-picking equipment. The cubic velocity is calculated by multiplying the quantity picked per item by the product's cubic dimensions. A product classified as slow moving might have a large cubic velocity. Similarly, a product classified as a fast mover might have a small cubic velocity. These two factors play a critical role when defining the picking equipment and potential product slotting.

Note that order profiling can be done by using simple calculations based on data collected for each product in the warehouse. Order profiling is an important input to the facility layout tool discussed in Section 4.1.2.

To select the most efficient picking equipment, the product activity, cubic velocity, and variety of products must be considered. The three basic types of picking are broken case, case, and pallet. The emergence of e-commerce has triggered the use of new technology in the picking process. Advanced picking technologies include radio frequency terminals, wireless speech technology, and pick- or put-to-light systems. In a radio frequency terminal system, the terminal is used to initiate and complete orders. The location and quantity of each product is displayed on the terminal screen. Wireless speech recognition terminals convert electronic text into voice commands that guide the operator during picking. The operator uses a laser or pen scanner to initiate and close orders. The information is received using the radio frequency terminal, which usually is secured around the operator's waist. In a pick- or put-to-light system, the operator uses a tethered scanner or a radio frequency scanner to initiate orders. Bay and location displays are illuminated to guide the operator through the picking process. The operator follows the lights and displays to complete orders. Typically, the right balance of technology and manual methods results in effective picking operations. Process flow diagrams (Section 4.1.2) and data-flow diagrams (Section 4.3) are useful in this phase of the design.

4.5 Summary

The knowledge and understanding of basic analysis tools for business process design is fundamental. Before one is able to employ more powerful tools to deal with the complexities of real business processes, it is essential to understand and apply the basic tools presented in this chapter. In some cases, these basic tools may be all that are needed to design a process. For instance, a process may be studied and viable designs may be developed using only flowcharts. Even in the cases when basic tools are not enough and analysts must escalate the level of sophistication, the concepts reviewed in this chapter are always needed as the foundation for the analysis.

Discussion Questions and Exercises

1. Develop a general process chart for the requisition process in Exercise 2 of Chapter 1.
2. Develop a general process chart for the receiving and delivering process in Exercise 3 of Chapter 1.
3. Develop a process activity chart for the IBM Credit example in Section 1.1.2.
4. Develop a flowchart for the claims-handling process in Section 1.2.2.
5. A department within an insurance company is considering the layout for a redesigned process. A computer simulation was built to estimate the traffic from each pair of offices. The load matrix in Table 4.14 summarizes the daily traffic.
   a. If other factors are equal, which two offices should be located closest to one another?

TABLE 4.14
Load Matrix for Exercise 5

From\To	B	C	E	F
A	10	75		140
B			95	
C			130	130
D			10	
E				95

b. Figure 4.21 shows one possible layout for the department. What is the total LD score for this plan using rectilinear distance? (*Hint*: The rectilinear distance between offices A and B is 3.)

c. Switching which two departments will most improve the total LD score?

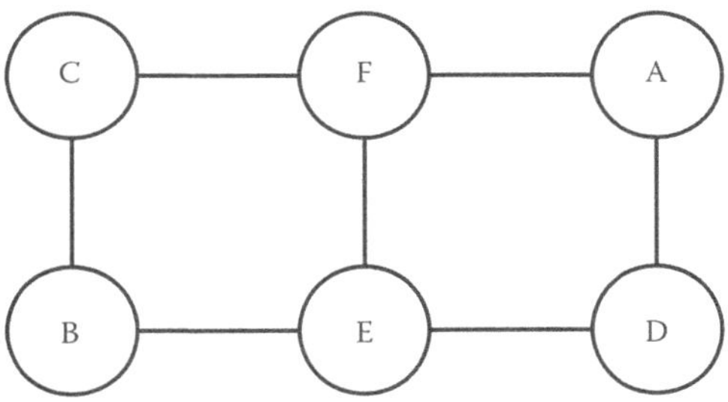

FIGURE 4.21
Current layout for Exercise 5.

6. A firm with four departments has the load matrix shown in Table 4.15 and the current layout shown in Figure 4.22.
   a. What is the LD score for the current layout? (Assume rectilinear distance.)
   b. Find a better layout. What is its total LD score?

TABLE 4.15
Load Matrix for Exercise 6

From\To	B	C	D
A	12	10	8
B		20	6

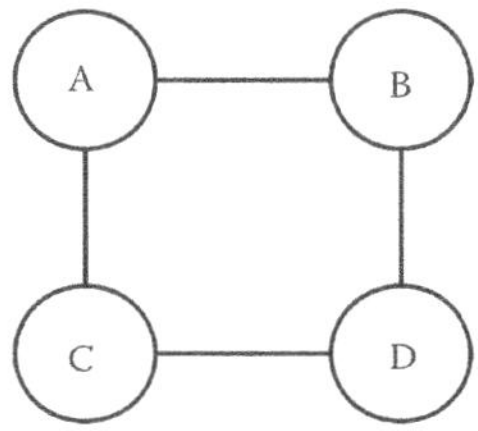

FIGURE 4.22
Current layout for Exercise 6.

7. A scientific journal uses the following process to handle submissions for publication:
   - The authors send the manuscript to the Journal Editorial Office (JEO).
   - The JEO sends a letter to the authors to acknowledge receipt of the manuscript. The JEO also sends a copy of the manuscript to the editor-in-chief (EIC).
   - The EIC selects an associate editor (AE) to be responsible for handling the manuscript and notifies the JEO.
   - The JEO sends a copy of the manuscript to the AE.
   - After reading the manuscript, the AE selects two referees who have agreed to review the paper. The AE then notifies the JEO.
   - The JEO sends copies of the manuscript to the referees.
   - The referees review the manuscript and send their reports to the JEO.
   - The JEO forwards the referee reports to the appropriate AE.
   - After reading the referee reviews, the AE decides whether the manuscript should be rejected, accepted, or revised. The decision is communicated to the JEO.
   - If the manuscript is rejected, the JEO sends a letter to the authors thanking them for the submission (and wishing them good luck getting the manuscript published somewhere else!).
   - If the manuscript is accepted, the JEO forwards the manuscript to production. The JEO also notifies the authors and the EIC.
   - If the manuscript needs revisions, the JEO forwards the referee reviews to the authors.
   - The authors revise the manuscript following the recommendations outlined in the referee reports. The authors then resubmit the manuscript to the JEO.
   - The JEO sends a resubmission directly to the responsible AE.

- After reading a resubmitted manuscript, the AE decides whether the revised version can now be accepted or needs to be sent to the referees for further reviewing.
    i. Construct a service system map for this process.
    ii. Identify opportunities for redesigning the process.

8. Calculate the efficiency of the line-balancing solution depicted in Figure 4.17.
9. Modify Ragsdale and Brown spreadsheet model[2] to solve the line-balancing problem described in Example 4.1.
10. Sola Communications has redesigned one of its core business processes. Processing times are not expected to vary significantly, so management wants to use the line-balancing approach to assign activities to workstations. The process has 11 activities, and the market demand is to process 4 jobs per 400 min working day. Table 4.16 shows the standard time and immediate predecessors for each activity in the process.
    a. Construct a precedence diagram.
    b. Calculate the cycle time corresponding to a market demand of 4 jobs per day.
    c. What is the theoretical minimum number of workstations?
    d. Use the longest activity time rule as the primary rule to balance the line.
    e. What is the efficiency of the line? How does it compare with the theoretical maximum efficiency?
    f. Is it possible to improve the line's efficiency? Can you find a way of improving it?

TABLE 4.16

Data for Exercise 10

Activity	Time (min)	Immediate Predecessor
A	70	—
B	15	A
C	8	—
D	32	—
E	47	C, D, G
F	25	B, E
G	61	—
H	52	—
I	29	G, H
J	42	I
K	50	F, J

11. A process consists of eight activities. The activity times and precedence relationships are given in Table 4.17. The process must be capable of satisfying a market demand of 50 jobs/day in a 400 min working day. Use the longest activity time rule to design a process line. Does the line achieve maximum efficiency?

TABLE 4.17

Data for Exercise 11

Activity	Time (min)	Immediate Predecessor
A	5	—
B	4	—
C	2	A
D	3	A and B
E	2	B
F	6	D and E
G	3	C and F
H	2	F

12. A process manager wants to assign activities to workstations as efficiently as possible and achieve an hourly output rate of four jobs. The department uses a working time of 56 min/h. Assign the activities shown in Table 4.18 (times are in minutes) to workstations using the "most followers" rule. Does the line of workstations achieve maximum efficiency?

TABLE 4.18

Data for Exercise 12

Activity	Time (min)	Immediate Predecessor
A	3	—
B	2	A
C	4	B
D	7	—
E	4	D and G
F	5	—
G	6	F
H	9	C and E
I	5	H

13. A business process has a market demand of 35 jobs per day. A working day consists of 490 min, and the process activity times do not exhibit significant amounts of variation. A process management

TABLE 4.19

Data for Exercise 13

Activity	Time (min)	Immediate Predecessor
A	3	—
B	6	—
C	7	A
D	5	A, B
E	2	B
F	4	C
G	5	F
H	7	D, E
I	1	H
J	6	E
K	4	G, I, J

team would like to apply the line-balancing technique to assign activities to workstations. The activity times and the immediate predecessors of each activity are given in Table 4.19.

a. Use the longest activity rule as the primary rule to assign activities to stations.

b. Compare the efficiency of the line with the theoretical maximum efficiency. Is it possible to improve the line efficiency?

14. Longform Credit receives an average of 1200 credit applications/day. Longform's advertising touts its efficiency in responding to all applications within hours. Daily application-processing activities, average times, and required preceding activities (activities that must be completed before the next activity) are listed in Table 4.20.

TABLE 4.20

Data for Exercise 14

Activity Description	Time (min)	Immediate Predecessor
AOpen and stack applications	0.20	—
BProcess enclosed letter; make note of and handle any special requirements	0.37	A
CCheck off form 1 for page 1 of application	0.21	A
DCheck off form 2 for page 2 of application; file original copy of application	0.18	A
ECalculate credit limit from standardized tables according to forms 1 and 2	0.19	C and D
FSupervisor checks quotation in light of special processing of letter and notes type of form letter, address, and credit limit to return to applicant	0.39	B and D
GSecretary types details on form letter and mails	0.36	E, F

15. Assuming an 8 h day, find the best assignment of activities to workstations using the longest activity time rule. Calculate the efficiency of your design. Has your design achieved maximum efficiency?
16. Suppose the jobs in Table 4.21 must be processed at a single facility. (All times are given in minutes.) Assume that processing starts after the last job arrives, that is, at time 20. Compare the performance of each of the following scheduling rules according to the average weighted tardiness, maximum tardiness, and number of tardy jobs:
    a. FIFO
    b. EDD
    c. SPT

TABLE 4.21

Data for Exercise 15

Job	Arrival Time (min)	Estimated Processing Time (min)	Due Date	Importance (Weight)
A	3	23	76	20
B	6	12	54	18
C	8	34	87	27
D	12	16	98	12
E	15	8	37	10
F	20	19	62	23

17. Consider the jobs in Table 4.22. Use Moore's algorithm to find the sequence that minimizes the number of tardy jobs. Assume that processing can start at time zero.

TABLE 4.22

Data for Exercise 16

Job	A	B	C	D	E	F
Due date	15	6	9	23	20	30
Processing time	10	3	4	8	10	6

18. Time commitments have been made to process seven jobs on a given day, starting at 9:00 AM. The manager of the process would like to find a processing sequence that minimizes the number of tardy jobs. The processing times and due dates are given in Table 4.23.

TABLE 4.23

Data for Exercise 17

Job	A	B	C	D	E	F	G
Due time	11:00	15:00	10:05	12:45	14:00	12:40	13:00
Processing time	0:30	1:15	1:00	0:20	0:45	2:00	1:10

19. A telecommunications company needs to schedule five repair jobs on a particular day in five locations. The repair times (including processing, transportation, and breaks) for these jobs have been estimated as shown in Table 4.24. Also, the customer service representatives have made due date commitments as shown in the table. If a due date is missed, the repair needs to be rescheduled, so the company would like to minimize the number of missed due dates. Find the optimal sequence, assuming that the first repair starts at 8:00 AM.

TABLE 4.24

Data for Exercise 18

Location	Repair Time	Due Time
Walnut	1:15	2:00 PM
Valmont	2:30	Noon
Topaz	1:45	10:00 AM
Wright	0:30	11:00 AM
Baseline	2:00	3:00 PM

Notes

1. The bottleneck is the resource with lowest capacity in the process.
2. http://archive.ite.journal.informs.org/Vol4No2/RagsdaleBrown/LineBalancing.xls

References

Chase, R.B., Aquilano, N.J., and Jacobs, F.J. 1998. *Production and operations management— Manufacturing and services.* 8th ed. New York: McGraw-Hill/Irwin.

Cross, K.F., Feather, J.J., and Lynch, R.L. 1994. *Corporate renaissance: The art of reengineering.* Cambridge, MA: Blackwell Business.

Goldratt, E.M., and Cox, J. 1992. *The goal.* 2nd revised ed. Great Barrington, MA: North River Press, Inc.

Hyer, N., and Wemmerlöv, U. 2002. *Reorganizing the factory: Competing through cellular manufacturing.* Portland, OR: Productivity Press.

Melnyk, S., and Swink, M. 2002. *Value-driven operations management: An integrated modular approach.* New York: McGraw Hill/Irwin.

Ragsdale, C.T., and Brown, E.C. 2004. On line balancing problems in spreadsheets. *INFORMS Transactions on Education* 4(2): 45–48.

Saenz, N. 2000. It's in the pick. *IIE Solutions* 32(7): 36–38

5

Managing Process Flows

A central idea in process dynamics is the notion of stocks and flows. Stocks are accumulations that are created due to the difference between the inflow to a process and its outflow. Everyone is familiar with stocks and flows. The finished goods inventory of a manufacturing firm is the stock of products in its warehouse. The number of people employed by a business also represents a stock—in this case, of resources. The balance in a checking account is a third example of a form of stock. Stocks are altered by inflows and outflows. For instance, a firm's inventory increases with the production flow and decreases with the flow of shipments (and possibly other flows due to spoilage and shrinkage). The workforce increases via a hiring rate and decreases via the rate of resignations, layoffs, and retirements. The balance of a bank account increases with deposits and decreases with withdrawals (Sterman, 2000).

This chapter uses the notion of stocks and flows in the context of business processes. Typically, three operational variables are used to study processes in terms of their stocks and flows: *throughput*, *work-in-process* (WIP), and *cycle time* (CT).[1] Also in this chapter, the relationship among these operational variables will be examined with what is known as Little's law. This chapter ends with an application of the *theory of constraints* (TOC) to capacity analysis.

5.1 Business Processes and Flows

Any business process (i.e., manufacturing or service) can be characterized as a set of activities that transform inputs into outputs (see Chapter 1). There are two main methods for processing inputs. The first, *discrete processing*, is familiar to most people in manufacturing and service industries. It is the production of goods (e.g., cars, computers, television sets, and so on) or services (e.g., a haircut, a meal, a hotel night, and so on) that are sold in separate, identifiable units.

The second method is nondiscrete or *continuous processing*, where products are not distinct units and typically involve liquids, powders, or gases. Examples include such products as gasoline, pharmaceuticals, flour, and paint and such continuous processes as the production of textiles and the

 DOI: 10.1201/9781032617237-5

generation of electricity. Note that ultimately, almost all products from a continuous process become discrete at either the packing stage or the point of sale, as is the case with electrical power.

Most business processes are discrete, where a unit of flow being transformed is often referred to as a *flow unit* or a *job* (see Section 1.1.2). Typical jobs consist of customers, orders, bills, or information. Resources perform the transformation activities.

As defined in Chapter 1, resources may be capital assets (e.g., real estate, machinery, equipment, or computer systems) or labor (i.e., the organization's employees and their expertise).

A job follows a certain routing within a process, determining the temporal order in which activities are executed. Routing provides information about the activities to be performed, their sequence, the resources needed, and the time standards. Routings are job-dependent in most business processes. In general, a process (architecture) can be characterized in terms of its jobs, activities, resources, routings, and information structure (see also Section 1.1.2).

Example 5.1 (Davis et al., 2003)

Speedy Tax Services offers low-cost tax-preparation services in many locations throughout New England. In order to expedite a client's tax return efficiently, Speedy's operations manager has established the following process. Upon entering a tax-preparation location, each client is greeted by a receptionist who asks a series of short questions to determine which type of tax service the client needs. This takes about 5 min. Clients are then referred to either a professional tax specialist if their tax return is complicated or a tax-preparation associate if the return is relatively simple. The average time for a tax specialist to complete a return is 1 h, and the average time for an associate to complete a return is 30 min. Typically, during the peak of the tax season, an office is staffed with six specialists and three associates, and it is open for 10 h/day. After the tax returns have been completed, the clients are directed to see a cashier (each location has two cashiers) to pay for having their tax return prepared. This takes about 6 min/client to complete. During the peak of the tax season, an average of 100 clients/day come into a location, 70% of which require the services of a tax specialist.

This example includes four resource types: receptionists, tax specialists, tax-preparation associates, and cashiers. The jobs to be processed are clients seeking help with their tax returns. The routing depends on whether the tax return is considered complicated. The routing also includes an estimate of the processing time at each activity. ■

As mentioned previously, the purpose of this chapter is to examine processes from a flow perspective. *Jobs* will be used here as the generic term for "units of flow." Jobs become process outflow after the completion of the activities in their specified routing. In manufacturing processes, industrial engineers are concerned with the flow of materials. In an order fulfillment process, the flow of orders is the main focus. A process has the following three types of flows:

- *Divergent flows* refine or separate the input into several outputs.
- *Convergent flows* bring several inputs together.
- *Linear flows* are the result of sequential steps.

One aspect of the process design is to determine the dominant flow in the process. For example, an order fulfillment process can be designed to separate orders along product lines (or money value), creating separate linear flows. As an alternative, the process may perform a number of initial activities in sequence until the differences in the orders require branching, creating divergent flows.

In manufacturing, material flows have been given the following names based on the shape of the dominant flow (Finch and Luebbe, 1995):

- *V-Plant*: A process dominated by divergent flows.
- *A-Plant*: A process dominated by converging flows.
- *I-Plant*: A process dominated by linear flows.
- *T-Plant*: A hybrid process that yields a large number of end products in the last few stages.

An important measure of flow dynamics is the *flow rate*, defined as "number of jobs per unit of time." Flow rates are not necessarily constant throughout a process over time. The notations $R_i(t)$ and $R_o(t)$ will be used to represent the inflow rates and the outflow rates at a particular time t. More precisely, we define the following:

$$R_i(t) = \text{rate of incoming jobs through all entry points into the process}$$

$$R_o(t) = \text{rate of outgoing jobs through all exit points from the process}$$

These definitions will be used to discuss key concepts that are the basis for modeling and managing flows (Anupindi et al., 2006).

5.1.1 Throughput Rate

Inflow rates and outflow rates vary over time, as indicated by the time-dependent notation of $R_i(t)$ and $R_o(t)$. Consider, for example, the inflow and outflow rates per time period t depicted in Figure 5.1.

The inflow rates during the first seven periods of time are larger than the outflow rates. However, during the eighth period (i.e., t = 8), the

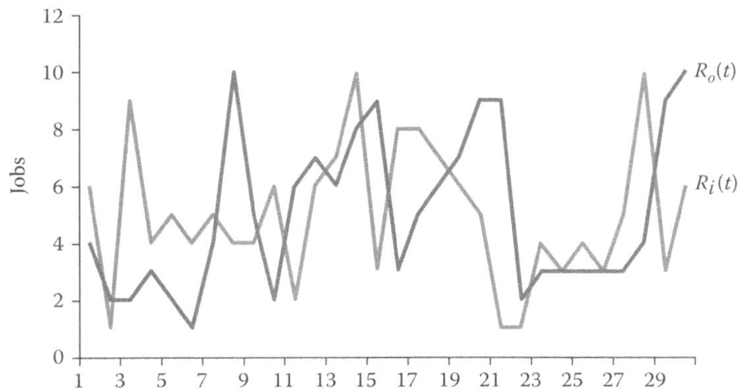

FIGURE 5.1
Inflow and outflow rates per time period.

outflow rate is 10 jobs per unit of time, and the inflow rate is only 4. Using the notation, we have

$$R_i(8) = 4$$

$$R_o(8) = 10$$

If we add all the $R_i(t)$ values and divide the sum by the number of time periods, we can calculate the average inflow rate R_i. Similarly, we can obtain the average outflow rate, denoted by R_o. In a stable process, $R_i = R_o$. Although the inflow and outflow rates of the process depicted in Figure 5.1 fluctuate considerably through time, the process can be considered stable over the time horizon of 30 periods because $R_i = R_o \approx 5$ jobs per unit of time.

Process stability, however, is defined in terms of an infinite time horizon. In other words, in a stable process, the average inflow rate matches the average outflow rate as the number of periods tends to infinity. Hence, when analyzing stable processes, it is not necessary to differentiate between the average inflow rate and the average outflow rate because both of them simply represent the average flow rate through the process. The Greek letter λ (lambda) denotes the average flow rate or throughput rate of a stable process:

$$\lambda = R_i = R_o$$

In queuing theory, λ typically denotes the average effective arrival rate to the system, that is, the arrivals that eventually are served per unit of time (see Section 6.2). In general, λ is often referred to as the process throughput (and is given in terms of jobs per unit of time).

5.1.2 Work-in-Process

If a snapshot of a process is taken at any given point in time, it is likely that a certain number of jobs would be found within the confines of the process. These jobs have not reached any of the exit points of the process because the transformation that represents the completion of these jobs has not been finished. As discussed in Section 4.2.2, the term *work-in-process* (WIP) originally was used to denote the inventory within a manufacturing system that is no longer raw material, but also not a finished product.

All jobs within the process boundaries are considered WIP, regardless of whether they are being processed or they are waiting to be processed. As discussed in Chapter 4, *batching* has a significant impact on the amount of WIP in the process. To take advantage of a particular process configuration, managers often prefer to process a large number of jobs before changing the equipment to process something else or before passing the items to the next processing step. Insurance claims, for example, are sometimes processed in batches in the same way that orders from online retailers are frequently batched. When an order arrives, the order is placed with a group of orders until the number of orders is considered sufficient to warrant sending them all to the warehouse to be filled and shipped. The accumulation of orders in some situations might increase the time that customers wait for a product because it might increase the amount of time the order spends in the process. A trend in manufacturing has been to reduce the batch sizes in order to become more responsive to a market where customers expect shorter waiting times (see also Section 4.2.3). The just-in-time (JIT) manufacturing philosophy dictates that production batches should be as small as possible in order to decrease the time each job spends in the process. Thereby each job spends less time waiting for a batch to be completed before it is moved to the next step. However, the key to implementing this philosophy is to reduce the time required to make necessary changes to the equipment to process a different batch of jobs.

This time is known as setup or changeover time. When setup time is lengthy, large batches are preferred because this implies fewer changeovers.

For instance, consider a bank that uses a machine to process checks, and suppose that the checks can be wallet size or book size. If changing the setup of the machine from one size to the other requires a considerable amount of time, then the manager in charge of the process might decide to run large batches of each job type to minimize the number of changeovers. Companies have developed a variety of strategies to reduce changeover time. Better design of processes and parts has resulted in greater standardization of components, fewer components, and fewer changeovers. In addition, better organization and training of workers have made changeovers easier and faster. Improved design of equipment and fixtures also has made the reduction in changeover time possible (Finch and Luebbe, 1995).

For instance, one strategy that is commonly used to reduce setup time is to separate setup into preparation and actual setup. The idea is to do as much as possible (during the preparation) while the machine or process is still operating. Another strategy consists of moving the material closer to the equipment and improving the material handling in general.

Because the inflow rate and the outflow rate vary over time, the WIP also fluctuates. We refer to the WIP at time t as WIP(t). The up-and-down fluctuation of WIP(t) obeys the following rules:

- WIP(t) increases when $R_i(t) > R_o(t)$. The rate of increase is $R_i(t) - R_o(t)$.
- WIP(t) decreases when $R_i(t) < R_o(t)$. The rate of decrease is $R_o(t) - R_i(t)$.

Figure 5.2 shows the WIP level observed over a period of time. From the beginning of the observation horizon to the time labeled as t_1, the outflow rate is larger than the inflow rate; therefore, the WIP is depleted at a rate that is the difference between the two flow rates. That is, the WIP decreases at a rate of $R_o(t) - R_i(t)$ during the beginning of the observation period until time t_1. During the time period from t_1 to t_2, the inflow rate is larger than the outflow rate; therefore, the WIP increases. The WIP stays constant from time t_2 to time t_3, indicating that the inflow and the outflow rates are equal during this period. In Figure 5.2, we consider that the inflow and outflow rates remain constant between any two consecutive time periods (e.g., between t_1 and t_2 or t_2 and t_3). For some processes, these time periods may be small. In the supermarket industry, for example, changes in the inflow and outflow rates are monitored every 15 min. These data are transformed into valuable information to make operational decisions, such as those related to labor scheduling.

The average WIP is also of interest. To calculate the average WIP when the periods of time are regular (i.e., they are all of the same length), we add the number of jobs in the process during each period of time and divide the

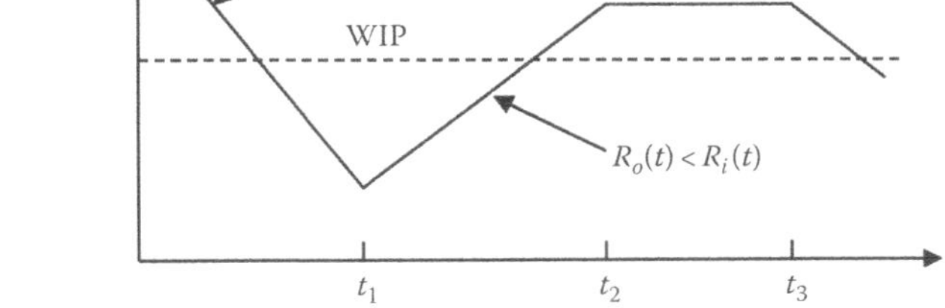

FIGURE 5.2
WIP level over time.

sum by the number of periods in the observed time horizon. For instance, consider an observation period of 1 h with four regular periods of 15 min in which 3, 6, 5, and 2 jobs were in process during each of the observed periods, respectively. The average WIP is given by

$$\text{Average WIP} = \frac{3+6+5+2}{4} = 4 \text{ jobs}$$

When the observation periods are irregular (i.e., they are not all of the same length), then the average WIP calculation must account not only for the number of jobs in each period but also for the length of the period. Suppose that the observation periods in our previous example were 10, 20, 20, and 10 min, respectively. In other words, the WIP was 3 jobs for 10 min, 6 jobs for 20 min, 5 jobs for 20 min, and 2 jobs for 10 min. Then, the average WIP is calculated as follows:

$$\text{Average WIP} = \frac{3\times 10 + 6\times 20 + 5\times 20 + 2\times 10}{10+20+20+10} = 4.5 \text{ jobs}$$

We will use WIP to denote the average (or expected) number of jobs in the process.[2] The dashed line in Figure 5.2 represents the average WIP during the observed period.

5.1.3 Cycle Time

Cycle Time, abbreviated CT (also known as throughput time) is one of the most important measures of performance of a business process. This value is frequently the main focus when comparing the performance of alternative process designs. The CT is the time that it takes to complete an individual job from start to finish. In other words, it is the time that it takes for a job to go from the entry point to the exit point of a process. This is the time that a customer experiences. For example, the CT at a bank may be the time that elapses from the instant a customer enters the facility until the same customer leaves. In an online ordering process, CT may be the time elapsed from the time a customer places an order until the order is received at home. Because jobs follow different routings in a process, the CT can be considerably different from job to job. For example, some jobs might have a routing through a set of activities performed by resources with large capacity and, therefore, would not have to wait to be processed, resulting in shorter CTs. On the other hand, long CTs can be the result of jobs having to compete for scarce resources.

The CT of any given job is the difference between its departure time from the process and its arrival time to the process. If a customer joins a queue in the post office at 7:43 AM and leaves at 7:59 AM, the customer's CT is 16 min. The average CT is the sum of the individual CTs associated with a set of jobs divided by the total number of jobs.[3] The CT depends not only on the arrival rate of jobs in a given time period but also on the routing and the availability of resources.

Because the CT is the total time a job spends in the process, the CT includes the time associated with value-adding and non-value-adding activities. The CT typically includes

- Processing time.
- Inspection time.
- Transportation time.
- Storage time.
- Waiting time (planned and unplanned delay time).

Processing time is often related to value-adding activities. However, in many cases, the processing time is only a small fraction of the CT. CT analysis is a valuable tool for identifying opportunities to improve process performance. For example, if an insurance company finds out that it takes 100 days (on average) to process a new commercial account and that the actual processing time is only about 2 days, then there might be an opportunity for a significant process improvement with respect to CT.

5.1.4 Little's Law

A fundamental relationship between throughput, WIP, and CT is known as Little's law. J. D. C. Little proposed a proof for this formula in connection with queuing theory (Little, 1961). The relationship, which has been shown to hold for a wide class of queuing situations, is

$$\text{WIP} = \lambda \times \text{CT}$$

The formula states that the average number of jobs in the process is proportional to the average time that a job spends in the process, where the factor of proportionality is the average arrival rate. Little's law refers to the average (or expected) behavior of a process. The formula indicates that if two of the three operational measures can be managed (i.e., their values are determined by conscious managerial decisions), the value of the third measure is also completely determined. Three basic relationships can be inferred from Little's law:

1. WIP increases if the throughput rate or the CT increases.
2. The throughput rate increases if WIP increases or CT decreases.
3. CT increases if WIP increases or the throughput rate decreases.

These relationships must be interpreted carefully. For example, is it true that in an order fulfillment process, more WIP inventory results in an increase in CT? Most people would argue that higher levels of inventory (either of the finished product or of WIP) should result in a shorter CT because the order can be filled faster when the product is finished (or in the process to be finished) than when it has to be produced from scratch. Is Little's law contradicting common sense? The answer is no. A closer examination of the order fulfillment process reveals that there are, in fact, two WIP inventories: one associated with purchasing orders and the other with products. In this case, Little's law applies only to the WIP inventory of orders and not to the inventory of products. Now it is reasonable to state that if the number of orders in the pipeline (i.e., the WIP inventory of orders) increases, the CT experienced by the customer also increases.

Finally, some companies use a performance measure known as *inventory turns* or *turnover ratio*. If WIP is the number of jobs in a process at any point in time, then the turnover ratio indicates how often the WIP is replaced in its entirety by a new set of jobs. The turnover ratio is simply the reciprocal of the CT; that is,

$$\text{Turnover ratio} = \frac{1}{\text{CT}}$$

Example 5.2

An insurance company processes an average of 12,000 claims/year. Management has found that on average, at any one time, 600 applications are at various stages of processing (e.g., waiting for additional information from the customer, in transit from the branch office to the main office, waiting for an authorization, and so on). If it is assumed that a year includes 50 working weeks, how many weeks (on the average) does processing a claim take?

$$\lambda = \frac{12{,}000}{50} = 240 \text{ claims/week}$$
$$\text{WIP} = 600 \text{ jobs}$$
$$\text{WIP} = \lambda \times \text{CT}$$
$$\text{CT} = \frac{\text{WIP}}{\lambda} = \frac{600}{240} = 2.5 \text{ weeks}$$

Little's law indicates that the average CT for this claim process is 2.5 weeks. Suppose management does not consider this CT acceptable (because customers have been complaining that the company takes too long to process their claims). What options does management have to reduce the CT to, say, 1 week? According to Little's law, in order to reduce the CT, either the WIP must be reduced or the throughput rate must be increased. A process redesign project is conducted, and it is found that most claims experience long, unplanned delays (e.g., waiting for more information). The new process is able to minimize or eliminate most of the unplanned delays, reducing the average WIP by half.

The CT for the redesigned process is then

$$\text{CT} = \frac{\text{WIP}}{\lambda} = \frac{300}{240} = 1.25 \text{ weeks}$$

■

5.2 Cycle Time and Capacity Analysis

In this section, the concepts of CT, throughput, and WIP inventory are used along with Little's law to analyze the capacity of processes. A process is still viewed as a set of activities that transforms inputs into outputs. Jobs are routed through a process, and resources perform the activities. The routing of jobs and the possibility of rework affect CT, and the amount of resources and their capabilities affect process capacity.

5.2.1 Cycle Time Analysis

CT analysis refers to the task of calculating the average CT for an entire process or a process segment. The CT calculation assumes that the time to complete each activity is available. The activity times are average values and include waiting time (planned and unplanned delays). Flow diagrams are used to analyze CTs, assuming that only one type of job is being processed. In the simplest case, a process may consist of a sequence of activities with a single path from an entry point to an exit point. In this case, the CT is simply the sum of the activity times. However, not all processes have such a trivial configuration. Therefore, the CT analysis needs to be considered in the presence of rework, multiple paths, and parallel activities.

5.2.1.1 Rework

An important consideration when analyzing CTs relates to the possibility of rework. Many processes use control activities to monitor the quality of the work. These control activities (or inspection points) often use specified criteria to allow a job to continue processing. That is, the inspection points act as an accept/reject mechanism. The rejected jobs are sent back for further processing, affecting the average CT and ultimately the capacity of the process (as will be discussed in the next section).

Example 5.3

Figure 5.3 shows the effect of rework on the average CT of a process segment. In this example, it is assumed that each activity (i.e., receiving the request and filling out parts I and II of the order form) requires 10 min (as indicated by the number between parentheses) and that the inspection (the decision symbol labeled "Errors?") is done in 4 min on the average. The jobs are processed sequentially through the first three activities, and then the jobs are inspected for errors. The inspection rejects an average of 25% of the jobs. Rejected jobs must be reworked through the last two activities associated with filling in the information in the order form.

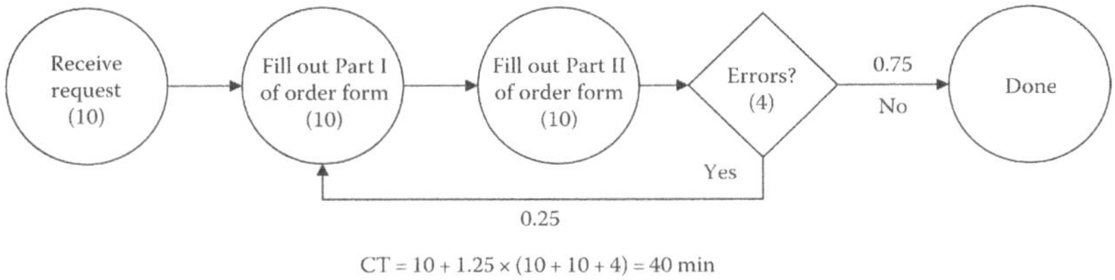

FIGURE 5.3
CT calculation with a rework loop.

Without the rework, the CT of this process segment from the entry point (i.e., receiving the request) to the exit point (out of the inspection activity) is 34 min, that is, the sum of the activity times and the inspection time. Because 25% of the jobs are rejected and must be processed through the activities that fill out the order form as well as being inspected once more, the CT increases by 6 min (24 × 0.25) to a total of 40 min on the average. In this case, the assumption is that jobs are rejected only one time. ■

If it is assumed that the rejection percentage after the inspection in a rework loop is given by r and that the sum of the times of activities within the loop (including inspection) is given by T, then the following general formula can be used to calculate the CT from the entry point to the exit point of the rework loop:

$$\text{CT} = (1 + r) \times T$$

This formula assumes that the rework is done only once. That is, it assumes that the probability of an error after the first rejection goes down to zero. If the probability of making an error after an inspection remains the same, then the CT through the rework loop can be calculated as follows:

$$\text{CT} = \frac{T}{1 - r}$$

In Example 5.3, the average CT for the entire process would be calculated as follows, taking into consideration that the probability of an error remains at 25% regardless of the number of times the job has been inspected:

$$\text{CT} = 10 + \frac{(10 + 10 + 4)}{0.75} = 42 \text{ min}$$

5.2.1.2 Multiple Paths

In addition to rework, a process might include routings that create separate paths for jobs after specified decision points. For example, a process might have a decision point that splits jobs into "fast track" and "normal track." In this case, a percentage must be given to indicate the fraction of jobs that follow each path.

Example 5.4

Figure 5.4 shows a flowchart of the process for Speedy Tax Services described in Example 5.1. All clients are received by the receptionist, after which a decision is made to send a fraction of the clients to a professional tax specialist and the remaining clients to a tax-preparation associate. On the average, 70% of the clients have complicated tax returns and need to work with a professional tax specialist. The remaining 30% of the clients have simple tax returns and can be helped by tax-preparation associates. The numbers between parentheses associated with each activity indicate the activity time (in minutes). With this information, the average CT associated with this process can be calculated. In this case, the CT represents the average time that it takes for a client to complete his or her tax return and pay for the service.

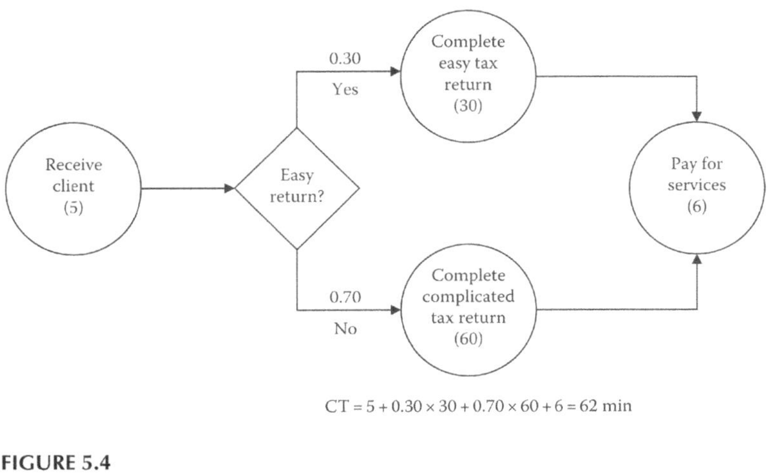

FIGURE 5.4
CT calculation with multiple paths.

The CT calculation in Figure 5.4 represents the sum of the contribution of each activity time to the total. All clients are received by a receptionist, so the contribution of this activity to the total CT is 5 min. Similarly, the contribution of the last activity to the CT is 6 min, given that all clients must pay before leaving the facility. The contribution of the other two activities in the process is weighted by the percentage of clients that are routed through each of the two paths. ■

A general formula can be derived for a process with multiple paths. Assume that m paths originate from a decision point. Also, assume that the probability that a job follows path i is p_i and that the sum of activity times in path i is T_i. Then the average CT across all paths is given by

$$\text{CT} = p_1 \times T_1 + p_2 \times T_2 + \cdots + p_m \times T_m$$

5.2.1.3 Parallel Activities

CT analysis also should contemplate routings in which activities are performed in parallel. For example, a technician in a hospital can set up the X-ray machine for a particular type of shot (as required by a physician), while the patient prepares (e.g., undresses). These two activities occur in parallel because it is not necessary for the technician to wait until the patient is ready before starting to set up the X-ray machine. Because one of these activities will require less time, the other has to wait for further processing. Typically, the patient will finish first and will wait to be called for the X-ray. The contribution to the CT is then given by the maximum time from all the parallel activities.

Example 5.5

Figure 5.5 depicts a process with five activities. The first activity consists of opening an envelope and splitting its contents into three different items: application, references, and credit history. Each of these items is processed in parallel, and a decision is made regarding this request after the parallel activities have been completed. The numbers between parentheses in Figure 5.5 are activity times in minutes.

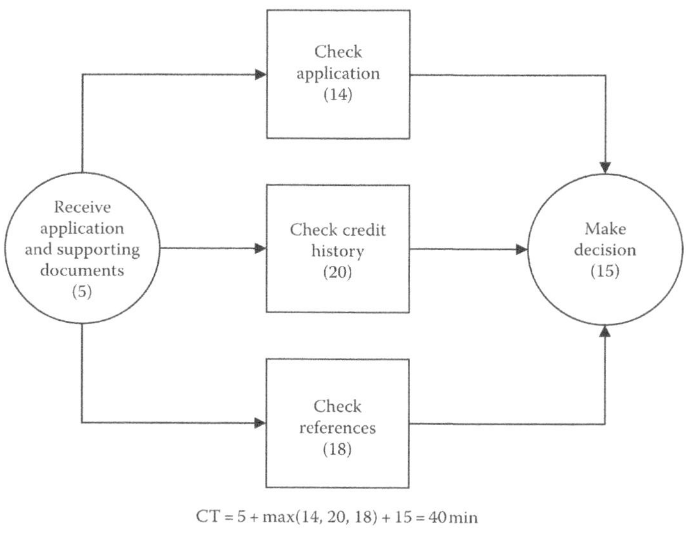

FIGURE 5.5
CT calculation with parallel activities.

The CT calculation in Figure 5.5 results in a total of 40 min for the process under consideration. This is the sum of 5 min for the first activity, 20 min for checking the credit history (which is the parallel activity with the longest time), and 15 min for making a decision. Note that the flow diagram in Figure 5.5 has no decision point after the first activity because all the jobs are split in the same way; that is, the routing of all the jobs is the same and includes the parallel processing of the "checking" activities. Note also that these are considered inspection activities, and therefore the square is used to represent them in the flowchart in Figure 5.5. ■

The general formula for process segments with parallel activities is a simplification of the one associated with multiple paths. Because there is no decision point when the jobs are split, it is not necessary to account for probability values. It is assumed that T_i is the total time of the activities in path i (after the split) and that the process splits into m parallel paths. Then the CT for the process segment with parallel paths is given by

$$\mathrm{CT} = \max(T_1, T_2, \ldots, T_m)$$

Example 5.6

Before concluding this section, let us apply the aforementioned principles for CT analysis to a small process. Assume that a process has been described and that the flow diagram in Figure 5.6 was drawn appropriately from these descriptions. The numbers in parenthesis shown in Figure 5.6 are activity times in minutes. The process consists of nine activities. A decision point is included after activity A because only 30% of the jobs are required to go through activity B. A rework loop sends jobs back from the inspection point I to activity D. This loop assumes that the reworked jobs are examined a second time and that they always pass the inspection; that is, jobs are reworked only one time. The process has two parallel activities, F and G, before the last activity is performed.

The CT calculation that corresponds to the process in Figure 5.6 starts with the time for activity A. Next, the contribution of 6 min (i.e., 0.3 × 20) from activity B is added. This is followed by the time of activity C (23 min) and the contribution of the activities in the rework loop. Note that the times within the rework loop are added, including the inspection time, and multiplied by 110%. This operation accounts for the 10% of jobs that are sent back after inspection. Finally, the contribution of the parallel activities is calculated as the maximum time between activities F and G, followed by the activity time for H. The average CT for this process is determined to be 92.5 min.

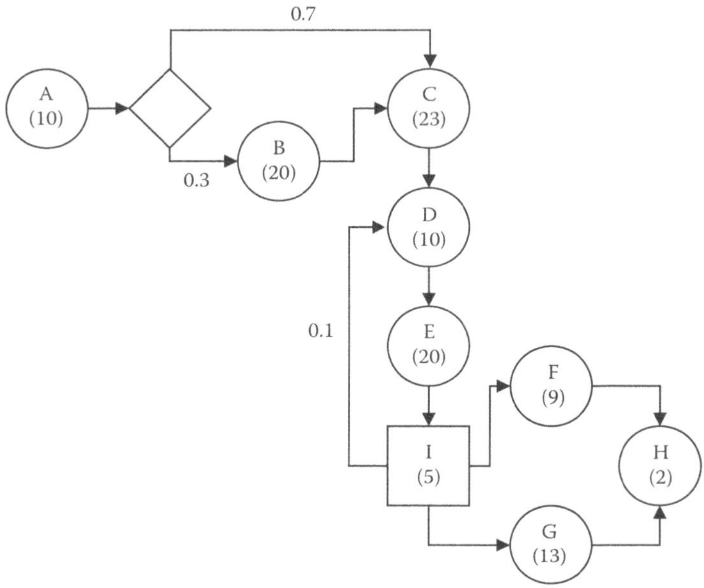

$$CT = 10 + 0.3 \times 20 + 23 + 1.1 \times (10 + 20 + 5) + \max(9, 13) + 2 = 92.5 \text{ min}$$

FIGURE 5.6
Example of CT analysis. ■

After calculating the CT of a process, the analyst should calculate what is known as the CT *efficiency*. Assuming that all processing times are value-adding, the CT efficiency indicates the percentage of time, from the actual CT, that is spent performing value-adding work. Mathematically, the CT efficiency is expressed as follows:

$$\text{CT efficiency} = \frac{\text{Process time}}{\text{CT}}$$

Process time also is referred to as the theoretical CT because theoretically a job can be completed in that time (i.e., if no waiting would occur). To calculate the process time, replace activity time with the processing time (i.e., the activity time minus the waiting time). It should be recognized that the CT efficiency calculated here is conceptually the same as the efficiency calculation in Chapter 4, which was connected with line balancing.

Example 5.7

Table 5.1 shows the time values associated with the activities in Example 5.6. The time values are broken down into processing time and waiting time. The activity time is the sum of these two values.

Using the processing times in Table 5.1 and the flow diagram in Figure 5.6, the process time (or theoretical CT) can be calculated as follows:

$$\text{Process time} = 2 + 0.3 \times 5 + 8 + 1.1 \times (3 + 4 + 4) + \max(2, 4) + 2 = 29.6 \text{ min}$$

The CT efficiency can now be found as follows:

$$\text{CT efficiency} = \frac{29.6}{92.5} = 0.32$$

This means that less than one-third of the actual CT is spent in processing, and the rest of the time is spent waiting.

TABLE 5.1

Processing Times, Waiting Times and Activity Times for Example 5.7

Activity	Processing Time (min)	Waiting Time (min)	Activity Time (min)
A	2	8	10
B	5	15	20
C	8	15	23
D	3	7	10
E	4	16	20
Inspection	4	1	5
F	2	7	9
G	4	9	13
H	2	0	2

■

Redesign projects in several industries (e.g., insurance, hospitals, and banking) indicate that it is not unusual to find CT efficiencies of less than 5%. This occurs, for example, in applications for a new insurance policy, where the process time is typically 7 min and the CT is normally 72 h.

5.2.2 Capacity Analysis

Capacity analysis complements the information obtained from a CT analysis when studying flows in a process. Flowcharts are important also for analyzing capacity because the first step in the methodology consists of estimating the number of jobs that flow through each activity. This step is necessary because resources (or pools of resources) perform the activities in the process and their availability limits the overall capacity. The number of jobs flowing through each activity is determined by the configuration of the process, which may include, as in the case of the CT analysis, rework, multiple paths, and parallel activities.

5.2.2.1 Rework

When a process or process segment has a rework loop, the number of jobs flowing through each activity varies according to the rejection rate.

Example 5.8

Figure 5.7 depicts a process with a rework loop. Requests are processed through three activities and then inspected for errors. Suppose 100 requests are received. Because an average of 25% of the requests are rejected, the second and third activities, which are inside the rework loop, end up processing 125 jobs in the long run. This is assuming that rejected requests always pass inspection the second time.

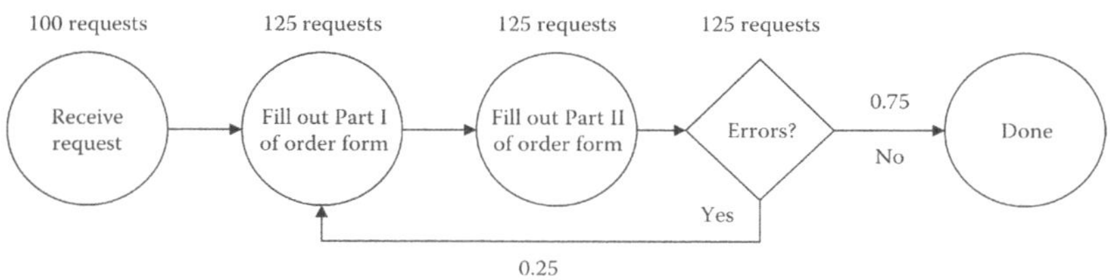

FIGURE 5.7
Number of requests flowing through each activity in a process with rework. ■

Next, a general formula is derived for calculating the number of jobs per activity in a rework loop. Assume that n jobs enter the rework loop and that the probability of rejecting a job at the inspection station is r. The number of jobs flowing through each activity in the loop, including the inspection station, is given by the following equation:

$$\text{Number of jobs} = (1 + r) \times n$$

When the rejection rate stays the same regardless of the number of times that a job has been reworked and inspected, then the number of jobs that are processed by activities inside the rework loop is given by the following equation:

$$\text{Number of jobs} = \frac{n}{1 - r}$$

According to this formula, the activities inside the rework loop in Example 5.8 (see Figure 5.7) process 133.33 requests on the average.

5.2.2.2 *Multiple Paths*

The job routings in a process may include decision points that create multiple paths. The flow through each path varies according to the frequency with which each path is selected (as indicated in the decision point).

Example 5.9

Figure 5.8 shows a flowchart of the process for Speedy Tax Services described in Example 5.1, which contains multiple paths. Assume that 100 clients enter the process, and we want to calculate the number of clients that are processed at each activity on the average.

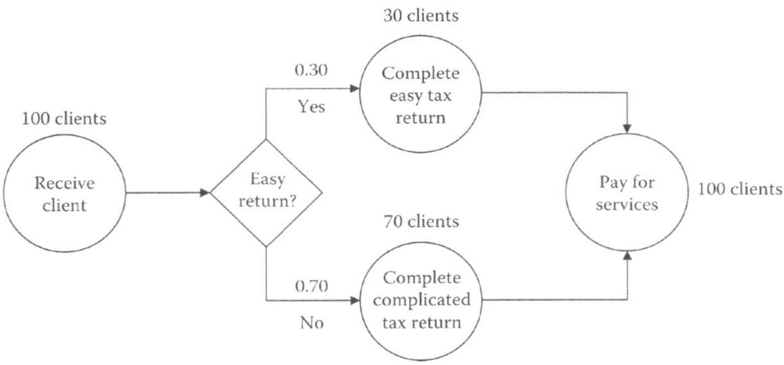

FIGURE 5.8
Number of clients flowing through each activity in a process with multiple paths.

Because 30% of the clients have simple tax returns, about 30 clients out of 100 are helped by tax-preparation associates. On the average, the remaining 70 clients are helped by professional tax specialists. It is important to note that all 100 clients must pay for the services rendered, so they are routed through the cashiers. ■

To derive a general formula for the number of jobs in each path, assume that the probability that a job follows path i is p_i. Also assume that n jobs enter the decision point. Then, the number of jobs in path i is given by the following equation:

$$\text{Number of jobs in path } i = p_i \times n$$

5.2.2.3 *Parallel Activities*

When jobs split into parallel activities, the number of jobs flowing through each activity remains the same as the number of jobs that enter the process (or process segment).

Example 5.10

Figure 5.9 shows a process with five activities, three of which are performed in parallel. Assuming that 100 applications are received, all activities experience the same load of 100 applications, as shown in Figure 5.9.

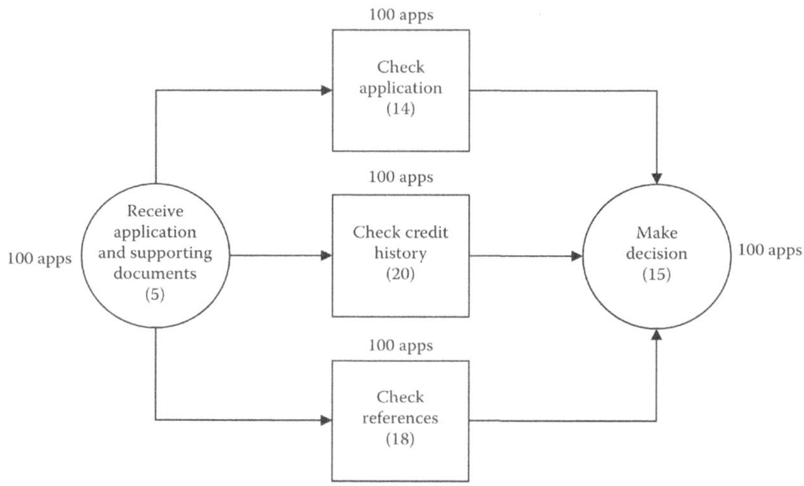

FIGURE 5.9
Number of applications flowing through each activity in a process with parallel activities.

In this case, no general formula is necessary because the number of jobs remains the same after a split into parallel activities or paths.

The next step in analyzing capacity is to determine the capacity of each resource or pool of resources. This calculation is illustrated using the process of Example 5.6, whose flowchart is depicted in Figure 5.6. ■

Example 5.11

For each activity in the process, the processing time, the type of resource required, and the number of jobs processed through each activity must be known. This information is summarized in Table 5.2.

The available resources of each type also must be known. Table 5.2 indicates that there are three types of resources, labeled R1, R2, and R3. Assume that there are two units of resource R1, two units of resource R2, and one unit of resource R3.

The resource unit and pool capacity can now be calculated as follows. For each resource type, find the *unit load*. To calculate the unit load for a given resource, first multiply the processing time by the number of jobs for each activity for which the resource is required and then add the

TABLE 5.2

Resource Capacity Data for Example 5.11

Activity	Processing Time (min)	Resource Requirements	Number of Jobs
A	2	R1	1
B	5	R1	0.3
C	8	R2	1
D	3	R2	1.1
E	4	R2	1.1
Inspection	4	—	1.1
F	2	R1	1
G	4	R3	1
H	2	R3	1

products. This is illustrated in the second column of Table 5.3. The *unit capacity* for each resource is the reciprocal of the unit load and indicates the number of jobs that each unit of resource can complete per unit of time. This is shown in the third column of Table 5.3.

TABLE 5.3

Pool Capacity Calculation for Example 5.11

Resource	Unit Load (min)	Unit Capacity (Jobs/min)	Available Resources (Units)	Pool Capacity (Jobs/min)
R1	$2 + 5 \times 0.3 + 2 = 5.5$	1/5.5	2	2/5.5 = 0.36
R2	$8 + 1.1 \times (3 + 4) = 15.7$	1/15.7	2	2/15.7 = 0.13
R3	$4 + 2 = 6$	1/6	1	1/6 = 0.17

Finally, to find the *pool capacity* associated with each resource type, multiply the number of resource units (i.e., the resource availability) by the unit capacity. The values corresponding to resources R1, R2, and R3 are shown in column 5 of Table 5.3 (labeled Pool Capacity). The pool capacities in Table 5.3 indicate that resource R2 is the bottleneck of the process because this resource type has the smallest pool capacity. The pool capacity of R2 is 0.13 jobs/min or 7.8 jobs/h, compared to 0.36 jobs/min or 21.6 jobs/h for R1 and 0.17 jobs/min or 10.2 jobs/h for R3. ■

It is important to realize that the bottleneck of a process refers to a resource or resource pool and not to an activity. In other words, capacity is not associated with activities but with resources. Also note that because the slowest resource pool limits the throughput rate of the process, the *process capacity* is determined by the capacity of the bottleneck. Therefore, the capacity of the process depicted in Figure 5.6 is 7.8 jobs/h.

The process capacity, as previously calculated, is based on processing times instead of activity times. Because processing times do not include waiting time, the process capacity calculation is a value that can be achieved only in theory. That is, if jobs are processed without any delays (either planned or unplanned), the process can achieve its theoretical capacity of 7.8 jobs/h. In all likelihood, however, the process actual throughput rate (or actual capacity) will not match the theoretical process capacity; therefore, a measure of efficiency can be calculated. The measure in this case is known as the *capacity utilization*, and it is defined as follows:

$$\text{Capacity utilization} = \frac{\text{Throughput rate}}{\text{Process capacity}}$$

With this mathematical relationship, the capacity utilization can be calculated for each resource type in a process.

Example 5.12

Assume that the throughput rate of the process in Example 5.11 is 6 jobs/h. This additional piece of information gives the following capacity utilization values for each resource in the process:

$$\text{Capacity utilization for R1} = \frac{6}{21.8} = 27.5\%$$

$$\text{Capacity utilization for R2} = \frac{6}{7.8} = 76.9\%$$

$$\text{Capacity utilization for R3} = \frac{6}{10.2} = 58.8\%$$

■

Similar to the way the capacity of the process is related to the capacity of the bottleneck resource, the capacity utilization of the process is related to the capacity utilization at the bottleneck. Thus, based on the calculations in Example 5.12, the capacity utilization of the process in Example 5.11 is 76.9%.

5.3 Managing Cycle Time and Capacity

The analysis of CT and capacity provides process managers with valuable information about the performance of the process. This information, however, is wasted if managers do not translate it into action. Through CT analysis, managers and process designers can find out the CT efficiency of the process and discover, for example, that the waiting time in the process is excessive for a desired level of customer service. This section discusses ways to reduce CT and strategies to increase capacity.

5.3.1 Cycle Time Reduction

Process CT can be reduced in five fundamental ways:

1. Eliminate activities.
2. Reduce waiting time.
3. Eliminate rework.
4. Perform activities in parallel.
5. Move processing time to a noncritical activity.

The first thing that analysts of business processes should consider is the elimination of activities. As discussed in earlier chapters, most of the activities in a process do not add value to the customer; therefore, they should be considered for elimination. Non-value-adding activities, such as those discussed in Chapter 1, increase the CT but are not essential to the process.

If an activity cannot be eliminated, then eliminating or minimizing the waiting time reduces the activity time. Recall that the activity time is the sum of the processing time and the waiting time. It is worth investigating ways to speed up the processing of jobs at an activity, but larger time reductions are generally easier to achieve when waiting time is eliminated. Waiting time can be reduced, for example, by decreasing batch sizes and setup times. Also, improved job scheduling typically results in less waiting and better utilization of resources.

Example 5.3 shows that rework loops add a significant amount of time to the process CT. As shown in that example, the CT when the activities are performed "right" the first time, and thus eliminating the need for the rework loop, is 85% (or 34/40) of the original CT. If the jobs always pass inspection, the inspection activity also can be eliminated, reducing the CT to 30 min. This represents a 25% reduction from the original design.

Changing a process design to perform activities in parallel has an immediate impact on the process CT. This was discussed in Chapter 4, where the notion of combining activities was introduced. (Recall that the combination of activities is the cornerstone of the process-design principle known as case management, as described in Chapter 3.) Mathematically, the reduction of CT from a serial process configuration to a parallel process configuration is the difference between the sum of the activity times and the maximum of the activity times. To illustrate, suppose that the "checking" activities in Figure 5.5 were performed in series instead of in parallel. The average CT for completing one application becomes

$$CT = 5 + 14 + 20 + 18 + 15 = 72 \text{ min}$$

The CT with parallel processing was calculated to be 40 min. The difference of 32 min also can be calculated as follows:

$$\Delta CT = (14 + 20 + 18) - \max(14, 20, 18) = 52 - 20 = 32 \text{ min}$$

Finally, CT decreases when some work content (or processing time) is shifted from a critical activity to a noncritical activity. In a process with parallel processing (i.e., with a set of activities performed in parallel), the longest path (in terms of time) is referred to as the critical path. The length of the critical path corresponds to the CT of the process.

Example 5.13

Consider, for instance, the activities depicted in Figure 5.5. This process has three paths.

The longest path is the second one with a length of 40 min. Then the critical activities are receive, check credit, and decide, and the path length of 40 min matches the CT calculated in Figure 5.5.

Path	Length
Receive → check application → decide	5 + 14 + 15 = 34 min
Receive → check credit → decide	5 + 20 + 15 = 40 min
Receive → check references → decide	5 + 18 + 15 = 38 min

■

Example 5.14

Figure 5.10 depicts a process with six activities. In this process, activities C and D are performed in parallel with activity E. Assume that the numbers between parentheses are activity times. The process has two paths.

Path	Length
A → B → C → D → F	10 + 20 + 15 + 5 + 10 = 60 min
A → B → E → F	10 + 20 + 12 + 10 = 52 min

The critical activities in the original process are A, B, C, D, and F. These activities belong to the critical path. In order to reduce CT, a redesigned process might move some work from a critical activity to a noncritical activity. In this case, activity E is the only noncritical activity. Suppose it is possible to move 4 min of work from activity C to activity

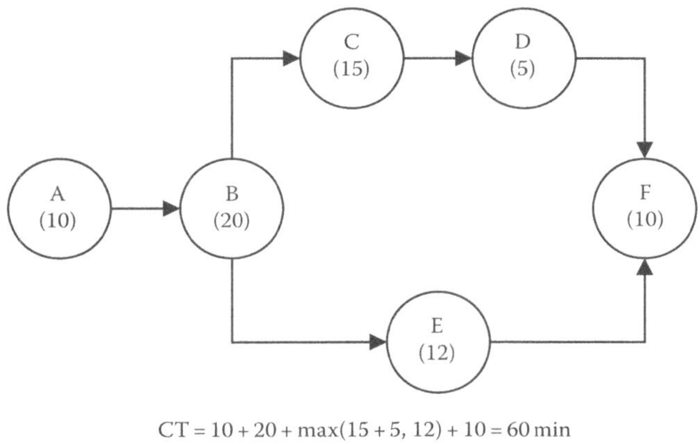

FIGURE 5.10
Original process in Example 5.13.

E. This change effectively reduces the CT by 4 min, making both paths critical. That is, both paths of activities in the process now have the same length of 56 min; therefore, all the activities in the process have become critical. Note also that in the following calculation of the CT, both of the time values compared in the "max" function equal 16 min: $CT = 10 + 20 + \max(11 + 5, 16) + 10 = 56$ min ■

We next examine ways to increase process capacity.

5.3.2 Increasing Process Capacity

Section 5.2.2 examined the direct relationship between process capacity and the capacity of the bottleneck resource. Given this relationship, it is reasonable to conclude that making the bottleneck resource faster increases the process capacity. Capacity of the process can be increased in two fundamental ways:

- Add resource availability at the bottleneck.
- Reduce the workload at the bottleneck.

Adding resources to the bottleneck might mean additional investment in equipment and labor or additional working hours (i.e., overtime). In other words, the available resources at the bottleneck can be increased with either more workers or with the same workers working more hours.

The approach of reducing workload at the bottleneck is more closely linked to the notion of process redesign. The reduction consists of either shifting activities from the bottleneck to a different resource pool or reducing the time of the activities currently assigned to the bottleneck. Shifting activities from one resource pool to another requires cross training so that workers in a non-bottleneck resource pool can perform new activities.

One must be careful when considering redesign strategies with the goal of reducing CT and increasing process capacity. Specifically, the decision must take into consideration the availability of resources and the assignment of activities to each resource pool.

Example 5.15

Assume that five resource types are associated with the process in Figure 5.10. Also assume that each type of resource has only one available unit. Finally, assume that activities A, B, C, and D are assigned to resources R1, R2, R3, and R4, respectively, and that activities E and F are assigned to resource R5. Given these assignments and assuming, for simplicity, that for all activities the processing times equal the activity times in Figure 5.10, the theoretical capacity of the original process can be calculated as shown in Table 5.4. The bottleneck is resource R5 with a capacity of 2.7 jobs/h, which also becomes the process capacity. The CT for this process was calculated to be 60 min.

TABLE 5.4

Capacity of the Original Process

Resource	Unit Load (min)	Unit Capacity (Jobs/min)	Available Resources (Units)	Pool Capacity (Jobs/h)
R1	10	1/10	1	60/10 = 6.0
R2	20	1/20	1	60/20 = 3.0
R3	15	1/15	1	60/15 = 4.0
R4	5	1/5	1	60/5 = 12.0
R5	12 + 10 = 22	1/22	1	60/22 = 2.7

If the process is modified in such a way that 4 min of work are transferred from activity C to activity E, then, as shown in Example 5.14, the CT is reduced to 56 min. Does the process capacity change as a result of this process modification? Table 5.5 shows the process capacity calculations associated with the modified process.

TABLE 5.5

Capacity of the Modified Process

Resource	Unit Load (min)	Unit Capacity (Jobs/min)	Available Resources (Units)	Pool Capacity (Jobs/h)
R1	10	1/10	1	60/10 = 6.0
R2	20	1/20	1	60/20 = 3.0
R3	11	1/11	1	60/11 = 5.4
R4	5	1/5	1	60/5 = 12.0
R5	16 + 10 = 26	1/26	1	60/26 = 2.3

The calculations in Table 5.5 indicate that the process capacity decreases as a result of the decision to transfer units from a critical activity to a noncritical activity associated with the bottleneck resource R5. At first glance, this result seems to contradict Little's law that specifies that the throughput rate increases when CT decreases. In this case, CT decreased from 60 to 56 min, and the process capacity (measured in units of throughput) decreased from 2.7 to 2.3 jobs/h.

The change in the level of WIP explains this apparent contradiction. Although it is true that a shorter CT should increase the throughput rate, this holds only if the WIP remains unchanged. In this example, the decision to move processing time from activity C to activity E decreases CT and throughput rate and as a result also decreases the WIP. ■

5.4 Theory of Constraints

Bottlenecks affect the efficiency of processes because they limit throughput, and in many cases, they also inhibit the achievement of value-based goals, such as quality, speed, cost, and flexibility. Bottlenecks often are responsible for the differences between what operations managers promise and what they deliver to internal and external customers (Melnyk and Swink, 2002). Due to the importance of bottlenecks, it is not sufficient to be able to identify their existence. That is, locating the bottlenecks in a process is just part of managing flows; the other part deals with increasing process efficiency despite these limitations. This section introduces a technique for managing bottlenecks that is known as Theory Of Constraints or TOC.

TOC provides a broad theoretical framework for managing flows. It emphasizes the need for identifying bottlenecks (or constraints) within a process, the entire firm, or even the business context. For example, an organization's information system represents a constraint if the order-entry function takes longer to prepare an order for release than the transformation process needs to make the product. Other constraints could be associated with resources that perform activities in areas such as marketing, product design and development, or purchasing. All of these constraints limit throughput and, therefore, affect the efficiency of the system.

As mentioned in Chapter 4, TOC draws extensively from the work of Eli Goldratt, an Israeli physicist. TOC leads to an operating philosophy that in some ways is similar to JIT.[4] TOC assumes that the goal of a business organization is to make money. To achieve this goal, the company must focus on throughput, inventory, and operating expenses. In order to increase profit, a company must increase throughput and decrease inventory and operating expenses. Then, companies must identify operations policies that translate into actions that move these variables in the right directions. However, these policies have to live within a set of relevant constraints.

TOC gets its name from the concept that a constraint is anything that prevents a system from achieving higher levels of performance relative to its goal (Vonderembse and White, 1994). Consider the following situations where decisions must be made in constrained settings.

Example 5.16

A company produces products A and B using the same process. The unit profit for product A is \$80, and the market demand is 100 units/week. The unit profit for product B is \$50, and the market demand is 200 units/week. The process requires 0.4 h to produce one unit of A and 0.2 h to produce one unit of B. The process is available 60 h/week. Because of this constraint, the process is unable to meet the entire demand for both products. Consequently, the company must decide how many products of each type to make. This situation is generally known as the product-mix problem.

If the objective of the company is to maximize total profit, one would be inclined to recommend the largest possible production of A because this product has the largest profit margin. This production plan can be evaluated as follows. First, calculate the maximum number of units of A that can be produced in 60 h of work:

$$\text{Units of A} = \frac{60 \text{ h/week}}{0.4 \text{ h/unit}} = 150 \text{ units}$$

Although 150 units can be produced per week, the market demands only 100. Therefore, the production plan would recommend 100 units of A. Because producing 100 units of A requires 40 h of process time (i.e., 0.4 × 100), 20 h are still available to produce product B. This process time translates into 100 units of product B (i.e., 20/0.2). The total profit associated with this production plan is

$$\text{Total profit} = \$80 \times 100 + \$50 \times 100 = \$13{,}000$$

Is this the best production plan? That is, does this production plan maximize weekly profits? To answer this question, first examine an alternative plan. If the company first attempts to meet the demand for product B, then the required process time would be 40 h (i.e., 200 × 0.2). This would leave 20 h to produce product A, resulting in 50 units (i.e., 20/0.4). The total profit associated with this plan would be

$$\text{Total profit} = \$80 \times 50 + \$50 \times 200 = \$14{,}000$$

The alternative plan increases total profit by \$1000/week. This example shows that when constraints are present, decisions cannot be made using simplistic rules. In this case, we attempted to solve the problem by simply maximizing the production of the product with the largest profit margin. This solution, however, takes into consideration neither the production constraints nor the market demand constraints. ■

Example 5.17

A department store chain has hired an advertising firm to determine the types and amounts of advertising it should have for its stores. The three types of advertising available are radio and television commercials and newspaper ads. The retail chain desires to know the number of each type of advertisement it should pursue in order to maximize exposure. It has been estimated that each ad or commercial will reach the potential audience shown in Table 5.6. This table also shows the cost associated with each type of advertisement.

TABLE 5.6

Resource Capacity Data for Example 5.11

Type of Advertisement	Exposure (People/Ad or Commercial)	Cost
Television	20,000	$15,000
Radio	12,000	$6000
Newspaper	9000	$4000

In order to make a decision, the company must consider the following constraints:

1. The budget limit for advertisement is $100,000.
2. The television station has time available for four commercials.
3. The radio station has time available for 10 commercials.
4. The newspaper has space available for seven ads.
5. The advertising agency has time and staff available for producing no more than a total of 15 commercials and/or ads.

The company would like to know how many and what kinds of commercials and ads to produce in order to maximize total exposure. If one follows the simple criterion of choosing advertising by its exposure potential, one would try to maximize TV commercials, then radio commercials, and finally newspaper ads. This would result in four TV commercials ($60,000), six radio commercials ($36,000), and one newspaper ad ($4000). The cost of the campaign would be $100,000 with an estimated exposure of 161,000 people. This problem, however, can be formulated as an *integer programming*[5] model and optimally solved, for instance, with Microsoft Excel's Solver. The optimal solution to this problem calls for two TV commercials, nine radio commercials, and four newspaper ads for a total cost of $100,000 and a total exposure of 184,000 people.

This example shows, once again, that simplistic solutions to constrained problems can lead to inferior process performance. In this case, a more comprehensive solution method is able to increase by 14% the total exposure that can be achieved with a fixed amount of money. ■

Examples 5.16 and 5.17 include constraints associated with production, demand, and capacity. In general, constraints fall into three broad categories:

- *Resource constraint*: A resource within or outside the organization, such as capacity, that limits performance.
- *Market constraint*: A limit in the market demand that is less than the organization's capacity.
- *Policy constraint*: Any policy that limits performance, such as a policy that forbids the use of overtime.

In Examples 5.16 and 5.17, the limit in the process hours and the budget for advertising can be identified as resource constraints. An external resource constraint is the limit on the number of ads or commercials the advertising agency is able to produce for the discount stores. The market constraint in Example 5.16 is characterized by a limit on the weekly demand for each product. One possible policy constraint in Example 5.17 might be to insist on a plan whereby the number of TV commercials is greater than the number of newspaper ads.

TOC proposes a series of steps that can be followed to deal with any type of constraint (Vonderembse and White, 1994):

1. Identify the system's constraints.
2. Determine how to exploit the system constraints.
3. Subordinate everything else to the decisions made in step 2.
4. Elevate the constraints so a higher performance level can be reached.
5. If the constraints are eliminated in step 4, go back to step 1. Do not let inertia become the new constraint.

Example 5.18

We now apply the first three steps of this methodology to a process with nine activities and three resource types. Three types of jobs must be processed, with each job following a different processing path. The routing for each job, the weekly demand, and the estimated profit margins are shown in Table 5.7.

TABLE 5.7

Job Data for Example 5.17

Job	Routing	Demand (Units/Week)	Profit Margin
A	4, 8, and 9	50	$20
B	1, 2, 3, 5, 6, 7, and 8	100	$75
C	2, 3, 4, 5, 6, 7, 8, and 9	60	$60

To make the analysis for this example simple, assume that activities 1, 2, and 3 require 10 min and that the other activities require 5 min of processing for each job. Also, assume that activities 1, 2, and 3 are performed by resource X; activities 4, 5, and 6 are performed by resource Y; and activities 7, 8, and 9 are performed by resource Z. Finally, assume that 2400 min of each resource are available per week.

1. *Identify the system's constraints*: This problem has two types of constraints: a resource constraint and a market constraint. The resource constraint is given by the limit on the processing time per week. The market constraint is given by the limit on the demand for each job. Next, identify which one of these two types of constraints is restricting the performance of the process. Table 5.8 shows the resource utilization calculations if the process were to meet the market demands. That is, these calculations assume that 50 A jobs, 100 B jobs, and 60 C jobs will be processed per week. The requirement calculations in Table 5.8 reflect the total time needed by each job type per resource. For example, resource Y performs activities 4, 5, and 6. Because 50 A jobs are routed through activity 4 and the activity time is 5 min, this requires 5 × 50 = 250 min of resource Y per week. Then 100 B jobs are routed through activities 5 and 6, adding 10 × 100 = 1000 min to resource Y. Finally, 60 C jobs are routed through activities 4, 5, and 6, resulting in 15 × 60 = 900 additional minutes of resource Y. Hence, the process requires 2150 min of resource Y per week. Because 2400 min are available, the utilization of resource Y is 2150/2400 = 90%. The utilization of resources X and Z is obtained in the same manner. In order to meet market demand, resource X is required at more than 100% utilization, so clearly the process is constrained by this resource.
2. *Determine how to exploit the system's constraint*: Next, determine how resource X can be utilized most effectively. Let us consider three different rules to process jobs, and for each rule, calculate the total weekly profit:
   a. *Rank jobs based on profit margin*: This rule would recommend processing jobs in the order given by B, C, and A.
   b. *Rank jobs based on their profit contribution per direct labor hour*: These contributions are calculated as the ratio of profit and

TABLE 5.8

Utilization Calculations for Example 5.17

Resource	Requirements (Min/Week)	Utilization
X	(30 × 100) + (20 × 60) = 4200	4200/2400 = 175%
Y	(5 × 50) + (10 × 100) + (15 × 60) = 2150	2150/2400 = 90%
Z	(10 × 50) + (10 × 100) + (15 × 60) = 2400	2400/2400 = 100%

total direct labor. For example, job A has a contribution of \$1.33 per direct labor minute because its profit margin of \$20 is divided by the total labor of 15 min (i.e., \$20/15 = \$1.33). Similarly, the contributions of B and C are \$1.50 and \$1.20, respectively. These calculations yield the processing order of B, A, and C.

c. *Rank jobs based on their contribution per minute of the constraint*: These contributions are calculated as the ratio of profit and direct labor in resource X (i.e., the bottleneck). For example, job B has a contribution of \$2.50 per direct labor minute in resource X because its profit margin of \$75 is divided by the total labor of 30 min in resource X (i.e., \$75/30 = \$2.50). Similarly, the contribution of C is \$3.00. The contribution of A in this case is irrelevant, because A jobs are not routed through any activities requiring the bottleneck resource X. The rank is then C and B, with A as a "free" job with respect to the constraint.

3. *Subordinate everything else to the decisions made in step 2*: This step involves calculating the number of jobs of each type to be processed, the utilization of each resource, and the total weekly profit. These calculations depend on the ranking rule used in step 2, so the results of using each of the proposed rules are shown as follows:
   a. Start by calculating the maximum number of jobs of each type that can be completed using ranking rule 2a, rule 2b and rule 2c above. The maximum number of B jobs is 80/week, which is the maximum that resource X can complete (i.e., 2400/30 = 80). If 80 B jobs are processed, then no C jobs can be processed because no capacity is left in resource X. However, type A jobs can be processed because they don't use resource X. These jobs use resources Y and Z. Resource Y does not represent a constraint because its maximum utilization is 90% (as shown in Table 5.8). After processing B jobs, 1600 min are left in resource Z (i.e., 2400 – 800). Each A job requires 10 min of resource Z, so a maximum of 1600/10 = 160 A jobs can be processed. Given that the demand is 50 A jobs/week, the entire demand can be satisfied. The utilization of each resource according to this plan is shown in Table 5.9. The total profit of this processing plan is \$75 × 80 + \$20 × 50 = \$7000.

TABLE 5.9

Utilization for Ranking Rules 2.1 and 2.2

Resource	Requirements (Min/Week)	Utilization
X	80 × 30 = 2400	2400/2400 = 100%
Y	(10 × 80) + (5 × 50) = 1050	1050/2400 = 44%
Z	(10 × 80) + (10 × 50) = 1300	1300/2400 = 54%

b. The ranking according to rule 2.2 is B, A, and C. Because A does not require resource X, the resulting processing plan is the same as the one for rule 2.1 (see Table 5.9). The total profit is also \$7000.
c. For the ranking determined by rule 2.3, first calculate the maximum number of C jobs that can be processed through the bottleneck. C jobs require 20 min of resource X, so a total of 120 C jobs can be processed per week (i.e., 2400/20 = 120). This allows for the entire demand of 60 C jobs to be satisfied. Now, subtract the capacity from the bottleneck and calculate the maximum number of B jobs that can be processed with the remaining capacity. This yields 1200/30 = 40 B jobs. Finally, the number of A jobs that can be processed is calculated. A jobs are "free" with respect to resource X but require 10 min of processing in resource Z. The updated capacity of resource Z is 1300 min (after subtracting 400 min for B jobs and 900 min for C jobs); therefore, the entire demand of A jobs can be satisfied. The utilization of the resulting plan is given in Table 5.10. The total profit of this processing plan is \$20 × 50 + \$75 × 40 + \$60 × 60 = \$7600.

TABLE 5.10

Utilization for Ranking Rule 2.c

Resource	Requirements (Min/Week)	Utilization
X	(30 × 40) + (20 × 60) = 2400	2400/2400 = 100%
Y	(5 × 50) + (10 × 40) + (15 × 60) = 1550	1550/2400 = 64%
Z	(10 × 50) + (10 × 40) + (15 × 60) = 1800	1800/2400 = 75%

■

Rule 2.c yields superior results in constrained processes where the goal is selecting the mix of products or services that maximizes total profit. Therefore, it is not necessary to apply rules 2.a and 2.b in these situations because their limitations have been shown by way of the previous example. The application of rule 2.c to the production and marketing examples (Examples 5.16 and 5.17) presented earlier in this section is left as exercises.

5.4.1 Drum–Buffer–Rope Systems

We finish our discussion of the connection between TOC and process capacity management with a brief description of drum–buffer–rope (DBR) systems. DBR is a planning and control system related to TOC that works by regulating the WIP at the bottleneck (also known as the capacity-constrained resource or CCR) in a process (Krajewski et al., 2010). The *drum* is the throughput rate of the bottleneck because it sets the beat for the entire process. The *buffer* ensures that the bottleneck does not starve and operates without disruptions. The *rope* is an information flow that controls the release of work into the process. In other words, the rope is a communication mechanism to manage the inflow rate. The rope attempts to balance the inflow rate and the CCR's throughput rate.

The main goal of a DBR system is to improve overall process capacity by a better utilization of the CCR. The implementation of DBR is related to two workflow design principles introduced in Chapter 4: Establish one-at-time processing (see Section 4.2.3) and balance flow to the bottleneck (see Section 4.2.4). DBR manages the size of transfer batches to allow workstations to start work before the completion of a process batch. Transfer batches at workstations upstream of the CCR may be as small as one unit, while the size may be significantly larger for workstations downstream of the bottleneck resource.

Although the ideas behind DBR were originally conceived with manufacturing systems in mind, the concepts have been adapted to business processes (Rhee et al., 2010). The main contribution of these authors is the analysis of business process capacity and the development of a DBR-based method for enhancing efficiency. The problem of managing capacity is viewed as one of assigning and scheduling work, and DBR is used as the base methodology to improve those decisions. Applied to a loan management process, the method monitors the pace of the CCR to make effective decisions on the number of loan applications to release into the process. A computer simulation (of the type that we will introduce in Chapter 7) was used to test the merit of the method and verify that it offered an improved performance in throughput, CT and WIP in the same work, particularly under heavy workloads.

5.5 Summary

This chapter has discussed two important aspects in the modeling and design of business processes: CT analysis and capacity analysis. CT analysis represents the customer perspective because most customers are concerned with the time that they will have to wait until their jobs are completed. Market demands are such that business processes must be able to compete in speed in addition to quality and other dimensions of customer service.

On the other hand, operations managers are concerned with the efficient utilization of resources. Capacity analysis represents this point of view.

Competitive processes are effective at balancing customer service and resource utilization. The techniques for managing flow discussed in this chapter are relevant in the search for the right balance.

Discussion Questions and Exercises

1. Explain in your own words the different types of flows in a process.
2. What is the relationship between WIP and the input and output rates over time?
3. A Burger King processes on average 1200 customers/day (over the course of 15 h). At any given time, 60 customers are in the store. Customers may be waiting to place an order, placing an order, waiting for the order to be ready, eating, and so on. What is the average time that a customer spends in the store?
4. A branch office of the University Federal Credit Union processes 3000 loan applications per year. On the average, loan applications are processed in 2 weeks. Assuming 50 weeks/year, how many loan applications can be found in the various stages of processing within the bank at any given time?
5. In Exercise 5.3, it is mentioned that at any given time, one can find 60 customers in the store. How often can the manager of the store expect that the entire group of 60 customers would be entirely replaced?
6. The process of designing and implementing a website for commercial use can be described as follows. First, the customer and the web design team have an informational meeting for half a business day. If the first meeting is successful, the customer and the web design team meet again for a full day to work on the storyboard of the site. If the first meeting is not successful, then the process is over, which means the customer will look for another web designer. After the storyboard is completed, the site design begins; immediately after that, the site is developed. The design of the site typically takes 10 business days. The development of the site requires 2 business days. While the site is being designed and developed, the contents are prepared. It takes 5 business days to complete an initial draft of the contents. After the initial draft is completed, a decision is made to have a marketing team review the initial draft. Experience shows that about 60% of the time the marketing review is needed and 40% of the time the final version of the contents is prepared without the marketing review. The marketing review requires 3 business days, and the preparation of the final version requires 4 business days.

The contents are then put into the site. This activity is referred to as building. (Note that before the building can be done, the development of the site and the contents must be completed.) Building the site takes about 3 business days. After the site is built, a review activity is completed to check that all links and graphics are working correctly. The review activity typically is completed in 1 business day. The final activity is the approval of the site, which is done in half a business day. About 20% of the time, the sites are not approved. When a site is rejected, it is sent back to the building activity. When the site is approved, the design process is finished.

a. Draw a flowchart of this process.
b. Calculate the CT.

7. Consider the process flowchart in Figure 5.11. The estimated waiting time and processing time for each activity in the process are shown in Table 5.11. All times are given in minutes.
   a. Calculate the average CT for this process.
   b. Calculate the CT efficiency.

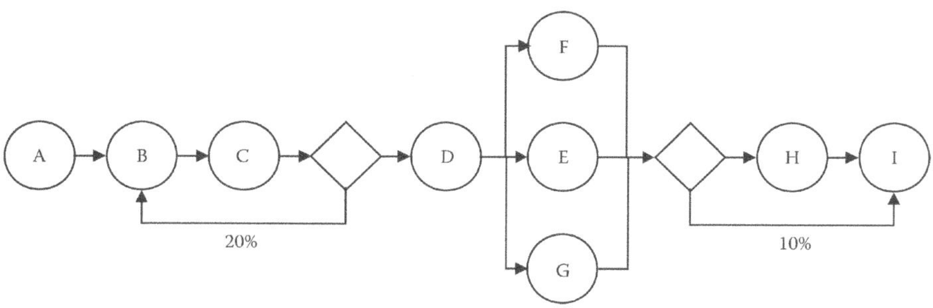

FIGURE 5.11
Flowchart of the business process for Exercise 5.7.

TABLE 5.11
Time Data for Exercise 5.7

Activity	Waiting Time (Min)	Processing Time (Min)
A	7	3
B	5	8
C	4	2
D	10	5
E	7	2
F	0	3
G	2	5
H	8	9
I	2	8

8. For the process in Exercise 5.7, assume that the resources in Table 5.12 are needed in each activity. Also assume that there are two units of R1, three units of R2, two units of R3, and two units of R4.
   a. Calculate the theoretical process capacity and identify the bottleneck.
   b. If the actual throughput has been observed to be 6 jobs/h, what is the capacity utilization?

TABLE 5.12

Resource Assignments for Exercise 5.8

Activity	Resource Type
A	R1
B	R2
C	R2
D	R2
E	R3
F	R3
G	R4
H	R4
I	R1

9. For the process flowchart in Figure 5.12, where the numbers between parentheses are the estimated activity times (in minutes), calculate the average CT.

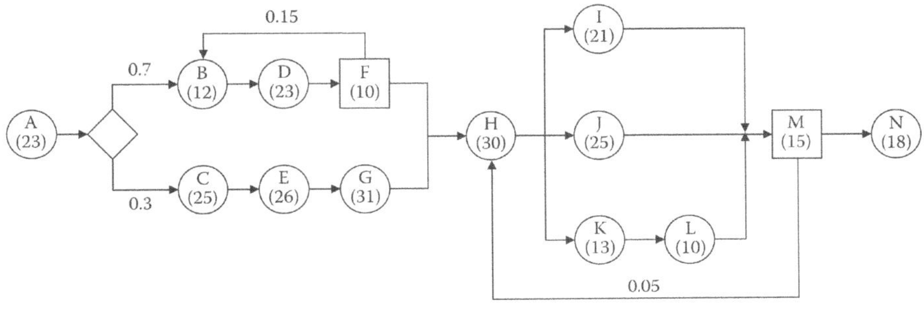

FIGURE 5.12
Flowchart for Exercise 5.9.

10. Assume that the processing times (in minutes) for the activities in Exercise 5.9 are estimated as shown in Table 5.13. Calculate the CT efficiency.

TABLE 5.13

Processing Times for Exercise 5.10

Activity	Processing Time (Min)
A	13
B	2
C	15
D	13
E	20
F	10
G	28
H	25
I	11
J	15
K	12
L	5
M	5
N	11

11. Assume that four resource types are needed to perform the activities in the process of Exercises 5.9 and 5.10. The resource type needed by each activity is shown in Table 5.14.
    a. Considering that there are two units of resource 1, two units of resource 2, three units of resource 3, and three units of resource 4, calculate the capacity of the process.
    b. If the actual throughput of the process is 2.5 jobs/h, what is the capacity utilization?

TABLE 5.14

Resource Assignments for Exercise 5.11

Resource	Activities
1	A, E, and G
2	B, D, and J
3	C, I, K, and M
4	F, H, L, and N

12. Three teams (T1, T2, and T3) work in the process depicted in Figure 5.13, where the numbers in each activity indicate processing times in minutes. Calculate the capacity utilization of the process assuming that the throughput is 1 job/h.

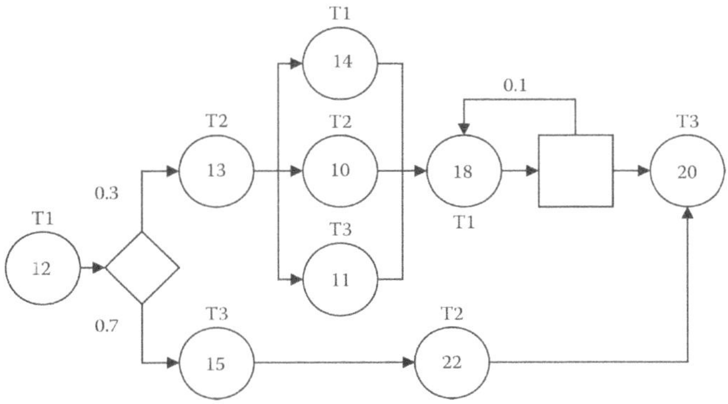

FIGURE 5.13
Flowchart for Exercise 5.12.

13. Consider the business process depicted in Figure 5.14 and the time values (in minutes) in Table 5.15. Use CT efficiency to compare this process with a redesigned version where the rework in activity G has been eliminated and activities D, E, and F have been merged into one with processing time of 10 min and zero waiting time.

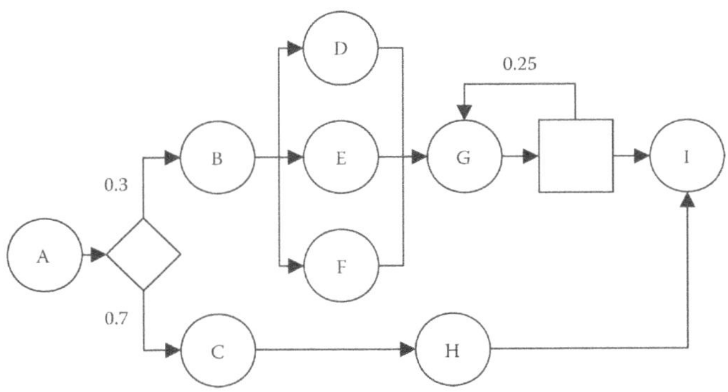

FIGURE 5.14
Flowchart for Exercise 5.13.

TABLE 5.15
Data for Exercise 5.13

Job	A	B	C	D	E	F	G	H	I
Activity time	12	13	15	14	10	11	18	22	20
Processing time	5	4	7	4	6	3	8	12	15

14. Nine people work in the process depicted in Figure 5.15. The numbers next to each activity are processing times in minutes. Table 5.16 shows the assignment of workers to activities.
    a. Calculate the capacity of the process in jobs per hour.
    b. Management is considering adding one worker to the process, but this would increase the operational cost by $23/h. Management knows that increasing the process capacity by 1 job/h adds $30/h to the bottom line. Based on this information, would you recommend adding a worker, and if so, with whom should the new person work?

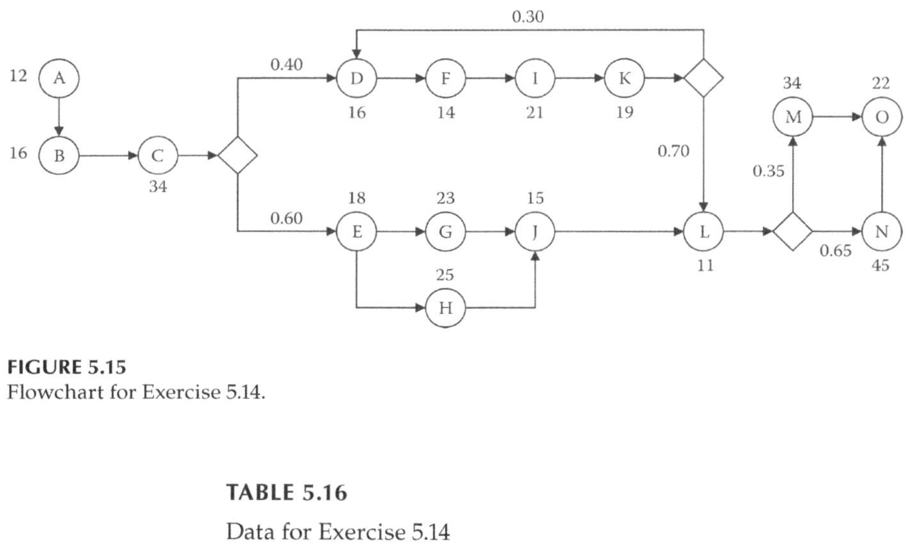

FIGURE 5.15
Flowchart for Exercise 5.14.

TABLE 5.16

Data for Exercise 5.14

Workers	Activities
David and Steve	A, B, C, and M
Laura	E and J
Debbie, Betty, and John	D, F, G, and N
Diane and Gary	I, K, and O
Fred	H and L

15. A process management team has studied a process and has developed the flowchart in Figure 5.16. The team also has determined that the expected waiting and processing times (in minutes) corresponding to each activity in the process are as shown in Table 5.17.
    a. Calculate the average CT for this process.
    b. Calculate the CT efficiency.

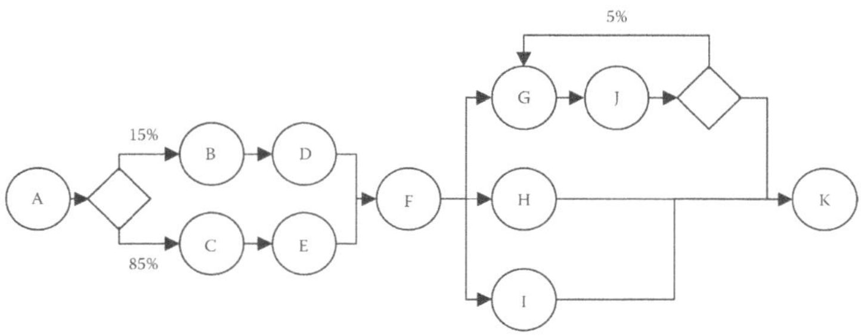

FIGURE 5.16
Flowchart for Exercise 5.15.

TABLE 5.17
Time Data for Exercise 5.15

Activity	Waiting Time (Min)	Processing Time (Min)
A	20	12
B	15	18
C	5	30
D	12	17
E	3	12
F	5	25
G	8	7
H	5	10
I	15	25
J	5	20
K	4	10

16. A process-design team is analyzing the capacity of a process. The team has developed the flowchart in Figure 5.17. The numbers between parentheses indicate the processing times in minutes, and the labels above or below each activity indicate the resource type (i.e., R1 = resource 1). The process has one unit of resource R1, two units of resource R2, and three units of resource R3. Assume that a reworked job has the same chance to pass inspection as a regular job.
    a. Calculate the theoretical process capacity and identify the bottleneck.
    b. If the actual throughput rate of the process is 1 job/h, what is the capacity utilization?

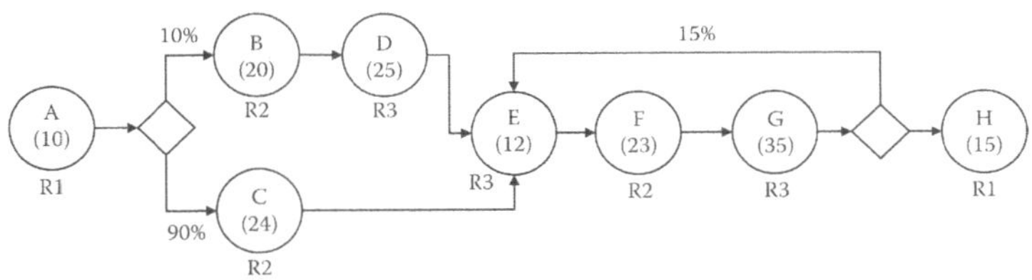

FIGURE 5.17
Flowchart for Exercise 5.16.

17. Use the TOC and the data in Tables 5.18 and 5.19 to determine how many units of each job type should be completed per week in order to maximize profits. Consider that the availability is 5500 min for resource R1, 3000 min for resource R2, and 8000 min for resource R3.

TABLE 5.18

Routing, Demand, and Profit Data for Exercise 5.17

Job	Routing	Demand (Units/Week)	Profit Margin ($)
A	1, 4, 7	80	10
B	2, 3, 5, 6	100	15
C	1, 3, 5, 6, 7	120	20

TABLE 5.19

Activity Time and Resource Assignment Data for Exercise 5.17

Activity	Time (Min)	Resource
1	20	R1
2	12	R2
3	7	R2
4	18	R1
5	9	R3
6	29	R3
7	8	R3

18. An order fulfillment process has demand for three order types during the next 4 weeks, as shown in Table 5.20. The assignment of activities to workers and processing time for each activity are shown in Table 5.21. All workers have 40 h/week available to work on this process. Use the TOC principles to find the number of orders of each type that should be processed to maximize total profit.

TABLE 5.20

Routing, Demand, and Profit Data for Exercise 5.18

Order Type	Activities	Demand	Profit Margin ($)
Basic	1, 2, and 5	150	125
Special	1, 3, 4, and 6	110	175
Deluxe	2, 4, 5, and 6	90	200

TABLE 5.21

Activity Time and Resource Assignment Data for Exercise 5.18

Activity	Time (Min)	Employee
1	24	Anne
2	18	Anne
3	25	Meg
4	27	Joe
5	15	Joe
6	14	Meg

Notes

1. CT also is called production lead time or lead time in manufacturing environments.
2. In queuing theory, which will be introduced in Chapter 6, L is used to denote the expected number of jobs in the process.
3. In queuing theory, the average CT is referred to as the expected time in the system and is denoted by W.
4. Originally developed by Toyota, JIT is now being used by many organizations around the world, including General Motors, Apple Computers, IBM, and others. JIT is mostly known for allowing companies to become more efficient by keeping low levels of inventory. However, JIT is more than just another inventory control system because its underlying philosophy for managing operations influences all parts of a company.
5. Integer programming is an optimization technique that allows decision problems to be modeled as a set of linear equations. There is one equation to model the objective function and one equation for each constraint in the problem. The solution method is based on a tree search that involves solving a series of linear programming problems, where decision variables are allowed to take on non-integer values.

References

Anupindi, R., Chopra, S., Deshmukh, S.D., Van Mieghem, J.A., and Zemel, E. 2006. *Managing business process flows: Principles of operations management*. 2nd ed. Upper Saddle River, NJ: Prentice Hall.

Davis, M.M., Aquilano, N.J., and Chase, R.B. 2003. *Fundamentals of operations management*. 4th ed. Boston, MA: McGraw-Hill Irwin.

Finch, B.J., and Luebbe, R.L. 1995. *Operations management: Competing in a changing environment*. Fort Worth, TX: Dryden Press.

Krajewski, L.J., Ritzman, L.P., and Malhotra, M.K. 2010. *Operations management: Processes and supply chains.* Upper Saddle River, NJ: Prentice Hall.
Little, J.D.C. 1961. A proof of the queuing formula $l = \lambda W$. *Operations Research* 9: 383–387.
Melnyk, S., and Swink, M.. 2002. *Value-driven operations management: An integrated modular approach.* New York: McGraw Hill/Irwin.
Rhee, S.H., Cho, N.W., and Bae, H. 2010. Increasing the efficiency of business processes using a theory of constraints. *Information Systems Frontiers* 12: 443–455.
Sterman, J.D. 2000. *Business dynamics: Systems thinking and modeling for a complex world.* Boston, MA: Irwin McGraw-Hill.
Vonderembse, M.A., and White, G.P. 1994. *Operations management: Concepts, methods, and strategies.* Ontario, Canada: John Wiley & Sons Canada Ltd.

6

Introduction to Queuing Modeling

The process models analyzed in Chapters 4 and 5 have one thing in common: they assume that activity times and demand are deterministic and constant; that is, they are known with certainty. In reality, some variability will always exist in the time it takes to perform a certain task and in the demand for service. In situations where the variability is small, deterministic models of the type studied in Chapters 4 and 5 might be an adequate way of describing a process. However, in situations with more accentuated variability, these models do not suffice. In fact, the variability itself is often one of the most important characteristics to capture in the model of a process.

Consider, for example, the checkout process at a grocery store where one of the main goals is to avoid long lines. Customers arrive at the checkout stations with their items to pay before leaving the store. The items need to be scanned, bagged, and paid for. These activities may be performed by the customer herself in a self-service checkout station or by a cashier who services the customer. The time it takes to service a customer depends on the amount and type of groceries and the form of payment used; therefore, it varies from customer to customer. In addition, the number of customers per time unit who arrive at the checkout station, meaning the demand for service, is highly uncertain and variable. Applying a deterministic model to describe this process, using only the average service time and the average number of customer arrivals per time unit, fails to capture the variability and explain why queues are forming. This is because the deterministic model assumes that it always takes exactly the same amount of time to perform the checkout activities and that customers arrive at constant intervals. Under these assumptions, it is easy to determine the number of checkout stations and cashiers needed to avoid queues. However, the fact that everyone has waited in line at a checkout station indicates that such a simplistic model is insufficient to describe the complex reality. Variability makes it difficult to match demand and capacity in such a way that queues are avoided.

The focus of this chapter is the incorporation of variability into models of business processes. These types of models, known in the literature as *stochastic models*, enable us to evaluate how the process design decisions affect waiting times, queue lengths, and service levels. An important difference compared with the deterministic models that have been considered so far is that in the stochastic models, the waiting time is no longer an input parameter; it is the result of the specified process time and the demand pattern. Because eliminating non-value-adding waiting time is

DOI: 10.1201/9781032617237-6

an important objective when designing a new business process (see, for example, Chapter 1), the importance of stochastic models in process design is obvious.

The modeling approaches that will be explored in this chapter belong to the area of *analytical queuing models* stemming from the field of mathematical queuing theory. Computer-based *simulation models,* which are experimental in nature, constitute flexible alternatives to analytical models. Simulation modeling of business processes is investigated in Chapters 7 through 10. The reason for considering both queuing models and simulation is that they offer different advantages to the process designer. The analytical queuing models express important performance characteristics in mathematical formulae and are convenient to use when the necessary conditions are in place. The drawback is that these conditions can be restrictive with regard to the process structure and the representation of variability in the models. The simulation models are more flexible, but they usually take more time to set up and run. They also require a computer with the appropriate software. Moreover, because the simulation models are experimental, they require statistical analysis of input and output data before the appropriate information on performance characteristics sought by the process designer can be obtained. This chapter describes some commonly used queuing models and how they can be applied. However, before getting into the modeling details, it is important to understand why variability in process parameters is such an important issue from an operational and economic perspective.

From an *operational perspective,* the main problem with variability in processing times, demands, and capacity is that it leads to an unbalanced use of resources over time, causing the formation of waiting lines. This will be loosely referred to as a *capacity-planning problem.* The core of this problem is that queues (or waiting lines) can arise at any given time when the demand for service exceeds the resource capacity for providing the service. This means that even if the average demand falls well below the average capacity, high variability will lead to instances when the demand for service exceeds the capacity, and a queue starts to form. This explains the formation of waiting lines in the grocery store example. On the other hand, in some instances, the service capacity will greatly exceed the demand for service. This reduces an existing queue, and if no queue exists, the resource providing the service will be idle. At a grocery store, even if a checkout station is empty for long periods during the day, the arrival of a customer with a full cart could initiate a queue.

At this point, it is important to recognize that queues concern individuals as well as documents, products, or intangible jobs such as blocks of information sent over the Internet. However, the issues with waiting in line are particularly important in service industries where the queue consists of people waiting for service. The reason is that, as opposed to objects or pieces of information, people usually find waiting frustrating, and they often react quickly and voice their dissatisfaction.

From an *economic perspective*, the capacity-planning problem caused by high variability in demand and process times comes down to balancing the cost of occasionally having excess capacity against the cost associated with long waiting lines at other times. The cost of delays and waiting can take on many different forms, including, for example,

- The *social costs* of not providing fast enough care at a hospital
- The *cost of lost customers* who go elsewhere because of inadequate service
- The *cost of discounts* because of late deliveries
- The *cost of goodwill loss* and bad reputation affecting future sales
- The *cost of idle employees* who have to wait for some task to be completed before they can continue their work

Ultimately, these costs will affect the organization, even though the impact sometimes might be hard to quantify. The cost of too much capacity is usually easier to identify. Primarily, it consists of the fixed and variable costs of additional and unused capacity, including increased staffing levels. Returning to the grocery store example, the store must balance the cost of hiring additional cashiers and installing more checkout stations against the cost of lost customers and lower sales revenues due to long queues and waiting times at the checkout stations.

Figure 6.1 depicts the economic trade-off associated with the capacity-planning problem. The x-axis represents the service capacity expressed as the number of jobs per unit of time the system can complete on average (see process calculations in Section 5.2.2). The y-axis represents the total costs associated with waiting and providing service. The waiting cost reflects the cost of having too little service capacity, and the service cost reflects the cost of acquiring and maintaining a certain service capacity.

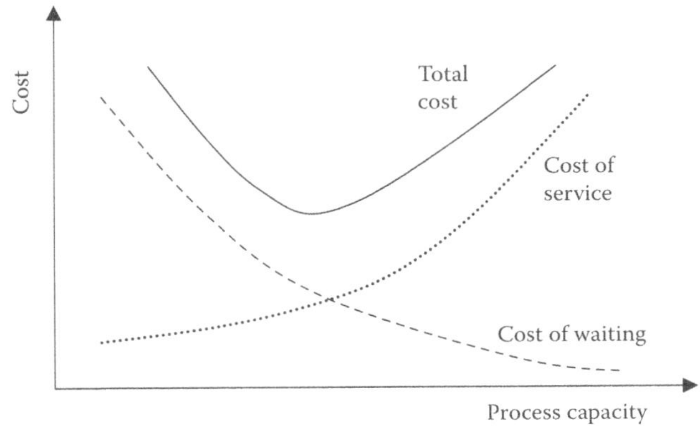

FIGURE 6.1
Economic trade-off between service capacity and waiting times.

To arrive at design decisions that will minimize the total costs, an important first step is to quantify the delay associated with a certain capacity decision. Second, this delay needs to be translated into monetary terms to compare this with the cost of providing a certain service capacity. This chapter investigates models that deal with both of these issues.

Section 6.1 specifies a conceptual model of the basic queuing process and defines what a queuing system is. This section also discusses strategies that are often used in service industries to mitigate the negative effects of making people wait in line. The conceptual model for the basic queuing process serves as a basis for the analytical queuing models investigated in Section 6.2. Finally, a summary and some concluding remarks are provided in Section 6.3.

6.1 Queuing Systems, the Basic Queuing Process, and Queuing Strategies

Queuing processes, queuing systems, and queuing models might appear to be abstract concepts with limited practical applicability, but nothing could be further from the truth. Wherever people go in their daily lives, they encounter queuing systems and queuing processes (e.g., at the bank, in the grocery store, or when calling for a taxi, going to the doctor, eating at a restaurant, buying tickets to the theater, or taking the bus). In simple terms, an elementary queuing system consists of a service mechanism with servers providing service and one or more queues of customers or jobs waiting to receive service. The queuing process, on the other hand, describes the operations of the queuing system; that is, how customers arrive at the queuing system and how they proceed through it. The queuing system is an integral part of the queuing process. Noting the subtle distinction between a queuing process and a queuing system will prove useful when discussing detailed modeling issues. However, in practical situations and higher-level discussions, these two terms are often interchangeable. Going back to the definition of a business process as a network of connected activities and buffers (see Section 1.1), a business process can be interpreted as a network of queuing processes or elementary queuing systems.

To expand the frame of reference regarding queuing systems and to make this concept more concrete, one can look at a number of examples of real-world queuing systems that can be broadly classified as commercial, transportation, business-internal, and social service systems. Many of the queuing systems we encounter daily are *commercial service systems,* whereby a commercial organization serves external customers. Often, the service is a personal interaction between the customer and an employee, but in some cases, the service provider might be a machine. Typically, customers go to a fixed location to seek service. Examples include a dentist's office (where the server is the dentist), banks, checkout stations at supermarkets or other stores, gas stations, and automated teller machines. However, the server also can travel to the customer. Examples include a plumber or cable company technician visiting the customer's home to perform a service.

The class of *transportation service systems* represents situations where vehicles are either customers or servers. Examples where vehicles are customers include cars and trucks waiting at tollbooths, railroad crossings, or traffic lights; trucks or ships waiting to be loaded; and airplanes waiting to access a runway. Situations where vehicles constitute the servers in a queuing system include taxicabs, fire trucks, elevators, buses, trains, and airplanes transporting customers between locations.

In *business-internal service systems*, the customers receiving service are internal to the organization providing the service. This definition includes inspection stations (the server), materials handling systems such as conveyor belts, maintenance systems where a technician (the server) is dispatched to repair a broken machine (the customer), and internal service departments such as computer support (the server) servicing requests from employees (the customers). It also includes situations where machines in a production facility or a computing center process jobs.

Finally, many *social service systems* are queuing systems. In the judicial process, for example, the cases awaiting trial are the customers, and the judge and jury are the servers. Other examples include an emergency room at a hospital, the family doctor making house calls, and waiting lists for organ transplants or student dorm rooms.

This broad classification of queuing systems is not exhaustive, and it does have some overlap. Its purpose is just to give a flavor of the wide variety of queuing systems facing a process designer.

A conceptual model of the basic queuing process describes the operation of the queuing systems just mentioned and further explains the subtle distinction between a queuing process and a queuing system. It also provides the basis for exploring the mathematical queuing models in Section 6.2 and help to conceptualize the simulation models in Chapters 7 through 9. Following the discussion of the basic queuing process, Section 6.1.2 looks at some pragmatic strategies used for mitigating the negative economic impact of long waiting lines.

6.1.1 The Basic Queuing Process

The basic queuing process describes how customers arrive at and proceed through the queuing system. This means that the basic queuing process describes the operations of a queuing system. The following major elements define a basic queuing process: the calling population, the arrival process, the queue configuration, the queue discipline, and the service mechanism.

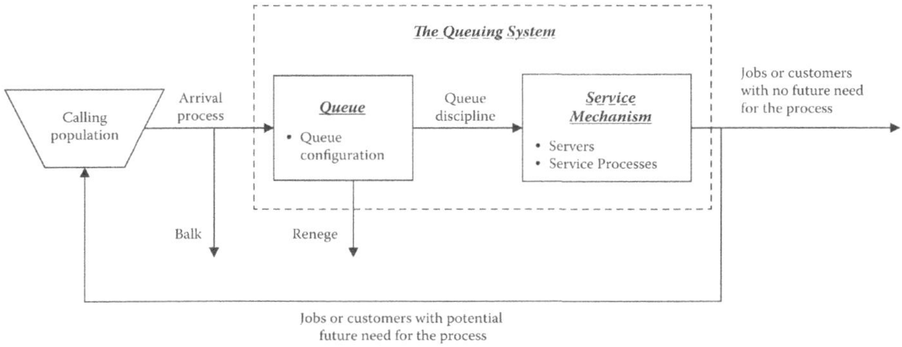

FIGURE 6.2
The basic queuing process and the queuing system.

These elements of the basic queuing process are illustrated in Figure 6.2, which also clarifies the distinction between the basic queuing process and the queuing system.

Customers or jobs from a calling population arrive at one or several queues (buffers) with a certain configuration. The arrival process specifies the rate at which customers arrive and the particular arrival pattern. Service is provided immediately if the queue is empty and a server (resource) is available. Otherwise, the customer or job waits in the queue for service.

In service systems where jobs are customers (i.e., people), some may choose not to join the queue when confronted with a potentially long wait. This behavior is referred to as *balking*. Other customers may consider, after joining the queue, that the wait is intolerable and *renege* (i.e., they leave the queue before being served). From a process performance perspective, an important distinction between balking and reneging is that the reneging customers take up space in the queue and customers that balk do not. For example, in a call center, the reneging customers will accrue connection costs as long as they remain in the queue.

When servers (resources) become available, a job is selected from the queue, and the service (activity) is performed. The policy governing the selection of a job from the queue is known as the *queue discipline*. Finally, the service mechanism can be viewed as a network of service stations where activities are performed. These stations may need one or more resources to perform the activities, and the availability of these resources will result in queues of different lengths.

This description of a queuing process is closely related to the definition of a general business process given in Section 1.1. In fact, as previously mentioned, a general business process can be interpreted as a network of basic queuing processes (or queuing systems).

Due to the importance of the basic queuing process, the remainder of this section is devoted to a further examination of its components.

6.1.1.1 The Calling Population

The calling population can be characterized as homogeneous (i.e., consisting of only one type of job) or heterogeneous, consisting of different types of jobs. In fact, most queuing processes will have heterogeneous calling populations. For example, the admissions process at a state university has in-state and out-of-state applicants. Because these applications are treated differently, the calling population is not homogeneous. In this case, the calling population consists of two subpopulations.

The calling population also can be characterized as infinite or finite. In reality, few calling populations are truly infinite. However, in many situations, the calling population can be considered infinite from a modeling perspective. The criterion is that if the population is so large that the arrival process is unaffected by how many customers or jobs are currently in the queuing system (in the queue or being served), the calling population is considered infinite. An example of this is a medium-sized branch office of a bank. All potential customers who might want to visit the bank define the calling population. It is unlikely that the number of customers in the bank would affect the current rate of arrivals. Consequently, the calling population could be considered infinite from a modeling perspective.

In other situations, the population cannot be considered infinite. For example, the mechanics at a motorized army unit are responsible for the maintenance and repair of the unit's vehicles. Because the number of vehicles is limited, a large number of broken-down vehicles in need of repair implies that fewer functioning vehicles remain that can break down. The arrival process—vehicles breaking down and requiring the mechanic's attention—is most likely dependent on the number of vehicles currently awaiting repair (i.e., in the queuing system). In this case, the model must capture the effect of a finite calling population.

6.1.1.2 The Arrival Process

The arrival process refers to the temporal and spatial distribution of the demand facing the queuing system. If the queuing process is part of a more general business process, there may be several entry points for which the arrival processes must be studied. The demand for resources in queuing processes that are not directly connected to the entry points to the overall business process is contingent on how the jobs are routed through this process. That is, the path that each type of job follows determines which activities are performed and, therefore, which resources are needed.

The arrival process is characterized by the distribution of interarrival times, meaning the probability distribution of the times between consecutive arrivals. To determine these distributions in practice, data need to be collected from the real-world system. Statistical methods can then be used to analyze the data and estimate the distributions. The approach can be characterized in terms of the following three-step procedure:

- Collect data by recording the actual arrival times into the process.
- Calculate the interarrival times for each job type.
- Perform statistical analysis of the interarrival times to fit a probability distribution.

Specific statistical tools to perform these steps are described in Chapter 9.

6.1.1.3 The Queue Configuration

The queue configuration refers to the number of queues, their location, their spatial requirements, and their effect on customer behavior. Figure 6.3 shows two commonly used queue configurations: the single-line versus the multiple-line configuration.

For the multiple-line alternative (Figure 6.3a), the arriving job must decide which queue to join. Sometimes, the customer, as in the case of a supermarket, makes the decision. However, in other cases, this decision is made as a matter of policy or operational rule. For example, requests for technical support in a call center may be routed to the agent with the fewest requests at the time of arrival. In the single-line configuration (Figure 6.3b), all the jobs join the same line. The next job to be processed is selected from the single line.

Some advantages are associated with each of these configurations. These advantages are particularly relevant in settings where the jobs flowing through

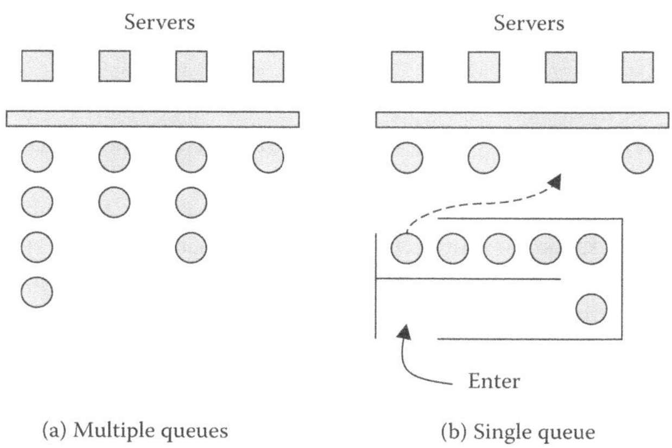

FIGURE 6.3
Alternative queue configurations. (Adapted from Fitzsimmons and Fitzsimmons, 1998.)

the process are people. The multiple-line configuration, for instance, has the following advantages:

1. The service provided can be differentiated. The use of express lanes in supermarkets is an example. Shoppers who require a small amount of service can be isolated and processed quickly, thereby avoiding long waits for little service.
2. Labor specialization is possible. For example, drive-in banks assign the more experienced tellers to the commercial lane.
3. The customer has more flexibility. For example, the customer in a grocery store with several checkout stations has the option of selecting a particular cashier of preference.
4. Balking behavior may be deterred. When arriving customers see a long, single queue snaked in front of a service, they often interpret this as evidence of a long wait and decide not to join the line.

The following are the advantages of the single-line configuration:

1. The arrangement guarantees "fairness" by ensuring that a first-come-first-served (FCFS) rule is applied to all arrivals.
2. Because only a single queue is available, no anxiety is associated with waiting to see whether the chosen line is the fastest.
3. With only one entrance at the rear of the queue, the problem of cutting in is resolved. Often, the single line is roped into a snaking pattern, which makes it physically more difficult for customers to leave the queue, thereby discouraging reneging. Reneging occurs when a customer in line leaves the queue before being served, typically because of frustration over the long wait.
4. Privacy is enhanced because the transaction is conducted with no one standing immediately behind the person being served.
5. This arrangement is more efficient in terms of reducing the average time that customers spend waiting in line.
6. Jockeying is avoided. Jockeying refers to the behavior of switching lines. This occurs when a customer attempts to reduce his or her waiting time by switching lines as the lines become shorter.

6.1.1.4 The Queue Discipline

The queue discipline is the policy used to select the next job to be served. The most common queue discipline is First-Come-First-Serve (FCFS), also known as *First-In-First-Out* (FIFO). This discipline, however, is not the only possible policy for selecting jobs from a queue. In some cases, it is possible to estimate the processing time of a job in advance. This estimate can then be used to implement a policy whereby the fastest jobs are processed first. This queue discipline is known as the *shortest processing time first* (SPT) rule, as discussed in Chapter 4. Other well-known disciplines are last-in-first-out (LIFO) and longest processing time first (LPT).

In addition to these rules, queue disciplines based on priority are sometimes implemented. In a medical setting, for example, triage (i.e., medical screening) is used to give priority to those patients who would benefit the most from immediate treatment. An important distinction in terms of priority disciplines is between *non-preemptive* and *preemptive* priorities. Under a non-preemptive priority discipline, a customer or job that is currently being served is never sent back to the queue to make room for an arriving customer or job with higher priority. In a preemptive discipline, on the other hand, a job that is being served will be thrown back into the queue immediately to make room for an arriving job with higher priority.

The preemptive strategy makes sense, for example, in an emergency room, where the treatment of a sprained ankle is interrupted when an ambulance brings in a patient with cardiac arrest, and the physician must choose between the two. The drawback with the preemptive discipline is that customers with low priorities can experience extremely long waiting times.

The non-preemptive strategy makes sense in situations where interrupting the service before it is finished means that all the work put in is wasted, and the service needs to be started from scratch when the job eventually rises in importance again. For example, if a limousine service is driving people to and from an airport, it makes little sense to turn the limousine around to pick up a VIP client at the airport just before it reaches the destination of its current passenger.

In many cases in which the customers are individuals, a preemptive strategy can cause severe frustration for low-priority customers. A non-preemptive approach is often more easily accepted by all parties. Note that queue disciplines in queuing theory correspond to dispatching rules in sequencing and scheduling (see Section 4.2.6).

6.1.1.5 The Service Mechanism

The service mechanism consists of one or more service facilities, each containing one or more parallel service channels referred to as *servers*. A job or a customer enters a service facility, where a server provides the service. In cases where multiple service facilities are set up in series, the job or customer might be served by a sequence of servers. The queuing model must specify the exact number of service facilities, the number of servers in each facility, and possibly the sequence of service facilities a job must pass through before leaving the queuing system. The time spent in a service facility, meaning time in one particular server or service station, is referred to as the *service time*. The *service process* refers to the probability distribution of service times associated with a certain server (possibly different across different types of customers or jobs). Often, it is assumed that the parallel servers in a given service facility have the same service time distribution. For example, in modeling the parallel tellers in a bank, it is usually reasonable to assume that the service process is the same for all tellers. Statistically, the estimation of the service times can be done in a similar way to the estimation of interarrival times (see Chapter 9).

The design of the service mechanism and queues, together with the choice of queuing disciplines, results in a certain service capacity associated with the queuing system. For example, the service mechanism may consist of one or more service facilities with one or more servers. How many is a staffing decision that directly affects the ability of the system to meet the demand for service. Adding capacity to the process typically results in decreasing the probability of long queues and, therefore, decreases the average waiting time of jobs in the queue. Often, the arrival process is considered to be outside the decision-maker's control. However, sometimes, implementing so-called *demand management strategies* is an option. This can involve working with promotional campaigns, such as "everyday low prices" at Walmart to encourage more stable demand patterns over time. These types of activities also go by the name of *revenue management.*

6.1.2 Strategies for Mitigating the Effects of Long Queues

In some situations, it might be unavoidable to have long waiting lines occasionally. Consider, for example, a ski resort. Most people arrive in the morning so that they can spend the whole day on the slopes. As a result, early in the day, lines will form at the ticket offices and ski lifts that everyone has to take to get on the mountain. A relevant question is, therefore, how the process designer or process owner can mitigate the negative economic effects of queues or waiting lines when they form. This is of particular interest in service systems that want to avoid balking and reneging. Commonly used strategies are based on the ideas of concealing the queue, using the customer as a resource, making the wait comfortable, distracting the customer's attention, explaining the reasons for the wait, providing pessimistic estimates of remaining waiting time, and being fair and open about the queuing discipline.

An often-used strategy for minimizing or avoiding lost sales due to long lines is to conceal the queue from arriving customers (Fitzsimmons and Fitzsimmons, 1998). Restaurants achieve this by diverting people to the bar, which in turn has the potential to increase revenues. Amusement parks such as Disneyland require people to pay for their tickets outside the park, where they are unable to observe the waiting lines inside. Casinos "snake" the waiting line for nightclub acts through the slot machine area, in part to hide its true length and in part to foster impulsive gambling.

In some cases, a fruitful approach is to consider the customer as a resource with the potential to be part of the service delivery. For example, a patient might complete a medical history record while waiting for a doctor, thereby saving the physician valuable time. This strategy increases the capacity of the process while minimizing the psychological effects of waiting.

Regardless of other strategies, making the customer's wait comfortable is necessary to avoid reneging and loss of future sales due to displeased and annoyed customers. This can be done in numerous ways: for example, by providing complementary drinks at a restaurant or by having a nicely decorated waiting room with interesting reading material in a doctor's office.

Closely related to this idea is distracting the waiting customers' attention, thereby making the wait seem shorter than it is. Airport terminals, for instance, often have public TV monitors, food courts, and shops.

If an unexpected delay occurs, it is important not only to inform the customer about this as soon as possible but also to explain the reason for the delay. The passengers seated on an airplane will be more understanding if they know that the reason for the delay is faulty radio equipment. Who wants to fly on a plane without a radio? However, if no explanation is given, attitudes tend to be much less forgiving, because the customers feel neglected and that they have not been treated with the respect they deserve. Keeping the customers informed is also a key concept in all industries that deliver goods. Consider, for example, the tracking systems that FedEx and other shipping companies offer their customers.

To provide pessimistic estimates of the remaining waiting time is often also a good strategy. If the wait is shorter than predicted, the customer might even leave happy. On the other hand, if the wait is longer than predicted, the customers tend to lose trust in the information and in the service.

Finally, it is important to be fair and open about the queue discipline used. Ambiguous VIP treatment of some customers tends to agitate those left out; for example, think about waiting lines at popular restaurants or nightclubs.

6.2 Analytical Queuing Models

This section will examine how analytical queuing models can be used to describe and analyze the basic queuing process and the performance characteristics of the corresponding queuing system. The mathematical study of queues is referred to as *queuing* or *queuing theory*.

The distributions of interarrival and service times largely determine the operational characteristics of queuing systems. After all, considering these distributions instead of just their averages is what distinguishes the stochastic queuing models from the deterministic models in Chapters 4 and 5. In real-world queuing systems, these distributions can take on virtually any form, but they are seldom thought of until the data is analyzed. However, when modeling the queuing system mathematically, one must be specific about the type of distributions being used. When setting up the model, it is important to use distributions that are realistic enough to capture the system's behavior. At the same time, they must be simple enough to produce a tractable mathematical model. Based on these two criteria, the exponential distribution plays an important role in queuing theory. In fact, most results in elementary queuing theory assume exponentially distributed interarrival and service times. Section 6.2.1 discusses the importance and relevance of the exponential distribution as well as its relation to the Poisson process.

The wide variety of specialized queuing models calls for a way to classify these according to their distinct features. A popular system for classifying queuing models with a single service facility with parallel and identical servers, a single queue, and a FIFO queuing discipline uses a notational framework with the structure $A1/A2/A3/A4/A5$. Attributes $A1$ and $A2$ represent the probability distribution of the interarrival and service times, respectively. Attribute $A3$ denotes the number of parallel servers. Attribute $A4$ indicates the maximum number of jobs allowed in the system at the same time. If there is no limitation, $A4 = \infty$. Finally, attribute $A5$ represents the size of the calling population. If the calling population is infinite, then $A5 = \infty$. Examples of symbols used to represent probability distributions for interarrival and service times—that is, attributes $A1$ and $A2$—include the following:

M = Markovian, meaning that the interarrival or service times follow exponential distributions.

D = Deterministic, meaning that the interarrival or service times are deterministic and constant.

G = General, meaning that the interarrival or service times may follow any distribution.

Consequently, $M/M/c$ refers to a queuing model with c parallel servers for which the interarrival times and the service times are both exponentially distributed. Omitting attributes $A4$ and $A5$ means that the queue length is unrestricted, and the calling population is infinite. In the same fashion, $M/M/c/K$ refers to a model with a limitation of at most K customers or jobs in the system at any given time. $M/M/c/\infty/N$, on the other hand, indicates that the calling population is finite and consists of N jobs or customers.

Exponentially distributed interarrival and service times are assumed for all the queuing models in this chapter. In addition, all the models will have a single service facility with one or more parallel servers, a single queue, and a FIFO queuing discipline. It is important to emphasize that the analysis can be extended to preemptive and non-preemptive priority disciplines; see, for example, Hillier and Lieberman (2010). Moreover, there exist results for many other queuing models not assuming exponentially distributed service and interarrival times. Some basic results for the specialized $M/G/1$ and $M/G/\infty$ models with general service time distributions are available in Appendix 6A.

Often, though, the analysis of these models tends to be much more complex and is beyond the scope of this book.

The exploration of analytical queuing models in this chapter is organized as follows. Section 6.2.1 discusses the exponential distribution, its relevance, important properties, and its connection to the Poisson process. Section 6.2.2 introduces some notation and terminology used in Sections 6.2.3 through 6.2.8. Important concepts include steady-state analysis and Little's law. Based on the fundamental properties of the exponential distribution and the basic notation introduced in Section 6.2.2, Section 6.2.3 focuses on general birth-and-death processes and their wide applicability in modeling queuing processes. Although it is a powerful approach, the method for analyzing general birth-and-death processes can be tedious for large models. Therefore, Sections 6.2.4 through 6.2.7 explore some important specialized birth-and-death models with standardized expressions for determining important characteristics of these queuing systems. Section 6.2.8 illustrates the usefulness of analytical queuing models for making process design–related decisions, particularly by translating important operational characteristics into monetary terms. Finally, a chapter summary is provided in Section 6.3.

6.2.1 The Exponential Distribution and Its Role in Queuing Theory

The exponential distribution has a central role in queuing theory for two reasons. First, empirical studies show that many real-world queuing systems have arrival and service processes that follow exponential distributions. Second, the exponential distribution has some mathematical properties that make it relatively easy to manipulate. In this section, the exponential distribution is defined, and some of its important properties and their implications when used in queuing modeling are discussed. The relationship between the exponential distribution, the Poisson distribution, and the Poisson process is also explained. This discussion assumes some prior knowledge of basic statistical concepts such as random variables, probability density functions, cumulative distribution functions, mean, variance, and standard deviation. Basic books on statistics contain material related to these topics, some of which are reviewed in Chapter 9. To provide a formal definition of the exponential distribution, let T be a random (or stochastic) variable representing either interarrival times or service times in a queuing process. T is said to follow an exponential distribution with parameter α if its probability density function (or frequency function), $f_T(t)$, is

$$f_T(t) = \begin{cases} \alpha e^{-\alpha t} & \text{for } t \geq 0 \\ 0 & \text{for } t < 0 \end{cases}$$

The shape of the density function $f_T(t)$ is depicted in Figure 6.4, where t represents the realized interarrival or service time.

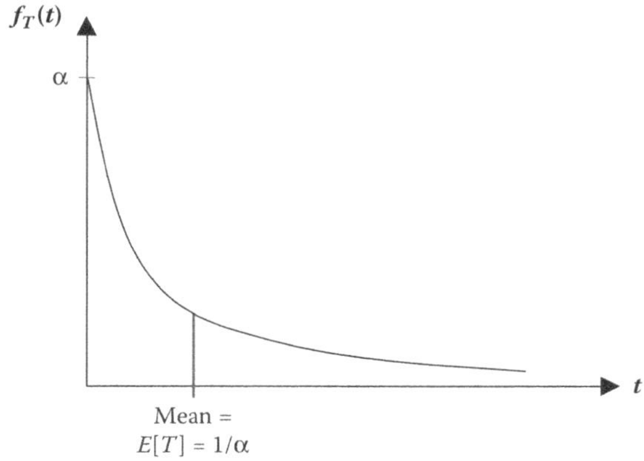

FIGURE 6.4
The probability density function for an exponentially distributed random variable T.

The expression for the corresponding cumulative distribution function, $F_T(t)$, is then

$$F_T(t) = \int_{-\infty}^{t} f_T(x)dx = \int_{0}^{t} \alpha e^{-\alpha x}dx = 1 - e^{-\alpha t}$$

To better understand this expression, recall that per definition $F_T(t) = P(T \leq t)$, where $P(T \leq t)$ is the probability that the random variable T is less than or equal to t. Consequently, $P(T \leq t) = 1 - e^{-\alpha t}$. This also implies that the probability that T is greater than t, $P(T > t)$, can be obtained as $P(T > t) = 1 - P(T \leq t) = e^{-\alpha t}$. Finally, we can conclude that the mean, variance, and standard deviation of the exponentially distributed random variable T, denoted $E(T)$, $\text{Var}(T)$, and σ_T, respectively, are:

$$E(T) = \frac{1}{\alpha}, \qquad \text{Var}(T) = \frac{1}{\alpha^2}, \qquad \sigma_T = \frac{1}{\alpha}$$

Note that the standard deviation is equal to the mean, which implies that the relative variability is quite high.

Next, the implications of assuming exponentially distributed interarrival and service times will be examined. Does it make sense to use this distribution? To provide an answer, it is necessary to look at a number of important properties characterizing the exponential distribution, some of which are intuitive and some of which have a more mathematical flavor.

First, from looking at the shape of the density function $f_T(t)$ in Figure 6.4, it can be concluded that it is decreasing in t (mathematically, it can be proved that it is *strictly decreasing in t*, which means that $P(0 \leq T \leq \delta) > P(t \leq T \leq t + \delta)$ for all positive values of t and δ).

Property 1: The Density Function $f_T(t)$ Is Strictly Decreasing in t

The implication is that exponentially distributed interarrival and service times are more likely to take on small values. For example, there is always a 63.2 percent chance that the time (service or interarrival) is less than or equal to the mean value $E(T) = 1/\alpha$. At the same time, the long right-hand tail of the density function $f_T(t)$ indicates that large time values occasionally occur. This means that the exponential distribution also encompasses situations when the time between customer arrivals, or the time it takes to serve a customer, can be very long.

For interarrival times, plenty of empirical evidence indicates that the exponential distribution in many situations is a reasonable way of modeling arrival processes of external customers to a service system. For service times, the applicability might sometimes be more questionable, particularly in cases of standardized service operations where all the service times are centered around the mean.

Consider, for example, a machining process of a specific product in a manufacturing setting. Because the machine does exactly the same work on every product, the time spent in the machine is going to deviate only slightly from the mean. In these situations, the exponential distribution does not offer a close approximation of the actual service time distribution. Still, in other situations, the specific work performed in servicing different customers is often similar but occasionally deviates dramatically. In these cases, the exponential distribution may be a reasonable choice. Examples of where this might be the case could be a bank teller or a checkout station in a supermarket. Most customers demand the same type of service—a simple deposit or withdrawal, or the scanning of fewer than 15 purchased items. Occasionally, however, a customer requests a service that takes a lot of time to complete. For example, in a bank, a customer's transaction might involve contacting the head office for clearance; in a supermarket, a shopper might bring two full carts to the cashier.

To summarize, the exponential distribution is often a reasonable approximation for interarrival times and of service times in situations where the required work differs across customers. However, its use for describing processing (or service) times in situations with standardized operations performed on similar products or jobs might be questionable. In these situations, one may need models that allow for more general service time distributions, such as those discussed in Appendix 6A.

Property 2: Lack of Memory

Another interesting property of the exponential distribution, which is less intuitive, is its lack of memory. This means that the probability distribution of the remaining service time, or time until the next customer arrives, is always the same. It does not matter how long a time has passed since the service started or since the last customer arrived. Mathematically, this means that $P(T > t + \delta \mid T > \delta) = P(T > t)$ for all positive values of t and δ (a mathematical derivation is available in Appendix 6A).

This property is important for the mathematical tractability of the queuing models, which will be considered later. However, the main question here is whether this feature seems reasonable for modeling service and interarrival times.

In the context of interarrival times, the implication is that the next customer arrival is completely independent of when the last arrival occurred. In the case of external customers arriving to the queuing system from a large calling population, this is a reasonable assumption in most cases. When the queuing system represents an internal subprocess of a larger business process, so that the jobs arrive on some kind of schedule, this property is less appropriate. Consider, for example, trucks shipping goods between a central warehouse and a regional distribution center. The trucks leave the central warehouse at approximately 10 a.m. and 1 p.m. and arrive at the regional distribution center around noon and 3 p.m., respectively. In this case, it is not reasonable to assume that the time remaining until the next truck arrives is independent of when the last arrival occurred.

For service times, the lack of memory property implies that the time remaining until the service is completed is independent of the elapsed time since the service began. This may be a realistic assumption if the required service operations differ among customers (or jobs). However, if the service consists of the same collection of standardized operations across all jobs or customers, it is expected that the time elapsed since the service started will help to predict how much time remains before the service is completed. In these cases, the exponential distribution is not an appropriate choice.

The exponential distribution's lack of memory implies another result. Assume as before that T is an exponentially distributed random variable with parameter α and that it represents the time between two events; that is, the time between two customer arrivals or the duration of a service operation. It can be asserted that no matter how much time has elapsed since the last event, the probability that the next event will occur in the following time increment δ is $\alpha\delta$. Mathematically, this means that $P(T \le t + \delta \mid T > t) = \alpha\delta$ for all positive values of t and small positive values of δ, or more precisely,

$$\lim_{\delta \to 0} \frac{P(T \le t + \delta | T > t)}{\delta} = \alpha$$

The result implies that α can be interpreted as the mean rate at which new events occur. This observation will be useful in the analysis of birth-and-death processes in Section 6.2.3.

Property 3: The Minimum of Independent Exponentially Distributed Random Variables Is Exponentially Distributed

A third property characterizing the exponential distribution is that *the minimum of independent exponentially distributed random variables is exponentially distributed* (a mathematical derivation is available in Appendix 6A). To explain this in more detail, assume that $T_1, T_2, \ldots, T_n$ are n independent, exponentially distributed random variables with parameters $\alpha_1, \alpha_2, \ldots, \alpha_n$, respectively. (Here, *independent* means that the distribution of each of these variables is independent of the distributions of all the others.) Furthermore, let $T_{\min}$ be the random variable representing the minimum of $T_1, T_2, \ldots, T_n$, that is, $T_{\min} = \min\{T_1, T_2, \ldots, T_n\}$. It can be asserted that $T_{\min}$ follows an exponential distribution with the parameter

$$\alpha = \sum_{i=1}^{n} \alpha_i$$

To interpret the result, assume that $T_1, T_2, \ldots, T_n$ represent the remaining service times for n jobs currently being served in n parallel servers operating independently of each other. $T_{\min}$ then represents the time remaining until the first of these n jobs has been fully serviced and can leave the service facility. Because $T_{\min}$ is exponentially distributed with parameter α, the implication is that currently, when all n servers are occupied, this multiple-server queuing system performs in the same way as a single-server system with service time $T_{\min}$. The result becomes even more transparent if we assume that the parallel servers are identical. This means that the remaining service times $T_1, T_2, \ldots, T_n$ are all exponential with parameter μ, that is, $\mu = \alpha_1 = \alpha_2 = \cdots = \alpha_n$, and consequently, $T_{\min}$ is exponential with parameter $\alpha = n\mu$. This result is very useful in the analysis of multiple-server systems.

In the context of interarrival times, this property implies that if we have a calling population with n customer types, all displaying exponentially distributed interarrival times but with different parameters $\alpha_1, \alpha_2, \ldots, \alpha_n$, the time between two arrivals is $T_{\min}$. Consequently, the arrival process of undifferentiated customers has interarrival times that are exponentially distributed with the parameter

$$\alpha = \sum_{i=1}^{n} \alpha_i$$

Consider, for example, a highway where connecting traffic arrives via an on-ramp. Assume that just before the on-ramp, the time between consecutive vehicles passing a given point on the highway is exponentially distributed with a parameter $\alpha_1 = 0.5$. Furthermore, assume that the time between consecutive vehicles on the ramp is exponentially distributed with parameter $\alpha_2 = 1$. After the two traffic flows have merged (i.e., after the on-ramp), the time between consecutive vehicles passing a point on the highway is exponentially distributed with a parameter $\alpha = \alpha_1 + \alpha_2 = 1.5$.

6.2.1.1 The Exponential Distribution, the Poisson Distribution, and the Poisson Process

Consider a simple queuing system, say a bank, and assume that the time T between consecutive customer arrivals is exponentially distributed with parameter λ. An important issue for the bank is to estimate how many customers might arrive during a certain time interval t. As it turns out, this number will be Poisson distributed with a mean value of λt. More precisely, let $X(t)$ represent the number of customers that has arrived by time t ($t \geq 0$). If one starts counting arrivals at time 0, then $X(t)$ is Poisson distributed with mean λt; this is often denoted $X(t) \in Po(\lambda t)$. Moreover, the probability that exactly n customers have arrived by time t is

$$P(X(t)=n)=\frac{(\lambda t)^n e^{-\lambda t}}{n!} \quad \text{for } n=0,1,2,\ldots$$

Note that if $n = 0$, the probability that no customers have arrived by time t is $P(X(t)=0)=e^{-\lambda t}$, which is equivalent to the probability that the arrival time of the first customer is greater than t, that is, $P(T>t)=e^{-\lambda t}$.

Every value of t has a corresponding random variable $X(t)$ that represents the cumulative number of customers that have arrived at the bank by time t. Consequently, the arrival process to the bank can be described in terms of this family of random variables $\{X(t);\, t \geq 0\}$. In general terms, such a family of random variables that evolves over time is referred to as a *stochastic* or *random process*. If, as in this case, the times between arrivals are independent, identically distributed, and exponential, $\{X(t);\, t \geq 0\}$ defines a Poisson process. It follows that for every value of t, $X(t)$ is Poisson distributed with mean λt. An important observation is that because λt is the mean number of arrivals during t time units, the average number of arrivals per time unit, or equivalently, the mean arrival rate, for this Poisson process is λ. This implies that the average time between arrivals is $1/\lambda$. This should come as no surprise because we started with the assumption that the interarrival time T is exponentially distributed with parameter λ or equivalently with mean $1/\lambda$. To illustrate this simple but sometimes confusing relationship between mean rates and mean times, consider an arrival (or counting) process with interarrival times that are exponentially distributed with a mean of 5 minutes. This means that the arrival process is Poisson, and on average, one arrival occurs every 5 minutes. More precisely, the cumulative number of arrivals describes a Poisson process with the arrival rate λ equal to 1/5 jobs per minute or 12 jobs per hour.

An interesting feature of the Poisson process (or, equivalently, the exponential distribution) is that aggregation or disaggregation results in new Poisson processes. Consider a simple queuing system, for example, the bank discussed earlier, with a calling population consisting of n different customer types, where each customer group displays a Poisson arrival process with arrival rate $\lambda_1, \lambda_2, \ldots, \lambda_n$, respectively. Assuming that all these Poisson processes are independent, the aggregated arrival process, where no distinction is made between customer types, is also a Poisson process but with mean arrival rate $\lambda = \lambda_1 + \lambda_2 + \cdots + \lambda_n$.

To illustrate the usefulness of Poisson process disaggregation, imagine a situation where the aforementioned bank has n different branch offices in the same area, and the arrival process to this cluster of branch offices is Poisson with the mean arrival rate λ. If every customer has the same probability p_i of choosing branch i, for $i = 1, 2, \ldots, n$, and every arriving customer must go somewhere, $\Sigma_{i=1}^{n} p_i = 1$, it can be shown that the arrival process to branch office j is Poisson with arrival rate $\lambda_j = p_j\lambda$. Consequently, the total Poisson arrival process has been disaggregated into branch office–specific Poisson arrival processes.

To conclude the discussion thus far, the exponential distribution is a reasonable way of describing interarrival times and service times in many situations. However, as with all modeling assumptions, it is important to verify their validity before using them to model and analyze a particular process design. To investigate the assumption's validity, data on interarrival and service times must be collected and analyzed statistically (see Chapter 9). For the remainder of this section on analytical queuing models, it is assumed that the interarrival and service times are exponentially distributed.

6.2.2 Terminology, Notation, and Little's Law Revisited

This section introduces some basic notation and terminology used in setting up and analyzing the queuing models to be investigated in Sections 6.2.3 through 6.2.8. An important concept is that of steady-state analysis as opposed to transient analysis. Also, Little's law (introduced in Chapter 5) will be revisited, and its applicability for obtaining average performance measures will be discussed.

When modeling the operation of a queuing system, the challenge is to capture the variability built into the queuing process. This variability means that we need some way to describe the different situations or states that the system will face. If one compares the deterministic models discussed in Chapters 4 and 5, the demand, activity times, and capacity were constant and known with certainty, resulting in only one specific situation or state that needs to be analyzed. As it turns out, a mathematically efficient way of describing the different situations facing a queuing system is to focus on the number of customers or jobs currently in the system (i.e., in the queue or in the service facility). Therefore, the state of a queuing system is defined as the number of customers in the system. This is intuitively appealing because when describing the current status of a queuing system, such as the checkout station in a supermarket, the first thing that comes to mind is how many customers are in that system—both in line and being served. It is important to distinguish between the queue length (number of customers or jobs in the queue) and the number of customers in the queuing system. The former represents the number of customers or jobs waiting to be served. It excludes the customers or jobs currently in the service facility. Defining the state of the system as the total number of customers or jobs in it makes it possible to model situations where the mean arrival and service rates, and the corresponding exponential distributions, depend on the number of customers currently in the system. For example, tellers might work faster when they see a long line of impatient customers, or balking and reneging behavior might be contingent on the number of customers in the system.

General notation:

$N(t)$ = The number of customers in the queuing system (in the queue and service facility) at time t ($t \geq 0$), or equivalently, the state of the system at time t.

$P_n(t)$ = The probability that there are exactly n customers in the queuing system at $t = P(N(t) = n)$

c = The number of parallel servers in the service facility.

λ_n = The mean arrival rate (i.e., the expected number of arrivals per time unit) of customers and jobs when there are n customers present in the system, that is, at state n.

μ_n = The mean overall service rate (i.e., the expected number of customers leaving the service facility per time unit) when there are n customers in the system.

In cases where λ_n is constant for all n, this constant arrival rate will be denoted λ. Similarly, in cases where the service facility consists of c identical servers in parallel, each with the same constant service rate, this constant rate will be denoted by μ. This service rate is valid only when the server is busy. Furthermore, note that from Property 3 of the exponential distribution discussed in Section 6.2.1, it is known that if k ($k \leq c$) servers are busy, the service rate for the entire service facility at this instance is $k\mu$. Consequently, the maximum service rate for the facility when all servers are working is $c\mu$. It follows that under constant arrival and service rates, $1/\lambda$ is the mean interarrival time between customers, and $1/\mu$ is the mean service time at each of the individual servers.

Another queuing characteristic related to λ and μ is the utilization factor, ρ, which describes the expected fraction of time that the service facility is busy. If the arrival rate is λ, and the service facility consists of one server with constant service rate μ, then the utilization factor for that server is $\rho = \lambda/\mu$. If c identical servers are in the service facility, the utilization factor for the entire service facility is $\rho = \lambda/c\mu$. The utilization factor can be interpreted as the ratio between the mean demand for capacity per time unit (λ) and the mean service capacity available per time unit ($c\mu$). To illustrate, consider a switchboard operator at a law firm. On average, 20 calls per hour come into the switchboard, and it takes the operator on average 2 minutes to figure out what the customer wants and forward the call to the right person in the firm. This means that the arrival rate is $\lambda = 20$ calls per hour, the service rate is $\mu = 60/2 = 30$ calls per hour, and the utilization factor for the switchboard operator is $\rho = 20/30 = 67\%$. In other words, on average, the operator is on the phone 67 percent of the time.

As reflected by the general notation, the probabilistic characteristics of the queuing system are in general dependent on the initial state and changes over time. More precisely, the probability of finding n customers or jobs in the system (or equivalently, that the system is in state n) changes over time and depends on how many customers or jobs were in the system initially. When the queuing system displays this behavior, it is said to be in a transient condition or transient state. Fortunately, as time goes by, the probability of finding the system in state n will in most cases stabilize and become independent of the initial state and the time elapsed (i.e., $\lim_{t \to \infty} P_n(t) = P_n$). When this happens, it is said that the system has reached a steady-state condition (or often just that it has reached steady state). Furthermore, the probability distribution $\{P_n; n = 0, 1, 2, \ldots\}$ is referred to as the *steady-state* (or *stationary*) *distribution*. Most results in queuing theory are based on analyzing systems in steady state. This so-called steady-state analysis is also the focus of this chapter. For a given queuing system, Figure 6.5 illustrates the transient and

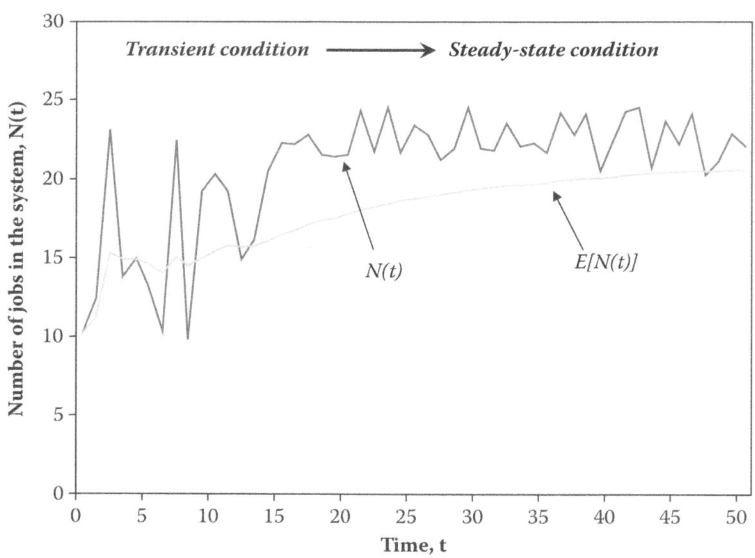

FIGURE 6.5
Illustration of transient and steady-state conditions for a given queuing process. $N(t)$ is the number of customers in the system at time t, and $E[N(t)]$ represents the expected number of customers in the system.

steady-state behavior of the number of customers in the system at time t, $N(t)$. It also depicts the expected number of customers in the system.

$$E[N(t)] = \sum_{n=0}^{\infty} nP_n(t)$$

(Recall that $P_n(t)$ is the probability of n customers in the system at time t, i.e., $P_n(t) = P(N(t) = n)$.)

Both $N(t)$ and $E[N(t)]$ change dramatically with the time t for small t values; that is, while the system is in a transient condition. However, as time goes by, their behavior stabilizes, and a steady-state condition is reached. Note that per definition, in steady state $P_n(t) = P_n$ for every state $n = 0, 1, 2, \ldots$. Consequently,

$$E[N(t)] = \sum_{n=0}^{\infty} nP_n$$

This implies that as a steady-state condition is approached, the expected number of customers in the system, $E[N(t)]$, should approach a constant value independent of time t. Figure 6.5 shows that this is indeed the case; as t grows large, $E[N(t)]$ stabilizes and approach a constant value of 21.5 customers. $N(t)$, on the other hand, will not converge to a constant value; it is a random variable for which a steady-state behavior is characterized by a stationary probability distribution (P_n, $n = 1, 2, 3, \ldots$), independent of the time t.

Due to the focus on steady-state analysis, a relevant question is whether all queuing systems are guaranteed to reach a steady-state condition. Furthermore, if this is not the case, how can one know which systems eventually will reach steady state? The answer to the first question is no; there may certainly be systems where the queue grows to an infinite size, implying that a steady-state condition is never reached. As for the second question, a sufficient criterion for the system to eventually reach a steady-state condition is that $\rho < 1$. In other words, if the mean demand for service is less than the mean available service capacity, it is guaranteed that the queue will not grow infinitely large. However, in situations where restrictions are placed on the queue length ($M/M/c/K$ in Section 6.2.6) or the calling population is finite ($M/M/c/\infty/N$ in Section 6.2.7), the queue can never grow to infinity. Consequently, these systems will always reach a steady-state condition even if $\rho > 1$.

For the steady-state analysis that will follow in Sections 6.2.3 through 6.2.7, the following notation will be used:

P_n = The probability of finding exactly n customers or jobs in the system or, equivalently, the probability that the system is in state n.

L = The expected number of customers or jobs in the system, including the queue and the service facility

L_q = The expected number of customers or jobs in the queue.

W = The expected time customers or jobs spend in the system, including waiting time in the queue and time spent in the service facility.

W_q = The expected time customers or jobs spend waiting in the queue before they get served.

Using the definition of *expected value*, the expected number of customers or jobs in the system can be determined as follows:

$$L = \sum_{n=0}^{\infty} nP_n$$

Furthermore, by recognizing that in a queuing system with c parallel servers, the queue will be empty as long as c or fewer customers are in the system, the expected number of customers in the queue can be obtained as follows:

$$L_q = \sum_{n=c}^{\infty} (n-c)P_n$$

To determine W and W_q, one can turn to an important relationship that is used extensively in steady-state analysis of queuing systems—namely, Little's law, which also was discussed in Chapter 5. In the following, this basic relationship will be recapitulated, and it will be shown how it can be extended to a more general situation with state-dependent arrival rates.

Assuming first that the mean arrival rate is constant and independent of the system state, $\lambda_n = \lambda$ for all n, Little's law states that

$$L = \lambda W \text{ and } L_q = \lambda W_q$$

A proof is found in Little (1961). For an intuitive explanation, think of a queuing system as a pipeline where customers arrive at one end, enter the pipeline (the system), work their way through it, and then emerge at the other end and leave. Now, consider a customer just before he or she leaves the pipeline (the system), which is such that customers cannot pass one another. Assume that 2 hours have gone by since the customer entered it. All the customers currently in the pipeline must have entered it during these 2 hours, so the current number of customers in the pipeline can be determined by counting how many have arrived during these 2 hours. Little's law builds on the same logic but states the relationships in terms of expected values: the mean number of customers in the system (or the queue) can be obtained as the mean number of arrivals per time unit multiplied by the mean time a customer spends in the system (or the queue).

If the arrival rate is state dependent, that is, λ_n is not the same for all n, Little's law still applies, but the average arrival rate $\bar{\lambda}$ has to be used instead of λ. The general expressions are then

$$L = \bar{\lambda} W \text{ and } L_q = \bar{\lambda} W_q$$

where

$$\bar{\lambda} = \sum_{n=0}^{\infty} \lambda_n P_n$$

Note that if $\lambda_n = \lambda$ for all n, it means that $\bar{\lambda} = \lambda$.

Another useful relationship in cases where the service rate at each server in the service facility is state independent and equal to μ is

$$W = W_q + \frac{1}{\mu}$$

This relationship follows since W_q is the expected time spent in the queue, $1/\mu$ is the expected service time, and W is the expected time spent in the queue and service facility together. An important observation is that by using this expression together with Little's law, it suffices to know $\bar{\lambda}$ and one of the mean performance measures L, L_q, W, and W_q, and all the others can be easily obtained.

6.2.3 Birth-and-Death Processes

All the queuing models considered in this chapter are based on the general birth-and death process. Although the name sounds ominous, in queuing, a birth simply represents the arrival of a customer or job to the queuing system, and a death represents the departure of a fully serviced customer or job from that same system. Still, queuing is just one of many applications for the general birth-and-death process, with one of the first being models of population growth, hence the name.

A birth-and-death process describes probabilistically how the number of customers in the queuing system, $N(t)$, evolves over time. Remember from Section 6.2.2 that $N(t)$ is the state of the system at time t. Focusing on the steady-state behavior, analyzing the birth-and-death process enables determination of the stationary probability distribution for finding 0, 1, 2 … customers in the queuing system. By using this stationary distribution, it is straightforward to determine mean performance measures such as the expected number of customers in the system, L, or in the queue, L_q. With Little's law, it is then easy to determine the corresponding average waiting times W and W_q. The stationary distribution also allows an evaluation of the likelihood of extreme situations to occur; for example, the probability of finding the queuing system empty or the probability that it contains more than a certain number of customers.

In simple terms, the birth-and-death process assumes that births (arrivals) and deaths (departures) occur randomly and independently of each other. However, the average number of births and deaths per time unit might depend on the state of the system (number of customers or jobs in the queuing system). A more precise definition of a birth-and-death process is based on the following four properties:

1. For each state, $N(t) = n$; $n = 0, 1, 2, \ldots$, the time remaining until the next birth (arrival), T_B, is exponentially distributed with parameter λ_n. (Remember that the state of a queuing system is defined by the number of customers or jobs in it.)
2. For each state, $N(t) = n$; $n = 0, 1, 2, \ldots$, the time remaining until the next death (service completion), T_D, is exponentially distributed with parameter μ_n.
3. The time remaining until the next birth, T_B, and the time remaining until the next death, T_D, are mutually independent.
4. For each state, $N(t) = n$; $n = 1, 2, \ldots$, the next event to occur is either a single birth, $n \longrightarrow n + 1$, or a single death, $n \longrightarrow n - 1$.

The birth-and-death process can be illustrated graphically in a so-called *rate diagram* (also referred to as a *state diagram*) (see Figure 6.6). In this diagram, the states (the number of customers in the system) are depicted as numbered circles or nodes, and the transitions between these states are indicated by arrows or directed arcs. For example, the state when zero customers are in the system is depicted by a node labeled 0. The rate at which the system moves from one state to the next is indicated in connection to the arc in question. Consider, for example, state n. The interarrival times and the service times are exponential with parameters λ_n and μ_n, respectively, so it is known from the properties of the exponential distribution that the mean arrival rate (or birth rate) must be λ_n, and the service rate (or death rate) is μ_n. This means that the process moves from state n to state $n - 1$ at rate μ_n and from state n to state $n + 1$ at rate λ_n. The rate diagram also illustrates the fact that the

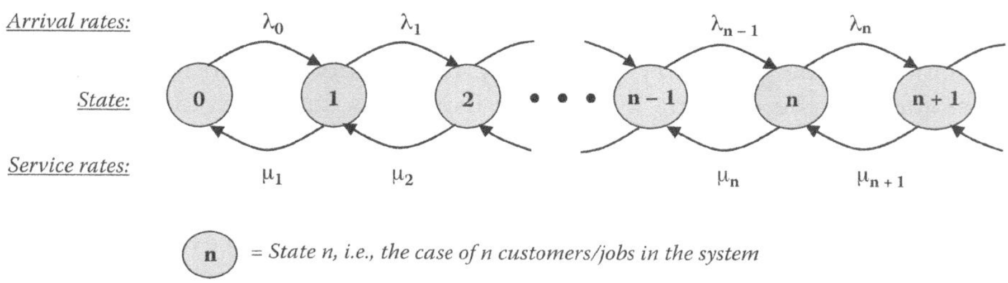

FIGURE 6.6
Rate diagram describing a general birth-and-death process.

birth-and-death process by definition allows transitions involving only one birth or death at a time ($n \longrightarrow n+1$ or $n \longrightarrow n-1$).

The rate diagram is an excellent tool for describing the queuing process conceptually. However, it is also useful for determining the steady-state probability distribution of finding 0, 1, 2, ... customers in the system $\{P_n; n = 0, 1, 2 \ldots\}$. A key observation in this respect is the so-called *rate in = rate out principle*, which states that when the birth-and-death process is in a steady-state condition, the expected rate of entering any given state n ($n = 1, 2, 3, \ldots$) is equal to the expected rate of leaving the state. To conceptualize this principle, think of each state n as a water reservoir, where the water level represents the probability P_n for the water level to remain constant (stationary); the average rate of water flowing into the reservoir must equal the average rate of water flowing out from it.

The Rate in = Rate out Principle

For every state n = 0, 1, 2, ..., the expected rate of entering the state = the expected rate of leaving the state. (A mathematical derivation of this result is available in Appendix 6A.)

The following example illustrates how the rate diagram and the rate in = rate out principle can be used to determine the stationary distribution for a queuing process.

Example 6.1: Queuing Analysis at TravelCall Inc.

TravelCall Inc. is a small travel agency that only services customers over the phone. Currently, only one travel agent is available to answer incoming customer calls. The switchboard can accommodate two calls on hold in addition to the one currently being answered by the travel agent. Customers who call when the switchboard is full get a busy signal and are turned away. Consequently, at most three customers can be found in the queuing system at any time. The calls arrive to the switchboard according to a Poisson process with a mean rate of nine calls per hour. The calls will be answered in the order they arrive; that is, the queuing discipline is FIFO. On average, it takes the travel agent 6 minutes to service a customer, and the service time follows an exponential distribution. The management of TravelCall wants to determine what the probability is of finding zero, one, two, or three customers in the system. If the probability of finding two or three customers in the system is high, it indicates that maybe the capacity should be increased.

The first observation is that the queuing system represented by TravelCall fits the aforementioned Criteria 1 through 4 that define a birth-and-death process. (Using the previously defined framework for classifying queuing models, it also represents an $M/M/1/3$ process). Second, it should be recognized that the objective is to determine the stationary probability distribution $\{P_n, n = 0, 1, 2, 3\}$ describing the probability of finding zero, one, two, or three customers in the system. Now that the system and the objective of the analysis are clear, it is time for some detailed calculations.

The first step in the numerical analysis is to decide on a time unit and to determine the mean arrival and service rates for each of the states of the system. The arrival rate and the service rate are constant and independent of the number of customers in the system. They are hereafter denoted by λ and μ, respectively. All rates are expressed in the number of occurrences per hour. However, it does not matter which time unit is used as long as it is used consistently. From the aforementioned data, it follows that $\lambda = 9$ calls per hour and $\mu = 60/6 = 10$ calls per hour.

Because the number of states, the mean arrival rate, and the service rate for each state are known, it is possible to construct the rate diagram (Figure 6.7) to describe the queuing process. For State 0, the mean rate at which the system moves to State 1 must be λ because this is the mean rate at which customers arrive to the system. On the other hand, the service rate for State 0 must be zero because at this state, the system is empty, and there are no customers who can leave. For State 1, the mean rate at which the system moves to State 2 is still λ because the mean arrival rate is unchanged. The service rate is now μ because this is the mean rate at which the one customer in the system will leave. For State 2, the mean arrival and service rates are the same as in State 1. At State 3, the system is full, and no new customers are allowed to enter. Consequently, the arrival rate for State 3 is zero. The service rate, on the other hand, is still μ because three customers are in the system, and the travel agent serves them with a mean service rate of μ customers per time unit.

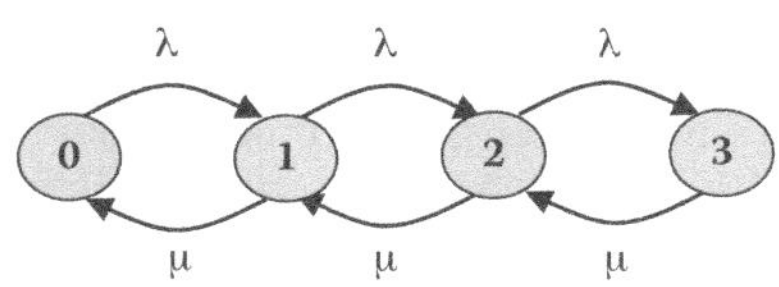

FIGURE 6.7
Rate diagram for the single-server queuing process at TravelCall.

To determine the probabilities P_0, P_1, P_2, and P_3 for finding zero, one, two, and three customers in the system, use the rate in = rate out principle applied to each state together with the necessary condition that all probabilities must sum to 1 ($P_0 + P_1 + P_2 + P_3 = 1$). The equation expressing the rate in = rate out principle for a given state n is often called the *balance equation* for State n. Starting with State 0, the balance equation is obtained by recognizing that the expected rate into State 0 is μ if the system is in State 1 and zero otherwise (see Figure 6.7). Because the probability that the system is in State 1 is P_1, the expected rate into State 0 must be μ times P_1 or μP_1. Note that P_1 can be interpreted as the fraction of the total time that the system spends in State 1. Similarly, the rate out of State 0 is λ if the system is in State 0 and zero otherwise (see Figure 6.7). Because the probability for the system to be in State 0 is P_0, the expected rate out of State 0 is λP_0. Consequently, we have

$$\text{Balance equation for State 0: } \mu P_1 = \mu P_0$$

Using the same logic for State 1, the rate into State 1 is λ if the system is in State 0 and μ if the system is in State 2 but zero otherwise (see Figure 6.7). The expected rate into State 1 can then be expressed as $\lambda P_0 + \mu P_2$. On the other hand, the rate out of State 1 is λ, going to State 2, and μ, going to State 0, if the system is in State 1 and zero otherwise. Consequently, the expected rate out of State 1 is $\lambda P_1 + \mu P_1$, and the corresponding balance equation can be expressed as follows:

$$\text{Balance equation for State 1}: \lambda P_0 + \mu P_2 = \lambda P_1 + \mu P_1$$

By proceeding in the same fashion for States 2 and 3, the complete set of balance equations summarized in Table 6.1 is obtained.

TABLE 6.1

Balance Equations for the Queuing Process at TravelCall

State	Balance Equations: Expected Rate in = Expected Rate Out
0	$\mu P_1 = \lambda P_0$
1	$\lambda P_0 + \mu P_2 = \lambda P_1 + \mu P_1$
2	$\lambda P_1 + \mu P_3 = \lambda P_2 + \mu P_2$
3	$\lambda P_2 = \mu P_3$

The balance equations represent a linear equation system that can be used to express the probabilities P_1, P_2, and P_3 as functions of P_0. (The reason why the system of equations does not have a unique solution is that we have four variables but only three linearly independent equations). To find the exact values of P_0, P_1, P_2, and P_3, we then use the condition that their sum must equal 1.

Starting with the balance equation for State 0, one finds that $P_1 = (\lambda/\mu)P_0$. Adding the equation for State 0 to that of State 1 results in $\lambda P_0 + \mu P_2 + \mu P_1 = \lambda P_1 + \mu P_1 + \lambda P_0$, which simplifies to $\mu P_2 = \lambda P_1$. This means that $P_2 = (\lambda/\mu)P_1$, and using the previous result that $P_1 = (\lambda/\mu)P_0$, one finds that $P_2 = (\lambda/\mu)^2 P_0$. In the same manner, by adding the simplified equation $\mu P_2 = \lambda P_1$ to the balance equation for State 2, one finds that $\lambda P_1 + \mu P_3 + \mu P_2 = \lambda P_2 + \mu P_2 + \lambda P_1$, which simplifies to $\mu P_3 = \lambda P_2$. Note that this equation is identical to the balance equation for State 3, which then becomes redundant. P_3 can now be expressed in P_0 as follows: $P_3 = (\lambda/\mu)P_2 = (\lambda/\mu)^3 P_0$. Using the expressions $P_1 = (\lambda/\mu)P_0$, $P_2 = (\lambda/\mu)^2 P_0$, and $P_3 = (\lambda/\mu)^3 P_0$, together with the condition that $P_0 + P_1 + P_2 + P_3 = 1$, the following is obtained:

$$P_0\left(1+\frac{\lambda}{\mu}+\frac{\lambda^2}{\mu^2}+\frac{\lambda^3}{\mu^3}\right)=1, \quad \text{equivalent to} \quad P_0\left(\frac{\mu^3+\mu^2\lambda+\mu\lambda^2+\lambda^3}{\mu^3}\right)=1,$$

which results in

$$P_0=\frac{\mu^3}{\mu^3+\mu^2\lambda+\mu\lambda^2+\lambda^3}$$

Because it is known that $\lambda = 9$ and $\mu = 10$, one can obtain $P_0 = 0.2908$. Because $P_1 = (\lambda/\mu)P_0$, $P_2 = (\lambda/\mu)^2 P_0$, and $P_3 = (\lambda/\mu)^3 P_0$, it follows that $P_1 = (9/10)P_0 \approx 0.2617$, $P_2 = (9/10)^2 P_0 \approx 0.2355$, and $P_3 = (9/10)^3 P_0 \approx 0.2120$.

These calculations predict that in steady state, the travel agent will be idle 29 percent of the time (P_0). This also represents the probability that an arbitrary caller does not have to spend time in the queue before talking to an agent. At the same time, it is now known that 21 percent of the time, the system is expected to be full (P_3). This means that all new callers get a busy signal, resulting in losses for the agency. Moreover, the probability that a customer who calls in will be put on hold is $P_1 + P_2 \approx 0.2617 + 0.2355 \approx 0.50$.

Although the explicit costs involved have not been considered, it seems reasonable that TravelCall would like to explore the possibilities of allowing more calls to be put on hold, that is, to allow a longer queue. In this way, they can increase the utilization of the travel agent and catch some customers who now take their business elsewhere. Another more drastic option to reduce the number of lost sales might be to hire one more travel agent. ■

The procedure for determining the stationary distribution of a general birth-and-death process with state-dependent arrival and service rates is analogous to the approach explained in Example 6.1. The balance equations are constructed in the same way, but their appearance is slightly different, as shown in Table 6.2. (See also the corresponding rate diagram in Figure 6.6.)

A complicating factor is that a large state space—one with many states to consider—means that it is tedious to find a solution following the approach in Example 6.1. For example, to handle the cases of infinite queue lengths, some additional mathematics are needed to obtain the necessary expressions. This is the motivation for studying the specialized models in Sections 6.2.4 through 6.2.7, where expressions are available for the steady-state

TABLE 6.2

Balance Equations for the General Birth-and-Death Process

State	Balance Equations: Expected Rate in = Expected Rate Out
0	$\mu_1 P_1 = \lambda_0 P_0$
1	$\lambda_0 P_0 + \mu_2 P_2 = \lambda_1 P_1 + \mu_1 P_1$
⋮	⋮
$n - 1$	$\lambda_{n-2} P_{n-2} + \mu_n P_n = \lambda_{n-1} P_{n-1} + \mu_{n-1} P_{n-1}$
n	$\lambda_{n-1} P_{n-1} + \mu_{n+1} P_{n+1} = \lambda_n P_n + \mu_n P_n$
⋮	⋮

probabilities and for certain important performance measures, such as average number of customers in the system and queue, as well as the corresponding waiting times. Still, for smaller systems, the general approach offers more flexibility.

As previously mentioned, interesting performance measures for queuing processes are the average number of customers or jobs in the system (L) and in the queue (L_q), as well as the corresponding average waiting times (W and W_q). If the stationary probability distribution $\{P_n; n = 1, 2, 3, \ldots\}$ is available, it is straightforward to obtain all these measures from the definition of expected value and Little's law. (See Section 6.2.2 for the general expressions.) Next is an explanation of how to do this in the context of TravelCall Inc.

Example 6.2: Determining Average Performance Measures at TravelCall

In Example 6.1, the steady-state probability distribution was determined for zero, one, two, and three customers in the queuing system describing the operations at TravelCall. The probabilities were $P_0 = 0.2908$, $P_1 = 0.2617$, $P_2 = 0.2355$, and $P_3 = 0.2120$. Although this is interesting information, TravelCall also wants know what it means in terms of average performance measures:

The expected number of customers in the system = L.
The expected number of customers waiting in the queue = L_q.
The expected time a customer spends in the system = W.
The expected time a customer spends in the queue before being served = W_q.

Because the probabilities of having zero, one, two, and three customers in the system are known, L can be calculated by applying the definition of an expected value as follows:

$$L = \sum_{n=0}^{\infty} nP_n = (0)P_1 + (1)P_1 + (2)P_2 + (3)P_3$$
$$= 0.2617 + (2)0.2355 + (3)0.2120 = 1.3687 \approx 1.37 \text{ customers}$$

To calculate L_q, proceed in a similar way but subtract the one customer that is being served in all those situations (states) where one or more customers are in the system. To clarify, when one customer is in the system, this customer is being served by the travel agent, so the queue is empty. If two customers are in the system, $n = 2$, then one is being served by the travel agent, and one is waiting in the queue. Finally, if three customers are in the system, then two of them are in the queue. The expected number of customers in the queue is obtained in the following way:

$$L_q = \sum_{n=1}^{\infty}(n-1)P_n = (1-1)P_1 + (2-1)P_2 + (3-1)P_3$$
$$= P_2 + (2)P_3 = 0.2355 + (2)0.2120 = 0.6595 = 0.66 \text{ customers}$$

To determine the mean time spent in the system, W, and the mean time spent in the queue, W_q, use Little's law as discussed in Section 6.2.2.

$$L = \bar{\lambda}W \quad \text{and} \quad L_q = \bar{\lambda}W_q, \quad \text{where } \bar{\lambda} = \sum_{n=0}^{\infty}\lambda_n P_n$$

First, determine the average arrival rate, which is not λ, since when three customers are in the system, no more customers can be accommodated, and the arrival rate for State 3 is effectively zero. For all other states, the arrival rate is λ (see Figure 6.7).

$$\bar{\lambda} = \sum_{n=0}^{3} P_n\lambda_n = \lambda P_0 + \lambda P_1 + \lambda P_2 + 0P_3 = \lambda(P_0 + P_1 + P_2) = \lambda(1 - P_3)$$
$$= 9(1 - 0.2120) = 7.0920 \approx 7.09 \text{ customers per hour}$$

W and W_q can now be obtained from Little's formula.

$$W = \frac{L}{\bar{\lambda}} = \frac{1.3687}{7.0920} = 0.1930 = 11.6 \text{ minutes}$$

$$W_q = \frac{L_q}{\bar{\lambda}} = \frac{0.6595}{7.0920} = 0.0930 \text{ hours} = 5.6 \text{ minutes}$$

Note that $W - W_q = 11.6 - 5.6 = 6$ minutes, which is the expected service time or $1/\mu$. With these average performance measures, TravelCall is in a better position to decide whether something needs to be done with respect to the system design. The waiting time of 5.6 minutes on the average might not seem long; however, for the customer waiting at the other end of the phone line, this might be perceived as a very long time and could lead to reneging (customer leaving the queue without before being served; see Section 6.1.1). ■

Thus far, the examples have assumed that the service facility has a constant mean service rate. To illustrate a more general model, Example 6.3 extends Examples 6.1 and 6.2 to a situation with two travel agents and state-dependent mean service rates. By relating back to the results in Examples 6.1 and 6.2, this example also shows how these types of queuing models can be used for better-informed design decisions.

Example 6.3: TravelCall Inc. Revisited

Based on the queuing analysis performed on the existing process in Examples 6.1 and 6.2, the management of TravelCall is contemplating hiring a second travel agent. However, before proceeding, they want to investigate the effects such a design change might have regarding the following:

The fraction of time (or equivalently, the probability) that the system is empty.
The fraction of time that the system is full and potentially loses customers.
The probability that a customer gets to talk to a travel agent immediately.
The probability that a customer will be put on hold.
The average performance measures L, L_q, W, and W_q.

The second travel agent is presumed to be just as efficient as the existing one, meaning that the service time for the new agent is also exponentially distributed with the parameter μ =10 customers per hour. The customer arrival rate is assumed to be unaffected; that is, Poisson distributed with mean rate $\lambda = 9$ customers per hour. Moreover, the travel agents work independently of each other and fully service one customer at a time. (In fact, TravelCall will encourage them to work from their homes and thereby reduce the overhead costs for office space, although still using a common switchboard.) As a result, the travel agency can be modeled as a queuing system with two parallel and identical servers. Recall from Section 6.1 that this means the mean service rate for the service facility is 2μ when both servers are busy and μ when only one server is busy. This observation is key when constructing the correct rate diagram shown in Figure 6.8. Starting in State 0, no customers are in the system, so the service rate is zero, and the arrival rate is λ. In State 1, one customer is in the system and is currently being served by one of the travel agents at the service rate μ. The arrival rate is unaffected and equals λ. In State 2, two customers are in the system, each being serviced by a travel agent. As discussed previously, the service rate for the service facility with the two parallel agents is then 2μ. The arrival rate is still λ. In State 3, three customers are in the system, one in the queue and two being serviced in parallel. The combined service rate for the two agents is, therefore, still 2μ. The arrival rate, on the other hand, is zero because the system is full, and any new arrivals are turned away.

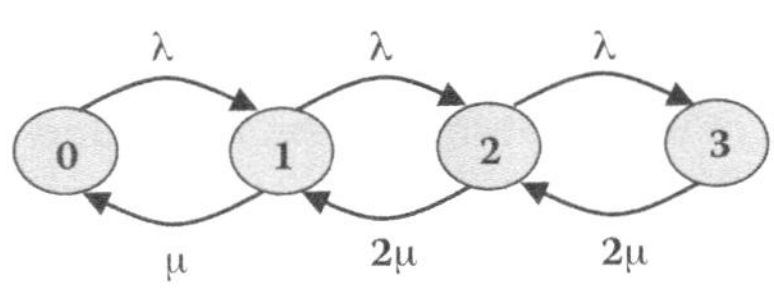

FIGURE 6.8
Rate diagram for the two-server queuing process at TravelCall. ■

TABLE 6.3

Balance Equations for the Two-Server Queuing Process at TravelCall

State	Balance Equations: Expected Rate in = Expected Rate Out
0	$\mu P_1 = \lambda P_0$
1	$\lambda P_0 + 2\mu P_2 = \lambda P_1 + \mu P_1$
2	$\lambda P_1 + 2\mu P_3 = \lambda P_2 + 2\mu P_2$
3	$\lambda P_2 = 2\mu P_3$

Based on the rate diagram, the balance equations can be formulated as in Table 6.3, ensuring that for each state the expected rate in equals the expected rate out.

Solving the equation system in the same way as described in Example 6.1, we obtain P_1, P_2, and P_3 expressed in P_0: $P_1 = (\lambda/\mu)P_0$, $P_2 = (\lambda^2/2\mu^2)P_0$, $P_3 = (\lambda^3/4\mu^3)P_0$. Using these expressions in conjunction with the condition that the steady-state probabilities must sum to 1, that is, the system must always be in one of the attainable states, $P_0 + P_1 + P_2 + P_3 = 1$, one gets the following:

$$P_0\left(1+\frac{\lambda}{\mu}+\frac{\lambda^2}{2\mu^2}+\frac{\lambda^3}{4\mu^3}\right)=1$$

equivalent to

$$P_0\left(\frac{4\mu^3+4\mu^2\lambda+2\mu\lambda^2+\lambda^3}{4\mu^3}\right)=1$$

which results in

$$P_0=\frac{4\mu^3}{4\mu^3+4\mu^2\lambda+2\mu\lambda^2+\lambda^3}.$$

With $\lambda = 9$ and $\mu = 10$, this means $P_0 = (1/2.48725) = 0.4021$, and consequently $P_1 = (\lambda/\mu)\ P_0 = (0.9)P_0 = 0.3618$, $P_2 = (\lambda^2/2\mu^2)P_0 = (0.9^2/2)P_0 = 0.1628$, and $P_3 = (\lambda^3/4\mu^3)P_0 = (0.9^3/4)P_0 = 0.0733$. It can then be concluded that in the new design

- The fraction of time that the system is empty = $P_0 \approx 40\%$.
- The fraction of time that the system is full and potentially loses customers = $P_3 \approx 7\%$.
- The probability that a customer gets to talk to a travel agent immediately = $P_0 + P_1 \approx 76\%$.
- The probability that a customer will be put on hold = $P_2 \approx 16\%$.

With a steady-state probability distribution in place, the expected number of customers in the system and in the queue, L and L_q, respectively, can be determined. (When comparing the following expressions with the general formulae, note that in the present case, $c = 2$ and $P_n = 0$ for $n > 3$.)

$$L = \sum_{n=0}^{\infty} nP_n = (0)P_0 + (1)P_1 + (2)P_2 + (3)P_3$$
$$= 0.3618 + (2)0.1628 + (3)0.0733$$
$$= 0.9073 \approx 0.91 \text{ customers}$$

$$L_q = \sum_{n=2}^{\infty} (n-2)P_n = (2-2)P_2 + (3-2)P_3$$
$$= (1)0.0733 \approx 0.07 \text{ customers}$$

The expected time in the system, W, and in the queue, W_q, can now be obtained from Little's law.

$$\bar{\lambda} = \sum_{n=0}^{\infty} \lambda_n P_n = \lambda P_0 + \lambda P_1 + \lambda P_2 + 0P_3 = \lambda(P_0 + P_1 + P_2)$$
$$= 9(0.4021 + 0.3618 + 0.1628) = 8.3403$$
$$= 8.34 \text{ customers per hour}$$

$$W = \frac{L}{\bar{\lambda}} = \frac{0.9073}{8.3403} = 0.1088 \text{ hours} = 6.5 \text{ minutes}$$

$$W_q = \frac{L_q}{\bar{\lambda}} = \frac{0.0733}{8.3403} = 0.0088 \text{ hours} = 0.5 \text{ minutes}$$ ■

Table 6.4 contrasts the results obtained for the system with two travel agents in Example 6.3 with those obtained for the current single-server system evaluated in Examples 6.1 and 6.2.

Hiring a second agent dramatically increases the customer service levels. For example, the probability that a customer will get immediate access to an agent goes up from 29 percent in the single-server system to 76 percent when two travel agents are working the phones. At the same time, the average waiting time on hold decreases from 5.6 minutes in the current single-server process to 0.5 minutes in the new design. Moreover, in terms

TABLE 6.4

Comparison of the Process Performance when There Are Two Travel Agents at TravelCall Instead of One

	System Design	
Performance Characteristic	**One Travel Agent**	**Two Travel Agents**
Fraction of time that the system is empty	29%	40%
Fraction of time that the system is full	21%	7%
Probability that a customer gets immediate access to a travel agent	29%	76%
Probability that a customer will be put on hold	50%	16%
Expected number of customers in the system (L)	1.37 customers	0.91 customers
Expected number of customers in the queue (L_q)	0.66 customers	0.07 customers
Expected time spent in the system (W)	11.6 minutes	6.5 minutes
Expected waiting time in the queue (W_q)	5.6 minutes	0.5 minutes

of potentially lost sales, customers in the two-server system encounter a busy signal only 7 percent of the time compared with 21 percent in the single-server system.

This improved service comes at the expense of lower resource utilization, indicated by the fact that the fraction of time the system is empty goes up from 29 percent in the single-server system to 40 percent in the system with two agents. Hiring the second travel agent also increases the operational costs significantly. To make an informed decision on whether to pursue the new design, these service costs must be quantified and compared against the expected revenue increase due to the improved customer service. Models dealing with these issues will be examined in Section 6.2.8.

Hiring a second agent is just one possibility for changing the process design. Another option is to allow customers to contact the travel agent via other channels, such as through a website or e-mail. In this way, the existing travel agent can work with these requests between phone calls. Most likely, it also will lead to fewer customers calling, implying shorter waiting times for telephone customers and a lower risk of encountering a busy signal. A further analysis of this option is beyond the scope of this example but illustrates the potential advantages of thinking about doing things in new ways. Another example of creative design ideas, which relates to the design principle of treating geographically dispersed resources as if they were centralized, as discussed in Section 3.6.2, is letting the travel agents work from their homes.

Examples 6.1 through 6.3 assume that customers who have the opportunity to enter the queue also will choose to join it. However, some

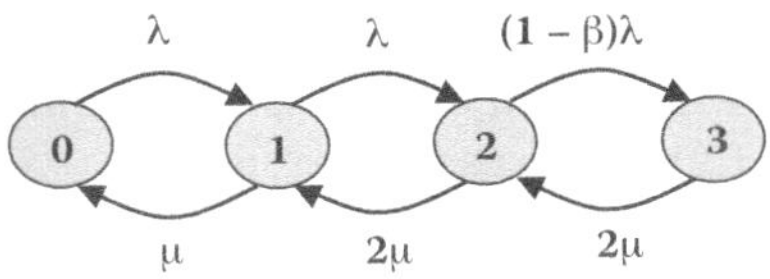

FIGURE 6.9
Rate diagram for the two-server queuing process at TravelCall if customers balk with probability β when they do not get immediate access to a travel agent.

customers, after realizing that they will have to wait before getting served, will choose to leave; in other words, they will balk. Balking can be modeled as state-dependent arrival rates using the fact that Poisson processes can be disaggregated into new Poisson processes, as seen in Section 6.2.1. Assume, for example, that customers calling TravelCall in Example 6.3 will hang up with probability β if they don't get immediate access to a travel agent. This means that as soon as both travel agents are busy, the arrival process will be Poisson with mean arrival rate $(1-\beta)\lambda$ as opposed to a mean arrival rate of λ for the other states. Figure 6.9 shows the corresponding rate diagram. It might be hard to assess an accurate value of β because customers who hang up are not registered. Most likely, a market research effort would have to be launched to explore and better understand customer attitudes.

Given the rate diagram in Figure 6.9, the model with balking can be analyzed using the same approach as in Examples 6.1 through 6.3.

The discussion on balking and state-dependent arrival rates concludes this investigation of the general birth-and-death process. Sections 6.2.4 through 6.2.7 focus on some important birth-and-death processes with specialized structures that are often used in queuing analysis. Their importance stems from their wide applicability and the expressions that make it straightforward to evaluate important performance measures quickly. Still, it is important to remember that all of these models are specialized birth-and-death processes that can be analyzed using the general approach explored in this section. It is also worth noting that this approach may also be used for analyzing more general continuous time Markov chains where state transitions are not restricted to a single birth ($n \longrightarrow n+1$) or a single death ($n \longrightarrow n-1$).

6.2.4 The *M/M/1* Model

The *M/M/1* model is a special case of the general birth-and-death process described in Section 6.2.3. Because of its simplicity, it has found applications in a wide range of areas, and it is one of the most commonly used queuing models. It describes a queuing system with a single server, exponentially distributed service and interarrival times, constant arrival and service rates, an infinite calling population, and no restrictions on the queue length. More precisely, going back to the major elements of the basic queuing process, the model is based on the following assumptions:

- *Calling population*: Jobs arrive from such a large population that it can be considered infinite for modeling purposes. The jobs are independent of each other; that is, the arrival of one job does not influence the arrival of others. Also, the arrival of jobs is not influenced by the system (e.g., there are no scheduled appointments).
- *Arrival process*: The interarrival times are exponentially distributed with a constant parameter λ, independent of the number of customers in the system. In other words, the arrival process is Poisson with the mean rate λ, independent of the state of the system.
- *Queue configuration*: A single waiting line is used with no restrictions on length. The model does not consider balking or reneging.
- *Queue discipline*: The queue is managed using a FIFO rule.
- *Service mechanism*: The service facility consists of a single server, whose service times are exponentially distributed with a constant parameter μ, independent of the number of customers in the system. Consequently, the mean service rate is μ customers per time unit for all states of the system.

The *M*/*M*/1 process can be illustrated by the rate diagram in Figure 6.10. Recall from Section 6.2.3 that in this diagram, the states, that is, the number of customers in the system, are depicted as numbered circles or nodes (node 0 indicates State 0 with zero customers in the system), and the transitions between these states are indicated by arrows or directed arcs. Furthermore, the rate at which the system moves from one state to the next is displayed in connection with each arc. As discussed previously, customers or jobs arrive to the system at the mean rate λ, independently of the number of customers currently in the system. Consequently, the rate at which the process moves from State n to State $n + 1$ is λ for all $n = 0, 1, 2, \ldots$. Similarly, the mean service

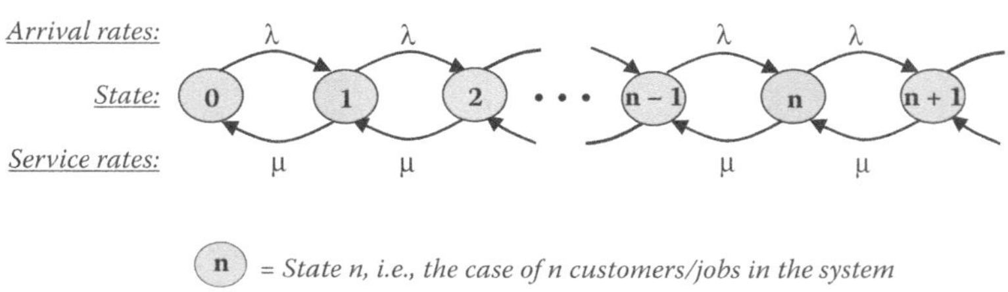

FIGURE 6.10
Rate diagram for the *M*/*M*/1 model.

rate at which the server services customers is μ for all states where there is someone to serve; that is, all states except State 0, when the system is empty, and the service rate is zero. As a result, the rate at which the process moves from State n to $n - 1$ is μ for all $n = 1, 2, \ldots$.

For the steady-state analysis of the $M/M/1$ model, the following notation, first introduced in Section 6.2.2, is used.

P_n = The probability of finding exactly n customers or jobs in the system, or equivalently, the probability that the system is in state n.

λ = The mean arrival rate (the mean number of arrivals per time unit); the same for all system states.

μ = The mean service rate (the mean number of service completions per time units in a busy server); constant for all relevant system states.

$\rho = \lambda/\mu$ = utilization factor.

L = The expected number of customers or jobs in the system including the queue and the service facility.

L_q = The expected number of customers or jobs in the queue.

W = The expected time customers or jobs spend in the system including waiting time in the queue and time spent in the service facility.

W_q = The expected time customers or jobs spend waiting in the queue before they get served.

A necessary and sufficient condition for the $M/M/1$ process to reach steady state—that is, for the existence of a steady-state probability distribution $\{P_n, n = 0, 1, 2, \ldots\}$—is that the mean arrival rate λ is strictly less than the mean service rate μ or equivalently $\rho < 1$. The analysis and expressions that will be explored in this section therefore assume that $\rho < 1$. If $\rho \geq 1$, it means that the average demand for service equals or exceeds the average service capacity, and the queue will grow infinitely large as the time progresses.

The steady-state operating characteristics of the $M/M/1$ system can be calculated from the following set of expressions. (The expressions can be derived using the general approach for analyzing birth-and-death processes in Section 6.2.3. The mathematical derivations are available in Appendix 6A.)

Probability of zero jobs in the system: $P_0 = 1 - \rho$.

Probability that the number of jobs in the system is exactly n: $P_n = P_0\rho^n = (1 - \rho)\rho^n$.

Probability that the number of jobs in the system is k or more: $P(n \geq k) = \rho^k$.

Probability that the number of jobs in the system is at least one: $P(n \geq 1) = 1 - P_0 = \rho$.

Because $P(n \geq 1)$ also represents the probability that the server is busy, it explains why ρ is known as the utilization factor for the server in this model.

Mean number of jobs in the system: $L = \dfrac{\rho}{1-\rho} = \dfrac{\lambda}{\mu-\lambda}$

Mean number of jobs in the queue: $L_q = L - \rho = \dfrac{\rho^2}{1-\rho} = \dfrac{\rho\lambda}{\mu-\lambda}$

Mean time a job spends in the system: $W = \dfrac{L}{\lambda} = \dfrac{1}{\mu-\lambda}$

Mean time a job spends in the queue: $W_q = \dfrac{L_q}{\lambda} = \dfrac{\rho}{\mu-\lambda}$

Example 6.4 illustrates how the previous expressions can be used to analyze a business process that can be modeled as an $M/M/1$ system.

Example 6.4: Policy Requests at an Insurance Company

An insurance company receives an average of 40 requests for new auto policies per week. Through statistical analysis, the company has been able to determine that the time between two consecutive requests arriving to the process is approximately exponentially distributed. A single team handles the requests and is able to complete, on average, 50 requests per week. The actual times to complete requests also are considered to follow an exponential distribution. The requests have no particular priority; therefore, they are handled on an FCFS basis. It also can be assumed that requests are not withdrawn.

What is the probability that an arriving request must wait for processing? That is, what is the probability that the team will be busy working on a request when a new request arrives?

Because $\lambda = 40$ and $\mu = 50$, then $\rho = 40/50 = 0.8$, and $P(n \geq 1) = \rho^1 = (0.8)^1 = 0.8$.

What is the probability that an arriving request does not have to wait? That is, what is the probability that the arriving request will find the team idle?

$$P_0 = 1 - \rho = 1 - 0.8 = 0.2$$

What is the expected number of requests in the process?

$$L = \frac{\lambda}{\mu-\lambda} = \frac{40}{50-40} = 4 \text{ requests}$$

What is the expected number of requests in the queue?

$$L_q = \frac{\rho\lambda}{\mu - \lambda} = \frac{(0.8)(40)}{50 - 40} = 3.2 \text{ requests}$$

What is the average time that a request spends in the process? In other words, what is the average cycle time? (It is assumed that a week has 5 working days and that each day has 8 hours.)

$$W = \frac{1}{\mu - \lambda} = \frac{1}{50 - 40} = 0.1 \text{ week} = 0.5 \text{ days} = 4 \text{ hours}$$

What is the average waiting time?

$$W_q = \frac{\rho}{\mu - \lambda} = \frac{0.8}{50 - 40} = 0.08 \text{ week} = 0.4 \text{ days} = 3.2 \text{ hours}$$

From these calculations, it can be predicted that the team will be busy, in the long run, 80 percent of the time. Therefore, 80 percent of the time, an arriving request will have to join the queue instead of receiving immediate attention. The average cycle time is estimated to be 4 hours, of which 3.2 hours are expected to be waiting time. This calculation reveals that only 20 percent of a request's cycle time is spent in value-adding activities, and the other 80 percent consists of a non-value-adding waiting time.

Another operating characteristic that can be calculated is the probability that a given number of jobs are in the process (either waiting or being served) at any point in time. This calculation is performed using the equation for P_n, where n is the number of jobs in the process. Figure 6.11 shows a plot of the probability values associated with a range of n-values from 0 to 20.

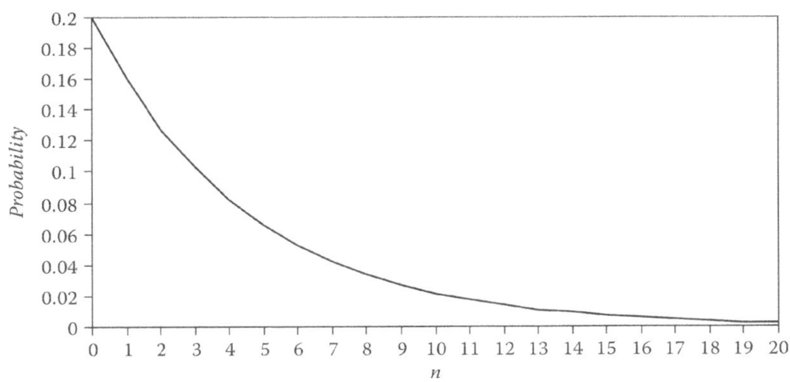

FIGURE 6.11
Probability of n jobs in an $M/M/1$ system with $\rho = 0.8$.

Figure 6.11 shows that the probability of finding n jobs in the process significantly decreases for $n \geq 10$. In fact, the exact value for the probability of finding 10 jobs or more in the process is $P(n \geq k) = \rho^k = (0.8)^{10} = 0.1074$. This means that the probability of finding fewer than 10 jobs in the process is almost 90 percent. ■

In some processes where space needs to be provided for jobs in the queue, the ability to estimate the probability of the number of jobs in the system is critical. For example, suppose that technical support for a popular piece of software is provided via a toll-free telephone number. The process designer needs to know the expected number of customers that typically would be put on hold to provide the necessary "space" in the telephone system. In the same way, it is important to estimate the waiting time that customers would experience with different levels of staffing. These and other extensions to the $M/M/1$ model will be explored in Sections 6.2.5 through 6.2.7.

Remark: Some steady-state results for the more general $M/G/1$ model with service times that can take on any distribution are available in Appendix 6A.

6.2.5 The *M/M/c* Model

The $M/M/c$ model is a generalization of the $M/M/1$ model that allows for multiple parallel servers in the service facility. More precisely, the added feature is that the service facility can consist of c identical parallel servers each with a constant service rate, μ, and exponentially distributed service times working independently of each other. Still, with the exception of the server configuration, the assumptions defining the $M/M/c$ model are the same as for the $M/M/1$ model, implying that the $M/M/c$ model also represents a birth-and-death process. Figure 6.12 describes the $M/M/c$ process in terms of its rate diagram. Note that for State 1, 2, ..., c, all customers or jobs in the system will be under service, and the queue will be empty. This means that these states will have 1, 2, ..., c independent servers working in parallel, each with the service rate μ. From Property 3 of the exponential distribution discussed in Section 6.2.1, it can be concluded that the service rates for the service facility in state 0, 1, 2, ..., c are 0, μ, 2μ, ..., $c\mu$, respectively. For states $c + 1$, $c + 2$, ..., all c servers are busy, and the service rate for the service facility is $c\mu$, so the additional customers have to wait in the queue. The arrival rate is independent of the system state and equal to λ.

In order for the $M/M/c$ model to reach steady state, the mean arrival rate λ must be less than the mean service rate when all c servers are busy, that is, $c\mu$. Because the utilization factor of the service facility ρ is defined as $\rho = (\lambda/c\mu)$

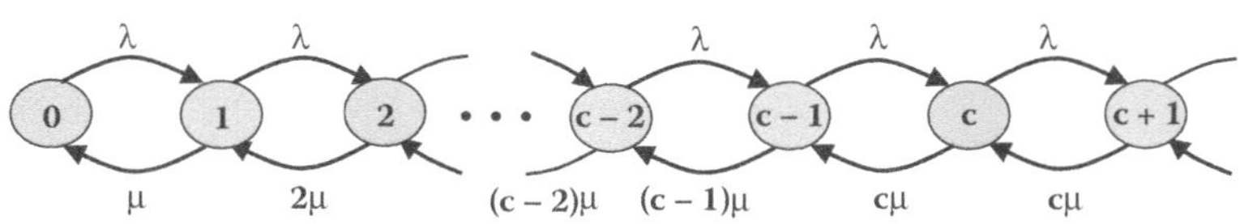

FIGURE 6.12
Rate diagram for the $M/M/c$ model.

(see Section 6.2.2), the following relationship must hold for the $M/M/c$ queue to reach steady state:

$$\rho = \frac{\lambda}{c\mu} < 1$$

In other words, the utilization of the service facility, ρ, must be less than 100 percent for the $M/M/c$ system to reach steady state; otherwise, the queue will explode in size and grow infinitely large.

As with the $M/M/1$ process, a set of expressions can be identified for evaluating the operating characteristics of this system. See, for example, Kleinrock (1975) and Hillier and Lieberman (2010). These equations can be derived from the approach used for analyzing the general birth-and-death process, but the algebra poses more of a challenge. Therefore, the mathematical derivations of the following expressions are omitted.

Probability of zero jobs in the system: $P_0 = \dfrac{1}{\sum_{n=0}^{c-1} \dfrac{(\lambda/\mu)^n}{n!} + \dfrac{(\lambda/\mu)^c}{c!(1-\rho)}}$

Probability that all servers are busy: $P(n \geq c) = \dfrac{(\lambda/\mu)^c}{c!(1-\rho)} P_0$

Probability of exactly n jobs or customers in the system:

$$P_n = \begin{cases} \dfrac{(\lambda/\mu)^n}{n!} P_0 & \text{for } n = 1,2,\ldots,c \\ \dfrac{(\lambda/\mu)^n}{c!c^{n-c}} P_0 & \text{for } n = c, c+1, \ldots; \end{cases}$$

Mean number of jobs in the system: $L = \dfrac{(\lambda/\mu)^c \rho}{c!(1-\rho)^2} P_0 + \dfrac{\lambda}{\mu}$

Mean number of jobs in the queue: $L_q = \dfrac{(\lambda/\mu)^c \rho}{c!(1-\rho)^2} P_0 = L - \dfrac{\lambda}{\mu}$

Note that if the steady-state criterion $\rho < 1$ is violated and $\rho = 1$, L and L_q will become infinite because their expressions include a division with zero.

Mean time a job spends in the system: $W = \dfrac{L}{\lambda} = \dfrac{L_q}{\lambda} + \dfrac{1}{\mu}$

Mean time a job spends in the queue: $W_q = \dfrac{L_q}{\lambda}$

A useful observation is that when L_q is known, we can easily find the expressions for W_q, W, and L using Little's law and the relationship $W = W_q + (1/\mu)$

discussed in Section 6.2.2. The expressions are $W_q = (L_q/\lambda)$; $W = W_q + (1/\mu)$; and $L = \lambda W = \lambda(W_q + 1/\mu) = L_q + \lambda/\mu$.

To illustrate how these formulae can be used to analyze a queuing system with parallel servers, Example 6.5 extends Example 6.4 to a situation with two servers.

Example 6.5: Policy Requests at an Insurance Company Revisited

Suppose the insurance company in Example 6.4 contemplates hiring a second team. However, before making a decision, they want to investigate the effects this will have on several important operating characteristics. A realistic assumption is that the new team will be just as efficient as the existing team, and for modeling purposes, they can be treated as parallel servers with identical service rates. Because the only difference from the situation described in Example 6.4 is that the new process will have two independent teams handling the arriving requests, the new process conforms to an $M/M/c$ queuing model with $c = 2$.

Recall from Example 6.4 that $\lambda = 40$ requests per week and $\mu = 50$. The new system has two parallel servers, so $c = 2$. The corresponding utilization factor for the service facility is then

$$\rho = \frac{\lambda}{c\mu} = \frac{40}{2(50)} = 0.4$$

Also note that just as in Example 6.4, $(\lambda/\mu) = 0.8$.

The first question posed by the insurance company concerns the probability that an arriving request must wait for processing. That is, what is the probability that both teams will be busy working on requests when a new request arrives?

To provide an answer, first calculate the probability of zero requests in the system. (Recall that $n! = n(n-1)(n-2)\cdots(2)(1)$ where by definition $0! = 1$.)

$$P_0 = \frac{1}{\sum_{n=0}^{c-1}\frac{(\lambda/\mu)^n}{n!} + \frac{(\lambda/\mu)^c}{c!(1-\rho)}} = \frac{1}{\sum_{n=0}^{1}\frac{(0.8)^n}{n!} + \frac{(0.8)^2}{2!(1-0.4)}}$$

$$= \frac{1}{1.8 + \frac{0.64}{1.2}} \approx 0.4286$$

Then for $c = 2$, $P(n \ge c)$ is calculated as follows:

$$P(n \ge c) = \frac{(\lambda/\mu)^c}{c!(1-\rho)}P_0 = \frac{(0.8)^2}{2!(1-0.4)}(0.4286)$$

$$\approx 0.2286 \approx 0.23$$

A related question concerns the probability that an arriving request gets processed immediately. That is, what is the probability that the arriving request will find at least one team idle?

$$P(n<2)=P_0+P_1=P_0+\frac{(\lambda/\mu)^1}{1!}P_0$$

$$=0.4286+\frac{(0.8)^1}{1!}0.4286=0.4286(1+0.8)\approx 0.77$$

Note that this probability is the complement of the probability of finding two or more requests in the process. Mathematically, this means that $P(n < 2) = 1 - P(n \geq 2) \approx 1 - 0.23 = 0.77$.

An important operating characteristic is the average work in process (WIP) or equivalently: what is the expected number of requests in the process?

$$L=\frac{(\lambda/\mu)^c\rho}{c!(1-\rho)^2}P_0+\frac{\lambda}{\mu}=\frac{(0.8)^2\,0.4}{2!(1-0.4)^2}(0.4286)+0.8\approx 0.95 \text{ requests}$$

A related performance measure is the expected number of requests in the queue.

$$L_q=L-(\lambda/\mu)=0.95-0.8=0.15 \text{ requests}$$

Finally, we need to analyze the time measures W and W_q. The first question is about the average time that a request spends in the process. In other words, what is the average cycle time W? (As before, we assume that a week has 5 working days and that each day has 8 hours.)

$$W=\frac{L}{\lambda}=\frac{L_q}{\lambda}+\frac{1}{\mu}=\frac{0.15}{40}+\frac{1}{50}=0.02375 \text{ weeks}$$
$$=0.11875 \text{ days}=0.95 \text{ hours}$$

Second, what is the average waiting time for a job before it is processed? That is, what is the value of W_q?

$$W_q=\frac{L_q}{\lambda}=\frac{0.15}{40}=0.00375 \text{ weeks}=0.01875 \text{ days}=0.15 \text{ hours}$$

To summarize the results, the average utilization of the two teams is ρ = 40 percent. The average cycle time is estimated to be 57 minutes (i.e., 0.95 hours), of which only 9 minutes are expected to be waiting time. The cycle time reduction is significant when compared with the 4 hours achieved with a single server. This reduction translates into 84.2 percent (i.e., 48/57) of value-adding activity time and 15.8 percent (i.e., 9/57) of non-value-adding activity time in the two-server system. In the single-server system, only 20 percent of a request's cycle time is spent in value-adding activities, and the other 80 percent consists of non-value-adding waiting time.

The decision about whether to increase the service capacity must be made based on the benefits associated with reducing non-value-adding time (in this case, waiting time) and the cost of achieving the reduction. Going from a one-server system to a configuration with two teams working in parallel reduces the total expected cycle time from 4 to 0.95 hours. However, at the same time, the utilization of the servers drops from 80 percent to 40 percent when the number of servers is increased. To assess this economic trade-off, the operating characteristics need to be translated into monetary terms, as explored in Section 6.2.8. ■

Remark: Steady-state results for the $M/G/\infty$ model with an infinite number of servers and generally distributed service times are available in Appendix 6A.

Next, two generalizations of the $M/M/c$ model will be considered: allowing limitations on the queue length and the size of the calling population.

6.2.6 The *M/M/c/K* Model

The $M/M/c/K$ model imposes the restriction that at most K customers can be in the $M/M/c$ system at the same time. This means that the system is full when K customers are in it, and customers arriving during this time are rejected. This type of queuing model is applicable in situations when the queuing system has a limited capacity to harbor customers. Examples include a limitation on the number of beds in a hospital, the size of a switchboard at a call center, the size of the waiting room at a barbershop, or the size of a car wash, a computer network, or a parking lot. The $M/M/c/K$ model also addresses the balking behavior of customers who choose to go elsewhere if they arrive when K customers are already in the system.

Apart from the constraint of no more than K customers in the system, the $M/M/c/K$ model is based on the same assumptions as the $M/M/c$ model: a single queue, exponential service and interarrival times, FIFO queuing discipline, and c identical servers in parallel working independently of one another, each with a mean service rate of μ. The difference is that because no more than K customers are allowed in the system, the arrival rate is zero for States $K, K+1, K+2, \ldots$, while it is λ for States $0, 1, 2, \ldots, K-1$. This is illustrated in the rate diagram for the $M/M/c/K$ model in Figure 6.13. It follows

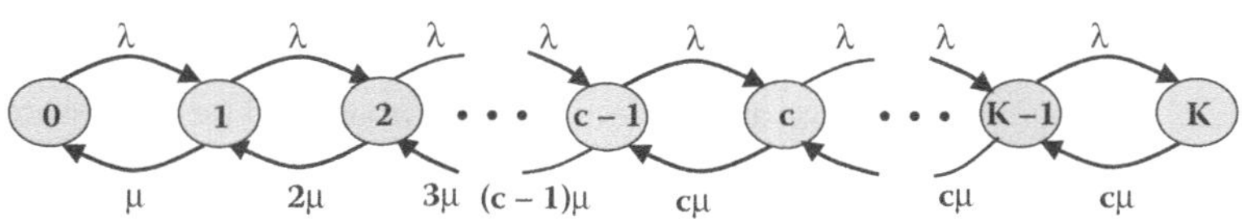

FIGURE 6.13
Rate diagram for the $M/M/c/K$ model.

that the $M/M/c/K$ model is a birth-and-death process with state-dependent service and interarrival rates.

$$\lambda_n = \begin{cases} \lambda & \text{for } n = 0,1,2,\ldots K \\ 0 & \text{for } n \geq K \end{cases} \quad \text{and} \quad \mu_n = \begin{cases} n\mu & \text{for } n = 0,1,2,\ldots,c \\ c\mu & \text{for } n \geq c \end{cases}$$

The queuing models analyzed in Examples 6.1 through 6.3 fall into the class of $M/M/c/K$ models.

An important observation is that the $M/M/c/K$ queue is guaranteed to reach steady state even if $\rho = (\lambda/c\mu) \geq 1$. This is so because the queue can never grow infinitely large. As a matter of fact, it can never be larger than $K - c$ customers or jobs (or equivalently, the number of customers in the system cannot exceed K, as shown in Figure 6.13). In this model, at most K servers can be busy at the same time. Therefore, without loss of generality, the following exposition assumes that $c \leq K$.

As in the case of $M/M/1$ and $M/M/c$ models, a set of expressions exists for calculating important performance characteristics for the $M/M/c/K$ model. Although the approach for analyzing general birth-and-death processes could be used, algebraically, it can be tedious for large K values. Therefore, the following expressions can be used to characterize the performance of $M/M/c/K$ systems. In addition to the assumption that $c \leq K$, the expressions use that $\rho = (\lambda/c\mu)$.

Probability of zero jobs in the system: $P_0 = \dfrac{1}{\sum_{n=0}^{c} \dfrac{(\lambda/\mu)^n}{n!} + \dfrac{(\lambda/\mu)^c}{c!} \sum_{n=c+1}^{K} \rho^{n-c}}$

Probability of exactly n jobs in the system:

$$P_n = \begin{cases} \dfrac{(\lambda/\mu)^n}{n!} P_0 & \text{for } n = 1,2,\ldots,c \\ \dfrac{(\lambda/\mu)^n}{c!c^{n-c}} P_0 & \text{for } n = c, c+1, \ldots, K \\ 0 & \text{for } n > K \end{cases}$$

Mean number of jobs in the queue:

$$L_q = \frac{(\lambda/\mu)^c \rho}{c!(1-\rho)^2} P_0 \left(1 - \rho^{K-c} - (K-c)\rho^{K-c}(1-\rho)\right)$$

Note that the expression for L_q is not valid for $\rho = 1$. However, this does not imply that steady state is never reached for $\rho = 1$. It is simply a result of how the expression is derived. The case of $\rho = 1$ can be handled separately, but this special case is not shown here.

Mean number of jobs or customers in the system:

$$L = L_q + \sum_{n=0}^{c-1} nP_n + c\left(1 - \sum_{n=0}^{c-1} P_n\right)$$

Observe that the last two terms in the expression for L represent the expected number of jobs in the service facility. In other words, the expression corresponds to the sum of the average number of jobs in the queue and the average number of jobs in the service facility.

Applying Little's law, we can now easily obtain the expected time spent in the system and in the queue. The average arrival rate in this case is

$$\bar{\lambda} = \sum_{n=0}^{\infty} \lambda_n P_n = \sum_{n=0}^{K-1} \lambda P_n = \lambda \sum_{n=0}^{K-1} P_n = \lambda(1 - P_K)$$

Mean time a job or a customer spends in the system:

$$W = \frac{L}{\bar{\lambda}} = \frac{L}{\lambda(1 - P_K)}$$

Mean time a job or a customer spends in the queue:

$$W_q = \frac{L_q}{\bar{\lambda}} = \frac{L_q}{\lambda(1 - P_K)}$$

In Example 6.6, the travel agency TravelCall Inc. is used to illustrate how the *M*/*M*/*c*/*K* model works. Example 6.6 studies the same process as in Example 6.3, but the performance measures are now determined from the expressions of the *M*/*M*/*c*/*K* model instead of using the general approach investigated in Section 6.2.3. For convenience, the TravelCall process is described again.

Example 6.6: TravelCall Inc. Revisited: A Streamlined Approach

Recall from Example 6.1 that TravelCall Inc. is a small travel agency that services customers over the phone. The agency employs only one travel agent to answer incoming customer calls. However, management is contemplating hiring a second agent and wants to investigate how the system would perform with two agents. It is assumed that these two agents would work independently on separate customers and that their service rates would be identical. The switchboard accommodates one call on hold in addition to the two routed to the travel agents. Customers who call when the switchboard is full will get a busy signal. Consequently, the maximum number of customers in the queuing system is three. The calls arrive to TravelCall according to a Poisson process with a mean rate of nine calls per hour, and the calls are answered in the order they arrive, so the queuing discipline is FIFO. On average, it takes a travel agent 6 minutes to service a customer, and the service time follows an exponential distribution. The travel agency wants to determine the probability of finding zero, one, two, or three customers in the system as well as the average performance measures L, L_q, W, and W_q.

The queuing system at TravelCall fits the $M/M/c/K$ model with $c = 2$ and $K = 3$, meaning a two-server system with at most $K = 3$ customers allowed in the system at the same time. Furthermore, the model can be described in terms of the rate diagram in Figure 6.14, which has the same structure as the rate diagram for the general $M/M/c/K$ process in Figure 6.13.

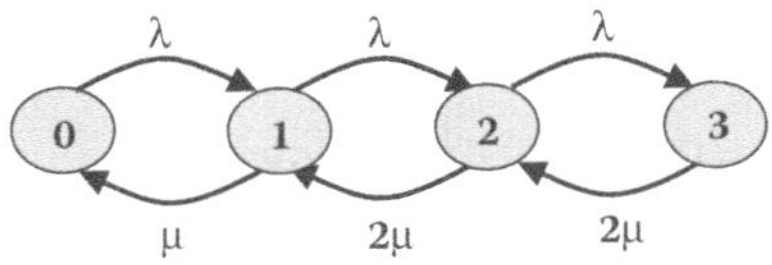

FIGURE 6.14
Rate diagram for the two-server queuing process at TravelCall Inc.

As before, the first step in the numerical analysis is to decide on a time unit and to determine the mean arrival and service rates. All rates will be expressed in the number of occurrences per hour. The mean arrival rate when no more than three customers are in the system is λ = nine calls per hour and zero otherwise (see Figure 6.14). Furthermore, the mean service rate for each of the two servers is $\mu = 60/6 = 10$ calls per hour. This means that the utilization factor is $\rho = \lambda/c\mu = 9/((2)(10)) = 9/20 = 0.45$.

First, the probability, P_0, that the system is empty is determined.

$$P_0 = \frac{1}{\sum_{n=0}^{c}\frac{(\lambda/\mu)^n}{n!} + \frac{(\lambda/\mu)^c}{c!}\sum_{n=c+1}^{K}\rho^{n-c}} = \frac{1}{\sum_{n=0}^{2}\frac{(0.9)^n}{n!} + \frac{(0.9)^2}{2!}\sum_{n=3}^{3}0.45^{n-2}}$$

$$= \frac{1}{(1+0.9+0.405)+0.405(0.45)} = \frac{1}{2.48725} \approx 0.4021$$

We can then obtain the probabilities P_1, P_2, and P_3 using

$$P_n = \begin{cases} \frac{(\lambda/\mu)^n}{n!}P_0 & \text{for } n = 1, 2, \ldots, c \\ \frac{(\lambda/\mu)^n}{c!c^{n-c}}P_0 & \text{for } n = c, c+1, \ldots, K \end{cases}$$

$$P_1 = \frac{(0.9)^1}{1!}P_0 = \frac{0.9}{2.48725} \approx 0.3618$$

$$P_2 = \frac{(0.9)^2}{2!}P_0 = \frac{0.81}{2}P_0 = \frac{0.405}{2.48725} \approx 0.1628$$

$$P_3 = \frac{(0.9)^3}{2!2^1}P_0 = \frac{0.729}{4}P_0 = \frac{0.18225}{2.48725} \approx 0.0733$$

These probability values are the same as those determined in Example 6.3. The next step is to determine the expected number of customers in the queue, L_q, and in the system, L.

$$L_q = \frac{(\lambda/\mu)^c \rho}{c!(1-\rho)^2} P_0\left(1-\rho^{K-c}-(K-c)\rho^{K-c}(1-\rho)\right)$$
$$= \frac{(0.9)^2 0.45}{2!(0.55)^2} \cdot \frac{\left(1-0.45^1-(1)0.45^1(0.55)\right)}{2.48725}$$
$$= 0.0733 \approx 0.07 \text{ customers}$$

$$L = L_q + \sum_{n=0}^{c-1} nP_n + c\left(1-\sum_{n=0}^{c-1} P_n\right)$$
$$= L_q + P_1 + 2(1-P_0-P_1) = 0.0733 + 0.3618$$
$$+2(1-0.4021-0.3618) = 0.9073 \approx 0.91 \text{ customer}$$

Finally, Little's law is used to determine the mean time a customer spends in the system, W, and the mean time he or she spends in the queue, W_q.

$$W = \frac{L}{\bar{\lambda}} = \frac{L}{\lambda(1-P_K)} = \frac{L}{\lambda(1-P_3)}$$
$$= \frac{0.9073}{9(1-0.0733)} \approx 0.1088 \text{ hours} \approx 6.5 \text{ minutes}$$

$$W_q = \frac{L_q}{\bar{\lambda}} = \frac{L_q}{\lambda(1-P_K)} = \frac{L_q}{\lambda(1-P_3)}$$
$$= \frac{0.0733}{9(1-0.0733)} \approx 0.0088 \text{ hours} \approx 0.5 \text{ minutes}$$

A comparison with Example 6.3 verifies that the results are identical. ■

6.2.7 The *M/M/c/∞/N* Model

The $M/M/c/\infty/N$ model is another generalization of the basic $M/M/c$ model, which deals with the case of a finite calling population of N customers or jobs. Apart from this feature, the $M/M/c/\infty/N$ model is based on the same assumptions as the $M/M/c$ model: a single queue, c independent, parallel and identical servers, exponentially distributed interarrival and service times, unlimited queue length, and a FIFO queuing discipline.

An important observation in understanding the $M/M/c/\infty/N$ model is that because the calling population is finite, each of the N customers or jobs must be either in the queuing system or outside, in the calling population. This implies that if n customers or jobs are in the system, meaning the queuing system is in State n, the remaining $N - n$ customers or jobs are found in the calling population. When constructing the model, it is assumed that for each of the customers or jobs currently in the calling population, the time remaining until it arrives to the queuing system is independent and exponentially distributed with parameter λ. As a result, if n ($n \leq N$) customers are in the queuing system, the arrivals at this system state occur according to a Poisson process with mean arrival rate $(N - n)\lambda$ (see Section 6.2.1). To obtain this, let T_j be the time remaining until job j, $j = 1, 2, \ldots, N - n$ in the calling population will arrive to the queue. The time remaining until the next arrival is then $T_{\min} = \min\{T_1, T_2, \ldots, T_{N-n}\}$. The conclusion follows because $T_{\min}$ is exponentially distributed with parameter $(N - n)\lambda$. Assuming, as before, that the mean service rate for each server is μ, the $M/M/c/\infty/N$ model can be described by the rate diagram in Figure 6.15. Moreover, because the service and interarrival times are mutually independent and exponentially distributed, the $M/M/c/\infty/N$ model describes a birth-and-death process with state-dependent interarrival and service rates.

$$\lambda_n = \begin{cases} (N-n)\lambda & \text{for } n = 0,1,2,\ldots,N \\ 0 & \text{for } n \geq N \end{cases}$$

$$\mu_n = \begin{cases} n\mu & \text{for } n = 0,1,2,\ldots,c \\ c\mu & \text{for } n \geq c \end{cases}$$

Figure 6.15 explicitly illustrates the fact that because only N jobs or customers are in the calling population, the maximum number of jobs or customers in the queuing system is N, which means that the maximum queue length is $N - c$. As a result, the $M/M/c/\infty/N$ system, like the $M/M/c/K$ system, is guaranteed to reach a steady state even if $\rho = (\lambda/c\mu) = 1$.

A common application area for the $M/M/c/\infty/N$ model is for analyzing machine maintenance or repair processes. These processes are characterized by c service technicians (or teams of service technicians) who are responsible for keeping N machines or service stations operational, that is, to repair them as soon as they break down. The jobs (or customers) in this queuing process are the machines or service stations. Those that are currently in operation

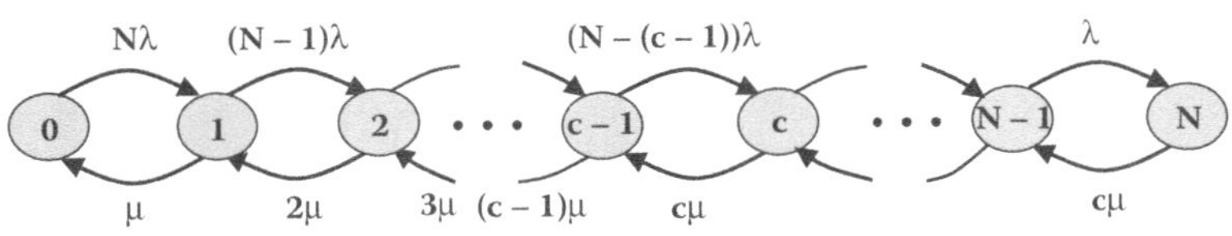

FIGURE 6.15
Rate diagram for the $M/M/c/\infty/N$ model.

are in the calling population, and those that are undergoing repair are in the queuing system. The service technicians or teams of technicians are the servers. For a maintenance process of this sort to be adequately described by an $M/M/c/\infty/N$ model, the time between breakdowns for each individual machine in operation needs to be exponentially distributed with a constant parameter, λ. Furthermore, the c service technicians (or service teams) should work independently of each other on separate machines all displaying exponentially distributed service times with the same service rate μ. Note that if n machines are undergoing repair, then $N - n$ machines are in operation.

To analyze the performance of a given $M/M/c/\infty/N$ system, the general approach in Section 6.2.3 can be used; however, as with the $M/M/c$ and $M/M/c/K$ systems, the algebra can be somewhat tedious for larger systems. As an alternative, the following expressions have been derived to aid in the analysis. See, for example, Kleinrock (1975) or Hillier and Lieberman (2010). It is assumed that $c \leq N$.

Probability of zero jobs in the system:

$$P_0 = \frac{1}{\sum_{n=0}^{c-1} \frac{N!}{(N-n)!n!}\left(\frac{\lambda}{\mu}\right)^n + \sum_{n=c}^{N} \frac{N!}{(N-n)!c!c^{n-c}}\left(\frac{\lambda}{\mu}\right)^n}$$

Probability of exactly n jobs in the system:

$$P_n = \begin{cases} \frac{N!}{(N-n)!n!}\left(\frac{\lambda}{\mu}\right)^n P_0 & \text{for } n = 0,1,2,\ldots c \\ \frac{N!}{(N-n)!c!c^{n-c}}\left(\frac{\lambda}{\mu}\right)^n P_0 & \text{for } n = c, c+1,\ldots,N \\ 0 & \text{for } n > N \end{cases}$$

Mean number of jobs or customers in the queue:

$$L_q = \sum_{n=c}^{N} (n-c) P_n$$

Note that this expression is simply the definition of L_q.

Mean number of jobs or customers in the system:

$$L = L_q + \sum_{n=0}^{c-1} nP_n + c\left(1 - \sum_{n=0}^{c-1} P_n\right)$$

Observe that the last two terms in this expression for L represent the expected number of jobs in the service facility. Consequently, the expression simply adds the average number of jobs in the queue to the average number of jobs in the service facility.

To obtain the expected time spent in the system and in the queue, we can use Little's law as discussed in Section 6.2.2. The average arrival rate in this case is

$$\bar{\lambda} = \sum_{n=0}^{\infty} \lambda_n P_n = \sum_{n=0}^{N} (N-n)\lambda P_n = \lambda \left(N \sum_{n=0}^{N} P_n - \sum_{n=0}^{N} nP_n \right) = \lambda(N-L)$$

Mean time a job or a customer spends in the system:

$$W = \frac{L}{\bar{\lambda}} = \frac{L}{\lambda(N-L)}$$

Mean time a job or a customer spends in the queue:

$$W_q = \frac{L_q}{\bar{\lambda}} = \frac{L_q}{\lambda(N-L)}$$

Remark: The $M/M/c/\infty/N$ model is based on the assumption that the time an individual job or customer spends in the calling population outside the queuing system is exponentially distributed. However, it has been shown (see Bunday and Scraton, 1980) that the expressions for P_0 and P_n (and consequently those for L, L_q, W, and W_q) also hold in more general situations. More precisely, the time that a job or customer spends outside the system is allowed to have any probability distribution as long as this distribution is the same for each job or customer and the average time is $1/\lambda$. Note that these situations fall outside the class of birth-and-death processes into a $G/M/c/\infty/N$ model.

Example 6.7 illustrates the analysis of the $M/M/c/\infty/N$ model in the context of a machine repair process as discussed previously.

Example 6.7: The Machine Repair Process at PaperCo Inc.

PaperCo Inc. is a manufacturer of fine paper. Its most important product is a special, environmentally friendly type of bleached paper used for laser printers and copiers. PaperCo has one production facility consisting of three large paper-milling machines. These machines are expensive and need to be kept operational to the largest extent possible. At the same time, their complexity makes them sensitive to malfunctions. Currently, PaperCo has one repair team on call 24 hours a day, 7 days a week. However, management is contemplating increasing this to two teams and would like a performance analysis of this double-team scenario. They already know the performance of the current process with one team. Each team is configured so that all competencies needed are available within the team, and having two teams working on the same machine provides no advantages. Therefore, it can be assumed that the two teams work independently of one another on different machines. Furthermore, their service times are assumed to be identical and exponentially distributed with a mean of 4 hours. Empirical data show that the time between malfunctions in each of the three machines when they are operating is exponential with a mean of 8 hours. The machines are equally important, so broken-down machines are serviced in a FIFO sequence.

PaperCo wants to know the steady-state probabilities that zero, one, two, or three machines are nonoperational. They also want to determine the average time before a machine that breaks down is operational again and how long time, on average, a broken machine has to wait before a service team starts to work on the problem.

The first step in analyzing this problem is to recognize that the PaperCo repair process can be modeled as an $M/M/c/\infty/N$ system with $c = 2$ servers (the service teams) and $N = 3$ jobs in the calling population (the machines). Furthermore, the mean service rate for each of the two teams is $\mu = 1/4 = 0.25$ jobs per hour, and the failure rate for each machine in operation is $\lambda = 1/8 = 0.125$ machines per hour. The rate diagram for the repair process is depicted in Figure 6.16. (See also Figure 6.15 for the general structure.) It should be noted that this problem can be analyzed using the general approach for birth-and-death processes discussed in Section 6.2.3, although a different method is used here.

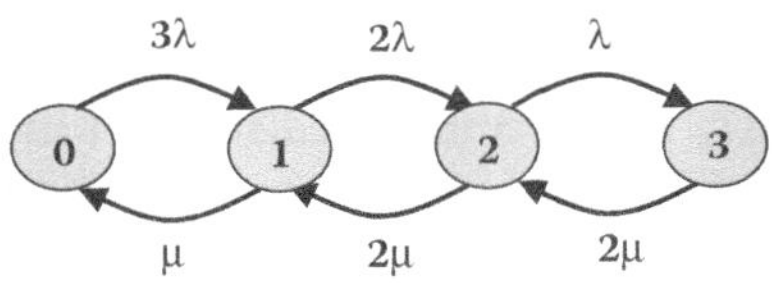

FIGURE 6.16
Rate diagram for the $M/M/2/\infty/3$ model describing the machine repair process at PaperCo.

Using the expressions given previously, one can determine the steady-state probabilities $\{P_n, n = 0, 1, 2, 3\}$ of zero, one, two, and three machines malfunctioning. Note that $N = 3$, $c = 2$, $\lambda = 0.125$, $\mu = 0.25$, and $(\lambda/\mu) = 0.5$.

Probability of zero jobs in the system:

$$P_0 = \frac{1}{\sum_{n=0}^{c-1} \frac{N!}{(N-n)!n!}\left(\frac{\lambda}{\mu}\right)^n + \sum_{n=c}^{N} \frac{N!}{(N-n)!c!c^{n-c}}\left(\frac{\lambda}{\mu}\right)^n}$$

$$= \frac{1}{\left(\frac{3!}{3!} + \frac{3!}{2!}(0.5)^1\right) + \left(\frac{3!}{2!}(0.5)^2\right) + \frac{3!}{2!2^1}(0.5)^3}$$

$$= \frac{1}{1 + 1.5 + 0.75 + 0.1875} = \frac{1}{3.4375} \approx 0.2909$$

Probability of exactly n jobs in the system:

$$P_n = \begin{cases} \dfrac{N!}{(N-n)!n!}\left(\dfrac{\lambda}{\mu}\right)^n P_0 & \text{for } n = 0,1,2,\ldots c \\ \dfrac{N!}{(N-n)!c!c^{n-c}}\left(\dfrac{\lambda}{\mu}\right)^n P_0 & \text{for } n = c, c+1,\ldots, K \\ 0 & \text{for } n > K \end{cases}$$

$$P_1 = \frac{3!}{(3-1)!1!}0.5^1 P_0 = \frac{1.5}{3.4375} \approx 0.4364$$

$$P_2 = \frac{3!}{(3-2)!2!}0.5^2 P_0 = \frac{0.75}{3.4375} \approx 0.2182$$

$$P_3 = \frac{3!}{(3-3)!2!2^1}0.5^3 P_0 = \frac{0.1875}{3.4375} \approx 0.0545$$

$$P_n = 0 \text{ for } n \geq 4$$

Mean number of jobs or customers in the queue:

$$L_q = \sum_{n=c}^{N}(n-c)P_n = (2-2)P_2 + (3-2)P_3 = P_3 \approx 0.0545 \approx 0.05 \text{ machines}$$

Mean number of jobs or customers in the system:

$$L = L_q + \sum_{n=0}^{c-1} nP_n + c\left(1 - \sum_{n=0}^{c-1} P_n\right) = L_q + (1)P_1 + 2(1 - P_0 - P_1)$$

$$\approx 0.0545 + 0.4364 + 2(1 - 0.2909 - 0.4364) = 1.0363 \approx 1.04 \text{ machines}$$

Little's law can be used to determine the expected downtime for a broken machine, W, and the expected time a broken machine has to wait for a service team, W_q.

$$W = \frac{L}{\bar{\lambda}} = \frac{L}{\lambda(N-L)} \approx \frac{1.0363}{0.125(3-1.0363)}$$
$$\approx 4.22 \text{ hours} \approx 4 \text{ hours and } 13.3 \text{ minutes}$$

$$W_q = \frac{L_q}{\bar{\lambda}} = \frac{L_q}{\lambda(N-L)}$$
$$\approx \frac{0.0545}{0.125(3-1.0363)} \approx 0.22 \text{ hours} \approx 13.3 \text{ minutes}$$

Consequently, with two service teams, all machines are operational 29 percent of the time, two out of three are working 44 percent of the time, two machines are down 22 percent of the time, and 5 percent of the time, all three machines are standing still. Moreover, it takes on average 4 hours and 13 minutes from the moment a machine breaks down until it is up and running again. Of this total downtime, 13 minutes is waiting time before a team is available.

It seems that PaperCo needs to do something about this process. Even with two teams, a significant amount of costly downtime is occurring. ■

6.2.8 Queuing Theory and Process Design

The analytical queuing models studied in Sections 6.2.3 through 6.2.7 have proved useful in analyzing the steady-state behavior of various queuing systems. The analysis focuses on the probability distribution for the number of customers in the queuing system and on the average performance measures, such as the mean queue length and the mean time customers spend in the queue and in the queuing system as a whole. Examples also have been provided of how this type of information can help the process designer better understand the impact that capacity-related decisions have on service to the customers. However, as discussed in the introduction to this chapter, an optimal design decision involves an economic trade-off between the cost of additional capacity and the cost of poor service in terms of long delays. This section explores these issues further, particularly by investigating how waiting costs (or shortage costs) can be used to translate the service performance into monetary terms. Challenges related to appropriately quantifying waiting costs also will be discussed. First, however, the decisions facing the process designer need to be examined.

When designing basic queuing systems such as those in Sections 6.2.3 through 6.2.7—single-queue, parallel-server systems—two design parameters are usually under the decision-maker's control:

- The number of parallel servers in the service facility; for example, the number of physicians at a hospital ward, the number of checkout stations in a grocery store, or the number of operators in a call center.
- The service time or speed of individual servers; for example, the size and thereby the service time of a repair team, the type of materials handling system (conveyor belts, forklift trucks, and so on) in a production facility, or the speed of an airplane transporting people.

In the context of the models in Sections 6.2.3 through 6.2.7, these decisions involve determining c, the number of parallel and identical servers, and μ, the service rate for an individual server (assuming that the service time for individual servers is state independent). Another design parameter that might be under the decision-maker's control is K, the maximum number of customers in the system. This could involve decisions regarding the number of calls that can be placed on hold at a call center, the size of the waiting area at a restaurant, the number of beds in a hospital, and so on. The effect of increasing K is that the system can accommodate more customers, and fewer customers are denied service, but at the same time, the waiting time for those in the system tends to increase. Translating these effects into monetary terms typically requires a somewhat more complex waiting cost structure than when considering changes only in c and μ. The following analysis assumes that $K = \infty$ and focuses on design changes in c and μ, meaning the focus is on designing $M/M/c$ systems.

In addition to the design parameters, c and μ, which relate to the queuing system's available capacity, the process design team sometimes is in a position to affect the demand for capacity through the arrival process. This could involve demand management initiatives to reduce the variability in the demand pattern, making the arrival process more stable. Note, however, that under the assumption of exponentially distributed arrival times, it is not possible to reduce the demand variability without changing the mean demand. Consequently, the effects of such demand management initiatives cannot be properly analyzed using the $M/M/c$ models that have been introduced. However, demand management decisions regarding the size of a limited calling population may be analyzed using the $M/M/c/\infty/N$ model. A service example of this type will be considered at the end of this section.

Given the available design parameters, the design team must decide how to set these parameters, meaning the amount and type of capacity to acquire to satisfy the customers' service requirements. A fundamental design issue is therefore to quantify these service requirements. The two standard approaches are to specify an acceptable service level and to specify a waiting cost (or shortage cost).

For the service level approach, the decision problem is to minimize the capacity-related costs under the service level constraint; that is, to determine c, μ, and if applicable K, to minimize the expected cost of operating the queuing system given that the service level is met. Different ways of measuring the service level exist, such as the average waiting time, the percentage of customers receiving service immediately, or the percentage of customers waiting less than, say, 5 minutes.

The focus in this section is on the waiting cost approach. In this case, the corresponding decision problem is to determine c or μ to minimize the total cost of waiting and operating the queuing system. To investigate this further, some additional notation needs to be introduced.

WC = Expected waiting cost per time unit for the queuing system.

SC = Expected service cost per time unit for operating the queuing system.

TC = Expected total cost per time unit for the queuing system = SC + WC.

C_w = Waiting cost rate = waiting cost per customer and time unit.

$C_s(\mu)$ = Service cost per time unit for an individual server with the service rate μ.

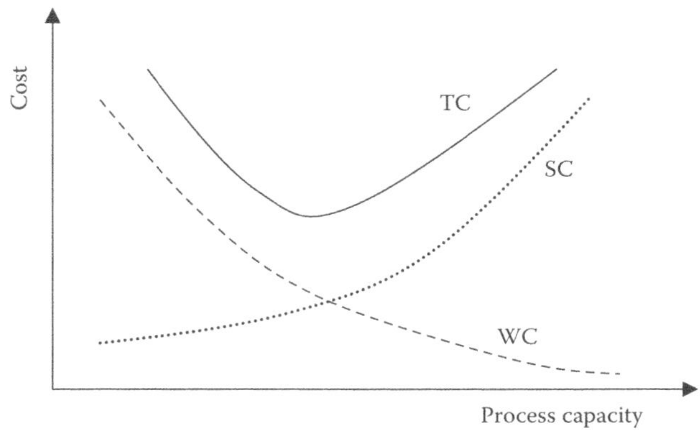

FIGURE 6.17
Schematic illustration of the cost functions *TC*, *SC*, and *WC*.

It is worth noting that although a constant waiting cost rate, C_w, is assumed, it is straightforward to extend the analysis to state-dependent waiting cost rates; see, for example, Hillier and Lieberman (2010).

The decision problem can be stated as:

$$\text{Minimize}\, TC = SC + WC$$

The corresponding cost functions, *TC*, *SC*, and *WC*, are illustrated in Figure 6.17.

6.2.8.1 Determining WC

To determine an expression for the expected waiting cost per time unit for the system, *WC*, note that if n customers are in the system, the waiting cost per time unit is $C_w n$. From the definition of expected value and the steady-state probability distribution for finding n customers in the system, an expression for *WC* can be found.

$$WC = \sum_{n=0}^{\infty} C_w n P_n = C_w \sum_{n=0}^{\infty} n P_n = C_w L$$

The expected waiting cost per time unit for the queuing system is obtained by multiplying the waiting cost per customer and per time unit with the expected number of customers in the system. This expression assumes that every customer in the queuing system incurs the same waiting cost, whether the customer is waiting in the queue or is in the service facility. In cases where only the customers in the queue incur waiting costs, the expression for *WC* is

$$WC = C_w L_q$$

If not otherwise specified, we will assume that the waiting costs are based on the number of customers in the system and not just those in the queue.

6.2.8.2 Determining SC

The expected service cost per unit of time for the system depends on the number of servers, c, and the speed of these servers. When all c servers are identical with service rates μ, the expected service cost can be expressed as follows:

$$SC = cC_s(\mu)$$

6.2.8.3 A Decision Model for Designing Queuing Systems

Based on the expressions for *WC* and *SC*, a decision model for determining the values of the design parameters c and μ can be formulated as follows. The objective is to minimize the total costs per time unit for the entire queuing system.

$$\underset{c,\mu}{\text{Minimize}}\ TC = SC + WC = cC_s(\mu) + C_w L$$

Examples 6.8 and 6.9 illustrate how the process design team can use this general decision model for optimizing the performance of queuing systems.

Example 6.8: Process Design at CopyCo Inc.—The Option of a Second Machine

CopyCo is a small print shop located near a large university campus. Currently, all jobs involving paper-to-paper copies are carried out on one copy machine. For some time, CopyCo has experienced backlog problems with jobs submitted to the paper-to-paper copier. Management is worried that the long waiting times might be detrimental to future sales, so they are contemplating the acquisition of a second identical copy machine. The current machine is leased at a cost of \$50 per day. Other costs for operating the machine, such as labor and electricity, amount to \$150 per day. The second copier also would be leased, and it is estimated that it would accrue the same costs as the existing one—\$200 per day. Empirically, it has been shown that the service and interarrival times are exponentially distributed. Furthermore, on average, CopyCo receives 95 jobs per day involving paper-to-paper copying, and the mean service time on the current machine is 6 minutes. CopyCo is open 10 hours per day, and the number of jobs that can be in the store at the same time is subject to no immediate space restrictions. The hourly waiting cost per customer is estimated to be $5. Management wants to design the paper-to-paper copying process to minimize the total expected costs while considering the option of acquiring a second machine.

First, a suitable queuing model needs to be selected for this situation. It is known that the service and interarrival times are exponentially distributed and that no restrictions are placed on the queue length. In the case of a single copier, the queuing process at CopyCo fits an $M/M/1$ model. In the case of two identical copiers working in parallel, the appropriate model is an $M/M/2$ model.

Because the objective is to find the design alternative that minimizes the total expected costs, an appropriate decision model is

$$\underset{c=\{1,2\}}{\text{Minimize}}\ TC = SC + WC = cC_s(\mu) + C_w L$$

The only decision variable is the number of servers, $c = 1$ or 2. The service rate μ for each individual server is fixed.

The known cost, demand, and service information is:

$\lambda = 95$ jobs per day $= 95/10 = 9.5$ jobs per hour
$\mu = 60/6 = 10$ jobs per hour
$C_s(\mu) = \$200$ per day $= 200/10 = \$20$ per hour
$C_w = \$5$ per customer and hour

All parameters should be expressed in the same time units, and in this case, hours have been chosen.

The only unknown values in our decision model are the expected number of jobs in the system for both the single and the double copier alternative. Let L_1 be the expected number of jobs in the $M/M/1$ system and L_2 be the expected number of jobs in the $M/M/2$ system. Using the formulae from Sections 6.2.4 and 6.2.5 and the decision model, one gets

$M/M/1$ system:

$$\rho = \lambda/\mu = 9.5/10 = 0.95$$

$$L_1 = \frac{\rho}{1-\rho} = \frac{(0.95)}{1-(0.95)} = 19 \text{ jobs}$$

$$TC_1 = SC_1 + WC_1 = 1(20) + 5(19) = \$115 \text{ per hour}$$

$M/M/2$ system:

$$\rho = \lambda/(2\mu) = 9.5/20 = 0.475$$

$$P_0 = \frac{1}{\sum_{n=0}^{c-1} \frac{(\lambda/\mu)^n}{n!} + \frac{(\lambda/\mu)^c}{c!(1-\rho)}} = \frac{1}{1 + 0.95 + \frac{(0.95)^2}{2!(1-0.475)}} \approx 0.3559$$

$$L_2 = \frac{(\lambda/\mu)^c \rho}{c!(1-\rho)^2} P_0 + \frac{\lambda}{\mu} \approx \frac{(0.95)^2\,0.475}{2!(1-0.475)^2} 0.3559 + 0.95 \approx 1.23 \text{ jobs}$$

$$TC_2 = SC_2 + WC_2 = 2(20) + 5(1.23) = \$46.15 \text{ per hour}$$

Because the expected total cost is considerably lower with a second copy machine in place ($TC_2 < TC_1$), it is recommended that the CopyCo management team chooses this alternative even though the service cost is twice as high. ■

Example 6.9: Process Design at CopyCo Inc.—The Option of a Single Faster Machine

As the management team at CopyCo contacted the leasing company to sign up for a second machine, the sales person informed them about another alternative. A brand new model of their old copier just came on the market. The new model is twice as fast as the old one, which for CopyCo would mean an estimated average service time of 3 minutes. The service times are assumed to be exponential, and the jobs arriving to the print shop are the same as before. The cost to lease this new model is \$150 per day. Operating the new model efficiently will require additional personnel and more electricity, rendering an estimated operating cost of about \$250 per day. Consequently, the total service cost of operating the new machine is \$400 per day, which is the same as the cost of operating two of the old machines. The question is whether leasing one fast machine is better than leasing two slower copiers.

Because service and interarrival times are exponential, and the queue length has no restrictions, the new alternative with one fast copier can be modeled as an *M*/*M*/1 queue.

From Example 6.8, it is known that

λ = 95 jobs per day = 95/10 = 9.5 jobs per hour
C_w = \$5 per customer and hour

From the information about the new model of the copy machine, it is known that

μ_{new} = 60/3 = 20 jobs per hour
$C_s(\mu_{\text{new}})$ = \$250 + \$150 per day = 400/10 = \$40 per hour

Because the objective is to find the design alternative that minimizes the total expected costs, an appropriate decision model is

$$\underset{c,\mu}{\text{Minimize}}\ TC = SC + WC = cC_s(\mu) + C_w L$$

The decision variables in this case are the number of servers, c = 1, 2, and their individual speed, μ = 10 or 20. However, only two alternatives are being compared: the alternative with two copiers of the old model ($c = 2$ and $\mu = 10$ jobs per hour) and the alternative with one copier of the new model ($c = 1$ and $\mu = 20$ jobs per hour). Furthermore, for the alternative with two copiers of the old model, the expected cost has already been determined in Example 6.8: $TC_2 = \$46.15$ per hour.

For the new alternative with one fast copier, the costs are obtained as follows:

M/*M*/1 system:

$$\rho = \lambda/\mu_{\text{new}} = 9.5/20 = 0.475$$

$$L_{\text{new}} = \frac{\rho}{1-\rho} = \frac{(0.475)}{1-(0.475)} = 0.9048 \text{ jobs}$$

$$TC_{\text{new}} = SC_{\text{new}} + WC_{\text{new}} = 1(40) + 5(0.9048) = \$44.52 \text{ per hour}$$

The cost comparison indicates that the alternative with one fast copier of the new model is preferable to the option of leasing two copiers of the old model (\$44.52 < \$46.15). ■

Remark: The conclusion in Example 6.9 is a consequence of a more general result regarding single- and multiple-server facilities with the same mean capacity. It can be shown that a queuing system with one fast server with service rate equal to $c\mu$ will always render shorter expected waiting times and fewer customers in the system than a service facility with c parallel servers, each with a service rate of μ. This also implies that the expected total cost TC is lower for the single-server system as long as the service cost SC is such that $Cs(c\mu) \leq cCs(\mu)$.

Thus far, it has been assumed that a relevant waiting cost rate C_w is readily available. Unfortunately, this is not always the case. In fact, finding an appropriate waiting cost rate is often a major challenge in the practical application of decision models based on waiting costs.

Because of the wide variety of queuing situations, there exists no single procedure for estimating the customers' waiting cost rates or waiting cost functions. In fact, the entire waiting cost concept is situation dependent. For example, the waiting costs for long delays at an emergency room are very different from those experienced by customers in a supermarket. Essentially, the waiting cost rate should reflect the monetary impact that a delay of a single customer or job has on the organization or society where the queuing system in question resides. This implies that, first, it is necessary to understand the consequences of time delays for the customers, and to understand customer requirements and behavior. Second, these consequences must be translated into monetary effects for the queuing system or the organization to which it belongs. This has to be done on a case-by-case basis. However, in broad terms, one can distinguish between situations in which the customers are external to the organization and ones in which they are internal. Furthermore, there is typically a difference between profit-maximizing organizations and nonprofit organizations.

In the case of external customers arriving to profit-maximizing organizations, the waiting costs are related primarily to lost profits and bad reputation affecting future sales. For example, long lines at a grocery store might discourage a customer from coming back or from entering the store in the first place. For nonprofit organizations, on the other hand, the waiting costs are typically related to future social costs. To estimate these costs, it is necessary to understand how waiting affects the individual customers and society as a whole. For example, what is the cost to society of having long waiting times at an emergency room? Social costs and costs related to loss of future profits can be hard to quantify in monetary terms. In some cases, it might be easier to estimate an acceptable service level and use the service level approach. Again, in other cases, the costs associated with delays and waiting might be easier to identify. Consider, for example, the drug approval process of a pharmaceutical company, which has a cost associated with its cycle time. If the process requires 1 year to be completed, and 80 percent of this time is due to delays, then the cost of waiting is 9.6 months of lost sales. In this case, loss of market share due to late market introduction also should be added to the lost sales cost.

In situations where the arriving customers are internal to the organization where the queuing system resides, it is usually somewhat easier to estimate the waiting costs. The customers (or jobs) in these cases are typically machines or employees of the organization. Therefore, it is often possible to identify the immediate consequences associated with having these customers idle. The main component of the waiting costs associated with this idleness is, in most cases, the lost profit due to the reduced productivity.

Estimating waiting costs is often a challenging task based on customer requirements and the consequences of poor service. However, as discussed in Chapters 1 through 3, gaining an understanding of the customer requirements is at the core of effective business process design.

As a final note, Example 6.10 illustrates how decisions about a limited client base may be included in the cost analysis and process design of a service system.

Example 6.10: Process Design at MedAid Inc.—Choice of Client Base Size

MedAid Inc. is a health care company that offers a mobile doctor's service to nursing homes, service home complexes for senior citizens, and other institutions. The service is such that a client (e.g., a nursing home) calls MedAid when they have a need for a doctor/nursing team. MedAid then promises to immediately send a team by car. If they are not able to immediately dispatch a unit because it is occupied with another client, MedAid pays a penalty fee to the client proportional to the time it has to wait before a unit is dispatched. The penalty fee is set to $1000 per client and hour.

Each client pays a flat fee of $3000 per day for MedAid's services, which are available 24 hours a day, 7 days a week. Of these revenues, $2400 is spent on salaries and other expenses. It should be noted that real emergency situations should still be handled by calling an ambulance for transportation to the nearest hospital. MedAid typically deals with issues where an ambulance would not be dispatched. An important feature of MedAid's service is that each client is associated with at most one mobile team at a time. The purpose is to establish good patient–doctor relationships. Because of the 24-hour service, and because each doctor's team works 8-hour shifts, each client has to deal with at most three different doctor's teams. This setup means that MedAid faces a problem of how many clients a doctor's team (or a group of three doctor's teams) should be assigned. The problem is clearly dependent on how often the clients demand service and how fast the doctor's team is in providing the service. MedAid has concluded from gathering a lot of data that it is reasonable to assume that both the time between service calls from each client and the service time for the doctors' units are exponentially distributed. In a particular instance, which they want help in analyzing, the average time between service calls from each client is 10 hours, and the service time (including the time it takes for the doctor's team to drive to the client) is 1 hour. Currently, the doctor's team in question has three clients to serve, but a request from a new potential client has been made, and MedAid wants to know if it is profitable to assign a fourth client to the team. As MedAid are trying to expand their business, they do not want to turn clients away. Thus, as long as they make a profit, they will accept the new client.

To analyze the problem and help MedAid to make a decision, one first needs to choose an appropriate model. Because of the exponentially distributed interarrival times and service times, the single server (doctor's team), and the limited client base (calling population), an appropriate queuing model is the $M/M/1/\infty/N$ model. In addition, an appropriate decision model for determining the expected total profit per time unit (TP) is

$$TP = NR - C_S(\mu) - C_W L_q$$

where

R = Revenue per client and hour = \$3000/24 = \$125 per hour
N = Number of clients = 4
$C_S(\mu)$ = Expenses for operating the server (the doctor's team) = \$2400/24 = \$100 per hour
C_W = Penalty fee (or waiting cost) per client and hour accruing from the moment a client requests a doctor's team until one is available and can be dispatched = \$1000 per client and hour
L_q = Expected number of clients waiting for a doctor's unit to be dispatched

It follows that the expected total profit is

$$TP = 4(125) - 100 - 1000L_q = 400 - 1000L_q$$

It remains to determine L_q. To do that, one can use the formulae for an $M/M/1/\infty/4$ model and

λ = Expected arrival rate = 1/10 = 0.1 service visits per client and hour,
$\rho\mu$ = Expected service rate = 1 service visit per hour.

$$L_q = \sum_{n=2}^{4} (n-1) P_n = (1) P_2 + (2) P_3 + (3) P_4$$

To calculate P_2, P_3, and P_4, one first needs to determine P_0, and then it is useful to note that $(\lambda/\mu) = 0.1$.

$$\begin{aligned} P_0 &= \frac{1}{\sum_{n=0}^{c-1} \frac{N!}{(N-n)!n!}\left(\frac{\lambda}{\mu}\right)^n + \sum_{n=c}^{N} \frac{N!}{(N-n)!c!c^{n-c}}\left(\frac{\lambda}{\mu}\right)^n} \\ &= \frac{1}{\frac{4!}{4!} + \left(\frac{4!}{3!}(0.1)^1 + \frac{4!}{2!}(0.1)^2 + \frac{4!}{1!}(0.1)^3 + \frac{4!}{0!}(0.1)^4\right)} \\ &= \frac{1}{1 + (0.4 + 0.12 + 0.024 + 0.0024)} = \frac{1}{1.5464} \approx 0.6467 \end{aligned}$$

The probability of exactly $n > 0$ jobs in the system can now be obtained from

$$P_n = \begin{cases} \frac{N!}{(N-n)!n!}\left(\frac{\lambda}{\mu}\right)^n P_0 & \text{for } n = 0,1,2,\ldots c \\ \frac{N!}{(N-n)!c!c^{n-c}}\left(\frac{\lambda}{\mu}\right)^n P_0 & \text{for } n = c, c+1, \ldots, K \\ 0 & \text{for } n > K \end{cases}$$

$$P_1 = \frac{4!}{(4-1)!1!} 0.1^1 P_0 = \frac{0.4}{1.5464} \approx 0.2587$$

$$P_2 = \frac{4!}{(4-2)!1!1} 0.1^2 P_0 = \frac{0.12}{1.5464} \approx 0.0776$$

$$P_3 = \frac{4!}{(4-3)!1!1^2} 0.1^3 P_0 = \frac{0.024}{1.5464} \approx 0.0155$$

$$P_4 = \frac{4!}{(4-4)!1!1^3} 0.1^4 P_0 = \frac{0.0024}{1.5464} \approx 0.0016$$

$$P_n = 0 \text{ for } n \geq 5$$

This renders

$$L_q = \sum_{n=2}^{4}(n-1)P_n = (1)P_2 + (2)P_3 + (3)P_4$$
$$= (1)\frac{0.12}{1.5464} + (2)\frac{0.024}{1.5464} + (3)\frac{0.0024}{1.5464} = \frac{0.1752}{1.5464} \approx 0.1133$$

The resulting total expected profit is

$$TP = 400 - 1000L_q \approx 400 - 1000(0.1133) = \$286.70 \text{ per hour}$$

Because the profit is positive, which was the decision criterion specified by MedAid, a contract should be signed with the fourth client. Note that the probability of zero or one client in the system (i.e., no clients in the queue waiting for service) is $P_0 + P_1 \approx 90.5\%$. Also note that if the MedAid objective had been to maximize its profit, a comparison should be made with the current profit gained when three clients are under contract, and the most profitable option should be chosen. ■

6.3 Summary

Variability in the time that it takes to perform certain activities as well as in the demand for these activities is an important concern when designing business processes. This variability causes unbalanced capacity use over time. This in turn manifests itself through queues and their associated non-value-adding waiting times. The economic implication is that the process design team is faced with the problem of balancing the cost of acquiring additional capacity against the cost of poor service.

This chapter investigated ways to explicitly incorporate variability in activity times and demand into process models. In particular, it explored analytical queuing models stemming from mathematical queuing theory. The analysis assumes an understanding of the elementary components defining a basic queuing process: the calling population, the arrival process, the queue configuration, the queue discipline, and the service mechanism.

Because even in well-designed processes occasional long queues might be unavoidable, it is important to consider strategies to mitigate the negative economic effects of these potentially long lines. This is particularly relevant in service situations in which the queue consists of people who are sensitive to waiting and might decide to leave. The pros and cons of several strategies that are often used in service industries were discussed, including (1) concealing the queue, (2) using the customer as a resource,

(3) making the wait comfortable, (4) keeping the customers informed of the reason for the wait, (5) providing pessimistic estimates of remaining waiting time, and (6) the importance of being fair and open about the queuing discipline used.

The exponential distribution plays a central role in mathematical queuing modeling. A thorough discussion about when using this distribution might be a realistic assumption for modeling service and interarrival times was provided. Of particular interest is to understand the implications of important properties, such as a strictly decreasing density function, the lack of memory and the relation to the Poisson process.

The analytical queuing models considered in this chapter all consist of a single queue with a FIFO queue discipline and a service mechanism with parallel servers. All models assumed exponentially distributed service and interarrival times. This represents an important class of queuing models, which are based on the general birth-and-death process. In the context of a queuing process, a birth represents a customer arrival, and a death represents a service completion. In addition to an approach for steady-state analysis of general birth-and-death processes, based on the so-called rate in = rate out principle and the corresponding balance equations, a number of specialized models also were explored. These models represent important special cases often seen in practice for which important performance characteristics can be obtained from standardized mathematical formulae. The specialized models covered in this chapter were $M/M/1$, $M/M/c$, $M/M/c/K$, and $M/M/c/\infty/N$.

In capacity-related design decisions, where a core issue is to balance the costs of additional capacity against the costs of poor service, waiting costs were introduced as a way to translate the effects of queues and waiting time into monetary terms. The corresponding decision model is based on the ability to analyze the behavior of queuing processes quantitatively, particularly in terms of expected number of customers in the system or in the queue. A major challenge in applying these types of decision models pertains to estimating relevant waiting cost rates—the waiting costs per customer and time unit. This task typically requires a good understanding of customer requirements and the economic consequences of delays.

Appendix 6A: Mathematical Derivations and Models with Generally Distributed Service Times

This appendix contains mathematical derivations of some key results used in Chapter 6 and supplemental material regarding queuing models with generally distributed service times.

6A.1 Mathematical Derivations of Key Results

6A.1.1 The Exponential Distribution (Section 6.2.1)

Derivation of Property 2: *Lack of Memory*

If T is an exponentially distributed stochastic variable (representing a service time or interarrival time), then for all positive values of t and δ,

$$P(T > t+\delta | T > \delta) = \frac{P(T > t+\delta, T > \delta)}{P(T > \delta)} = \frac{P(T > t+\delta)}{P(T > \delta)} = \frac{e^{-\alpha(t+\delta)}}{e^{-\alpha\delta}} = e^{-\alpha t} = P(T > t)$$

This means that the probability distribution of the remaining service time, or time until the next customer arrives, is always the same. It does not matter how long a time has passed since the service started or since the last customer arrived.

Derivation of Property 3: *The Minimum of Independent Exponentially Distributed Random Variables Is Exponentially Distributed*

If $T_1, T_2, \ldots, T_n$ are n independent, exponentially distributed random variables with parameters $\alpha_1, \alpha_2, \ldots, \alpha_n$, respectively, and $T_{\min} = \min\{T_1, T_2, \ldots, T_n\}$, it follows from the definition of the exponential distribution that

$$\begin{aligned} P(T_{\min} > t) &= P(T_1 > t, T_2 > t, \ldots, T_n > t) \\ &= P(T_1 > t)P(T_2 > t)\cdots P(T_n > t) \\ &= e^{-\alpha_1 t}e^{-\alpha_2 t}\cdots e^{-\alpha_n t} = e^{-\left(\sum_{i=1}^{n}\alpha_i\right)t} \end{aligned}$$

Thus, $T_{\min}$ is exponentially distributed with parameter $\alpha = \sum_{i=1}^{n}\alpha_i$.

6A.1.2 Birth-and-Death Processes (Section 6.2.3)

Derivation of the Rate in = Rate out Principle

Consider an arbitrary state n and count the number of times the process enters and leaves this state, starting at time 0.

Let

$I_n(t)$ = The number of times the process has entered State n by time $t \geq 0$

$O_n(t)$ = The number of times the process has left State n by time $t \geq 0$

It follows that for any t, the absolute difference between $I_n(t)$ and $O_n(t)$ is 0 or 1:

$$|I_n(t) - O_n(t)| \leq 1$$

Dividing this expression by t and letting t approach infinity ($t \to \infty$) that is, by considering what happens to the average number of arrivals and departures at State n in steady state, the following is obtained:

$$\left|\frac{I_n(t)}{t} - \frac{O_n(t)}{t}\right| \le \frac{1}{t} \quad \text{and} \quad \lim_{t\to\infty}\left|\frac{I_n(t)}{t} - \frac{O_n(t)}{t}\right| = 0$$

which implies that

$$\lim_{t\to\infty}\frac{I_n(t)}{t} = \lim_{t\to\infty}\frac{O_n(t)}{t}$$

The rate in = rate out principle follows because by definition

$\lim_{t\to\infty}\frac{I_n(t)}{t}$ = Mean rate (average number of times per time unit) at which the process enters State n = expected rate in

$\lim_{t\to\infty}\frac{O_n(t)}{t}$ = Mean rate at which the process leaves State n = expected rate out

6A.1.3 The M/M/1 Model (6.2.4)

Derivation of Steady-State Operating Characteristics—Probabilities and Mean Performance Measures

1. *Probability of exactly n jobs in the system:* P_n

   Using the rate diagram in Figure 6.10, it is straightforward to construct the balance equations for the $M/M/1$ queue using the approach in Section 6.2.3 (see also Table 6.2) with $\lambda_n = \lambda$ and $\mu_n = \mu$ for all n. From these equations, it follows that

   $$P_1 = (\lambda / \mu)P_0 = \rho P_0, P_2 = \rho^2 P_0, \ \ldots\text{and consequently}, P_n = \rho^n P_0$$

   To determine P_0, invoke the condition $\sum_{n=0}^{\infty} P_n = 1$, which results in $P_0 \sum_{n=0}^{\infty} \rho^n = 1$. Observing that $\sum_{n=0}^{\infty} \rho^n$ is the sum of a geometric series for which it is known that $\sum_{n=0}^{\infty} \rho^n = \frac{1}{1-\rho}$ for any $\rho < 1$ renders

   $$P_0 = 1 - \rho$$

2. *Probability of k or more jobs in the system:* $P(n \ge k)$

$$P(n \ge k) = \sum_{n=k}^{\infty} P_n = \sum_{n=k}^{\infty}(1-\rho)\rho^n = (1-\rho)\rho^k \sum_{n=k}^{\infty}\rho^{n-k}$$

$$= (1-\rho)\rho^k \sum_{n=0}^{\infty}\rho^n = (1-\rho)\rho^k \frac{1}{1-\rho} = \rho^k$$

where $\sum_{n=0}^{\infty} \rho^n$ is the sum of a geometric series for which it is known that $\sum_{n=0}^{\infty} \rho^n = \dfrac{1}{1-\rho}$ for any $\rho < 1$.

3. *Mean number of jobs in the system: L*

   From the definition of expected value

$$L = \sum_{n=0}^{\infty} nP_n = \sum_{n=0}^{\infty} n(1-\rho)\rho^n = (1-\rho)\rho \sum_{n=0}^{\infty} n\rho^{n-1}$$

$$= (1-\rho)\rho \sum_{n=0}^{\infty} \frac{d}{d\rho}\rho^n = (1-\rho)\rho \frac{d}{d\rho}\left(\sum_{n=0}^{\infty}\rho^n\right)$$

$$= (1-\rho)\rho \frac{d}{d\rho}\left(\frac{1}{1-\rho}\right) = (1-\rho)\rho\left(\frac{1}{(1-\rho)^2}\right) = \frac{\rho}{(1-\rho)}$$

where one uses that the derivative of x^n is nx^{n-1}, that is, $\dfrac{d}{dx}\left(x^n\right) = nx^{n-1}$, and the fact that $\dfrac{d}{d\rho}\left(\sum_{n=0}^{\infty}\rho^n\right) = \sum_{n=0}^{\infty}\dfrac{d}{d\rho}\left(\rho^n\right) = \sum_{n=0}^{\infty} n\rho^{n-1}$ for $\rho < 1$

4. *Mean number of jobs in the queue: L_q*

   From the definition of expected value and using that $P_0 = 1 - \rho$, one gets

$$L_q = \sum_{n=1}^{\infty}(n-1)P_n = \sum_{n=0}^{\infty} nP_n - \sum_{n=1}^{\infty} P_n = L - (1 - P_0)$$

$$= L - \rho = \frac{\rho}{1-\rho} - \rho = \frac{\rho - (1-\rho)\rho}{1-\rho} = \frac{\rho^2}{1-\rho}$$

5. *Mean time a job spends in the system and in the queue: W and W_q*

   The expressions follow directly from Little's law, $\rho = (\lambda/\mu)$, and the expressions for L and L_q shown earlier.

$$W = \frac{L}{\lambda} = \frac{\rho}{(1-\rho)\lambda} = \frac{\lambda}{\mu\left(1-\dfrac{\lambda}{\mu}\right)\lambda} = \frac{1}{\mu-\lambda};\ W_q = \frac{L_q}{\lambda} = \frac{\rho^2}{(1-\rho)\lambda} = \frac{\rho}{\mu\left(1-\dfrac{\lambda}{\mu}\right)} = \frac{\rho}{\mu-\lambda}$$

6A.2 Queuing Models with Generally Distributed Service Times

As noted in Chapter 6, the assumption of exponentially distributed service times may sometimes be questionable. In the following, two queuing models that allow the service times to follow any distribution will be considered. The models are the $M/G/1$ model and the $M/G/\infty$ model.

6A.2.1 The M/G/1 Queuing Model

The only difference in model assumptions between the *M/G/*1 model and the *M/M/*1 model studied in Section 6.2.4 is that the service time distributions in the *M/G/*1 model can take on any form. Thus, the *M/G/*1 model represents a queuing system with a single server and a single queue with no capacity restrictions, Poisson arrivals from an infinite calling population, and an FCFS queue discipline. The *M/G/*1 model is not a birth-and-death-process, as the remaining time until the next service completion (death) is not exponentially distributed (except for the special case of *M/M/*1). Typically, queuing models with generally distributed interarrival or service times are very difficult to analyze. Fortunately, for the *M/G/*1 model, some simple steady-state results are available. These results only require knowledge about the mean and variance of the service time. Assuming as before that μ denotes the service rate, the mean service time is $1/\mu$. The service time variance is in the following denoted by σ^2. As for the *M/M/*1 system, the *M/G/*1 queuing system will reach steady state if $\rho = \lambda/\mu < 1$. The steady state results for the *M/G/*1 model follow. The proofs are omitted, as they are beyond the scope of this book. Interested readers are referred to, for example, Kleinrock (1975).

Probability of zero jobs in the system: $P_0 = 1 - \rho$

Expected number of jobs in the queue: $L_q = \dfrac{\lambda^2\sigma^2 + \rho^2}{2(1-\rho)}$

Expected number of jobs in the system: $L = L_q + \rho$

Expected waiting time in the queue: $W_q = \dfrac{L_q}{\lambda}$

Expected time in the system: $W = W_q + \dfrac{1}{\mu}$

The expression for L_q is commonly referred to as the *Pollaczek–Khintchine formula* after the persons who originally derived it. Because of its simplicity and usefulness, it is often considered one of the most important results in queuing theory. An important observation is that L_q, L, W_q, and W are all increasing with the service time variance, σ^2, when the mean service time is kept constant. Thus, the performance of the queuing system is improved if the variability of the service times can be reduced!

6A.2.2 The M/G/∞ Queuing Model

The *M/G/*∞ model describes a queuing system with Poisson arrivals from an infinite calling population to a service station with an infinite number of identical servers. Thus, a queue will never form, as all customers will be assigned to a server immediately upon arrival. As there is an infinite server capacity, the *M/G/*∞ system always reaches steady state; moreover, it is not a birth-and-death process unless the service times are exponentially distributed. The *M/G/*∞

model has been used for the analysis of telecommunication systems. It is also useful, for example, when analyzing the number of customers in self-service facilities such as stores or amusement parks, or the amount of inventory in a storage facility. A key result for the analysis of this type of queuing system is Palm's theorem (Palm 1938), which states that the total occupancy level in an $M/G/\infty$ system is Poisson distributed with mean λ/μ, where the mean arrival rate is λ, and μ is the mean service rate for each individual server. This means that the probability of finding n customers in the system is

$$P_n = \frac{(\lambda / \mu)^n}{n!} e^{-(\lambda/\mu)} \text{ for } n = 0, 1, 2, \ldots$$

The expected values L and W are then easily obtained from their definitions (note: $L_q = 0$, $W_q = 0$, as no queue can form):

$$L = \sum_{n=0}^{\infty} nP_n = \frac{\lambda}{\mu} \quad W = \frac{L}{\lambda} = \frac{1}{\mu}$$

To illustrate the $M/G/\infty$ model's usefulness in modeling inventory systems, consider a single item stocked at a single location. All demands that cannot be satisfied immediately from stock are backordered. Demand follows a Poisson process with arrival rate λ, and the inventory is controlled by a base-stock policy with base-stock level S. This means that whenever a demand occurs (always for one unit), a replenishment of exactly the same amount is ordered from an external supplier. As a result, the inventory position (=outstanding orders on their way from the suppliers + inventory on hand – backorders) always equals S. It also means that the inventory level (=inventory on hand – backorders) equals S minus the outstanding orders. Assuming that the replenishment lead times are stochastic, with mean $1/\mu$, and independent (i.e., orders may cross each other in time), the number of outstanding orders corresponds to the occupancy level in the $M/G/\infty$ system and is therefore Poisson distributed with mean λ/μ. Thus, the probability that the inventory level is k, $P(IL = k)$, is

$$P(IL = k) = P(\text{number of outstanding orders} = S - k)$$

$$= P_{S-k} = \frac{(\lambda / \mu)^{S-k}}{(S-k)!} e^{-(\lambda/\mu)} \text{ for } k = S, S-1, \ldots, -\infty$$

Note that a negative inventory level of $-k$ means that there are k units backordered and no inventory on hand. On the other hand, a positive inventory level of k units means that there are k units on hand and no backorders.

It is now straightforward to calculate the expected inventory on hand and the expected number of backorders. If the holding cost rate and backorder cost rate per unit and time unit are known, the associated expected holding and backorder costs follow directly. It is then easy to optimize the base-stock level S to minimize these expected costs.

Discussion Questions and Exercises

1. Demonstrate that each of the following situations can be represented as a basic queuing process by identifying its components. How would you define the corresponding queuing system?
   a. The checkout stations in a supermarket
   b. The tollbooth at a road-toll checkpoint
   c. An auto shop or garage
   d. An emergency rescue unit
   e. A plumber
   f. A bus station
   g. The materials-handling equipment at a production facility
2. A mechanic, Bob, requires on average 3 hours to complete a repair. Furthermore, the repair time closely follows an exponential distribution. Bob realizes that by hiring his nephew Bill as an assistant, he can reduce the average repair time to 2 hours. The repair time will still be exponential because the job structure is unchanged. For each repair job, Bob gets paid $220 if he finishes in less than 2 hours and $180 if he cannot. Assume that Bob does not want to decrease his expected payment per repair job. What is the maximum amount per job that Bob would be willing to pay Bill?
3. Purchasing requests arrive to an agent at a rate of six per day. The time between arrivals is exponentially distributed. The agent typically requires 1 hour to process a request. The processing time is also exponentially distributed. Assuming 8-hour working days:
   a. What is the cycle time (the average time that a request spends in the process)?
   b. What is the WIP inventory (the average number of requests in the process)?
   c. How often is the WIP entirely replaced with a new set of requests?

4. A small branch office of a local bank has two tellers for which customers line up in a single queue. The customers are being served on an FCFS basis. It has been determined that the steady-state probability distribution for finding exactly n customers in the system, $\{P_n, n = 0, 1, 2 \ldots\}$, is $P_0 = (2/18)$, $P_1 = (5/18)$, $P_2 = (6/18)$, $P_3 = (3/18)$, $P_4 = (1/18)$, $P_5 = (1/18)$, and $P_n = 0$ for $n = 6, 7, \ldots$.
   a. Determine the expected number of customers in the system, L.
   b. Determine the expected number of customers in the queue, L_q.
   c. Determine the expected number of customers in the service facility.
   d. Assuming that the mean arrival rate is six customers per hour, determine the expected time a customer spends in the system, W, and in the queue, W_q.
   e. Determine the expected service time—the expected time a customer spends in the service facility—assuming that the two tellers are identical.
5. It has been concluded that a single-server queuing system with exponentially distributed service and interarrival times can be modeled as a birth-and-death process with state-dependent mean service and arrival rates, μ_n and λ_n, respectively.

$$\mu_n = \begin{cases} n & \text{for } n = 0, 1, 2, 3 \\ 0 & \text{otherwise} \end{cases}$$

$$\lambda_n = \begin{cases} 3 - n & \text{for } n = 0, 1, 2, 3 \\ 0 & \text{otherwise} \end{cases}$$

   a. Construct the corresponding rate diagram.
   b. Calculate the stationary probabilities for finding exactly n customers in the system, $\{P_n, n = 0, 1, 2, 3, \ldots\}$.
   c. Determine the expected number of customers in the queuing system, L, and in the queue, L_q. Also determine the expected time a customer spends in the system, W, in the queue, W_q, and in the service facility, W_s.
6. A queuing process is modeled as a birth-and-death process with mean arrival rates $\lambda_0 = 2$, $\lambda_1 = 4$, $\lambda_2 = 3$, $\lambda_3 = 1$, $\lambda_n = 0$ for $n > 3$ and mean service rates $\mu_1 = 2$, $\mu_2 = 4$, $\mu_3 = 1$, and $\mu_n = 0$ otherwise.
   a. Construct a rate diagram
   b. Develop the balance equations and solve them to obtain $P_n, n \geq 0$.
   c. Determine L, L_q, W, and W_q assuming that the queuing process has two parallel servers.

7. Consider a birth-and-death process with mean birth and death rates, λ_n and μ_n, shown in Table 6.5.
   a. Construct a rate diagram.
   b. Calculate the stationary probabilities for finding the process in state n, $n = 0, 1, 2, 3, \ldots$.
   c. If a queuing process with two identical servers fits this birth-and-death process, what are the mean arrival and service rates for each of the two servers when they are busy serving customers?

TABLE 6.5

Birth-and-Death Process for Exercise 6.7

State (n)	Mean Birth Rate (λ_n)	Mean Death Rate (μ_n)
0	2	0
1	2	2
2	2	4
3	0	4

8. The mean service and arrival rates for a birth-and-death process describing a queuing system with two parallel servers are given in Table 6.6.
   a. Construct the corresponding rate diagram.
   b. Develop the balance equations and solve them to determine the stationary probabilities for finding n customers in the system P_n, $n = 0, 1, 2, 3, 4, \ldots$.
   c. Determine the expected number of customers in the system, L, and in the queue, L_q, as well as the expected time spent in the system, W, and in the queue, W_q.
   d. Identify the type of queuing model that corresponds to this birth-and-death process. Use the formulae derived for this queuing model to calculate P_n, $n = 0, 1, 2, 3, 4, \ldots, L, L_q, W, W_q$.

TABLE 6.6

Birth-and-Death Process for Exercise 6.8

State	Mean Birth Rate (λ_n)	Mean Death Rate (μ_n)
0	4	0
1	4	1
2	4	2
3	4	2
4	0	2

9. A gas station has a single automated carwash. Cars arrive to the gas station according to a Poisson process with an average of 30 cars per hour. One-third of the customers also want a carwash; that is, for every customer, there is a 33.33 percent chance it needs a carwash. The mean service rate is 10 cars per hour. Although there are several different washing programs to choose from, it is a bit of a stretch to say that the service times follow an exponential distribution. Still, as a first cut analysis, management has decided to make this assumption and model the queuing system as a birth-and-death process. It has been observed that customers balk from the carwash queue when it increases in length. More precisely, the probability that a customer will balk is $n/3$ for n =1, 2, 3 customers in the carwash system. If there are more than three customers in the system, no customers will join the queue.
   a. Construct a rate diagram for this queuing process.
   b. Formulate the balance equations and solve them to determine the steady-state probability distribution for the number of cars in the carwash.
   c. What is the expected utilization of the carwash?
   d. Determine the expected waiting time in line for those customers who join the queue.
10. A small machine shop consists of three sensitive machines that break down frequently but independently of each other. The company has two service technicians on standby to repair the machines as soon as they break down. For *each* fully functional machine, the breakdown rate is 0.1 times per hour, and the time between breakdowns is exponentially distributed. The service technicians work independently of one another, and due to restricted space around the machines, only one technician at a time can work on a machine. For each technician, the average service rate is 0.2 machines per hour, and the service times follow an exponential distribution. The operations manager has realized that this system can be modeled as a birth-and-death process with state-dependent arrival and service rates. He has also figured out what these arrival and service rates are, expressed in λ = the machine breakdown rate (the same for all machines) and μ = the service rate for each of the technicians. Table 6.7 shows a summary of what he has discovered.
   a. Unfortunately, the operations manager lacks training in analyzing queuing systems, and his analysis of the arrival and service rates is not entirely correct. Show the correct service and arrival rates for the system by drawing a state/rate diagram.

   The operations manager is very stubborn and does not admit that he is wrong, so he orders you to continue the analysis based on the arrival rates he has specified.

TABLE 6.7

Summary Information for Exercise 6.10

Number of Broken-Down Machines	Break Down Rate (=Arrival Rate)	Service Rate
0	3λ	0
1	2λ	2μ
2	λ	2μ
3	0	2μ

b. Construct the state/rate diagram corresponding to the given arrival and service rates. Develop the balance equations and solve them to determine the steady-state probability distribution. What is the probability that all the machines are functioning?

c. Calculate the expected number of broken machines and the expected time for a machine to be operating again after it has broken down.

11. Consider an $M/M/2$ model. Derive the following expressions by constructing the rate diagram and solving the balance equations.

$$\rho = \lambda/(2\mu)$$
$$P_0 = (1 - \rho)/(1 + \rho)$$
$$P_n = 2\rho^n P_0$$
$$L = 2\rho/(1 - \rho^2)$$
$$L_q = 2\rho^3/(1 - \rho^2) = L - 2\rho$$

(*Hint*: Look at the technique used to derive the corresponding results for the $M/M/1$ queue in Appendix 6A.)

12. A pharmaceutical company has formed a team to handle FDA applications for approval of new drugs. Requests for applications arrive at a rate of one every year. The arrivals follow a Poisson process. On average, the team processes an application in 9 months. The company estimates that the average cost of revenue losses associated with waiting for a new drug to be approved is $100,000 per drug per month. The cost of a team is $50,000 per month.

a. Estimate the total cost due to revenue losses in the current process.

b. Should the company add a second team?

13. A telecommunications company receives customer calls at a rate of 25 per hour. The interarrival times are exponentially distributed. Each call requires, on the average, 20 minutes. The times for each call also follow an exponential distribution.

a. What is the minimum number of customer service agents needed for this process to reach steady state?

b. Create a plot that shows the expected waiting time as a function of the number of agents.

c. The telecommunications company would like to limit the average waiting to 2 minutes or less. How many agents should the company hire?

14. A mechanic is responsible for keeping two machines in working order. The time until a working machine breaks down is exponentially distributed with a mean of 12 hours. The mechanic's repair time is exponentially distributed with a mean of 8 hours.

a. Show that this queuing process is a birth-and-death process by defining the states, $n = 0, 1, 2, 3 \ldots$, specifying the state-dependent mean arrival and service rates, λ_n and μ_n, for $n = 0, 1, 2, 3, \ldots$, and constructing the rate diagram. Also, specify the criteria defining a birth-and-death process and make sure that this process satisfies these criteria.

b. Specify the balance equations and use them to determine the steady-state probability distribution for finding n customers in the system P_n, $n = 0, 1, 2, 3, \ldots$.

c. Use the definitions and Little's law to determine L, L_q, W, and W_q.

d. Determine the fraction of time that at least one machine is working.

15. Consider the description of Problem 14.

a. Identify an appropriate queuing model that adequately describes the queuing process. Use the corresponding formulae to determine P_n, $n = 0, 1, 2, 3, \ldots, L, L_q, W, W_q$.

b. Determine the fraction of time that the mechanic is busy.

c. Determine the fraction of time that both machines are operational.

16. A workstation has enough storage space for storing three jobs in addition to the one being processed. Excess jobs are routed to another workstation, which is used solely to handle this overflow of jobs from the regular workstation. Jobs arrive to the regular workstation according to a Poisson process with a mean of eight jobs per day. The process time in the regular workstation is exponentially distributed with a mean of 30 minutes. A workday consists of 8 hours.

a. Construct the rate diagram describing this queuing process.

b. Develop the balance equations and use them to determine the steady-state probability distribution for finding n customers in the system $\{P_n, n = 0, 1, 2, 3, \ldots\}$.

c. Identify a specialized queuing model that describes this queuing process and use the corresponding formulae to determine the steady-state probability distribution for finding n customers in the system $\{P_n, n = 0, 1, 2, 3, \ldots\}$. Compare the results with Part (b).

d. Determine the fraction of time that the storage space at the regular work center is adequate to handle the demand.

e. What is the average number of jobs stored at the regular work center (excluding the job being processed), and what is the average waiting time in line before processing?

17. Plans are being made to open a small gas station in a central location in Springfield. The owner must decide how much space should be provided for waiting cars. This is an important decision, since land prices are high. It is assumed that customers (cars) arrive according to a Poisson process with a mean rate of 20 per hour. However, if the waiting area is full, customers will go elsewhere. The time it takes to service one customer is exponentially distributed with a mean of 4 minutes. The gas station will have two gas pumps. Before making a decision about how much land to acquire, the owner wants to analyze the situation further using queuing modeling.

    a. Identify an appropriate queuing model to describe this queuing process. Use the corresponding formulae to calculate the fraction of time that sales will be lost for the following options regarding spaces for waiting cars (excluding the ones being served, i.e., filling gas or paying):

        I. 0 spaces for waiting cars

        II. 2 spaces for waiting cars

        III. 4 spaces for waiting cars

    b. Assuming the gas station is open 24 hours a day, what is the expected number of lost customers per day for alternatives (i), (ii), and (iii) in Part (a)? If, on average, a customer generates a profit of $3 for the owner of the gas station, what is the expected lost profit per day under alternatives (i), (ii), and (iii)?

18. A facility management company has recently acquired a number of new commercial properties within the city of Greenfield. They are faced with the problem of hiring a number of on-call janitors, who are responsible for fixing emergency problems that arise at these different facilities. The question is how many locations to assign to each janitor. The janitors work independently on the properties that they are responsible for serving. When the janitor gets a call from a location other than the one he/she is at, the request is placed in a queue and served on a FIFO basis. When visiting a certain location, the janitor will attend to all matters at this facility. This means that a location that has called for the janitor will not call again even if new matters arise. We can, therefore, consider a facility that has placed a call to the janitor as "non-operational." If there are no matters that need janitorial attention, the facility is considered fully operational. For each fully operational facility, the times between emergency calls to

the janitor are exponentially distributed with a mean of 2.5 hours. The time it takes for the janitor to fix the problems at a location is exponentially distributed with a mean of 20 minutes, including travel time. The facility management company has determined a service level criterion, specifying that at least 75 percent of the time the facilities should receive immediate emergency service from a janitor.

a. Identify a queuing model that adequately describes this process of emergency janitorial assistance. Use the corresponding steady-state formulae to determine the largest number of facilities that can be assigned to a janitor without violating the specified service requirement.
b. Given the maximum number of facilities assigned to each janitor in Part (a), what is the expected fraction of time that a janitor is busy with emergency calls? What is the expected number of facilities not fully operational? What is the expected waiting time before a facility gets emergency assistance?

19. A company has a central document-copying service. Arrivals can be assumed to follow a Poisson process, with a mean rate of 15 per hour. It can be assumed that service times are exponentially distributed. With the present copying equipment, the average service time is 3 minutes. A new machine can be leased that has a mean service time of 2 minutes. The average wage of the people who bring the documents to be copied is $8 an hour.
    a. If the machine can be leased for $5 per hour more than the old machine, should the company replace the old machine?
    b. Suppose that the new machine is leased. How much space (e.g., number of chairs) must be provided for people to wait to guarantee that at least 90 percent of the time this space will be sufficient?
20. The manager of a movie theater would like to predict the consequences of adding a second ticket clerk. Data show that arrivals to the theater are Poisson distributed at a rate of 250 per hour, and service times are exponentially distributed with a mean of 12 seconds. The manager has also estimated that the loss in concession revenue when customers wait in the ticket line is $1.50 per waiting hour. This revenue is not recovered once the customers enter the theater. Given that it costs $5 per hour to have a ticket clerk, should the manager add a second server?
21. A process has a bottleneck resource that consists of specialized equipment. Jobs arrive to this machine at a rate of 40 per hour (according to a Poisson arrival process). The processing times average 1 minute and are exponentially distributed. Compare the performance (e.g., average cycle time through the bottleneck) of the current process with the following alternatives:
    a. Add a second identical machine in the bottleneck.
    b. Replace the current machine with one that is twice as fast.

22. Most arrivals to a hospital emergency room are not considered emergencies, in that the patients can wait to see a doctor until they complete the proper forms. At County Hospital, emergency patients arrive at a rate of six per hour. This process is, to no one's surprise, a Poisson arrival process. It takes the admission clerk approximately 10 minutes to fill out the patient's form. The length of time is not exact and, in fact, follows an exponential distribution. As soon as the form is filled out, the patient is examined. The chief of staff is concerned with the quality of the operations and wants to know the expected performance of the current process. Use queuing theory to evaluate the performance of this process.
23. A fast food chain is opening a new restaurant in a shopping mall. The restaurant needs to hire a cashier and has two prime candidates. Candidate 1 is experienced and fast but demands a higher salary. Candidate 2 is inexperienced and slower but also has more modest salary claims. The question is which one the restaurant manager should choose. Both candidates can be considered to have exponentially distributed service times, Candidate 1 with a mean of 1 minute and Candidate 2 with a mean of 1.5 minutes. The customers arrive according to a Poisson process with a mean of 30 customers per hour. The waiting cost has been estimated at $6 per minute for each customer until they have been fully served and have paid for their meal. Determine the maximum difference in monthly salary that would justify hiring Candidate 1 instead of Candidate 2 when the objective for the restaurant manager is to minimize the expected total cost. Assume that there are 30 workdays in a month, and each workday is 8 hours long.
24. Consider a cashier at a grocery store. Customers arrive at this cashier's station according to a Poisson process with a mean of 25 customers per hour. The cashier's service times are exponentially distributed with a mean of 2 minutes. By hiring a person to help the cashier bag the groceries and occasionally assist customers by bringing their groceries to their car, the mean service time for the checkout station can be reduced to 1 minute. The service times are still exponentially distributed. Hiring a cashier's aid would incur a cost of $10 per hour for the store. Furthermore, it has been estimated that the store is facing a waiting cost (lost profit and future lost sales) of $0.10 per minute that a customer spends in line or in the checkout station. The store manager wants to determine whether proceeding to hire a cashier's aid is a good idea or not. The criterion is that the manager wants to minimize the expected total costs. What is your recommendation after analyzing the situation using an appropriate queuing model?

25. The arrival and service rates in Table 6.8 pertain to telephone calls to a technical support person in a call center on a typical day. Both the interarrival times and the service times are exponentially distributed.
    a. Determine the average time the callers wait to have their calls answered for each period.
    b. What should be the capacity of the switchboard so that it is capable of handling the demand 95 percent of the time? The capacity of the switchboard is measured as the number of calls that can be placed on hold plus the call that the technical support person is answering.

TABLE 6.8

Arrival Rate and Service Times for Exercise 6.25

Period	Incoming Rate (Calls per Hour)	Service Time (in Minutes)
Morning	3.2	8
Afternoon	3.5	10
Evening	7.8	5

26. Truckloads of seasonal merchandise arrive to a distribution center within a 2-week span. Because of this, merchandise-filled trucks waiting to unload have been known to back up for a block at the receiving dock. The increased cost due to unloading delays, including truck rental and idle driver time, is of significant concern to the company. The estimated cost of waiting and unloading for truck and driver is $18 per hour. During the 2-week delivery time, the receiving dock is opened 16 hours per day, 7 days per week, and can on average unload 35 trucks per hour. The unloading times closely follow an exponential distribution. Full trucks arrive during the time the dock is opened at a mean rate of 30 per hour, with interarrival times following an exponential distribution. To help the company get a handle on the problem of lost time while trucks are waiting in line or unloading at the dock, find the following measures of performance:
    a. The average number of trucks in the unloading process
    b. The average cycle time
    c. The probability that there are more than three trucks in the process at any given time
    d. The expected total daily cost of having the trucks tied up in the unloading process
    e. It has been estimated that if the storage area were to be enlarged, the sum of waiting and unloading costs would be cut in half next year. If it costs $9000 to enlarge the storage area, would it be worth the expense to enlarge it?

27. A case team completes jobs at a rate of two per hour, with actual processing times following an exponential distribution. Jobs arrive at a rate of about one every 32 minutes, and the arrival times are also considered exponential. Use queuing theory to answer the following questions.
    a. What is the average cycle time?
    b. What is the cycle time efficiency? (*Hint*: remember that cycle time includes both processing time and waiting time.)
28. The manager of a grocery store is interested in providing good service to the senior citizens who shop in his store. The manager is considering the addition of a separate checkout counter for senior citizens. It is estimated that the senior citizens would arrive at the counter at an average of 30 per hour, with interarrival times following an exponential distribution. It is also estimated that they would be served at a rate of 35 per hour, with exponentially distributed service times.
    a. What is the estimated utilization of the checkout clerk?
    b. What is the estimated average length of the queue?
    c. What is the estimated average waiting time in line?
    d. Assess the performance of this process.
    e. What service rate would be required to have customers average only 8 minutes in the process? (The 8 minutes includes both waiting and service time.)
    f. For the service rate calculated in Part (e), what is the probability of having more than four customers in the process?
    g. What service rate would be required to have only a 10 percent chance of exceeding four customers in the process?
29. A railroad company paints its own railroad cars as needed. The company is about to make a significant overhaul of the painting operations and needs to decide between two alternative paint shop configurations.

    Alternative 1: Two "wall to wall" manually operated paint shops, where the painting is done by hand (one car at a time in each shop). The *annual* joint operating cost for each shop is estimated at $150,000. In each paint shop, the average painting time is estimated to be 6 hours per car. The painting time closely follows an exponential distribution.

    Alternative 2: An automated paint shop at an *annual* operating cost of $400,000. In this case, the average paint time for a car is 3 hours and exponentially distributed.

    Regardless of which paint shop alternative is chosen, the railroad cars in need of painting arrive to the paint shop according to a Poisson process with a mean of 1 car every 5 hours (=the interarrival time is 5 hours). The cost for an idle railroad car is $50 per hour. A car is considered idle as soon as it is not in traffic; consequently, all the time spent in the paint shop is considered idle time. For efficiency reasons, the paint shop operation is running 24 hours, 365 days a year, for a total of 8760 hours per year.

a. What is the utilization of the paint shops in Alternatives 1 and 2, respectively? What are the probabilities for Alternatives 1 and 2, respectively, that no railroad cars are in the paint shop system?
b. Provided that the company wants to minimize the total expected cost of the system, including operating costs and the opportunity cost of having idle railroad cars, which alternative should the railroad company choose?

30. At Martha's café, Martha herself operates the espresso machine. Customers arrive and demand cups of espresso according to a Poisson process with an average rate of 30 cups per hour. The time it takes for Martha to make a cup of espresso is exponential with a mean of 75 seconds.
    a. Determine L, L_q, W, and W_q for this queuing system. Show that the results are the same if you use the expressions available for $M/M/1$ and $M/G/1$ systems, respectively.
    b. Assume that Martha buys a new fully automated espresso machine that makes a cup of espresso in exactly 75 seconds always. Determine L, L_q, W, and W_q under these new conditions.
    c. Determine the ratio of the mean time in the queue, L_q, for the systems in Parts (a) and (b). What are your conclusions?
31. Consider a bank office where customers arrive according to a Poisson process with an average arrival rate of λ customers per minute. The bank has only one teller servicing the arriving customers. The service time is exponentially distributed, and the mean service rate is μ customers per minute. It turns out that the customers are impatient and are only willing to wait in line for an exponentially distributed time with a mean of $1/\mu$ minutes. Assume that there is no limitation on the number of customers that can be in the bank at the same time.
    a. Construct a rate diagram for the process and determine what type of queuing system this corresponds to on the form $A1/A2/A3$.
    b. Determine the expected number of customers in the system when $\lambda = 1$ and $\mu = 2$.
    c. Determine the average number of customers per time unit that leave the bank without being served by the teller when $\lambda = 1$ and $\mu = 2$.
32. The BlockBlaster store has only one cashier working at a time. Assume that the customers arrive to the cashier according to a Poisson process with average rate of one customer per minute. When comparing the service times of two cashiers that work different shifts, some differences in service time distributions were noted. Cashier A has an exponentially

distributed service time with mean of 0.5 minutes per customer. Cashier B also has a mean service time of 0.5 minutes per customer, but the time distribution is uniform on the interval [0.2, 0.8] minutes. The management of BlockBlaster wants to know whether these differences in service time distributions affect the expected number of customers waiting in line and the associated expected time in the queue. If there are differences, the management team wants to know which cashier performs better so that the other cashier may learn from the best practice performance, and the overall process can be improved.

33. The engineering firm AllDesign is reviewing the process for handling customer requests for a certain type of assignment. The requests arrive according to a Poisson process with a mean arrival rate λ = one request per day, and the company considers two different process alternatives. The first one is to let anyone in the office that is available take on the assignment when it arrives. The processing time is then assumed to be identical and exponentially distributed with a rate μ = three requests/day for all employees. It is also assumed that someone at the firm can start working on the assignment immediately when it arrives. The second alternative is to have one of the employees specialize in these assignments and handle all the requests of this type that arrive to the firm. This person is then expected to process on average four requests per day, and the standard deviation of the service time is estimated to be 0.15 days. The distribution of the service time is unknown.
    a. What queuing models adequately describe the two alternative process designs?
    b. Which process design should the firm choose if it wants to have as few open requests as possible?

References

Bunday, B.D., and Scraton, R.E. 1980. The G/M/r machine interference model. *European Journal of Operational Research* 4: 399–402.

Fitzsimmons, J.A., and Fitzsimmons, M.J. 1998. *Service management: Operations, strategy, and information technology*. New York: Irwin/McGraw-Hill.

Hillier, F.S., and Lieberman, G.J. 2010. *Introduction to operations research*. 9th ed. New York: McGraw-Hill.

Kleinrock, L. 1975. *Queuing systems—Volume I: Theory*. New York: John Wiley and Sons.

Little, J.D.C. 1961. A proof for the queuing formula L = λW. *Operations Research* 9: 383–387.

Palm, C. 1938. Analysis of the Erlang traffic formula for busy signal assignment. *Ericsson Technics* 5: 39–58.

7

Introduction to Simulation

Chapter 6 explored how analytical queuing models offer powerful means for understanding and evaluating queuing processes. However, the use of these analytical models is somewhat restricted by their underlying assumptions. The limitations pertain to the structure of the queuing system, the way variability can be incorporated into the models, and the focus on steady-state analysis. Because many business processes are cross-functional and characterized by complex structures and variability patterns, a more flexible modeling tool is needed. Simulation offers this flexibility and represents a powerful approach to the analysis and quantitative evaluation of business processes.

In general, to simulate means to mimic reality in some way. Simulation can be done, for example, through physical models such as wind tunnels, through simulators where pilots or astronauts train by interacting with a computer in virtual or artificial reality, or through computer-based models for the evaluation of a given technical system or process design. In the last case, simulation software is used to create a computer model that mimics the behavior of the real-world process.

The rapid development of computer hardware and software in recent years has made computer simulation an effective tool for process modeling and an attractive technique for predicting the performance of alternative process designs. It also helps in optimizing their efficiency. Simulation is useful in this context because business process design is a decision-making problem for which the following is true:

- Developing analytical mathematical models in many cases might be too difficult or perhaps even impossible.
- The performance of a process design typically depends heavily on the ability to cope with variability in interarrival and processing times (implying a need for a modeling tool that can incorporate several random variables).
- The dynamics are often extremely complex.
- The behavior over a period of time must be observed to validate the design.
- The ability to show an animation of the operation is often an important way to stimulate the creativity of the design teams.

Despite its many virtues, simulation is sometimes met with skepticism from practitioners and managers. Much of the reluctance toward using simulation

DOI: 10.1201/9781032617237-7

stems from the misconception that simulation is extremely costly and time-consuming. This is despite the many success stories showing that the savings from using simulation to improve process designs have far exceeded its costs (see Examples 7.1 and 7.2). In fact, with the advanced modeling tools that are currently available, the model development and experimentation phase might take only a few days or weeks, representing only a small fraction of the overall project development time. One of the most resource-consuming efforts that go into building a valid simulation model is understanding how the process operates. However, as discussed in Chapter 3, this process understanding is a necessity for achieving an effective process design and must be done whether or not simulation is used. Savings from the use of simulation are well documented, as illustrated in Examples 7.1 and 7.2.

Example 7.1: Improving Customer Service Levels

Due to an acquisition to support growth plans, a financial institution offering a range of mortgages, loans, and other financial products through responsible lending principles was reviewing its operational efficiency. In particular, the institution wanted to ensure that its new business and servicing capability for savings and mortgages was capable of absorbing the increase in workload from its acquisition. The primary business motivation was the need to maintain competitive customer service levels. Predicting customer contact workload can be challenging because it is often influenced by external factors such as market conditions and seasonal consumer behavior patterns. The institution was looking to gain insights into its existing back-office business processes. It wanted to understand the impact on its customer service levels caused by unexpected surges in workload or resource downtime. It also needed to understand how quickly the business could recover to normal operating levels after such events. A group of consultants worked with the institution's service delivery team to develop a clear understanding of the current processes and service levels. A business process simulation model was built and benchmarked against historical performance measures. The model was then used to analyze (1) the impact on service levels when the business experiences a surge in new business volumes for a predetermined period of time, (2) the length of time that would be typically taken to clear the resulting backlog at business-as-usual staffing levels, and (3) the effect on business-as-usual workload during system downtimes. This analysis demonstrated the benefits of implementing workload prioritization policies. ■

Example 7.2: Capacity Expansion in Connection with Facility Relocation

When a company decided to relocate to a new facility and expand its production capacity, it needed to analyze the impacts of changes in future demand. The relocation also provided the company with an opportunity to analyze its current production process and to seek improvements before increasing its capacity. Process simulation was used for capacity analysis and process improvement. Outputs from the process simulation were the primary inputs in determining the equipment and work-in-process inventory. The expansion alternatives from the process simulation identified the type and number of each piece of equipment needed to meet future capacity requirements. The queue lengths (average and maximum) were used to estimate the work-in-process inventory that needed to be accommodated at the new facility. Therefore, the facility layout design was fashioned along the recommended feasible alternatives and using other outputs from the process simulation, given the prescribed space constraints. The suggested layout design generated a throughput that exceeded the target production goal by 50 percent (Eneyo and Pannirselvam, 1999). ■

Simulation analysts must be effective in building the right model to represent the system under study. This means that the model must be of the right size and complexity to answer the important questions without including unnecessary details. The key for the analyst is to develop the ability to distill the idiosyncrasies of real processes and extract their essence so that they can be modeled in the simplest way possible but still retain the dynamic behavior needed to solve relevant problems. In that sense, and in the words of George E. P. Box, "all models are wrong, but some are useful.", the main advantage of simulation is that it is a tool that compresses time and space and thus enables a robust validation of ideas for process design and improvement. Successful business process implementations withstand the test of time and solve real problems. Their performance through months or years of operation makes them valid, with a demonstrated return-on-investment. The risk of new implementations of these processes is low because analysts know that with high probability, these processes will work as expected. Although predicting the performance and the associated return-on-investment of process innovations is difficult, a well-conceived simulation model can help and substantially reduce the risk of deploying a new process.

7.1 Simulation Models

Simulation models in general and computer-based models in particular can be classified in three different ways according to their attributes:

- Static or dynamic
- Deterministic or stochastic
- Discrete or continuous

A static model is used when time does not play a role in the actual system. For example, a model of a bridge does not depend on time. A deterministic model is such that the outputs are fully determined after the inputs are known. Take a computer model for calculating the water pressure of a pipe network as an example. The pressure is known once the designer selects the pipe diameters for all the pipe segments. A discrete model considers that individual units (i.e., the transient entities of the system) are important. Most manufacturing, service, and business processes are discrete. Business processes in general are represented as computer-based dynamic, stochastic, and discrete simulation models.

A computer-based simulation model is an abstraction of the actual business process, represented in the computer as a network of connected activities and buffers (or equivalently, a network of basic queuing systems) through which jobs or customers flow. To provide a correct representation of the process, the model also must capture the resources and various inputs needed to perform the activities.

Because process modeling is only one application area for simulation modeling, a more general terminology for describing a simulation model is to refer to it as an *abstraction of a system* rather than a process. Conceptually, a system is defined as a collection of entities that interact with a common purpose according to sets of laws and policies. Entities are either transient or resident. In the process terminology defined in Chapter 1, transient entities are the jobs that flow through the system, and the resident entities are the buffers, workstations, and resources that make up the process network. Laws are not under the process designer's control. Laws generally are represented by parameters. Sensitivity analysis is the experimentation used to determine the effect of changes in parameter values. Policies, on the other hand, are under the designer's control. A policy typically is implemented by changing input factors. Design of Experiments (DOE), an area of statistics, is used to determine the effect of changes in input factors and system structure.

Example 7.3: An Order-Fulfillment Process

Consider an order-fulfillment process in which customers call a toll-free number to place an order. In this process, the orders are transient entities, and the sales representatives answering phone calls are resident entities. The arrival pattern of phone calls is a law because it is beyond the process designer's immediate control. The number of sales representatives can be controlled; therefore, it is considered a policy. In a simulation model of this process, the number of sales representatives can be expressed as an input factor, so the analyst is able to experiment with different values, meaning that a sensitivity analysis can be carried out, and the performance of the process can be tested (as measured, e.g., by the cycle time). ■

In process design, models are used to study the behavior of a process. A process can be modeled symbolically, analytically, or with simulation. Symbolic models (or graphical tools) were introduced in Chapter 4. These models include process activity charts, process diagrams, and flowcharts. Symbolic models are quick and easy to develop and are easily understood by others. The main disadvantage of symbolic models is that they fail to capture the dynamics of the process. Various analytical models have been investigated in Chapters 4 through 6. Deterministic models were introduced in Chapters 4 and 5, and stochastic models were introduced in Chapter 6. If their underlying assumptions are valid for a system under study, these models can be a convenient way to evaluate quantitatively different process designs. The main disadvantage is that these underlying assumptions are often restrictive. Furthermore, because these tools offer no obvious graphical representation, they might appear abstract to the process designer.

Modern simulation software in a sense combines the descriptive strength of the symbolic models with the quantitative strength of the analytical models. It offers a graphical representation of the model through graphical interfaces as well as a graphical illustration of the system dynamics through plots of output data and animation of process operations. At the same time, it enables the estimation of quantitative performance measures through statistical analysis of output data. The main disadvantage of simulation is the time spent learning how to use the simulation software and how to interpret the results.

To summarize, the following are some of the main attributes that make simulation powerful:

- Simulation, like analytical modeling, provides a quantitative measure of performance (e.g., resource use or average waiting time).
- Simulation, unlike analytical and symbolic models, is able to take into consideration any kind of complex system variation and statistical interdependencies.
- Simulation is capable of uncovering inefficiencies that usually go undetected until the system is in operation.

An interesting distinction between a simulation model and an optimization model is that simulation is a tool for evaluating a given design, and an optimization model is a tool used to search for an optimal solution to a decision problem. That is, a simulation model is by nature descriptive, and an optimization model is by nature prescriptive, because it provides an optimal solution, prescribing a course of action to the user.

Until recently, simulation software packages could be used only as what-if tools. This means that given a simulation model, the designer would experiment with alternative designs and operating strategies to measure system performance. Consequently, in such an environment, the model becomes an experimental tool that is used to find an effective design.

However, modern simulation software packages merge optimization technology with simulation. The optimization consists of an automated search for the best values (near-optimal values) of input factors (the decision variables).

This valuable tool allows designers to identify critical input factors that the optimization engine can manipulate to search for the best values. The best values depend on the measure of performance that is obtained after one or several executions of the simulation model. The notion of optimizing simulations will be expanded in Chapter 10.

7.2 Monte Carlo Simulation

A model is an abstraction whose simplifying assumptions help us understand the complexity of a real system. Within the large spectrum of modeling paradigms, mathematical models are the most commonly used for decision-making in business. For instance, decisions regarding new products or services are often made based on profitability. The mathematical model to calculate profit is relatively simple:

$$\text{Profit} = \text{Revenue} - \text{Cost}$$

Both revenue and cost could be treated as known quantities if it is assumed that good predictive models are available to estimate their values. Basing decisions on expected values has some severe limitations. The most important one is that it ignores risk, which is typically defined as *the likelihood of an undesirable outcome*. Assessing risk entails evaluating the probability that an undesirable outcome will occur as well as its severity. The uncertainty of an outcome on its own does not mean that a decision is particularly risky. For example, purchasing a drink from a machine carries the risk that the machine will not deliver the product. However, most people would not consider the decision of inserting money into a soft-drink machine to carry great risk. First, past experiences indicate that it is likely that the machine will deliver. Second, the loss if the machine does not deliver is relatively small for most people. This indicates that the risk involved is a function of both the probability of what is undesirable and the loss associated with it.

Many models used for decision-making contain estimated values for input variables that are uncertain. Random variables emerge when it is not possible in a model to determine with certainty the value of a particular input. Analysts attempt to make informed guesses about the values such inputs will assume. The logic is that inserting the expected, or most likely, values for all the uncertain inputs in a model will provide the most likely value for the output. This logic, however, has two problems. First, the relationship between the inputs and the output of a model may not be linear (in the mathematical sense). In this situation, using expected values for the inputs does not generally result in the expected value of the output. Second, even if an accurate estimate of the expected value of the output is obtained, there is no information on its expected variability. For example, a financial analyst might determine that an investment opportunity of $1000 will result in an outcome of $10,000 within two years. However, how much variability exists in this predicted outcome? If after consideration of all the uncertainty, the potential outcomes are close to $10,000 (for instance, between $9000 and $11,000), then an investor might find this opportunity attractive and with low risk. If, on the other hand, the potential outcome exhibits large variability around $10,000 (for instance, between –$30,000 and $50,000), then a risk-averse investor might find this opportunity unattractive. This indicates that while two scenarios might have the same expected or average value, the risks involved could be quite different. Therefore, even if it is possible to determine the expected value of an outcome, it is equally important, or perhaps even more important, to assess the risk involved. To be able to measure risk, decision models must be able to produce the data necessary to evaluate the variability of an output.

The most common techniques to analyze risk are best-case/worst-case analysis, what-if analysis, and simulation. Of these approaches, simulation is the most complete because it contains the other two. Relevant questions that simulation helps address as a technique for assessing risk include:

- What is the probability of financial loss?
- What are the probabilities of several levels of potential losses?
- What is the probability of running out of inventory?
- What are the chances that a software development project will be completed on time?

The input variables in the profit calculation shown earlier can be decomposed into pieces that are more fundamental. For instance, suppose that sales depend on leads and that only a fraction of those leads (a so-called conversion rate) result in sales. Also, suppose that there is a cost for generating a lead, a unit cost that includes operating costs for producing and delivering the product, and a fixed cost. The profit equation now looks like this:

$$\text{Profit} = \text{Sales} * \text{Price} - \text{Leads} * \text{Cost per Lead} - \text{Sales} * \text{Unit Cost} - \text{Fixed Cost}$$

where Sales = Leads * Conversion Rate. This model now reveals some sources of uncertainty. Clearly, the conversion rate cannot be known with certainty. The unit cost could also exhibit variability as well as the number of leads. Price may be considered known, and the variability around the fixed cost could be negligible. The cost per lead could also be considered known; for example, in online advertising, where the cost per click on a banner may be fixed. In this case, a click is considered a lead. Therefore, the model has three known inputs (price, cost per lead, and fixed cost) and three uncertain inputs (leads, conversion rate, and unit cost). Monte Carlo simulation is an effective risk analysis tool for a model like this.

Monte Carlo simulation is the process of generating random values for uncertain inputs in a model, computing the output variable of interest, and repeating this process for many trials to obtain a distribution of the output. Monte Carlo refers to the casino in Monaco. Generating a number of random numbers between zero and one and assigning values less than or equal to 0.50 as heads and greater than 0.50 as tails is a Monte Carlo simulation of tossing a coin. There are three main reasons to perform risk analysis with Monte Carlo simulation: (1) formulation of the mathematical model; (2) data analysis of the uncertain inputs to determine their probability distributions; and (3) generation of simulation trials and analysis of results.

Example 7.4: Simple Monte Carlo Simulation on a Spreadsheet

Risk analysis with a Monte Carlo simulation process can be illustrated in a spreadsheet model. Microsoft Excel is used to illustrate the process. The task is to build a Monte Carlo simulation of the mathematical model to calculate profit, which was introduced above. Therefore, the first step has already been completed; there is a mathematical model. The known inputs into the model are

Price: $50
Fixed cost: $5000
Cost per lead: $0.65

It is assumed that historical records are available and that they were analyzed to determine the following probability distributions for the uncertain inputs:

Number of leads: Discrete uniform between 1200 and 2000
Conversion rate: Continuous uniform between 0.1 and 0.3
Cost per unit: Normal with an average of $30 and a standard deviation of $5

What is the probability that this product will be profitable? This is the type of risk-analysis question that the Monte Carlo simulation is equipped to address. Table 7.1 shows a spreadsheet Monte Carlo simulation of the profit model.

The known inputs, price, fixed cost, and cost per lead are in Cells B1, B2, and B3, respectively. The parameters of the probability distributions of the unknown inputs are in Cells E1 to F3. Each simulation trial (starting in Row 6) is modeled in the same way. The formulae in Row 6 are

Leads = RANDBETWEEN(E1, F1)
Conversion rate = E2 + (F2 – E2) * RAND()
Cost per unit = NORM.INV(RAND(), E3, F3)
Revenue = B6 * C6 * B1
Cost = B6 * C6 * D6
Profit = E6 – F6 – B2

TABLE 7.1

Spreadsheet Model of a Monte Carlo Simulation

	A	B	C	D	E	F	G
1	Price	$50		Number of leads	1,200	2,000	
2	Fixed cost	$5,000		Conversion rate	0.1	0.3	
3	Cost per lead	$0.65		Cost per unit	$30	$5	
4							
5	Trial	Lead i	Conversion rate	Cost per unit	Revenue	Cost	Profit
6	1	1,604	0.216	$19.32	$17,331.51	$6,697.29	$5,634.22
7	2	1,875	0.117	$41.32	$10,960.14	$9,057.16	($3,097.02)
8	3	1,373	0.267	$32.27	$18,357.54	$11,849.66	$1,507.88
9	4	1,742	0.247	$22.03	$21,523.45	$9,482.01	$7,041.45
10	5	1,565	0.167	$36.28	$13,099.59	$9,506.13	($1,406.54)
11	6	1,227	0.275	$25.29	$16,883.37	$8,538.27	$3,345.11
12	7	1,520	0.202	$33.73	$15,336.44	$10,347.07	($10.64)
13	8	1,881	0.154	$25.11	$14,479.64	$7,271.64	$2,208.00
14	9	1,806	0.274	$36.93	$24,739.50	$18,272.15	$1,467.36
15	10	1,944	0.224	$32.31	$21,809.97	$14,095.06	$2,714.91

The RANDBETWEEN() function generates discrete uniform random numbers between the specified parameters, in this case Cells E1 and F1. This is a useful function to generate integer values that are uniformly distributed, such as the number of leads. The RAND() function generates uniformly distributed values between zero and one. The conversion rate formula scales the random numbers to fall between 0.1 and 0.3. These parameter values are in Cells E2 and F2. The cost per unit follows a normal distribution. These random numbers are calculated with the inverse function of the normal distribution. The functions for revenue, cost, and profit follow the mathematical model introduced earlier.

Table 7.1 shows only the results of 10 simulation trials. All the following rows have the same equations, where the use of the dollar sign in Microsoft Excel indicates that the cell is locked and that the formula should always refer to that cell. Cells that are not locked are in relative positions and change accordingly when the formulae are copied. Various statistics related to the profit could be calculated with the value in Table 7.1. However, a sample with 10 trials is generally considered too small in a simulation study. The complete spreadsheet associated with Table 7.1 has 1000 trials. The output of interest is profit, and some relevant statistics can be computed for this random variable:

Minimum profit = – ($3989.62)
Maximum profit = $13,046.82
Mean profit = $1356.37

TABLE 7.2
Histogram of Profit Values from a Simulation with 1000 Trials

Range	Count	Cumulative Frequency (%)
Profit < −2,500	35	3.5
−2,500 ≤ Profit < 0	304	33.9
0 ≤ Profit < 2,500	353	69.2
2,500 ≤ Profit < 5,000	215	90.7
5,000 ≤ Profit < 7,500	78	98.5
7,500 ≤ Profit < 10,000	12	99.7
10,000 ≤ Profit < 12,500	2	99.9
Profit ≥ 10,000	1	100.0

The range of profit values goes from a loss of almost 4000 dollars to a profit of more than 13,000. Since the mean value is closer to the minimum value, the distribution of profits must be skewed to the right. These are valuable insights, but additional analysis is necessary to answer the question regarding the probability of losing money. A histogram can be used to answer this question. Table 7.2 shows the values associated with a histogram with eight bins. The first column indicates the range, the second column contains the number of observations in each range, and the third column shows the cumulative frequency. This histogram estimates that the probability of a loss is 33.9 percent, which corresponds to 339 observations with a negative profit. ■

Monte Carlo simulation has been shown to be very useful in areas such as finance, engineering, computational biology, physical sciences, and even law. The main characteristic of these applications is the existence of a mathematical model for which some inputs are uncertain. Business processes, however, cannot be represented as a static mathematical model. Their dynamic nature forces the use of a different approach, namely, discrete-event simulation.

7.3 Discrete-Event Simulation

Business processes usually are modeled as computer-based, dynamic, stochastic, and discrete simulation models. The most common way to represent these models in a computer is using discrete-event simulation. In simple terms, discrete-event simulation describes how a system with discrete flow units or jobs evolves over time. Technically, this means that a computer program tracks how and when state variables such as queue lengths and resource availabilities change over time. The state variables change as a result of an event (or discrete event) occurring in the system. An important characteristic is that discrete-event models focus only on the time instances when these discrete events occur. This feature allows significant time compression because it makes it possible to skip through all time segments between events when the state of the system remains unchanged. Therefore, in a short period of time, a computer can simulate a large number of events corresponding to a long real-time span.

To illustrate the mechanics of a discrete-event simulation model, consider an information desk with a single server. Assume that the objective of the simulation is to estimate the average delay of a customer. The simulation then must have the following state variables:

- Status of the server (busy or idle)
- Number of customers in the queue
- Time of arrival of each person in the queue

As the simulation runs, two events can change the value of these state variables: the arrival of a customer or the completion of service.

The arrival of a customer either changes the status of the server from idle to busy or increases the number of customers in the queue. The completion of service, on the other hand, either changes the status of the server from busy to idle or decreases the number of customers in the queue.

Because the state variables change only when an event occurs, a discrete-event simulation model examines the dynamics of the system from one event to the next. That is, the simulation moves the "simulation clock" from one event to the next and considers that the system does not change in any way between two consecutive events. For example, if a single customer is waiting in line at a grocery store, and the next event is the completion of service of the customer who is currently paying for his groceries, then discrete-event simulation does not keep track of how the customer in the line spends her waiting time. In other words, the simulation keeps track of the time when each event occurs but assumes that nothing happens during the elapsed time between two consecutive events.

Figure 7.1 summarizes the steps associated with a discrete-event simulation. The simulation starts with initializing the current state of the system and an event list. The initial state of the system, for example, might include some jobs in several queues as specified by the analyst. It also could specify, for instance, the availability of some resources in the process. The most common initial state is to consider that no jobs are in the process and that all resources are currently available and ready. The event list indicates the time when the next event will occur. For example, the event list initially might

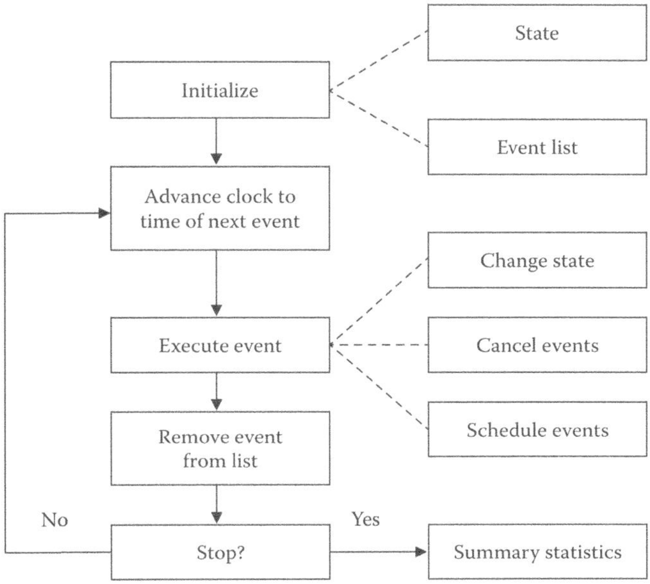

FIGURE 7.1
Discrete event simulation.

include the time of the first arrival to the process. Other events might be scheduled initially, as specified by the analyst.

Once the initialization step is completed, the clock is advanced to the next event in the event list. The next event is then executed. Three activities are triggered by the execution of an event. First, the current state of the system must be changed. For example, the executed event might be a job arriving to the process. If all the servers are busy, then the state change consists of adding the arriving job to a queue. Other state changes might require deleting a job from a queue or making a server busy.

The execution of an event might cause the cancellation of other events. For example, if the executed event consists of a machine breakdown, then this event forces the cancellation of the processing of jobs that are waiting for the machine. Finally, the execution of an event may cause the scheduling of future events. For example, if a job arrives and is added to a queue, a future event is also added to the event list indicating the time that the job will start processing.

When an event is executed, the event is removed from the event list. Then, the termination criterion is checked. If the criterion indicates that the end of the simulation has been reached, then raw data along with summary statistics are made available to the analyst. On the other hand, if the termination criterion indicates that the simulation has not finished (e.g., because more events remain in the event list), then the clock is advanced to the time of the next event.

Example 7.5: Discrete Events and Simulation Timeline in a Single-Server Process

A single-server queuing process can be represented by a timeline on which the time of each event is marked. Assume the following notation.

t_j: Arrival time of the *jth* job
$A_j = t_j - t_{j-1}$: Time between the arrival of job $j-1$ and the arrival of job j
S_j: Service time for job j
D_j: Delay time for job j
$c_j = t_j + D_j + S_j$: Completion time for job j
e_i: Time of occurrence of event i

Figure 7.2 shows a graphical representation of the events in a single-server process. This example has six events, starting with Event 0 and finishing with Event 5. Event 0, e_0, is the initialization of the simulation. Event 1, e_1, is the arrival of the first job, with arrival time equal to t_1. The arrival of the second job occurs at time t_2. Because $c_1 > t_2$, the second job is going to experience a delay. The delay D_2 is equal to the difference between c_1 and t_2 ($D_2 = c_1 - t_2$).

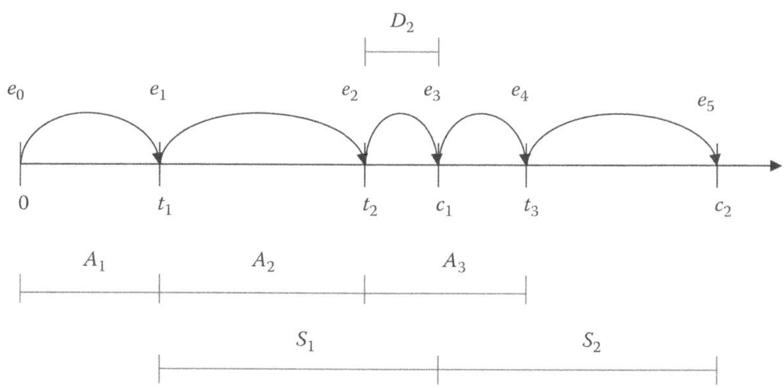

FIGURE 7.2
Events timeline for a single server.

In Figure 7.2, the completion time for Job 1 is calculated as $c_1 = t_1 + S_1$ because this job does not experience any delay. The last event in this figure, labeled e_5, is the completion time for Job 2. In this case, the calculation of the completion time c_2 includes the waiting time D_2 ($c_2 = t_2 + D_2 + S_2$). ■

7.4 Getting Started in Simulation Modeling

Simulation modeling is an art as well as a science that requires training and experience. While an advanced degree in simulation, industrial engineering, or some other related discipline is not necessary to be a competent modeler, there are several steps or guidelines that should be considered to produce effective models. The following steps have been adapted from Banks and Gibson (1996).

7.4.1 Step 1: Defining the Problem

Perhaps the most important, but often the most overlooked, aspect of simulation modeling is defining exactly the problem that needs to be solved. A common misconception among those unfamiliar with simulation is that once a model is built, it will be able to provide answers to any and all questions relating to the simulated process. Like other computer applications, a simulation model can only do what it was designed to do, and it is impractical to design it to do everything. Therefore, it is critical to carefully determine the scope of the model and its level of detail. A good way to begin is to construct a list of the questions that the model will answer. These questions should be as specific as possible. For example, it is better to ask "What is the maximum number of callers on-hold at a call center?" than "How big should the process be?" The questions should be open-ended. For example, ask, "How many sortation lanes are used?" rather than "Are 10 sortation lanes needed?"

The questions should be classified into two categories: key and desirable. Key questions are those that the model must be able to answer; desirable questions are those that we may want the model to answer, but they are not critical. The classification helps focus the model building effort and tells the model developer what may be left out if there is a need to meet scheduled deadlines and budget constraints.

During this step, judgements might have to be suspended until the process is better understood. If some parts of the process are well understood, and answers to relevant questions are known, those aspects should be considered not important or unrelated to the problem solution. It is also helpful at this step to quantify the benefits anticipated from obtaining answers to each question. This process will help to limit the list of questions to only the important ones. Too many questions will lead to a large and cumbersome model, which will be difficult to construct, validate, and use.

It's important to solve the right problem. If the completed model cannot answer relevant questions, the project may be considered a failure even though the model was "correct." This step might have to be revisited later (especially after Steps 2 and 3) to ensure that the model focus is still on target.

7.4.2 Step 2: Understanding the Process

After determining what the model is going to achieve, the next step is to gain a thorough understanding of the process or facility that will be simulated. It is important to understand that this step is different from Step 5 in Chapter 3. The "acquiring process understanding" step of Chapter 3 refers to the task of finding out what a process is supposed to do and what the customers expect from it to create the best possible design. In contrast, the assumption here is that a process exists either because it is already in operation or because there is a conceptual design for it. The goal now is to create a simulation model of this process to answer relevant questions and find improvements. Hence, understanding the process that will be simulated is essential because "it is not possible to model what is not understood." As trivial as it sounds, this simple axiom is often ignored in the rush to construct a "baseline" model.

The simulation model should be designed around the questions to be answered rather than trying to imitate the real process exactly. Pareto's Law—which states that in every group or collection of entities, there exist a vital few and a trivial many—seems to apply in simulation modeling because often 80 percent of a system's behavior can be explained by the action of 20 percent of its components. The problem, then, is to make sure that the vital few components are identified and included in the model. Reviewing the design with those responsible for it is a good start. It is important to compose a list of the process components, including equipment and human resources. Also important is to define (1) the entities that will flow through the process, (2) the stations where something will be done to the entities, (3) the basic flow patterns of the entities, (3) the routing logic, (4) the scheduling rules and operating policies, and (5) the alternative designs that will be considered. Flow diagrams and interviews with those involved in the process, particularly those in supervisory roles, are two of the most useful tools to accomplish this step.

Finally, common sense must be applied. Is there sufficient understanding of all aspects of the process that are relevant to answering the key questions? Is it possible to define a "paper" model design capable of addressing each question? For instance, if a key question relates to order completion time in a distribution center, and the order release and batching logic is not completely understood, then there is more homework to do before proceeding.

7.4.3 Step 3: Determining Goals and Objectives

Once there is an understanding of the process to be simulated, the next step is to plan the simulation project. The plan starts with setting explicit goals and objectives for the project. This is very closely related to determining which questions the model must answer, and these two steps may need to be repeated until they are consistent and accepted by management. The goal of the modeler should be to limit the size and scope of the model to only what is required to address the project objectives and answer the key questions. As with other computer applications, the effort required to build and validate a model grows exponentially with the requirements.

A decision must be made about what scenario(s) are to be simulated; for example, a peak demand day, or an average day with no exceptional conditions, or a third shift at the end of the month. The normal operating cycles of the process determine the period or length of time to be included for each scenario. For instance, if the objective is to determine capacity requirements for peak demand (a typical simulation scenario), then historical data could be used to identify the busiest period and limit the data collection effort (see Section 7.4.6. Additional scenarios can be specified, but each new scenario might require a careful review of the model capabilities and scope, and possibly a new input data set.

The evaluation criteria that will be used to analyze the model results must be determined. What are the key measures of performance? What level of resolution is required to be useful to management? Measures of performance may be separated into required and desirable categories, since some measures may require significant additional model complexity, and they may not be necessary. Finally, the project objectives must be summarized in writing and reviewed and agreed on by all involved with the project. Differences in expectations should be resolved at this point before continuing to the next step. Ignoring a lack of agreement may risk the model's credibility and ultimately the success of the project.

7.4.4 Step 4: Obtaining Support from Management

This seems like an obvious step, but it is often overlooked in the rush to move a project forward. Clear support from management must be obtained at the beginning of the project, or management will be reluctant to accept the results. If the results are not used, the simulation project will be deemed a failure.

A key component of this step is education. Management in most companies has little or no experience with simulation modeling. Many managers will have very unrealistic expectations regarding the project schedule and the model's scope or capabilities. The first task in obtaining support from management is to help them understand the modeling process and establish reasonable expectations for the project. This step should include an overview meeting to review the project goals and objectives, time frame, model scope, questions that will be answered, and measures of performance to be used. This meeting is also used to provide background information on the process and to discuss what models can and cannot do. A helpful technique is to review the highlights of a similar project at another company—what was learned, how long it took, and any management comments about the project that are appropriate.

After the initial meeting, it is advisable to hold regular but short meetings to keep management informed of the progress, while circulating a memo summarizing the key topics for those managers who are not available for a meeting. When the model is completed and verified, management should be included in the analysis process, thus making them part of the discovery and decision process. Having sat through some of these sessions, managers will gain insight into the process and understand and share in "ownership" of the results.

Attaining management support is a continuing requirement—it is not limited to just the beginning of the project. In theory, simulation models are built to answer difficult questions—questions in which management may have a keen interest. The more involved managers are, the more they will appreciate the process and learn from the project.

7.4.5 Step 5: Choosing Simulation Software

Making a choice from the vast amount of software that is available for discrete-event simulation may be bewildering, particularly to newcomers to the field. The search could be narrowed by focusing on software that specializes in business process modeling. Even within this subset, decisions must be made by considering software attributes pertaining to input capabilities, processing capabilities, output capabilities, the simulation environment, support, and cost. Powerful capabilities can make for increased simulation modeler productivity. Speed is also important because the greater expense is the modeler's waiting time. The bottom-line question is to choose the "right" simulation modeling tool. In some cases, it is possible to ask simulation vendors to run a model of a small version of the problem. It is also recommended to seek the opinion of consultants who use several products and companies with similar applications. Training courses for most simulation software are also available.

7.4.6 Step 6: Determining Data Requirements and Availability

There are several types of data that are typically used in simulation. These include time-based data such as interarrival times, demand rates, downloading times, uploading times, processing times, time to fail, time to repair, and travel times. Other types of data used in simulation include a fraction of the jobs failing review, a fraction requiring additional processing, and a fraction of each customer type. Data problems occur when availability is limited to small samples, summary data (e.g., the mean only), qualitative data, guesstimates (estimates that are only guesses), and data in the wrong format (e.g., last year's data instead of this year's data or ordering rather than shipping data).

Data sources include direct observations, data from time studies and history, and automated data. However, automated data may be flawed; for example, is a telephone line idle because it is down, or has the customer service representative gone to lunch? Sometimes, data from similar processes may be used, but making inferences from such a source may be risky. Employee estimates are another data source but they are highly suspect because humans are such poor estimators. We tend to forget the extremes and emphasize the present. Vendor claims or specifications are another source; however, there is a tendency for them to be highly optimistic; for instance, the reliability and speed of a particular piece of equipment might be overestimated.

It is often the case that no data exists; for instance, when a process is in the design phase, or when improper data have been collected. Nonetheless, there are techniques that may be used when no data or insufficient data exist. For example, if a process activity is truly random, under certain general conditions, the exponential distribution is a good way to capture the underlying uncertainty.

7.4.7 Step 7: Developing Assumptions about the Problem

For numerous reasons, the accomplishment of this step is a major factor in the success of the project. The assumptions must include the appropriate process scope. If the scope is too large, additional time and expense will be encountered in completing the simulation. If the scope is too small, the questions that are asked of the simulation may not be answered. Similarly, the complexity of the simulation model should be determined. The complexity should be sufficient to answer the questions that are asked, but not more than that.

The assumptions are also important at the end of the simulation. It is an unenviable position to be hit with criticism at the end of the simulation that the wrong process has been modeled. Such criticism can be halted by pointing to the assumptions that were agreed on at the outset of the project. The best situation is to have complete agreement of the assumptions and to verify that the assumptions are valid as the simulation project progresses. In the case of a difference of opinion, it may be necessary to negotiate an extension in time or resources to complete the project.

7.4.8 Step 8: Determining Desired Outputs

There are many possible outputs, including throughput, work-in-process, resource use, overtime, waiting time, cycle time, percentage of jobs that are late, and the time by which jobs are late, to mention just a few. These can be presented as questions to be answered as follows:

- Will the system meet throughput requirements?
- What happens to response time at peak periods?
- What is the recovery time when short-term surges cause congestion and queuing?
- What is the system capacity?
- What is the cause of specific problems?

The purpose is to gain understanding by involving statistical considerations. For this purpose, output analysis capabilities are built into many simulation software packages. There are also add-on packages or stand-alone software for statistical analysis of simulation output. While these software options provide the "number crunching" aspect of the statistical analysis, it is still the responsibility of the analyst to select the right data to be collected during the simulation run, choose the right statistical models, and then interpret the statistical analysis software output correctly.

7.4.9 Step 9: Building the Simulation Model

A decision must be made as to whether the simulation model will be made internally or externally. Actually, the decision could include an option whereby the simulation model is built via some combination of internal and external resources. For first-time users—that is, those firms without internal simulation expertise—building the simulation model internally may be unwise. There are too many decisions regarding modeling options and strategies that have to be made. Having an expert build that initial simulation model might be a useful learning exercise for the firm and an appropriate application of the technology. Another possibility is having an expert work with the firm on the first application of the simulation technology. The "jump start" is a variation of this, whereby the expert initiates the project and then turns over the reins to the internal group.

For infrequent users, consideration must be given to the time for relearning. For example, if simulation is used only once per year for a brief period of time, then the analyst (assuming that he/she is the same person) would practically have to relearn the software (or different software) every year. Additionally, the same software might change as a function of newer versions. When there is a short deadline for the completion of a simulation project, spending time relearning subtracts from the time available for the simulation activity. An expert, for instance, a simulation consultant, can certainly prove valuable in this case. Frequent users of simulation, on the other hand, typically have an internal group capable of handling demand for simulation within desired deadlines.

7.4.10 Step 10: Project Kickoff

The project should kick off with a formal, upbeat, fast-moving meeting that maintains the interest of management. Milestones should be presented and discussed because they are useful in keeping the decision-makers involved in the project. If they maintain their involvement, there is a much greater probability of implementation at project completion. Interaction should be encouraged to ensure the continued interest of all those involved. Responding to concerns and objections shows that there is interest in what decision-makers and others not on the simulation team have to say. Often, some valuable points are made when concerns and objections are raised. Great care must be taken not to alienate the decision-makers and others not on the simulation team. While the tone of the meeting should be that all those involved are going to give what is needed to make the project a success, wild promises of what can be accomplished, and how quickly and cheaply, can prove to be the undoing of a simulation project.

7.5 An Illustrative Example

A simple example of a service process adapted from Ingalls (2011) helps to illustrate what discrete simulation is and the type of questions that a model is able to answer. The flowchart in Figure 7.3 shows the activities and logic associated with a drive-through at a fast-food restaurant. As a car enters the parking lot of the fast-food restaurant, the customer decides whether to stay and enter the drive-through or leave. If the customer decides to leave the restaurant, s/he leaves as an unsatisfied customer, and the simulation model is able to capture that. That is, a simulation model of this process is able to answer questions related to unsatisfied customers. If the customer decides to stay, then s/he waits until the menu board is available to place an order. The customer moves forward as space becomes available, and at the same time, the order is transmitted to the kitchen. As soon as the customer reaches the pickup window, s/he pays and picks up the order if ready. The customer may experience some waiting time if the order is not ready. The customer leaves as soon as the order arrives.

The structural components of a discrete-event simulation include entities, activities and events, resources, global variables, a random number generator, a calendar, system state variables, and statistics collectors. The primary entities in the drive-through example are the cars arriving at the fast-food restaurant. The order is another type of entity that is created when the customer reaches the menu board. This entity has a relatively short life in the simulation, lasting only from the time the order is taken until it is delivered at the pickup window. Also, this entity starts as an information flow (the dashed line from the menu board to the kitchen) and is later transformed into a physical item (the food and/or beverages that go from the kitchen to the pickup window).

There are two attributes associated with a customer. The first is the time of day when the car enters the parking lot, and the second is the dollar value of the order. Both of these are unique to each customer (entity) flowing through the simulation model. The activities in the model are (1) placing an order, (2) making an order, and (3) paying and picking up. There are three resources in the process: (1) menu board, (2) kitchen, and (3) pickup window. The menu board is used from the time a customer moves to it until the customer moves

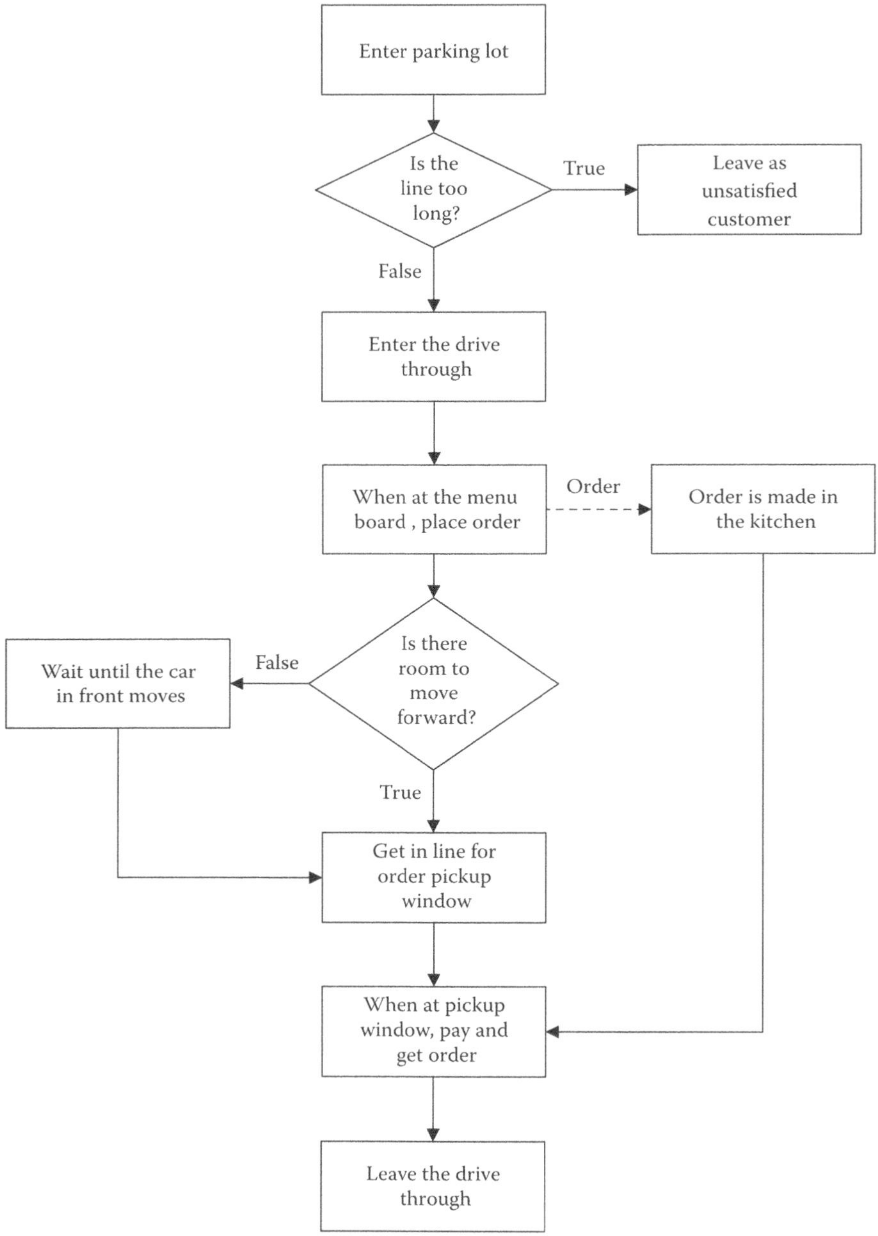

FIGURE 7.3
Drive-through flowchart.

away from it. In the model, the customer does not automatically move away from the menu board after the order has been placed. The customer waits until there is space to move forward. Hence, the menu board might be occupied for both productive time (when the order is being placed) and unproductive time (when there is no space for a car to pull forward). We assume that the kitchen processes one order at a time, and therefore, it is used from the time an order arrives until the order is finished. The order pickup window might also experience unproductive time, which occurs when a customer arrives at the window but the order has not arrived from the kitchen. The simulation is capable of tracking performance measures associated with these resources, including use and costs.

Because resources are limited, there is a queue associated with each resource. That is, as represented in Figure 7.3, customers might have to wait to reach the menu board and might also have to wait to reach the pickup window. Likewise, an order might have to wait for the kitchen. The flow of entities is determined by the logic in the model. For instance, the model must include logic to simulate how a potential customer answers the question of whether the line to the menu board is too long. One possible model consists of using a discrete probability distribution function for which the probability of entering the queue decreases with the number of cars waiting to reach the menu board.

Walking through a couple of steps in the model illustrates how a simulation keeps track of the state of the system, the entities on the calendar, the values of attributes and state variables, and the data collected for output statistics. The goal of the analysis is to determine the performance of the process when three spaces are provided for the menu board (i.e., one car can be at the board and up to two can be waiting) and three spaces for the order pickup window (i.e., one car being served and up to two waiting). The state of the system at noon on a given day is shown in Table 7.3. The calendar of the simulation consists of the entities that are scheduled to complete an activity.

TABLE 7.3

State of the System at 12:00:00

Entity	State
Car 1	At the order pickup window receiving its order
Car 2	Waiting for the order pickup window
Car 3	Waiting for the order pickup window
Car 4	Placing an order on the menu board
Car 5	Waiting for the menu board
Car 6	Waiting for the menu board
Order 2	Being cooked in the kitchen
Order 3	Waiting for the kitchen
Car 7	Scheduled to arrive at a later point

TABLE 7.4

Calendar of the Simulation at 12:00:00

Entity	Event	Event Time
Car 7	Arrive at restaurant	12:00:20
Car 1	Order picked up	12:00:40
Order 2	Completed	12:00:56
Car 4	Order placed	12:01:10

These entities are Car 1, Car 4, Order 2, and Car 7. The calendar for these four entities is shown in Table 7.4. In addition, the attributes of the "car" entities at noon are shown in Table 7.5. Since the current simulation time is 12:00:00, the attributes of Car 7 have not yet been set. This entity is scheduled to arrive at 12:00:20, as shown in Table 7.4.

Collecting data to calculate output statistics is a very important aspect of performing a simulation. For the purpose of this illustration, it is assumed that the simulation run started at 11:00:00, and therefore data have been collected for 1 simulated hour. It is also assumed that during that hour, 44 customers have completed the process; that is, they have entered the drive-through, placed an order, picked up the order, and left. The third column of Table 7.6 shows the statistics associated with the data collected during 1 simulated hour (i.e., from 11:00:00 until 12:00:00). The total revenue is the sum of all orders from 11:00:00 until the current simulation time. The lost revenue is the value of all orders of customers who leave because they find no space in the line for the menu board. One of the assumptions in the simulation model is that the order value follows a probability distribution function and uses this function to generate a specific order value for each arrival. The order value is generated whether or not the customer enters the system. Therefore, the simulation is able to estimate a value for lost revenue. It should be noted that in the long run, the lost revenue equals the average order value multiplied by the percentage of time that

TABLE 7.5

Attributes of Entities in the System at 12:00:00

Entity	Start Time	Order Value
Car 1	11:54:20	$10
Car 2	11:55:50	$6
Car 3	11:57:10	$4
Car 4	11:58:20	$14
Car 5	11:59:30	$14
Car 6	12:00:00	$10

TABLE 7.6

Value of Statistics of Interest

Quantity of Interest	Statistic	Value at 12:00:00	Value at 12:00:20
Revenue	Total	$504	$504
Lost revenue	Total	$74	$90
Menu board	Use	99.84%	99.84%
Kitchen	Use	70.06%	70.23%
Order pickup window	Use	96.78%	96.80%
Menu board queue (cars)	Average length	1.2306	1.2349
Kitchen queue (cars)	Average length	0.0822	0.0873
Order pickup window queue (cars)	Average length	1.0311	1.0365
Time in the system (min)	Average	5.60	5.60

the queue at the menu board is full. Other statistics of interest shown in Table 7.6 include the use of the three resources, the average length of the queues that are associated with these resources, and the total time in the system.

The event calendar determines the next step of the simulation. This is what differentiates discrete-event simulation from other types of simulation models. A simulation model operating in fixed time intervals might simulate time periods where nothing happens. The next event on the calendar shown in Table 7.4 is the arrival of Car 7. Discrete-event simulation does not simulate the 20 seconds between noon and the next event on the calendar; instead, the system time is moved directly to 12:00:20. Since the next event is a customer arrival, the simulation sets the values of the attributes that uniquely identify this entity; that is, the start time and the order value. The start time is simply the current system time. The order value, as mentioned earlier, is drawn from a probability distribution function provided by the user. For instance, the order value could be assumed to be a random number between $4 and $25. Or, the value could be assumed to follow a normal distribution with a mean of $10 and a standard deviation of $2. Historical sales data could be used to determine the right probability distribution function to model the value of an order. Assuming that the order value for Car 7 is $16, the statistic associated with lost revenue increases by this amount because at the time that this customer arrives, there are three cars at the menu board, two waiting and one placing an order. Since there is no room for Car 7, this customer leaves, and the associated revenue is lost. At this point, the simulation generates the next arrival and adds it to the event calendar. The time between arrivals is another assumption that the modeler (i.e., the person who built the simulation model) must make. While the time between arrivals may be deterministic, in most processes, the time follows a probability distribution function. As seen in Chapter 6, the exponential distribution is a function typically used for the purpose of

TABLE 7.7

Calendar of the Simulation at 12:00:20

Entity	Event	Event Time
Car 1	Order picked up	12:00:40
Order 2	Completed	12:00:56
Car 4	Order placed	12:01:10
Car 8	Arrive at restaurant	12:01:30

modeling the time between consecutive arrivals. For this illustrative example, it is simply assumed that the next arrival will occur 70 seconds later, that is, at 12:01:30, and therefore, the event calendar at 12:00:20 is the one shown in Table 7.7.

Since the simulation time has moved from 12:00:00 to 12:00:20, the time-dependent statistics have changed. In particular, the use of resources and the average length in the queue are calculated by taking into consideration the simulation time. For example, for 1 hour (i.e., from 11 until noon), the use of the kitchen has been 70.06 percent. In the last 20 seconds, however, the kitchen has been busy preparing Order 2. Therefore, the use of the kitchen at 12:00:20 is calculated as follows:

$$(70.06\% \times 3600 + 100\% \times 20) / 3620 = 70.23\%$$

There are 3600 seconds in 1 hour, and therefore, at 12:00:20, the total number of seconds is 3620, of which the kitchen was fully used for 20 seconds. The statistics associated with the queues can be updated in a similar fashion. For instance, the queue at the menu board had two cars in the last 20 seconds of the simulation and an average of 1.2306 cars in the first hour. The new average is then given by

$$(1.2306 \times 3600 + 2 \times 20) / 3620 = 1.2349 \text{ cars}$$

The updated statistics at 12:00:20 are shown in the last column of Table 7.6. The updated values show that due to the departure of Car 7, the lost revenue went from $74 to $90. There is also a small change in all the time-dependent values (i.e., resource use and average queue lengths), while there is no change in the revenue or time in the system. These two values change only when a customer leaves after paying and picking up an order.

The next event on the calendar shown in Table 7.7 is the completion of an order. At 12:00:40, Car 1 is scheduled to complete the process. At this point, the accumulated revenue changes from $504 to $514 because the order value for Car 1 is $10 (see Table 7.5). The statistic associated with the time in the system also changes. As shown in Table 7.5, Car 1 arrived at 11:54:20 and was leaving at 12:00:40, resulting in a time in the system of 6 minutes and 20 seconds (or 6.33 minutes). Since the assumption was that in the first simulation hour, 44 customers completed the process, this new completion is the 45th, and the "time in system" statistic is updated as follows:

$$(5.60 \times 44 + 6.33) / 45 = 5.62 \text{ minutes}$$

While the example is limited to average values, modern simulation software is capable of reporting a wide range of statistics that include both measures of centrality (such as the mean) and measures of dispersion (such as maximum and minimum values or the standard deviation). Assuming that the period of interest in the analysis is lunchtime, then a simulation run could be set to start at 11:00:00 and end at 14:00:00. Statistics are then collected for the three simulated hours, and these final values could be used to answer questions related to process performance. However, basing answers on a single simulation run is unwise because the output statistics depend on the random numbers that were generated to model the arrivals, the value of an order, the time to place an order at the menu board, the time to make an order at the kitchen, and the time spent at the order pickup window. The good news is that once a simulation model is built, performing more than one run and collecting data for statistics of interest is as trivial as changing an option in simulation software. The efficiency of the software and the processing speed of the computer equipment determine the amount of time that the process will take, but there is no work involved from the analyst's point of view.

The current operating environment of the drive-through is considered the base case. If the process is simulated 30 times, that is, the lunch hours are simulated for 30 days, the corresponding average values of the statistics at the end of each simulation run might be as shown in the second column of Table 7.8. These statistics show that, on average, 12.83 customers are lost during lunchtime on any given day, representing a potential average revenue of $155. The values also show the high use of both the menu board and the pickup window. Because the simulation was set assuming that the time to place and pick up an order is fixed at 70 seconds (or 2.33 minutes), the average time in the system of 5.31 minutes includes about 3 minutes of total waiting time (5.31 – 2.33 = 2.98 minutes). The simulation model could be used to try process changes and predict what would happen if these changes were implemented in the real system. Three scenarios are considered, and the simulation model is changed to calculate the statistics of 30 runs, as shown in Table 7.8.

In Scenario 1, the analyst is interested in measuring the impact of eliminating the spaces allowed for cars to form a line to wait for a resource (e.g., to wait for the menu board or the pickup window). The idea is to reduce drastically the waiting time for the customer by creating a totally balanced process with two main activities, both taking about the same amount of time. The change to the simulation model consists simply of eliminating the space for

TABLE 7.8

Value of Statistics for 30 Runs

Statistic	Original	Scenario 1	Scenario 2	Scenario 3
Total revenue	$1649	$989	$1690	$1822
Lost revenue	$155	$851	$99	$10
Lost customers	12.83	70.43	8.50	0.70
Menu board use	97%	68%	99%	81%
Kitchen use	73%	43%	73%	79%
Pickup window use	99%	84%	99%	97%
Menu board queue length (cars)	0.95	—	3.07	0.26
Kitchen queue length (cars)	0.12	—	0.13	0.23
Pickup window queue length (cars)	1.23	—	1.28	0.85
Average time in the system (min)	5.31	3.31	6.33	3.42

cars before the menu board and the pickup window. The resulting simulation model is run 30 times, and the statistics are shown in the third column of Table 7.8. The average time in the system has now been reduced by 37.7 percent from the base case. However, the number of customers leaving without entering the system has increased by a factor of 5.5, causing a 40 percent reduction in total revenues. Without space for queues, there is an excessive amount of *blocking*. Blocking occurs when something in a process does not allow a resource to become available even when service has been completed at the resource. In the model without queues, a car cannot move forward after placing an order until the pickup window is available. While the causes of blocking are immediately obvious in this simple example, they might not be when dealing with large, complex processes. This is why process analysis emphasizes the identification of bottlenecks in an attempt to reduce and manage them.

After analyzing the potential consequences of trying to eliminate waiting lines, a second scenario could be explored, consisting of adding spaces for cars entering the drive-through. The model is modified to allow six cars at the menu board, including the one placing the order (i.e., one being served and up to five waiting). The values of the performance statistics for this scenario are shown in the fourth column of Table 7.8. Under this scenario, revenue increases as the number of lost customers decreases. However, the time in the system increases by a minute, from 5.31 to 6.33 minutes on average. This is a very typical trade-off in business process management and is in agreement with Little's law (see Chapter 5), which establishes a direct proportional relationship between cycle time (i.e., time in the system) and work in process (i.e., entities in the system).

A third scenario is considered whereby a new technology cuts the average service time at the menu board and at the order window by 20 percent from 70 to 56 seconds. Implementing the technology requires an investment of $30,000, which must be paid by increased revenue at lunchtime. The change is made in the original model, and the simulation is run 30 times to obtain the statistics shown in the last column of Table 7.8. The values show that, on average, the time in the system decreases from 5.31 to 3.42 minutes. There are almost no lost customers, and the revenue has increased by $174 during the lunch hours. The investment would be amortized in less than 6 months (172 days), and the use of the resources is slightly more balanced, ranging from 70 percent to 97 percent. The results of this scenario lead to the obvious conclusion that investing in the new technology seems like a good idea, as long as the assumptions that were used to build the model are valid. In the drive-through example, these assumptions affect the pattern of customer arrivals, the service times at the different stations (i.e., menu board, kitchen, and pickup window), and the order values. The assumption of the effectiveness of the technology to reduce service time is also very important to base the decision on the results produced by the simulation model.

7.6 Spreadsheet Simulation of a Process

Most simulation models in practice are built using commercial software. However, it is possible to create some simple simulations—for example, a process with a single server—using an electronic spreadsheet program such as Microsoft Excel. The spreadsheet model must be capable of handling the following events:

1. A job arrives.
2. A job waits to be processed if the server is busy.
3. A job is processed.
4. A job leaves the system.

Next, a spreadsheet model will be created with the goal of estimating average waiting time and the percentage of time that the server is busy (the server use), which is actually calculated by computing the server's idle time. To calculate waiting times, the arrival time of each job and its completion time must be known. To calculate idle time, one must know the time when processing starts for the current job and the completion time of the previous job. The following definitions are needed in addition to the variables defined in Example 7.5.

b_j: Time when processing begins for job j

I_j: Idle time between job j and job $j-1$

In process-driven spreadsheet queuing simulation, it is assumed that for each job, the interarrival time A_j and its service time S_j can be generated using a random generation function. The following relationships can be established based on the interarrival and service values:

Arrival time of job *j*, where t_0 is the process start time: $t_j = A_j + t_{j-1}$
Starting time of job *j*, where $c_0 = t_0$: $b_j = \max(t_j, c_{j-1})$
Completion time of job *j*, where $c_0 = t_0$: $c_j = S_j + b_j$
Waiting time for job *j*: $W_j = b_j - t_j$
Total time for job *j*: $T_j = c_j - t_j$
Idle time associated with job *j*: $I_j = b_j - c_{j-1}$

Table 7.9 shows a spreadsheet simulation of a single server. The start time (Cell C4), interarrival times (Cells B5 to B14), and service times (Cells D5 to D14) are fixed values on the spreadsheet. The model simulates 10 customers (A5 to A14) arriving to a customer service desk, the average waiting time (Cell I15), and the server utilization (Cell G15). The formulae needed to create the simulation model are shown in Table 7.10. The table shows the formulae for the first customer (Row 5). All the other customers are modeled after the formulae in Row 5. To calculate the statistics of interest, the

TABLE 7.9

Process-Driven Spreadsheet of a Single-Server Simulation

1	A	B	C	D	E	F	G	H	I
2	Single-Server Queuing Simulation								
					Server		Customer		
3	Customer	Interarrival Time	Arrival Time	Service Time	Start Time	Completion Time	Idle Time	Cycle Time	Waiting Time
4	Start		9:00			9:00			
5	1	0:04	9:04	0:05	9:04	9:09	0:04	0:05	0:00
6	2	0:09	9:13	0:09	9:13	9:22	0:04	0:09	0:00
7	3	0:10	9:23	0:06	9:23	9:29	0:01	0:06	0:00
8	4	0:08	9:31	0:08	9:31	9:39	0:02	0:08	0:00
9	5	0:01	9:32	0:03	9:39	9:42	0:00	0:10	0:07
10	6	0:05	9:37	0:05	9:42	9:47	0:00	0:10	0:05
11	7	0:05	9:42	0:04	9:47	9:51	0:00	0:09	0:05
12	8	0:04	9:46	0:05	9:51	9:56	0:00	0:10	0:05
13	9	0:03	9:49	0:09	9:56	10:05	0:00	0:16	0:07
14	10	0:02	9:51	0:03	10:05	10:08	0:00	0:17	0:14
15						Use	83.8%	Average	0:04

TABLE 7.10

Formulae for customer 1 (row 5) in the Spreadsheet Simulation in Table 7.9

C	E	F	G	H	I
=B5 + C4	=MAX(C5, F4)	=D5 + E5	=E5 – F4	=F5 – C5	=E5 – C5

following formulae are used in Cells I15 (Average waiting time) and G15 (Utilization), respectively.

$$\text{Average waiting time:} = \text{AVERAGE}(\text{I5}:\text{I14})$$
$$\text{Server utilization:} = 1 - \text{SUM}(\text{G5}:\text{G14})/(\text{F14} - \text{F4})$$

The utilization is based on the percentage of idle time. The percentage of idle time is the sum of the idle times (Cells G5 to G14) divided by the simulation time, which is calculated as the completion time of the last job (Cell F14) minus the start time (Cell F4). The utilization is then calculated as the complement of the percentage of idle time.

One limitation of performing process-driven simulations on a spreadsheet is that estimates of the number of jobs in the queue cannot be calculated (Evans and Olson, 1998). These statistics are a function of time rather than the number of customers. Because one customer is processed at a time, there is no convenient way of capturing data related to the length of the queue within the spreadsheet model. Another serious limitation of spreadsheets for waiting-line simulation is that behaviors such as balking, reneging, and jockeying are difficult to include. Fortunately, commercial simulation software is capable of handling these and other model complexities.

7.7 Successful Simulation in Practice

Successful simulation projects require a combination of both quantitative and qualitative skills, strong support from multiple areas of the organization, and a suite of tools and techniques to enable effectiveness and innovation. The recommendations in this section are adapted from an article by Sadowski and Grabau (1999).

First, it is important to identify how success will be defined. In the best scenario, a successful simulation project is one that delivers useful information at the appropriate time to support a meaningful decision. The most important aspect of presenting the right information is to think about what the decision-makers need to know in the context of what they are going to do with the information to deliver value to their businesses. The nature of the information that may be of interest to the decision-makers may vary, even at different times within a project. Beyond the typical data associated with a simulation project—such as cycle times, costs, or resource uses—may lie other types of information; for instance, options that were unexplored or quickly discarded and the rationale behind those decisions.

The timing of when meaningful information can be delivered is also critical to a project's success. A high-fidelity answer that is too late to influence a decision is not nearly as useful as a timely rough-cut estimate. This applies throughout a study and not just at its completion. If preliminary insights into a system's behavior can be provided early in a project, the owners of the design may change the options they are considering or adjust the focus of the simulation efforts.

The third aspect of succeeding may be out of the simulation analyst's control, but is important to understand it. Namely, the success of a simulation project is linked to its role in influencing an important decision. Wonderful simulation work, advanced analysis, and eye-grabbing animation all completed on time still are of no value if they are not delivered to the right persons in the right context, that is, to those who have the power to make decisions within the organization.

To reach this goal of succeeding with simulation, two questions seem most relevant; namely, how can things go wrong, and what could be done to make them go right?

- *Tackling the Wrong Problem*—Sometimes, the biggest mistake is made at the outset of a simulation study. If the organization has picked the wrong problem to explore with simulation, the analyst might be put at a high risk of failure. One of the interesting situations where this occurs is when simulation is used to prove what is already "known" or decided. In some firms, certain thresholds of capital acquisition require a simulation study, although the study might be initiated well after firm commitments to a particular course of action have been made. Another situation relates to analysts falling into the trap of "when one has a hammer (simulation), everything looks like a nail." Certainly, simulation is the right tool for many problems. However, other problems can be readily solved using other tools, such as queuing analysis, optimization, or simple spreadsheet calculations (more on that in the next section). Analysts should step back and double-check that simulation is the best tool for the job before embarking on a simulation project. The cost of a simulation study must be avoided if something simpler is able to provide the same quality of results.
- *Working on the Right Problem at the Wrong Time*—To increase the chance of providing a good answer at the right time, the analyst may need to think carefully about when to start a simulation project. If the designers of a process are still considering widely differing ideas,

or are brainstorming on how to solve some of the fundamental problems in the system, then it may be premature to perform more than a rudimentary analysis. It is more difficult to identify timing problems once a project is under way. If there are regular and significant changes to the nature of the project, the analysts feel the effects, since they will have to rework the simulation model. It is hard, however, to know whether or when simulation work should be paused to let the project team do some more preliminary design. Sometimes, the best value comes from using simulation for rough-cut analysis and putting on hold a more detailed study that might have been initially commissioned.

- *Missing the Warning Signs of the "Data Woes"*—Experienced simulation analysts agree that data are the most aggravating, challenging, and dangerous aspect of a simulation project. Data management (i.e., gathering and analysis) is very time-consuming. Frequently, the problem is that there are not enough data. Service times, arrival rates, rework percentages, routing probabilities, and many other important aspects of the dynamics of a process may not be collected for other business purposes. Because gathering data may take a long time, it is critical to establish data needs as early in a study as possible and to assess the amount of effort required to obtain the necessary data. Estimated project duration might vary significantly based on what is found when the search for numbers starts. Sometimes, the problem with data is that there is too much. For instance, the relevant data may exist in large databases, and the extraction and formatting may not be trivial. Even in situations when the need for data matches the availability, the analyst must make sure that he/she understands what the data really mean. For instance, data labeled "cycle times" may have different meanings in different organizations.
- *Letting the Window of Opportunity Close*—The greatest and most widely discussed risk of failure of a simulation project is that the project is not finished on time. Creating a valid, useful simulation model is essentially a software development project, with similar risks and challenges. Returning to the earlier definition of success, the timing of the information can be as important as the quality of the information. While there are countless reasons why simulation projects are late in delivering results, four particular pitfalls seem worth special consideration: (1) getting lost in the detail, (2) leaving no time for analysis, (3) overdoing the animation, and (4) testing only at the end of the project. One of the common traps is getting too involved in modeling. As mentioned in Section 7.4, the art of simulation involves assessing what level of detail is required to support the project's goals. In successful simulation projects, all elements of the project (problem definition, model formulation, verification, validation, and analysis) are performed repeatedly throughout the effort,

> growing in scope as the model progresses. By the last 25 percent of the time allotted for the project, preliminary analysis should have been performed on the model for a number of different scenarios. Animation holds a similar attraction to "prettying up" PowerPoint presentations. Animation is a great tool for model validation and for engaging those who are not part of the simulation project team. However, a simulation should be enhanced with animation only to the point of what is needed to meet the project goals. Finally, as with analysis, verification and testing of the model must be performed throughout a project and not just at the end. Delivering a "state of the model" assessment, which grades the various segments of the model regarding quality and completeness, should be a standard part of your simulation project at each milestone.

Best practice in a simulation project consists of reviewing the model and other deliverables often and with more intensity as the project nears completion. Structured walkthroughs with colleagues and clients are ideal for discovering problems with logic or errors in the model. In a structured walkthrough, the modeler steps through the modeling constructs and explains model logic, how areas were abstracted, and what assumptions were made. Colleagues may point out easier ways to accomplish portions of the model or even point out an incorrect interpretation of the process under study. In addition, the simulation team should review the model specifications, data analyses, animation, output reports, and client presentations.

Flexibility is important throughout the duration of the project. The simulation team must search for ways to deal with situations such as changes in the scope of the project, data management issues, or lack of subject matter expert availability. More importantly, as these situations arise, they should be questioned regarding whether or not they are consistent with the true motivation for the project.

7.8 When Not to Simulate

Although most business process managers and analysts agree that simulation is a useful tool, they also agree that it is not the right tool in all situations. The observations in this section are adapted from an article by Banks and Gibson (1997).

Simulation modeling has become an essential tool for analyzing anticipated performance, validating designs, demonstrating and visualizing processes, testing hypotheses, and performing many other analyses. It is the preferred tool in a variety of industries, and, as mentioned above, in some industries, it is even required prior to any major capital investment. However, whether simulation modeling is the right tool for the problem is a question that is often overlooked. In the past, simulation modeling was reserved for only very large or specialized projects that required one or more programmers or analysts with specialized training and much experience. The proliferation of simulation software has led to a significant increase in applications—some by users without appropriate training or experience. It has also led to increasing dependence on simulation to solve a variety of problems. Although many of these projects are successful, the tool can be, and sometimes is, misapplied. An awareness of when quantitative problem requirements or qualitative project dynamics indicate that simulation may not be appropriate should help avoid this mistake.

Rule 1: The problem can be solved using common-sense analysis. Consider the design of an automobile tag facility. Customers arrive at random to purchase their automobile tags at a rate of 100 per hour. The time for a clerk to serve a customer varies but averages 5 minutes. What is the minimum number of clerks required? At least nine clerks will be needed to avoid a situation where lines would grow explosively. The more clerks, the shorter will be the average waiting time. This problem could have been analyzed by simulation, but that is unnecessary and would take longer to program and run than the solution shown here.

Rule 2: The problem can be solved analytically. There are steady-state queuing models (see Chapter 6), probabilistic inventory models, and others that can be solved using equations—that is, in a closed form—and this is a much less expensive method to use compared with simulation. In the license tag example in Rule 1, assume that all of the times are exponentially distributed. How long, on average, do the customers wait in the queue if there are 10 clerks? This is the $M/M/c$ model introduced in Chapter 6, and an equation can be used to determine the probability that the system is empty, from which the average number in the system can be determined. This is certainly a much faster analysis than using simulation.

Rule 3: It's easier to change or perform direct experiments on the real system. This is not always obvious. Cases have been documented where a model that has been commissioned to solve a problem has actually taken more time and money to complete than a simple direct experiment would have required. Consider the case where a detailed model of a drive-through fast-food restaurant was constructed and used to test improvements in customer service time of adding a second drive-up window. The model took weeks to complete, while a competitor tested the same concept by staging a second person with a remote handheld terminal and voice communication along the drive-up line, completing the entire study in a matter of days. The rule of thumb is that if the problem involves an existing process that can be perturbed or measured without undue consequences, it is wise to first look for a direct experiment to find answers to the questions. A direct experiment has the additional advantage of avoiding those questions related to whether the model was detailed enough or was properly validated.

Rule 4: The cost of the simulation exceeds the benefits. Although almost every simulation project has many qualitative benefits, the expense of the model, data collection, and analysis is usually justified by the expected quantitative stake. Accurately estimating the total costs of a simulation project requires some experience. Factors to be considered include

- Project planning, problem definition, and process documentation
- Model development and testing
- Data collection, review, and formatting
- Model validation
- Experimentation and analysis
- Possible updates or enhancements to the model
- Project documentation and presentations

Also to be considered are the costs of the simulation software (if not readily available) and computer resources. The estimated total simulation project costs should be compared with potential savings, cost avoidance, or additional revenue generation. If the cost–benefit analysis does not clearly favor embarking on a simulation project, building a simulation model may not be justified. On the other hand, some simulation projects are undertaken because of perceived risk for processes that are too complex to understand otherwise. The model provides a level of insurance to understand whether and where possible problems lurk. Can a price be calculated for this risk reduction?

Rule 5: Appropriate resources are not available for the project. Primary resources required to complete a successful simulation project include people, software/computers, and money. Experienced analysts who understand the problem, select the proper level of detail, translate it into a simulation model requirement, and program, test, and validate the model are the most critical component in any successful simulation project. Simulation is both an art and a science, with the art gained through experience and the science gained through proper training. Modern simulation software certainly helps, but it is not a substitute for the proper people resources for a project. A properly trained simulation modeler is critical because a poorly constructed model is worse than no model at all. In addition to people, funding is also critical. If the project cost estimate is much larger than the available project funding, it is best not to simulate. Otherwise, the project objectives would have to be compromised and corners cut in the model design and planned experimental analysis to come close to the budget. The project goals would be at risk because the resulting model would not be capable of providing the required results. Simulation, or the software selected, or both will be mistakenly held at fault.

Rule 6: There is not enough time for the model results to be useful. Time might be another type of insufficient resource. This is usually caused by one of three reasons: the project schedule is too short, model development and testing takes too long, or the window is too narrow. A frustrating, but not uncommon, problem in simulation projects is that decisions are made before their completion because "management did not have time to wait for the simulation results." Simulation studies that are commissioned at the last minute as a final check tend to have unrealistic schedules. If there is not sufficient time to conduct a proper project, the analyst might be forced to make coarser assumptions, skip details, or otherwise cut corners in an attempt to meet the schedule. It may be better not to use simulation if there is not enough time allowed in the overall project schedule to produce results and put them to use. This means allowing time to make changes to the process design and re-simulate if needed.

Rule 7: Lack of data. As discussed in the previous section, during the design phase of a simulation project, it is critical to determine the availability of the data required to meet project expectations and support the level of detail planned for the model. In some cases, the data may not be available; they may be impossible, impractical, or too expensive to collect. Committing to a project and building a model should not be done before verifying that the necessary data are available or obtainable. It is possible to perform sensitivity testing of a model using estimates of the data values, but this still requires estimates about the range of values for critical data items.

Rule 8: The model cannot be verified or validated. This problem is usually caused by a lack of one of three critical ingredients: people, data, and time. For instance, the project analyst may not understand how to properly verify the model (lacks sufficient training and/or experience). There may be a lack of useful performance data for comparing the model results against test scenarios to validate the model. Or, the project schedule may not allow for sufficient testing and/or validation activities. The correct procedure is to first complete a base case scenario by comparing the model results against those for the real process (or those expected from a new process) and then use this case to compare future test cases, as done in the illustrative example of Section 7.5. There are other methods that can be used even when data for the base case are not available. For instance, the *degeneracy test* checks whether the model behavior degenerates appropriately to extreme inputs. What happens when arrival rates get really high? How does the model respond? Do bottlenecks develop where expected? The *face validity* test applies common sense to analyze model outputs for the base case. Is the output reasonable? Can we explain model behavior based on experience with similar systems? The *sensitivity analysis* test analyzes model outputs for different inputs. For repeated test cases with different input values, do the outputs change in the direction anticipated? Do they all track together? These test procedures may help build confidence in the model, but it must still be questioned whether the model is sufficient to support the decisions that may be made based on simulation results. If the model is not properly verified and validated, results will be questioned and may not be accepted.

Rule 9: Project expectations cannot be met. Nine times out of ten, the failure to meet project expectations is due to a failure to properly educate the decision-makers about what is realistic and possible when using simulation modeling.

Management may have unreasonable expectations—usually, they expect too much, too fast. When it cannot be delivered, they may mistakenly blame simulation technology or the analyst. A manager with no experience in simulation may conclude that once a system is modeled, the model will be capable of answering any question. It can be difficult to explain, especially late in the project, that models are only capable of answering the explicit questions that they were designed to address. If the project expectations are unreasonable and cannot be modified or controlled, it will be very difficult to complete the project successfully, making it a "don't simulate" candidate.

Rule 10: System behavior is too complex or cannot be defined. The process to be simulated must be thoroughly understood before simulating, or the analyst will be forced to guess. Some processes are so complex that building an accurate model (within an acceptable schedule and budget) is not possible. This is often the case when (complex) human behavior is a critical part of the simulated process. For example, because modern automated distribution centers are complex, they are frequently simulated prior to implementation or modification. Most are driven by computerized warehouse management system software, which selects and combines orders to process. Almost all of the actual order processing (picking) is performed manually, and people run the facility, even in automated facilities. Typically, the scenario simulated is an average day, and the model results can be quite accurate. But in a real facility, when an unusual event occurs and the orders start falling behind schedule, people will change their normal behavior or activities to find a way around the system constraints in an attempt to meet the schedule. This behavior can be quite varied and virtually impossible to describe completely and simulate for all possible scenarios. Model results for these crash-case scenarios almost never match what occurs in the real system and are simply unreliable.

Simulation can be such a powerful analysis tool that it tends to be regarded in some industries as a universal solution. Perhaps in part because of the variety of success stories in recent years (for otherwise intractable problems), or perhaps in part because of the ready availability of sophisticated simulation software packages claiming user-friendliness, simulation is frequently the only tool considered. However, not every problem is a nail with simulation as the hammer.

7.9 Summary

In situations where the business process under study violates the assumptions of existing analytical models, and new mathematical models may be hard or even impossible to derive, an attractive modeling tool is computer-based, discrete-event simulation. This modeling technique allows the representation of processes, people, and technology in a dynamic computer model. The simulation model mimics the operations of a business process, including customer arrivals, truck deliveries, absent workers, or machine breakdowns. A characteristic of discrete-event simulation is that it focuses solely on the instances when discrete events occur and change the state of the simulated process. This allows considerable compression of time, enabling the analyst to look far into the future. While stepping through the events, the simulation software accumulates data related to the model elements and state variables, including capacity use of resources, number of jobs at different workstations, and waiting times in buffers. The performance of the process can then be evaluated through statistical analysis of the collected output data.

To illustrate the principles of discrete-event simulation, the chapter includes an example of a drive-through with three resources and two sequential queues. The example shows how the state of the system and the event calendar are managed during a simulation and how statistics are calculated and updated after each event. Additional insight into the mechanics of discrete-event simulation is obtained with an exploration of how a single-server queuing process could be simulated in a spreadsheet environment such as Microsoft Excel.

The chapter concludes with a set of observations from simulation practitioners. Their experiences are shared as a collection of "dos and don'ts" associated with successful simulation projects. Furthermore, a set of rules are presented that give a roadmap to determine when tools other than simulation might be more appropriate to deal with problems related to process design and improvement.

Discussion Questions and Exercises

1. List a few entities, attributes, activities, events, and state variables for the following processes:
   a. Check-out process at a grocery store
   b. Admission process at hospital
   c. Insurance claim process
2. Perform a web search of "business process simulation" and prepare a short report of your findings.
3. Perform the steps in the drive-through simulation that remain to empty the event calendar of Table 7.7. Calculate the statistics at the end of these events. Was Car 8 able to enter the drive-through?
4. Use the spreadsheet template in Table 7.11 to simulate a customer service desk with two servers. The start time of the process is 9:00 a.m. (Cell C4), and the interarrival times (Cells B5 to B14) are given in

TABLE 7.11

Spreadsheet for Exercise 7.4

1	A	B	C	D	E	F	G	H	I	J	K	L
2	Exercise 7.4: Two-Server Queuing Simulation											
3												
					Server 1			Server 2			Customer	
4	Customer	Interarrival Time	Arrival Time	Service Time	Start Time	Completion Time	Idle Time	Start Time	Completion Time	Idle Time	Cycle Time	Waiting Time
5	Start		9:00									
6	1	0:04		0:05								
7	2	0:03		0:09								
8	3	0:04		0:06								
9	4	0:08		0:08								
10	5	0:01		0:10								
11	6	0:05		0:05								
12	7	0:02		0:04								
13	8	0:04		0:05								
14	9	0:03		0:09								
15	10	0:02		0:03								
16						Use			Use		Average	

minutes. Assume that the process consists of a single queue operating under a first-in-first-out discipline. That is, customers join the queue and move to the first available customer service representative in the order in which they arrive. Calculate the average waiting time and the use of each server.

5. The chief of staff in the emergency room of Exercise 6.22 is considering the computerization of the admissions process. This change will not reduce the 10-minute service time, but it will make it constant. Develop a spreadsheet simulation to compare the performance of the proposed automated process with the performance of the manual existing process. *Hint*: Note that interarrival and service times that follow an exponential distribution can be generated with the Excel functions RAND() and LN() and the following expressions:

$$A_j = -\left(\frac{1}{\lambda}\right) \times LN(RAND())$$

$$S_j = -\left(\frac{1}{\mu}\right) \times LN(RAND())$$

In these expressions, μ is the mean service rate, and λ is the mean arrival rate. The Excel formula RAND() generates a random number between 0 and 1. The natural logarithm of the random number is found to transform this number into an exponentially distributed value. The transformation to an exponential distribution is completed when the value is divided by the appropriate mean rate and the negative sign is applied. (Simulate 500 patients in each case.)

6. At Letchworth Community College, one person, the registrar, registers students for classes. Students arrive at a rate of 10 per hour (Poisson arrivals), and the registration process takes 5 minutes on average (exponential distribution). The registrar is paid $5 per hour, and the cost of keeping students waiting is estimated to be $2 for each student for each hour waited (not including service time). Develop a process-driven spreadsheet simulation to compare the estimated hourly cost of the following three systems. (See the hint in Exercise 7.5 and simulate 500 students in each case.)
   a. The current system.
   b. A computerized system that results in a service time of exactly 4 minutes. The computer leasing cost is $7 per hour.
   c. Hiring a more efficient registrar. Service time could be reduced to an average of 3 minutes (exponentially distributed), and the new registrar would be paid $8 per hour.
7. A process manager is considering automating a bottleneck operation. The operation receives between three and nine jobs per hour in a random fashion. The cost of waiting associated with the jobs is estimated at $2.20 per hour. The team's equipment choices are specified in Table 7.12. Develop a spreadsheet simulation to compare the estimated hourly cost of the three alternatives. *Hint:* Uniform random numbers between a and b can be generated in Excel with the formula $= RAND() * (b - a) + a$. Random numbers from a normal distribution with a mean m and standard deviation s can be generated in Excel with the formula $= NORM.INV(RAND(), m, s)$. (Simulate 500 jobs in each case.)

TABLE 7.12

Equipment Choices for Exercise 7.7

Machine	Service Time	Hourly Cost
A	Uniform between 4 and 12 minutes	$12
B	Exponential with an average of 8 minutes	$8
C	Normal with an average of 8 minutes and a standard deviation of 1 minute	$15

8. An online retailer would like to predict total revenues for 2019 based on the 2018 revenues and the estimated growth in each of the five departments. The estimates are summarized in Table 7.13. The retailer believes that any of the growth rates between the minimum and the maximum are equally likely to occur. Create a simulation model on a spreadsheet to estimate the probability that total revenues in 2019 will be more than 5 percent higher than those in 2018. *Hint:* Uniform random numbers between a and b can be generated in Excel with the formula $= RAND() * (b - a) + a$.

TABLE 7.13

Revenue and Growth Rate Estimates for Exercise 7.8

		Growth Rates	
Department	**2018 Revenue**	**Minimum**	**Maximum**
Electronics	$6,342,213	2%	10%
Garden Supplies	$1,203,231	–4%	5%
Jewelry	$4,367,342	–2%	6%
Sporting Goods	$3,543,532	–1%	8%
Toys	$4,342,132	4%	15%

9. The Maestranza Hotel in downtown Málaga has 94 rooms for which it charges $143 per night. An occupied room costs the hotel $28 in operating costs. There is a 4.5 percent chance that a guest with a booked reservation does not show up. The number of booked guests that show up at the hotel follows a binomial distribution. The hotel uses overbooking and pays $200 to compensate guests whose reservations cannot be honored. What overbooking limit should the hotel use to maximize the average daily profit? *Hint:* Build a Monte Carlo simulation model in Excel to answer this question. The number of guests that show up can be simulated with the Excel function =BINOM.INV(Booked, Show-up probability, RAND()).
10. Through statistical analysis of historical data, Mile High Airlines has determined that the demand for one of its flights follows a Poisson distribution with a mean of 150 passengers. It has also determined that, out of the booked passengers, the number that actually show up for the flight follows a binomial distribution with a probability of success (i.e., probability of showing up) of 92 percent. The airline uses an Airbus 319 with 134 seats for this flight and has an overbooking limit of 13 passengers. That is, the

airline policy is to overbook the flight by at most 13 passengers. If the flight is overbooked and more than 134 passengers show up for boarding, some passengers must be bumped to another flight. Bumping more than two passengers creates customer service problems for the airline in addition to the cost of compensating the bumped passengers.

a. Create a Monte Carlo simulation to estimate the probability that more than two passengers are bumped with the current overbooking limit. Use at least 1000 simulation trials.

b. What should the overbooking limit be if the airline would like to keep the probability of bumping more than two passengers to no more than 5 percent?

Hint: The actual demand can be generated with the Random Number Generation option in the Data Analysis tool of Excel. The Data Analysis tool is in the Data tab of the Excel ribbon. Use this random number generator to generate 1000 demand values that follow a Poisson distribution with a mean (lambda value) of 150. The function =BINOM.INV(Booked, Show-up rate, RAND()) can be used to simulate the number of passengers that show up, where Booked is the number of passengers booked for the flight, the show-up rate is 92 percent, and RAND() is the Excel function that generates uniform random numbers between zero and one.

11. The owner of a ski apparel store in Winter Park, CO must make a decision in July regarding the number of ski jackets to order for the following season. Each ski jacket costs $86 and can be sold during the ski season for $245. Any unsold jackets at the end of the season are sold for $55. Historical data show that the demand follows a Poisson distribution with a mean value of 80 jackets. Jackets must be ordered in lot sizes of 10.

a. How many jackets should be ordered to maximize the expected profit?

b. What are the best and worst possible outcomes of implementing the solution in Part (a)?

c. What is the likelihood of making at least $12,000 with this apparel item?

Hint: The actual demand can be generated with the Random Number Generation option in the Data Analysis tool of Excel. The Data Analysis tool is in the Data tab of the Excel ribbon. Use this random number generator to generate 1000 demand values that follow a Poisson distribution with a mean (lambda value) of 80.

References

Banks, J., and Gibson, R. 1996. Getting started in simulation modeling. *IIE Solutions 28*(11): 34–37.

Banks, J., and Gibson, R. 1997. Don't simulate when: 10 rules for determining when simulation is not appropriate. *IIE Solutions 29*(9): 30–32.

Eneyo, E.S., and Pannirselvam, G.P. 1999. Process simulation for facility layout. *IIE Solutions 31*(11): 37–40.

Evans, J.R., and Olson, D.L. 1998. *Introduction to simulation and risk analysis*. Upper Saddle River, NJ: Prentice Hall.

Ingalls, R.G. 2011. *Introduction to simulation*. In Proceedings of the 2011 Winter Simulation Conference, editors S. Jain, R.R. Creasey, J. Himmelspach, K.P. White, and M. Fu, 1379–1393.

Sadowski, D.A., and Grabau, M.R. 1999. Tips for successful practice of simulation. In Proceedings of the 1999 Winter Simulation Conference, editors P.A. Farrington, H.B. Nembhard, D.T. Sturrock, and G.W. Evans, 60–66.

8

Modeling and Simulating Business Processes with ExtendSim

Chapter 7 introduced simulation as a flexible and general approach for analyzing business processes. This chapter continues to explore the use of simulation by focusing on how to build discrete-event simulation models of real business processes using a commercial simulation software package called ExtendSim.[1] By applying tools such as this (ExtendSim is only one of many available on the market), it is relatively easy to create discrete-event simulation models of complex processes. These models can then be used for analyzing process performance and for evaluating alternative designs. Hence, discrete-event simulation modeling can be a very useful tool in designing complex business processes that exhibit uncertainty and variation.

Recall that the process flow analysis in Chapter 5 assumed constant and deterministic activity times (i.e., the times were certain, and there was no variation). However, in practical settings, this is rarely the case. It is therefore important to have tools for assessing process performance that take variability into consideration. In general, three types of variability are relevant in business processes: (1) variation in the arrival of jobs, (2) variation in the processing time of each activity, and (3) variation in the availability of resources. Along with variability, the simulation models offer opportunities to add other realistic elements of business processes that the basic tools of analysis in Chapters 4 through 6 cannot easily incorporate, for example, more complex state-dependent decision rules.

Apart from exploring process modeling in ExtendSim, this chapter also discusses important general concepts and principles for modeling and simulating business processes independently of the software used.

Turning to the ExtendSim software, a first observation is that it is not industry specific. Thus, it can be used to model and simulate discrete-event or continuous systems in a wide variety of settings ranging from production processes to population growth models. For the modeling and simulation of business processes, it is the discrete-event Item library that contains most of the needed functionality. However, the libraries Value and Chart also contain features that are needed to build realistic models. The former contains tools for data processing and statistics, and the latter contains tools for displaying results graphically. Functionality from these three ExtendSim libraries will be used throughout Chapters 8–10.

DOI: 10.1201/9781032617237-8

The best way to learn the material discussed in this chapter is in front of a computer. Students should install ExtendSim and open the application to either create the models shown in this chapter or review the models that are available on the website associated with this book. For convenience, all the models used for illustration purposes are available on the website. The models are named Figurexx.xx.mox, where xx.xx is the number that corresponds to the figure number in the book. For instance, the first model discussed in this chapter is the one that appears in Figure 8.2, so the corresponding file is Figure08.02.mox.

8.1 Developing a Simulation Model—Principles and Concepts

The development of a simulation model entails the following steps, which may be more or less accentuated and challenging depending on the context:

1. Determine the goals of the model.
2. Understand the process to be modeled through the use of basic analysis and design tools, such as flowcharting.
3. Draw a block diagram using the appropriate blocks in the available libraries to represent each element of the model accurately.
4. Specify appropriate parameter values for each block.
5. Define the logic of the model and the appropriate connections between blocks.
6. Perform careful model verification and validation.
7. Add data collection and graphical analysis capabilities.
8. Analyze the output data and draw conclusions.

Before building a model, the analyst must think about what questions the model is supposed to answer. For example, if the goal is to examine the staffing needs of a process, many details that are not related to process capacity can be left out of the model. The level of detail and the scope of the model are directly related to the overall goal of the simulation effort. Because the model will help the analyst predict process performance, a good starting point is to define what constitutes "good" performance. This definition can be used later to enhance the process design via optimization.[2]

Before modeling processes within the simulator's environment, the analyst needs to understand the process thoroughly. This implies that the analyst knows the activities in the process, the sequence in which they are performed, the resources needed to perform the activities, and the logical relationships among activities. This understanding includes the knowledge of the arrival and service process for each activity. In particular, the analyst must be able to estimate the probability distributions that govern the behavior of arrivals to the process and the time needed to complete each activity. This step usually requires statistical analysis to determine appropriate probability distributions associated with the processing times in each activity and the time between job arrivals. Typically, data are collected for the processing times in each activity, and then a distribution is fitted to the data. Similarly, data are collected to figure out what distribution best represents the arrival pattern. A statistical test, such as the Chi-square, can be used for this purpose, as discussed in Chapter 9. ExpertFit (www.averill-law.com) and Stat::Fit (www.geerms.com) are two examples of software packages that can aid in determining what standardized probability distribution best represents a set of real-world data. These products are designed to fit continuous and discrete distributions and to provide relative comparisons among distribution types. The use of Stat::Fit is illustrated in Chapter 9.

ExtendSim, like most modern simulation packages, provides a graphical modeling environment in which the model of the process is represented by a block diagram. Blocks with different functionality are dragged and dropped into the model from available libraries. The blocks are easily connected to each other by drawing connection lines between them. Modeling in ExtendSim entails the appropriate selection of blocks and how they are connected. Blocks can model activities, resources, and the routing of jobs throughout the process. They can also collect data, calculate statistics, and display output graphically with frequency charts, histograms, and line plots. ExtendSim also includes blocks to incorporate databases and activity-based costs in the simulation model.

The most basic block diagram of a business process starts with the representation of the activities in the process. Then, the conditions that determine the routing of jobs through the process and, if appropriate, a representation of resources are included. In addition, rules are provided to define the operation of queues and the use of available resources. Finally, data collection and displays of summarized output data are added to the model.

To measure performance, the analyst must determine whether additional blocks are needed. If performance is measured by the waiting time in the queues, no additional blocks are necessary because the blocks that model queues collect waiting time data. The same goes for resource blocks and utilization data. However, if the analyst, for example, wants to count the number of jobs that are routed through a particular activity during a simulation run, an Information block (from the Information submenu of the Item library) needs to be included in the model. ExtendSim libraries also include blocks that can be used to collect other important data, such as cycle times.

One of the most important aspects of business process simulation is the definition and monitoring of resources. As discussed in Chapter 5, resources are needed to perform activities, and resource availability determines process capacity. Therefore, for a simulation model of a business process to be useful, resources must be defined, and the resource requirements for each activity must be specified. ExtendSim provides a Resource Pool block (or, alternatively, a Resource Item block), found in the Resources submenu of the Item library, which is used to define the attributes of each resource type, including availability and usage. A Resource Pool may represent, for instance, a group of nurses in a hospital. Every time a nurse starts or finishes an activity, the number of available nurses is adjusted in the Resource Pool. These blocks are particularly useful for determining staffing needs based on performance measures such as queue length, waiting time, and resource use.

The proper completion of Steps 1 through 8 above, results in a simulation model that describes the real process (at a level of detail that is consistent with the goals outlined in Step 1). Because, in all likelihood, some mathematical and logical expressions will be used in the model, it is necessary to check that the syntax employed to write these expressions is indeed correct. Checking the syntax allows the analyst to create a simulation model that is operational. A computer simulation cannot be executed until the syntax is correct.

Not all simulation models that run are correctly programmed and valid; that is, one can develop a simulation model where jobs arrive, flow through the process, and leave. However, this does not mean that the model is correct with respect to the modeler's intentions or that it represents the actual process accurately. Model verification and validation are essential steps in building simulation models. Verification of a model means ensuring that the programmed model does what the modeler intended. Validation of a model consists of ensuring that the operational rules and logic in the model are an accurate reflection of the real system. Model verification and validation must be done before the simulation results are used to draw conclusions about the behavior of the process.

8.1.1 Model Verification

Model verification is the process of debugging a programmed model to ensure that it operates as expected (i.e., that the model behaves as expected given the logic implemented). One way of verifying that the model is correct is to use an incremental building technique. This means building the model in stages and running each stage to verify that it behaves as expected. Another technique is to reduce the model to a simple case for which the outcome can be easily predicted. This simplification can be obtained as follows:

- Remove all the variability to make the model deterministic.
- Run the deterministic model twice to verify that the same results are obtained.
- For processes with several job types, run the model using one job type at a time.
- Reduce the size of the Labor Pool (e.g., to one worker).
- Uncouple interacting parts of the model to see how they run on their own.

Other techniques for verification include accounting for all the items in the model and adding animation. After the model is verified, it needs to be validated.

8.1.2 Model Validation

Model validation refers to determining whether the model represents the real process accurately. A valid model is a reasonably accurate representation of the real processes that conforms to the model's intended purpose. During validation, it is important that comparisons with the real process are made using the same metrics (i.e., that measures of performance are calculated in the same manner). Determining whether the model makes sense or not is often part of validating a model. The analyst also can ask someone familiar with the process to observe the model in operation and approve its behavior. Simulation results also can be compared with historical data to validate a model. If enough historical data are available (e.g., arrival and processing times), these data can be used to simulate the past and validate that the model resembles the actual process under the same conditions.

8.2 ExtendSim Elements

A simulation model in ExtendSim is an interconnected set of blocks. A block performs a specific function, such as simulating an activity, a queue, or a pool of resources. An item is a process element that is being tracked or used. For example, items being tracked (transient entities) could be jobs, telephone calls, patients, or data packets. Items being used (resident entities) could be workers, fax machines, or computers. Items are individual entities and can have unique properties (e.g., a rush order versus a normal order) that are specified by their attributes and priorities. An item can only be in one place at a time within the simulation model. Items flow in a process and states change when events occur. For example, a server could change from idle to busy when a customer arrives.

Values provide information about items and the state of the simulated process. Values can be used to generate output data such as the waiting time in a queue or the actual processing time of an activity. These values will be referred to as *output values*. Output values include statistics such as the average queue length and the average use of a resource. There are also *state values*, which indicate the state of the process. For example, the number of customers waiting in line at a given time indicates the state of the system. Most blocks in ExtendSim include connectors that can be used to track output or state values.

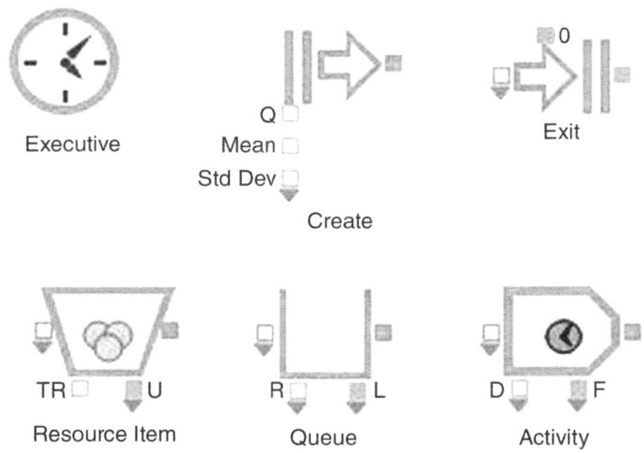

FIGURE 8.1
Basic ExtendSim blocks.

The blocks are available in libraries, which serve as repositories of blocks. For a particular block in a model to be used, the library where it is located must be open. The tutorial of Section 8.3 will show how to open libraries. A simple simulation model of a process can be constructed using six basic blocks (see Figure 8.1).

- The Executive block from the Item library is a special block that is automatically included in all new ExtendSim models. The block controls the timing and passing of events in the model. This block is the heart of a business process simulation and is located at the upper left corner of the simulation model.
- The Create block from the Routing submenu of the Item library is used to generate items at arrival times specified in its dialog window. When developing the model, one must determine what items will be generated, for example, jobs, customers, purchasing orders, insurance claims, or patients. The block may be configured to generate items randomly or according to predetermined schedules. If items are to be generated randomly, the modeler chooses the probability distribution of the interarrival times (constant time between arrivals is a special case, where the probability for the specified constant time is one). ExtendSim provides a long list of standardized distributions to choose from. Depending on the choice made, the Create block asks for the necessary parameters. For example, if Constant is chosen, it asks for the constant delay time; if Normal is chosen (corresponds to a normal distribution), it asks for the mean and standard deviation; if Uniform, Real is chosen (corresponding to a continuous uniform distribution), it asks

for the maximum and minimum values. These parameters may be entered directly in the block's dialog window where query boxes appear, or they may be sent in through value input connectors at the bottom of the block. This can be seen in Figure 8.1, where the Create block is configured for interarrival times following a normal distribution. Hence, the input value connectors *mean* and *Std Dev* appearing at the bottom of the block may be used to specify the necessary parameters. The available input connectors change with the choice of distribution. It is noteworthy that these input value connectors make it easy to change the distribution parameters during a simulation run. This can be useful when modeling business processes where the arrivals of jobs change over time, for example, during rush hours.

- The Exit block from the Routing submenu of the Item library is used to take care of items that leave the process. Because a business process can have several exit points, several Exit blocks might be needed in a simulation model. It is noteworthy that the Exit block can have many input connectors where items may be passed out of the model. The number of input connectors is chosen by a simple "drag and drop" procedure in the direction of the arrow on the input connector on the left side of the block.
- The Resource Item block from the Resources submenu of the Item library is similar to a holding area that can initially contain a specified number of items. This block can be used to model the "in box" of a clerk. In the dialog, the number of items can be specified at the beginning of the simulation. Properties such as attributes and a priority can be added to the items as they leave.
- The Queue block found in the Queues submenu of the Item library is similar to the Resource Item block in the sense that it serves as a holding area where items queue up while waiting to be processed. The Queue block can deal with different queuing disciplines, reneging, and other queuing features, while the Resource Item block is more restricted with regard to such features. On the other hand, the Queue block will always be empty when the simulation starts.
- The Activity block from the Activities submenu of the Item library has many applications; however, its most common use is to simulate an activity or a workstation with one or several parallel servers. The simulation of an activity is achieved by delaying an item passing through the Activity block by an amount of time that is equivalent to the processing time. The processing times specified in this block may be constant or specified by a long list of probability distributions reflecting the process time uncertainty. The processing times can also be specified as attributes of the items that are being processed, or they can be taken from a lookup table.

Figure 8.2 shows these basic elements in a simple credit application process. The block labeled "Apps In" generates the arrival of an application every 10 minutes. The application is placed in the block labeled "In Box" until the reviewer is available. The application stays in the block labeled "Reviewer" for 14 minutes and then moves to the block named "Apps Done." The In Box block shows the number of applications held in this repository, and the Apps Done block shows the number of completed applications. These numbers are updated during the simulation run. In addition to the blocks, the model in Figure 8.2 contains a text box with a brief description of the model. ExtendSim includes a drawing tool that can be used to add graphics and text to the model to enhance presentation and documentation.

Double-clicking on a block activates its dialog window, which is used to set the properties of a block. For example, Activity blocks are used to model activities in a business process. The basic property of an activity is its processing time. The dialog window of an Activity block allows one to set the processing time of the activity that the block is modeling. Figure 8.3 displays the dialog window of the Activity block labeled "Reviewer" in the credit application process shown in Figure 8.2. In the lower left corner of every dialog window is a convenient help button. Clicking on this button opens a help window, which explains all the features of this particular block and how to use them.

Typically, additional blocks are needed to model the logic associated with an actual process as well as to collect output data. A complete simulation model may contain one or more blocks that are not directly related to an activity in the process, but they help implement some necessary logical conditions, data collection, and reporting. These blocks will be introduced as this discussion moves from simple to more elaborate simulation models.

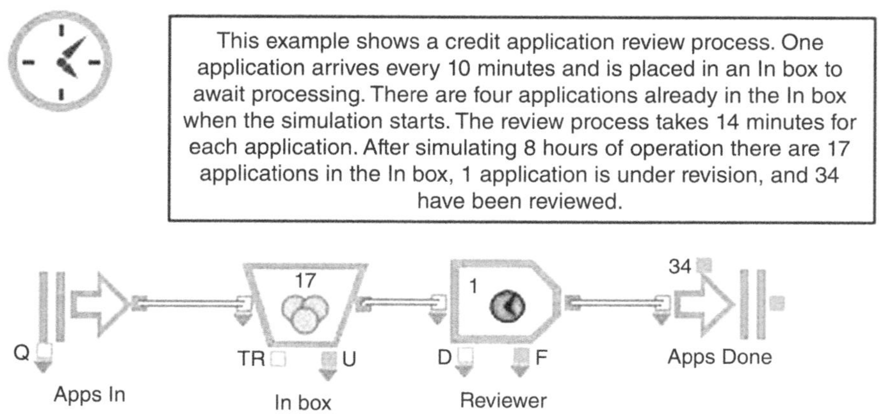

FIGURE 8.2
A simple credit application process.

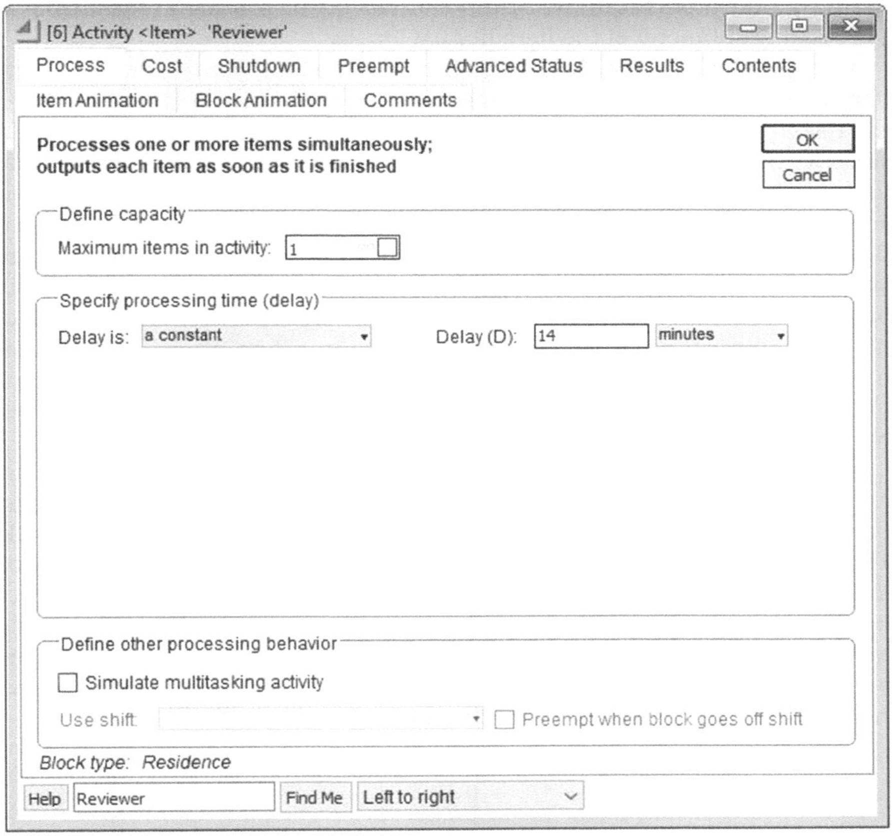

FIGURE 8.3
Dialog window of the activity block in the credit application process.

In simulation models, a path establishes the relationship between any pair of blocks. When more than one path originates from a block, it might be necessary to add a Select Item Out block from the Routing submenu of the Item library. Select Item Out blocks are used for routing jobs correctly through the process based on some defined decision rules. For example, if 90 percent of the credit approval requests can be processed by a group of generalists, and 10 percent of the requests require a specialist, the probabilistic selection that determines where the job should go can be implemented in the Select Item Out block. Figure 8.21 illustrates how a Select Item Out block is used for splitting the total flow of Requests In arriving to the system into Simple and Regular requests with different properties. For more complex decision rules, it may be necessary to model the decision logic outside the Select Item Out block and send information about the decision to this block, which then routes the items accordingly. Additional blocks for routing are available in the Routing submenu of the Item library.

A Queue block (in the Queues submenu of the Item library) or a Resource Item block (in the Resources submenu of the Item library) is needed when the resources required for performing an activity are limited. For example, consider the activity of a patient walking from the reception area to a laboratory in a hospital admissions process. The relevant resource to perform this activity is the hallway that connects the reception area with the laboratory. Because, the capacity of the hallway, for all practical purposes, can be considered unlimited, the block used to model this activity (e.g., the Activity block in the Item library with the option Maximum items in activity: ∞) does not need to be preceded by a Queue or Resource Item block. On the other hand, after the patient reaches the laboratory, a Queue or Resource Item block might be necessary to model the queue that is formed due to the limited number of technicians performing lab tests.

In many business processes, transportation activities such as mailing, e-mailing, or faxing documents are modeled with Activity blocks allowing parallel processing without a preceding Queue block or Resource Item block. This is because the only interest is in adding the time that it takes for the job to go from one place to another. The combination of a Queue or Resource Item block and an Activity block is used to model activities with limited capacity, such as recording data to a database or authorizing a transaction, because the resources needed to perform these activities are usually limited. An example of this situation is depicted in Figure 8.2.

8.3 ExtendSim Tutorial: A Basic Queuing Model

The basic queuing model consists of one server (e.g., a teller at a bank, an underwriter at an insurance company, and so on). The model also considers that jobs wait in a single line (of unlimited size) and that they are served according to a first-in-first-out (FIFO) discipline.[3] In this tutorial, a simulation model of this basic queuing system will be built following a set of simple steps. Although the resulting simulation model will represent a simple process, the methodology can be extended to model more elaborate systems. Students are strongly encouraged to build the models presented in this tutorial, even though it is possible to just load them into ExtendSim from the web page associated with this book.

It is assumed that the ExtendSim application has been launched. The main ExtendSim window that appears after the software is loaded is shown in Figure 8.4.

The default horizontal and vertical toolbars displayed in Figure 8.4 can be moved by the user who may also customize which toolbars that should be

FIGURE 8.4
Main ExtendSim window.

displayed under the tools tab on top. A short description of each button in the toolbars can be obtained by placing the mouse pointer directly on top of each button. Starting with the horizontal toolbar on top, the first four buttons (from left to right) correspond to the File tools menu: New Model Worksheet (Ctrl + N), Open (Ctrl + O), Save (Ctrl + S), and Print (Ctrl + P). The next nine buttons correspond to the Edit tools menu: Cut (Ctrl + X), Copy (Ctrl + C), Paste (Ctrl + V), and Undo (Ctrl + Z), followed by buttons for zooming and one for finding text, blocks, etc. in the model The next nine buttons are from the Model tools menu, followed by buttons for choosing different types of cursors from the Cursers tools menu and buttons for text editing from the Text tools menu. The last two buttons to the far right are from the Library tools menu, allowing library windows that are open to be visible in the window or hidden.

In the vertical toolbar to the left, the first ten buttons counting from the top are from the Alignment tools menu and the subsequent nine buttons are from the Shape tools menu. All buttons in the vertical toolbar to the right belong to the Dialog Item tools menu.

Before blocks are added to the simulation model, a new model worksheet and the libraries that contain the blocks to be added must be opened. Because the goal is to model business processes, the Item library always needs to be opened. To do this, choose Open Library from the Library menu (i.e., Library > Open Library…,[4] or Ctrl + L) and then choose Item from the Libraries folder. The Libraries folder is one of the folders created during the installation of ExtendSim and is located in the folder where the ExtendSim application resides. After the library has been added, Item should appear under the Library menu. A model can now be built using blocks in this library as illustrated in the following example.

Example 8.1

An insurance company receives an average of 40 requests for underwriting per week. Through statistical analysis, the company has been able to determine that the time between two consecutive requests arriving to the process is adequately described by an exponential distribution. A single team handles the requests and is able to complete on average 50 requests per week. The requests have no particular priority; therefore, they are handled on an FCFS basis. It can also be assumed that requests are not withdrawn and that a week consists of 40 working hours.

To model this situation, a time unit needs to be agreed on for use throughout the simulation model. Suppose *hours* is selected as the time unit. The "Global time units" option "Hours" should then be selected in the Run menu. This is done by opening the Setup dialog (i.e., Run > Simulation Setup and click on the Setup tab), choosing Hours in the Global time units pop-up menu, and clicking OK. If 40 requests arrive per week, the average time between arrivals is 1 hour. If the team is able to complete 50 requests in a week, the average processing time per request is 0.8 hours (i.e., 40/50). Perform the following steps to build an ExtendSim simulation model of this situation. The complete model is shown in Figure 8.6.

1. Open a new model either by clicking the button "New Model" in the upper left corner of the horizontal toolbar or by choosing File>New Model. Recall that this notation means clicking on the File menu and then choosing the New Model option from the list that appears. A new model worksheet then appears with an executive block in the upper left corner.
2. Add a block to generate request arrivals. Choose the Create block from the Routing submenu of the Item library. Double-click on the Create block and change the Distribution to Exponential. Make sure the mean value is 1 (and the location value is 0). Also, type "Requests In" in the text box next to the Help button (lower left corner of the dialog window). Press OK. The complete dialog window after performing this step is shown in Figure 8.5.
3. When a Create block is used to generate requests, it must be followed by a Resource Item or a Queue block because the Create block always pushes items out of the block regardless of whether or not the block directly after in the process flow is ready to receive them. Add a Queue block to the model to simulate the In Box of the underwriting team. Choose the Queue item from the Queues submenu of the Item library. Drag the Queue block and place it to the right of the Create block. Double-click on the Queue block and type "In Box" in the text box next to the Help button (lower left corner of the dialog window). Do not change any other field; it is assumed that requests arriving to the In Box are sorted and processed in a FIFO manner.
4. Add an Activity block to the model to simulate the underwriting activity. Choose the Activity item from the Activities submenu of the Item library. Drag the Activity block and place it to the right of the Queue block. Double-click on the Activity block, set "Maximum items in activity:" to 1, "Delay is:" *a constant,* and "Delay (D):" to 0.8. Also, type "Underwriting" in the text box next

to the Help button (lower left corner of the dialog window). No other field needs to be changed to model the current situation.

5. Add an Exit block to simulate the completion of a request. Choose the Exit item from the Routing submenu of the Item library. Drag the Exit block to the model window and place it to the right of the Activity block. Double-click on the Exit block and type "Requests Done" in the text box next to the Help button (lower left corner of the dialog window).
6. Connect the blocks in the model. To make a connection, make sure that the Block/Text Cursor is selected in the horizontal toolbar (this is the white arrow cursor); then, click on one connector and move the cursor to the other connector. Usually, the default is that the item input connectors are on the left-hand side of the blocks, and the item output connectors are on the right-hand side. For most blocks, this can be reversed by choosing "Right to left" at the bottom of the blocks dialog window. The item connectors are used to allow items (jobs) to move from one block to another. Value connectors on a block are used to input values and to output results. Under the Model tab and the submenu Connection Line Style in the main ExtendSim window, it is possible to choose between different styles for the connection lines. The options include: Smart, Right Angle, Straight, and Free Form. Now, make a connection between the Requests In block and the In Box block. Also, make a connection between the In Box block and the input connector of the Underwriting block. Finally, make a connection between the Underwriting block and the Requests Done block.

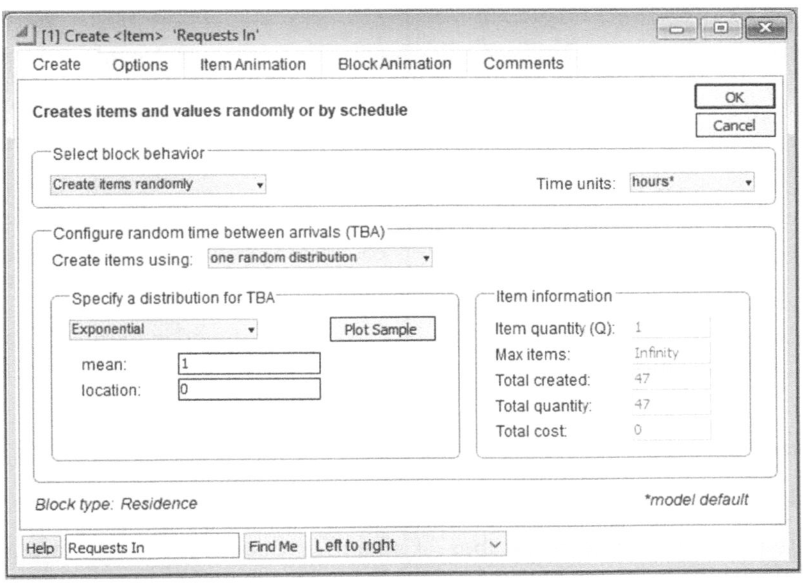

FIGURE 8.5
Create block dialog window.

The simulation model should look as depicted in Figure 8.6. Before running the simulation, the simulation time needs to be set to 40 hours so that the process can be observed for 1 working week. This is done in the Setup dialog (i.e., Run > Simulation Setup and click on the Setup tab) by setting the End time to 40, the Start time to 0 and Runs to 1 (default) and pressing OK. Now, make sure that Show 2D Animation and Add Connection Line Animation are chosen in the Run menu. Then, choose Run > Run Simulation. The number of requests in the In Box is updated as requests go in and out of this Queue block. Also, the Requests Done block shows the number of completed requests. If this model is run more than once, the number of completed requests changes from one run to another. As default, ExtendSim uses a different sequence of random numbers every time a simulation is run. To force ExtendSim to use the same sequence of random numbers every time, a random seed that is different from zero must be specified. Choose Run > Simulation Setup…, click on the Random Numbers tab, and change the Random Seed value to a number other than zero.

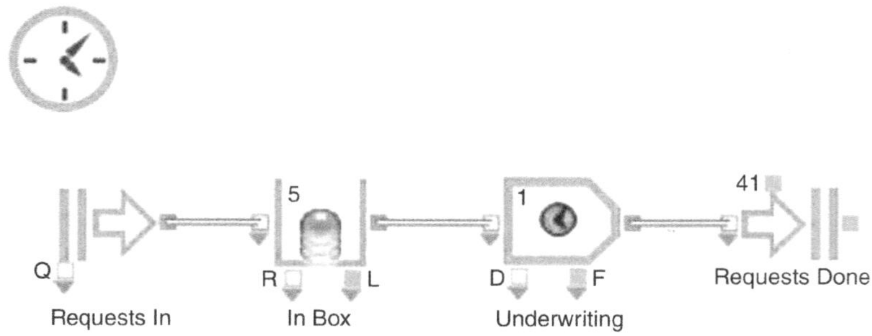

FIGURE 8.6
Underwriting process model with a single team.

During the simulation, the activity in any block can be monitored by double-clicking on the block to display its dialog window. For example, run the simulation either by choosing Run > Run Simulation or by pressing Ctrl + R and double-click on the Underwriting block. In the dialog window, click on the Results tab and monitor the use of the underwriting team. When the simulation is completed, the final use value is the average use of the underwriting team during the simulated week. The average number of requests in the In Box, and the average time a request waits in the In Box before being processed can be found in the Results tab of the dialog window to the Queue block labeled "In Box." ■

8.4 Basic Data Collection and Statistical Analysis

ExtendSim provides a number of tools for data collection and statistical analysis. The easiest way of collecting and analyzing output data in an ExtendSim model is to add Statistics blocks from the Statistics submenu of

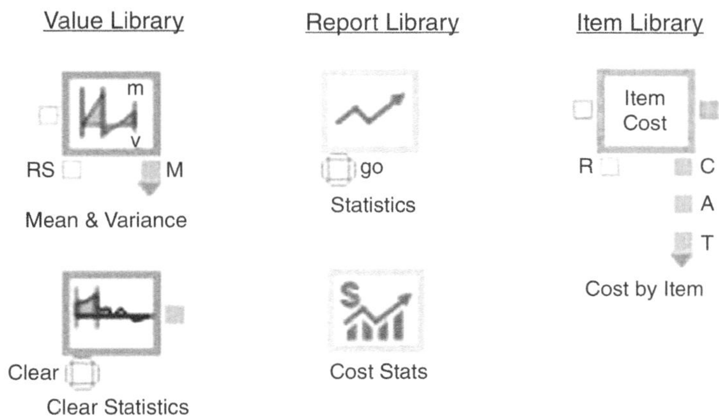

FIGURE 8.7
Blocks for statistical analysis of output data found in the Value Library, the Report Library, and the Information submenu of the Item library.

the Value library, from the Report library, and from the Information submenu of the Item library. The blocks in these libraries that are used most commonly in business process simulations are shown in Figure 8.7. The Cost Stats, Statistics, and Clear Statistics blocks can be placed anywhere in the model and do not need to be connected to any other blocks.

- The Mean & Variance block from the statistics submenu of the Value library can calculate the mean, variance, and standard deviation of the values sent into the block via the input connector (on the left side of the block) during the simulation. The choices in the dialog window include the options to calculate confidence intervals and to calculate statistics for multiple simulation runs. In business process simulations, throughput is often an interesting performance measure. A convenient way to calculate throughput statistics over multiple simulation runs is to use a Mean & Variance block and connect it to an Exit block where the number of completed jobs is available at one of the data output connectors. If an input is a NoValue, it is ignored and does not affect the statistics.
- The Clear Statistics block from the statistics submenu in the Value library can be used to clear the statistics calculations for specified blocks at a certain time or event. This provides a convenient way, for example, to eliminate warm-up (transient) periods from the statistical calculations when analyzing steady-state behavior.
- The Statistics block from the Report library provides a convenient method for collecting data and reporting statistics from all blocks of a specified type in a single table using a specified statistical method. In the dialog of this block, one can specify what block type should

be reported and whether confidence intervals should be computed. A variety of blocks can be chosen, including Activities, Queues, Mean & Variance, Resource Item, Resource Pool, Resource Manager, Mixed Blocks, Workstations, Convey Flow, and Tanks. Depending on the block type, different results are recorded and analyzed. The results can be copied and pasted into a word processing program, or they can be cloned to the model Notebook. The Statistics blocks can be placed anywhere in the model, and no connection lines need to be drawn.

- The Cost Stats block from the Report library provides a convenient method for collecting data from all cost blocks, such as an Activity block. The main statistics calculated in this block are cost per item, cost per unit of time, and total cost.
- The Cost by Item block from the Information submenu in the Item library collects data and calculates statistics associated with the cost of every item that passes through the block. The block must be connected to the process model to collect data on the items passing through it. In addition to displaying the cost of each item, the dialog of this block also shows the average cost and the total cost. The C, A, and T connectors can be used to plot the values of the current, the average, and the total cost, respectively.

Returning to the model of the underwriting process in Figure 8.6. Suppose that we are interested in analyzing the statistics associated with the requests that have to wait in the In Box before being processed by the underwriting team. Since we used a Queue block to model the In Box, we can obtain queue data with the Statistics block by choosing the option Queues for the "Block type to report" in the dialog window. Figure 8.8 shows the model with the Statistics block added.

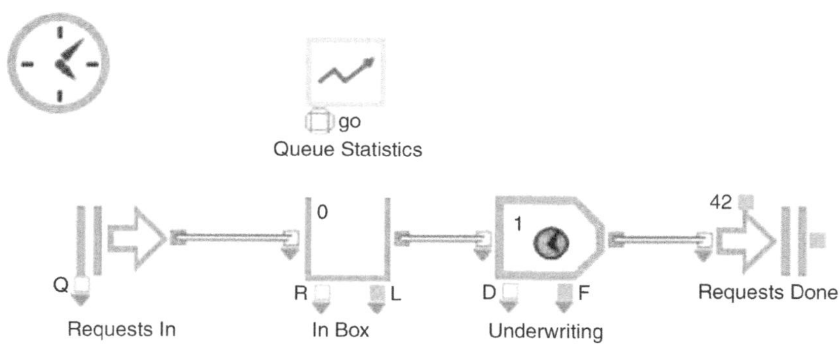

FIGURE 8.8
Underwriting process with a Statistics block added to gather and analyze Information from the Queue block modeling the In Box.

When running the model, the statistics associated with the In Box block can be analyzed. Using a Random Seed value of 1000 (a value chosen arbitrarily to make the experiment reproducible) the queue statistics shown in Figure 8.9 are found under the Results tab in the dialog window of the In Box block (the dialog window is opened by double-clicking on the block). The same information is available under the Statistics tab of the dialog window for the Queue Statistics block shown in Figure 8.10. From this figure, it may appear that some results are missing in the table, but on opening the block, one can scroll to the right of the table and find all the information shown in Figure 8.9. From Figures 8.9 and 8.10, it can be seen that for the simulated week, the average waiting time was 1.337 hours, and the maximum waiting time was 4.803 hours. Moreover, the average number of requests in the In Box was 1.438, and at no point during the simulation were more than seven requests waiting in the In Box of the underwriting team.

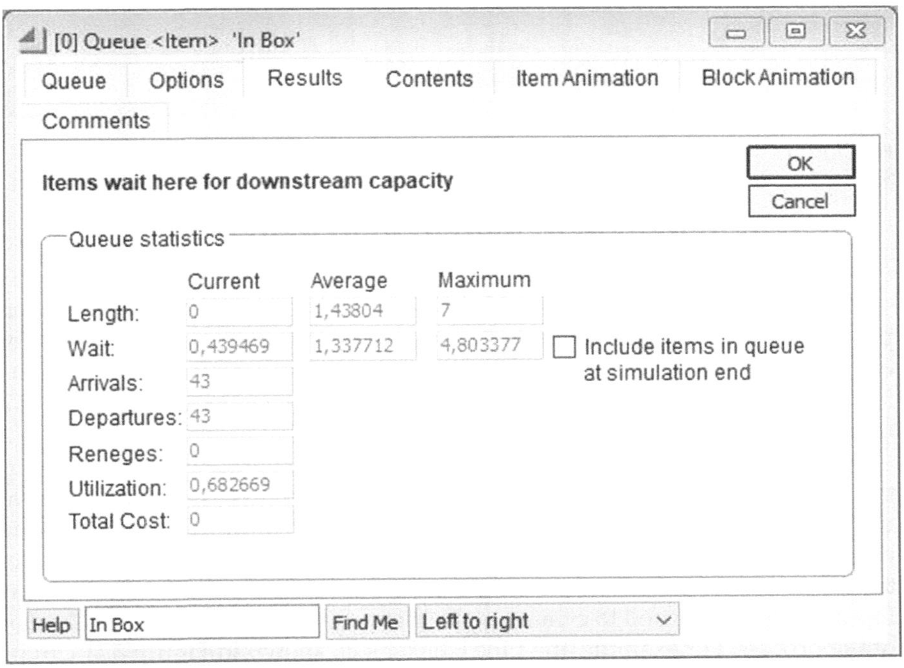

FIGURE 8.9
Statistics available under the results tab in the dialog of the In Box block.

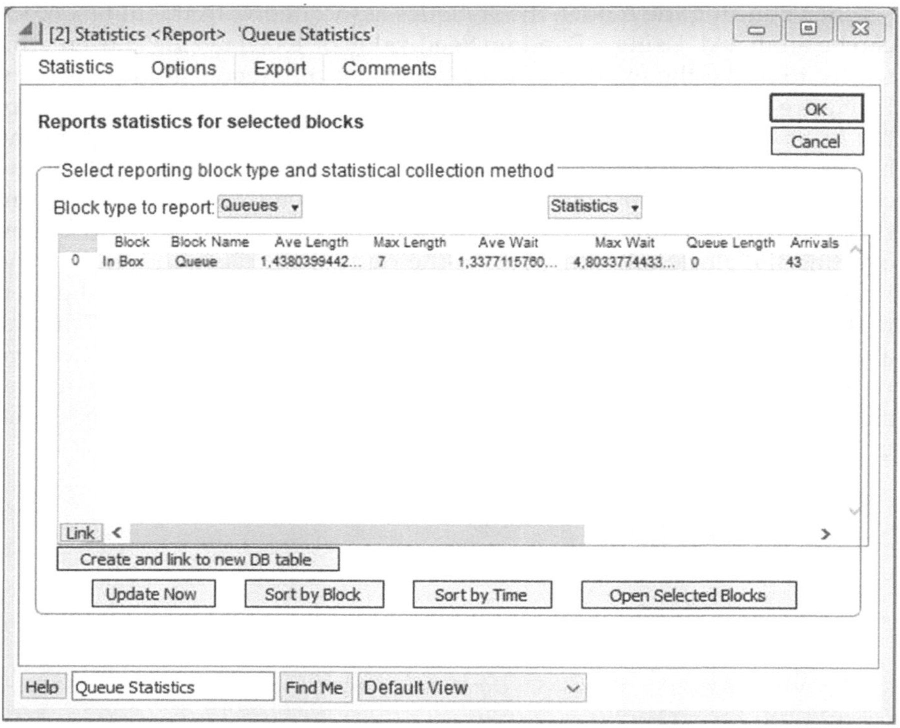

FIGURE 8.10
Statistics for the queue named In Box available in the Statistics block labeled "Queue Statistics."

The statistics that are relevant for measuring the performance of a given process vary according to the context in which the system operates. For instance, in a supermarket, the percentage of customers who find a queue length of more than three may be preferred to the average waiting time as a measure of customer service.

Although analyzing statistical values at the end of a simulation run is important, these values fail to convey the dynamics of the process during the simulation run. For example, knowing the final average and maximum waiting times does not provide information about when during the week the time of the week the waiting times are longer or shorter. ExtendSim includes a Chart library that can be used to create graphs that provide insight into the behavior of the process. For example, the Line Chart block shown in Figure 8.11 can be used to generate a waiting time vs. simulation time plot.

Next, we add a Line Chart block to the ExtendSim model of the underwriting process. If the Chart library is not already open, we first do that by choosing Open Library under the Library Tab and then double click on the Chart.

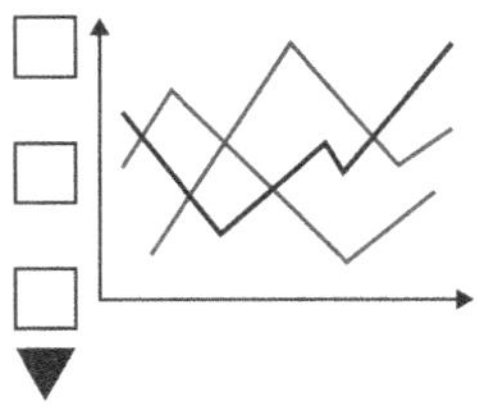

FIGURE 8.11
The Line Chart block found in the Chart library.

lbr icon in this directory. Now, choose Library > Chart > Line Chart to add the block to the model. Then connect the W output value connector from the In Box to the top input value connector of the Line Chart block. The modified model looks as depicted in Figure 8.12.

After the simulation is run, a waiting time vs. simulation time plot is produced by the Line Chart block, as shown in Figure 8.13.

The plot in Figure 8.13 discloses that the busiest time for the underwriting team was after 33.3 simulation hours, when the maximum waiting time of 4,803 hours occurred. In a real setting, the analyst must run the simulation model several times with different random seeds to draw conclusions about the process load at different points of the simulation horizon (i.e., 1 week for the purpose of this illustration). Next, model enhancements are examined, and additional blocks are introduced.

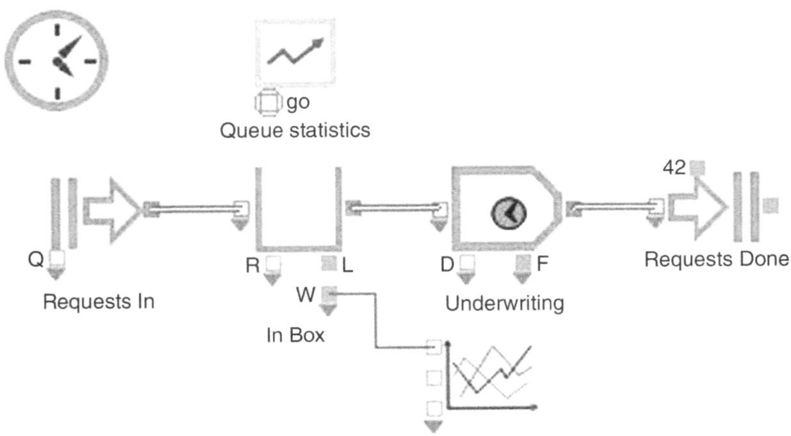

FIGURE 8.12
Underwriting process with a Line Chart block for plotting the waiting time during the simulation.

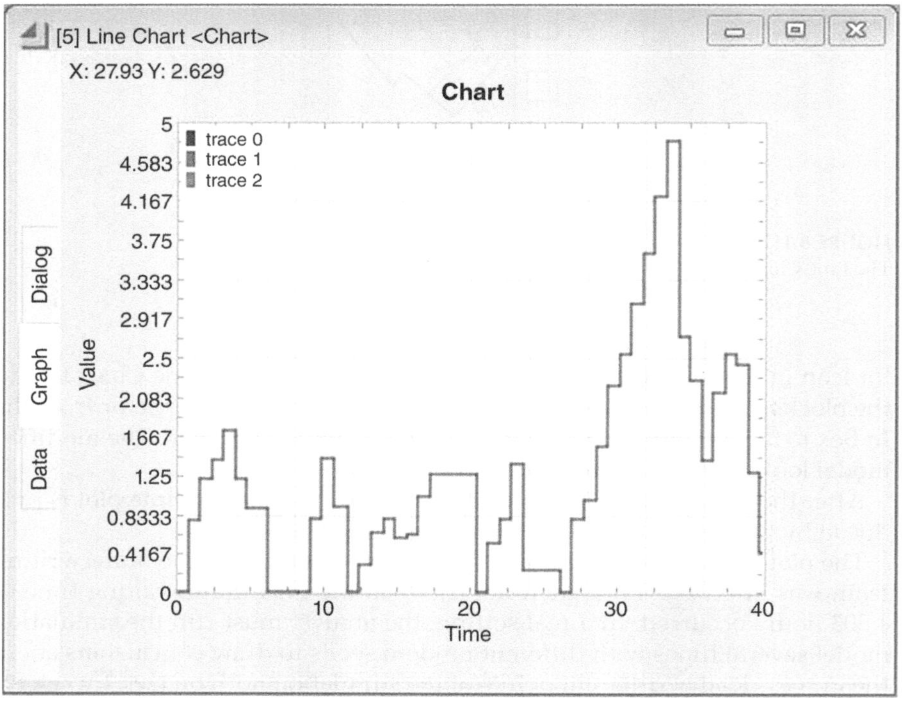

FIGURE 8.13
Waiting time versus simulation time plotted using the Line Chart block.

8.5 Adding Randomness to Processing Times and the Use of Attributes

So far, the models of the underwriting process in this chapter have used constant values to represent the time it takes the underwriting team to process a request. In real-world business processes, it is rare to find constant processing times. They usually inhibit some degree of variation and uncertainty. Sometimes, these variations are small enough that modeling the processing times as constant is a good approximation. However, if the variations are larger, they should be explicitly incorporated into the simulation model by the use of random numbers with specified probability distributions. In ExtendSim, random processing times can easily be incorporated into the model directly in the Activity block. The steps are as follows:

1. Double-click on the Activity block to open the dialog window.
2. Select the Process tab and set "Delay is:" to "specified by a distribution."

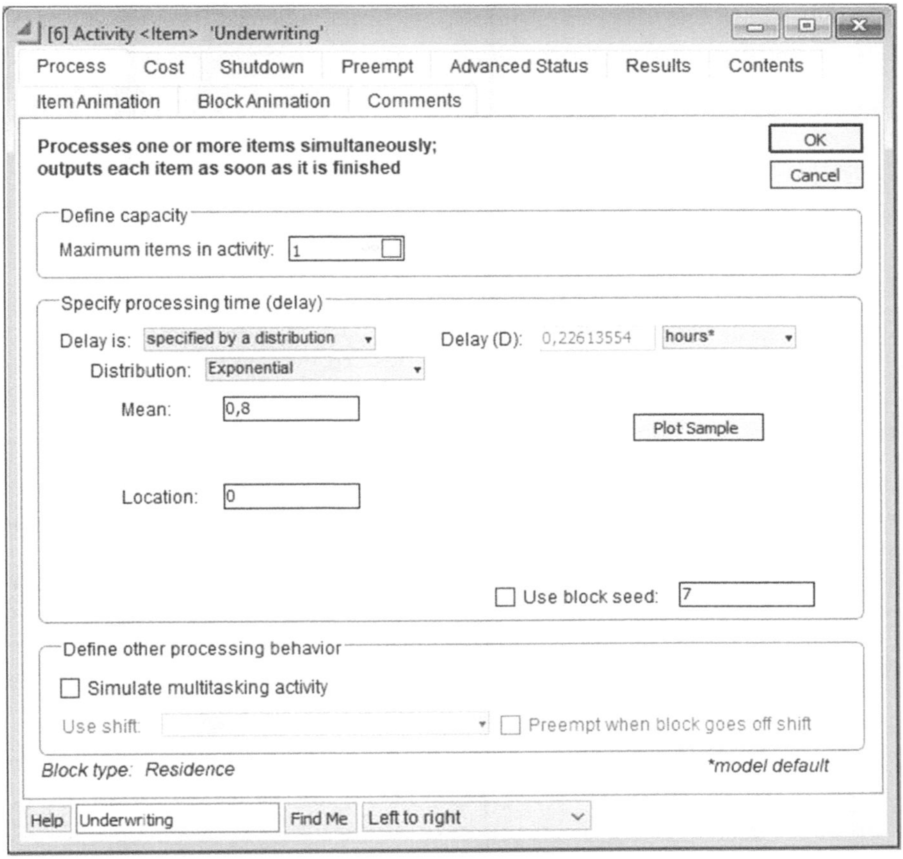

FIGURE 8.14
Dialog window of the Activity block in the underwriting process when processing times are exponential with mean 0.8.

3. Choose a distribution from the list that appears and specify the associated parameter values so that the distribution has a good fit with the actual real-world processing times.

For the underwriting process in this tutorial, it is assumed that the processing times follow an exponential distribution with a mean value of 0.8 hours. Figure 8.14 shows the dialog window of the underwriting block when these processing times have been implemented. The ExtendSim model that represents the current situation in the insurance company is shown in Figure 8.15.

Because the current model uses an exponential distribution with a mean value of 1 hour to model interarrival times and an exponential distribution with a mean of 0.8 hours to model processing times, the model should be able to approximate the theoretical values calculated in Chapter 6 for the same

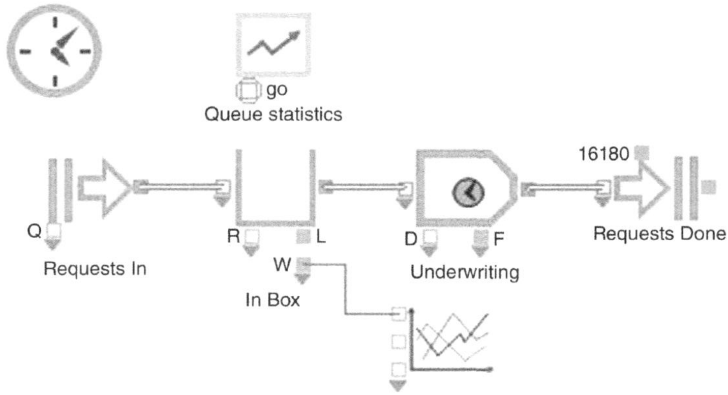

FIGURE 8.15
Underwriting process model with random processing times.

example. The current model is run for 400 weeks (16,000 hours), and queue statistics are collected in the In Box block (see Figure 8.16).

The empirical values for the average queue length (3.22 requests) and average waiting time (3.18 hours) in Figure 8.16 are close to the analytical values (3.2 requests and 3.2 hours, respectively) calculated with the $M/M/1$ model in Chapter 6.

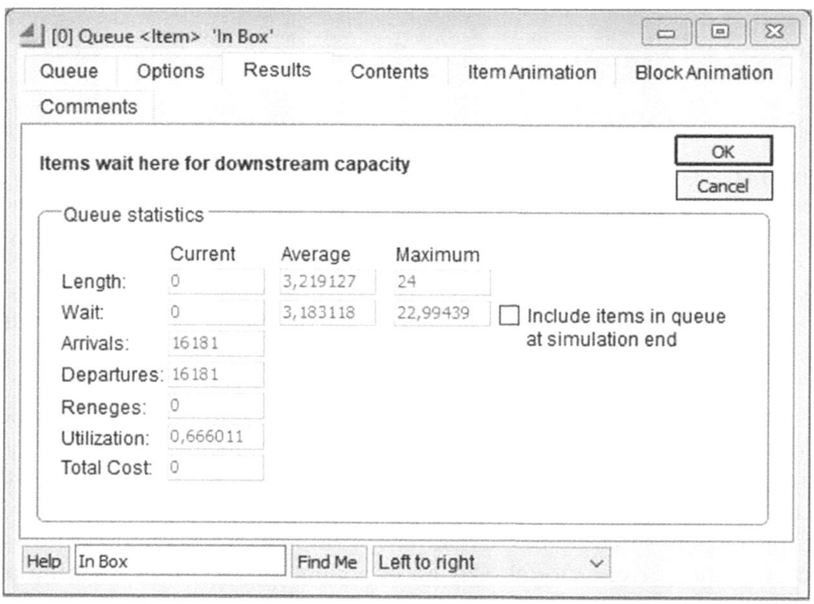

FIGURE 8.16
Queue statistics for the In Box block (400-week run).

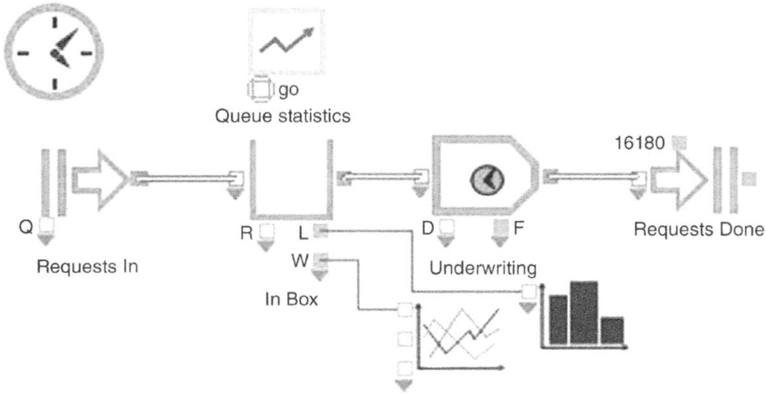

FIGURE 8.17
Underwriting process model with Histogram block added.

An interesting question regarding the performance of the current underwriting process relates to the percentage of time that the In Box block contains more than a specified number of requests. For example, one might wonder what percentage of time the In Box contains more than three requests (i.e., four or more requests). A Histogram block from the Chart library can help answer this question. Choose Library > Chart > Histogram. Connect the L output value connector at the In Box block to the top input value connector of the Histogram block. Then configure the block to collect time-weighted statistics in six bins. This is done by double-clicking on the Histogram block and choosing the Dialog option in the vertical menu bar. To specify the number of bins, choose Bins in the menu bar on top of the window and write 6 in the dialog box after "Number of bins:". To specify collection of time-weighted statistics, choose Data collection in the menu bar and check the option "Frequencies represented on the chart are time weighted." The resulting model is shown in Figure 8.17.

The Queue Length column in Table 8.1 represents the length of the queue, i.e., the number of requests in the In box. That is, of the total simulation time

TABLE 8.1
Frequency Table for the In Box Block

Queue Length (L)	Time with This Queue Length	Percentage of Total Time with This Queue Length
$0 \leq L < 4$	10322.42	64.5%
$4 \leq L < 8$	3415.47	21.3%
$8 \leq L < 12$	1562.53	9.8%
$12 \leq L < 16$	547.51	3.4%
$16 \leq L < 20$	131.71	0.8%
$20 \leq L < 24$	20.21	0.1%

of 16,000 time units, the In Box contains less than four requests 10322.42 time units (or 64.5% of the time). In terms of the original question, this means that 35.5% of the time (i.e., 100% – 64.5%), the In Box contained more than three requests (or equivalently four or more requests, as there is an integer number of requests in the queue at any given time).

In Chapter 6, an analytical queuing model was used to predict that the underwriting team would be busy 80 percent of the time in the long run. In other words, the prediction was made that the use of the underwriting team would be 80 percent. The simulation model can be used to compare the theoretical results with the empirical values found with the simulation model. Figure 8.18 shows the Results tab of the dialog window associated with the Underwriting block. For instance, the utilization of the underwriting team after 400 simulated weeks is 81.7 percent.

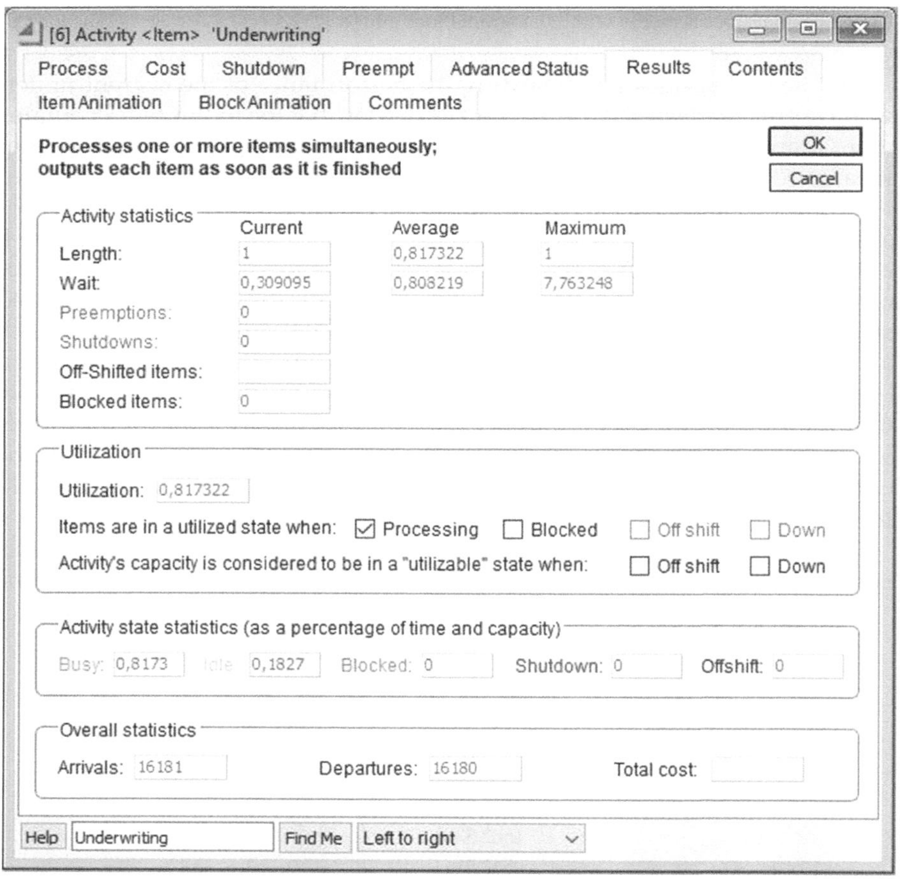

FIGURE 8.18
Dialog window of the Underwriting block.

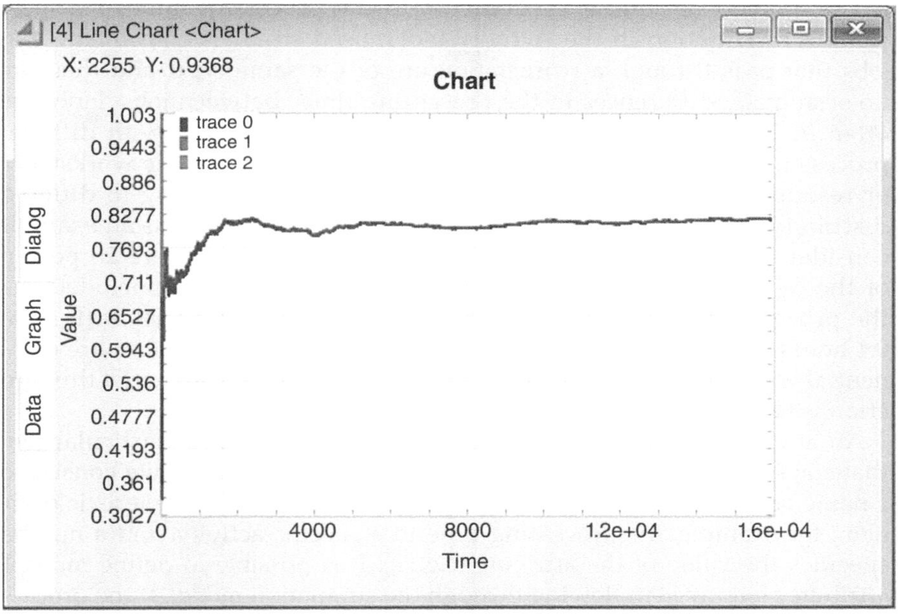

FIGURE 8.19
Average use of the underwriting team plotted against the simulation time using the Line Chart block.

The long-run average utilization of 80 percent is observed after 400 simulated weeks, but the utilization varies throughout the simulation and, in particular, within the first 10 simulated weeks. Figure 8.19 shows a plot of the average utilization of the underwriting team during the simulation.

The utilization plot can be obtained easily by connecting the u output connector of the Underwriting block to the Line Chart block. The same Line Chart block can be used to create Figures 8.13 and 8.19. Using the Line Chart block, up to 20 different data inputs can be turned on and off. Figure 8.19 reveals the importance of running a simulation for a sufficiently long time to predict the behavior of a process at steady state (after a period, known as *warm-up*, in which average values have not converged yet). If the simulation had been stopped after 400 hours (10 weeks), a team utilization of about 70 percent would have been predicted. Equally important as running simulations for a sufficiently long time is running the model several times. To run a model more than once, choose Run > Simulation Setup and change the number of runs in the Setup tab. Chapter 9 addresses issues associated with terminating and nonterminating processes as well as warm-up periods.

The method of adding randomness to the processing times by specifying a distribution in the Activity block used so far is convenient if all jobs that pass through a workstation are of the same type, and there are no systematic differences in the processing times between jobs. However, often in business processes, there are many types of jobs with different processing time characteristics that are processed in the same workstation or resource. If the processing times for different jobs belong to different distributions, the method described above does not suffice. As an example, consider a refined version of the underwriting process where 20 percent of the requests are much simpler to process. It has been concluded that the processing times for the simple requests are exponential with mean 0.4 hours and that the processing times for the regular requests are exponential with mean 0.8 hours as before. A convenient way to model this situation is to use attributes.

An attribute is a quality characteristic associated with a particular item that stays with it as it moves through the model. Each attribute consists of a name and a numeric value. The name identifies some characteristic of the item, for example, the processing time in a specific activity, and a number specifies the value of the attribute (e.g., 2). It is possible to define multiple attributes for any item flowing through the simulation model. If the processing time is available as an attribute value, the Activity block can read the named attribute and delay the item for the specified time. Thus, the modeling approach is to specify the processing time for each item according to the different distributions of interest, assign this information as a value to a named attribute (say Processing Time 1), and let the delay time in the activity be controlled by this attribute value for each item passing through the block. The new ExtendSim blocks used when modeling the refined underwriting process are as follows (see also Figure 8.20):

- The *Select Item Out block* from the Routing submenu of the Item library is used for separating the simple requests from the regular requests and routing them to separate blocks for setting the attributes.
- The processing times are generated using *Random Number blocks* from the Inputs submenu of the Value library. A Random Number block generates random numbers based on a catalog of probability distributions. The probability distribution with appropriate parameter values is specified in the Distributions tab of the dialog window.
- The *Set block* in the Properties submenu in the Item library is used for assigning the generated process time as a value for a specified attribute. The attribute name is determined by the modeler. In the present example, the attribute name *Process time* is used. The value

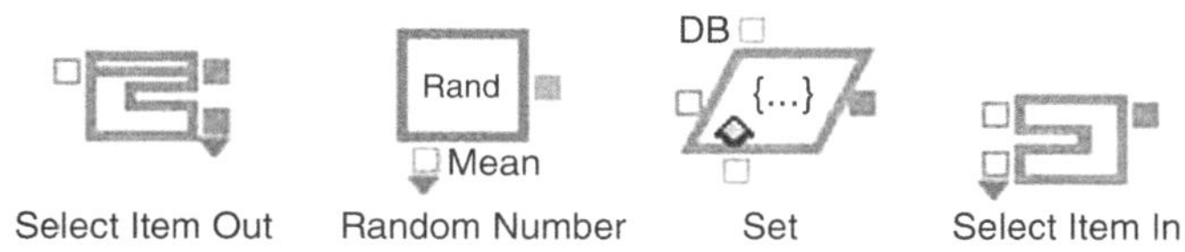

FIGURE 8.20
ExtendSim blocks modeling the process times in the refined underwriting process.

of the process time for each item passing through the block is sent in from the Random Number block through the input connector.

- The *Select Item In block* found in Routing submenu of the Item library is used to merge the flows of simple and regular requests into the same In Box.

An important step in completing the refined model shown in Figure 8.21 is to configure the Underwriting block by choosing the options "Delay is:" *an item's attribute value* and "Attribute:" *Process time* as depicted in Figure 8.22.

It is worth emphasizing that apart from specifying heterogeneous processing times, attributes are commonly used for specifying routing instructions, item types, calculation of cycle times, and many other things.

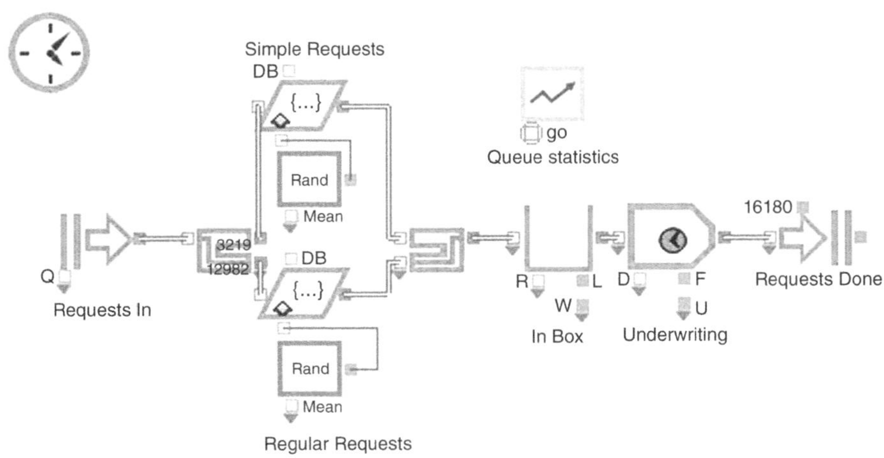

FIGURE 8.21
ExtendSim model of the refined underwriting process.

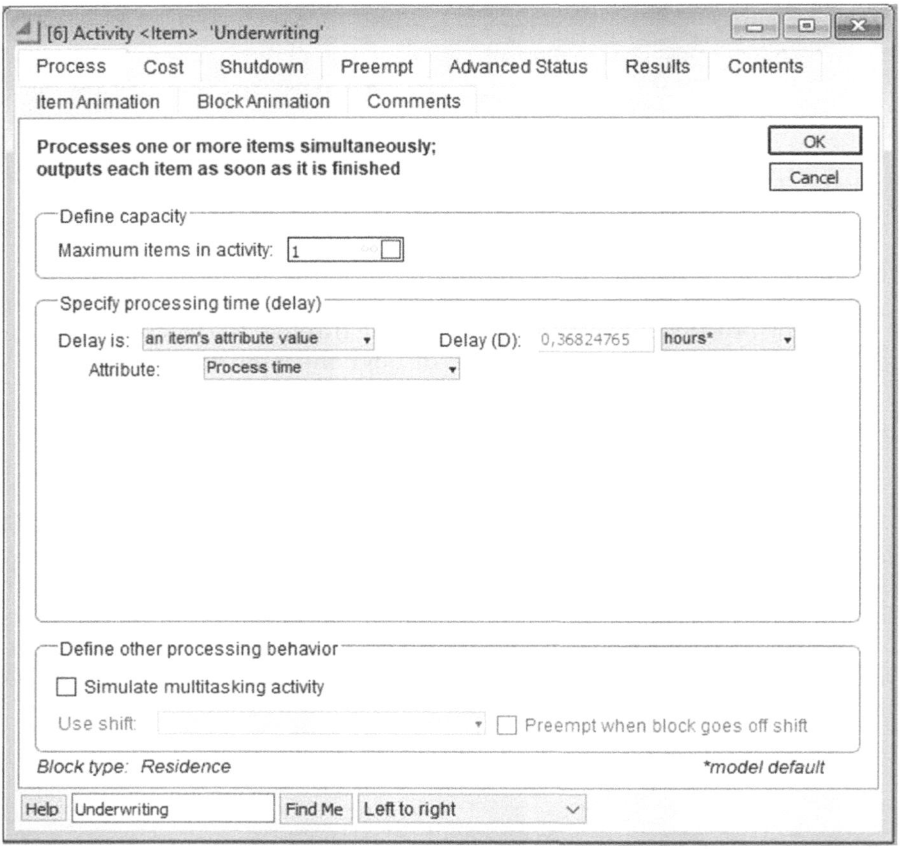

FIGURE 8.22
Dialog window of the Underwriting block in the refined Underwriting process.

8.6 Adding a Second Underwriting Team

The original model of the underwriting process illustrated a common method in modeling; namely, making simple assumptions first and then expanding the model to incorporate additional elements of the actual process. Suppose one would like to predict the effect of adding another underwriting team. If the two teams are identical and work in parallel, the simplest way to model this in ExtendSim is to change the capacity in the Activity block labeled "Underwriting" so that the "Maximum items in activity" is 2 instead of 1, as shown in Figure 8.23. This method means that no additional blocks are needed, and it is very easy to evaluate the effects of adding processing capacity.

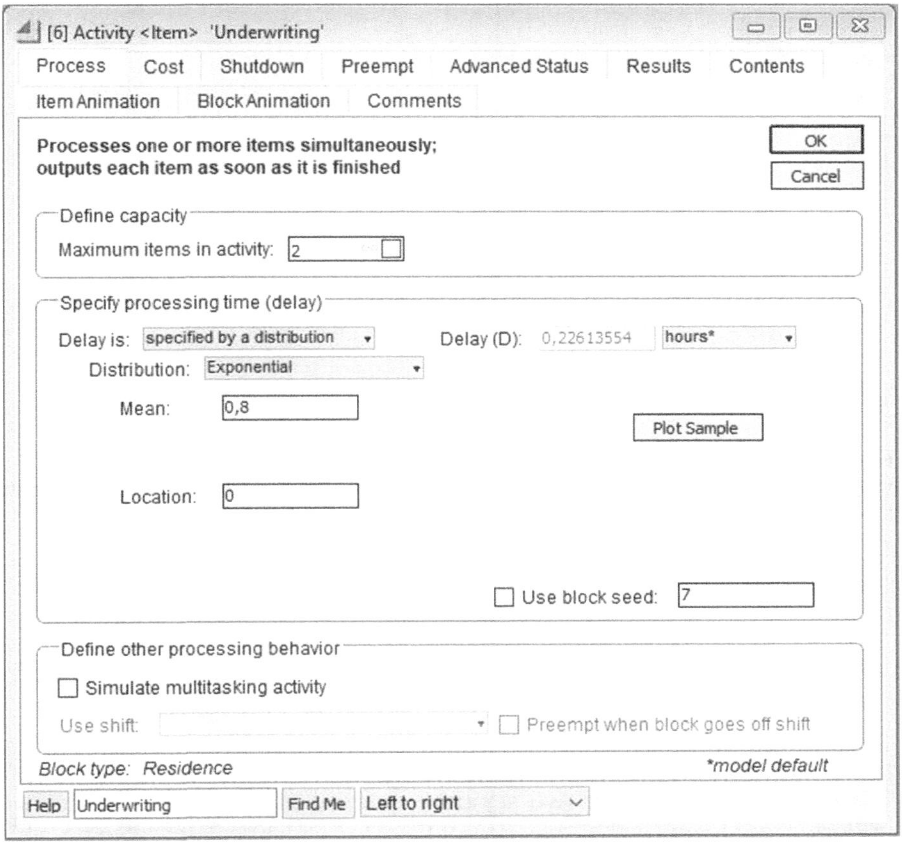

FIGURE 8.23
Activity block underwriting with two teams working in parallel.

An alternative way to model the second underwriting team in ExtendSim is, of course, to duplicate the original Underwriting block, which only allows one item in the block, and expand the Exit block with one more input connector as shown in Figure 8.24. The steps to arrive at this model are as follows:

1. Select the Activity block Underwriting.
2. Move it up to make room for the second underwriting team.
3. Choose Edit > Copy (Ctrl + C).
4. Click below the original block.
5. Choose Edit > Paste (Ctrl + V). Open the dialog window of the new activity block and change the block name to Underwriting 2.

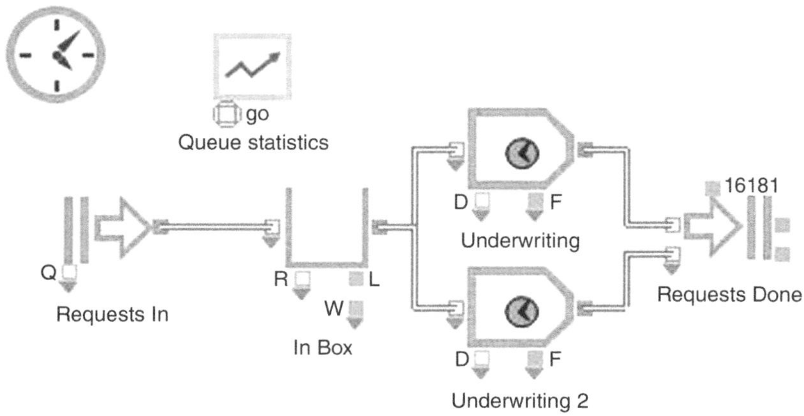

FIGURE 8.24
Underwriting process with two underwriting teams modeled with two separate Activity blocks.

6. Expand the Exit block to include two input connectors by placing the marker on the arrow attached on the lower side of the existing input connector and dragging downward until a second input connector appears.
7. Attach the input of the new Activity block to the output of the Queue block and attach the output connector of the new Activity block to the new input connector of the Exit block.

An advantage of using two separate Activity blocks to model the two underwriting teams is that they do not have to be identical. It is also possible to specify the logic for routing the requests to Team 1 or Team 2. The teams may, for example, be specialized in handling different types of requests.

After the model in Figure 8.24 has been run with two underwriting teams, the resulting queue statistics can be compared with those in Figure 8.16 (which correspond to the model with a single underwriting team). Table 8.2 shows the direct comparison.

TABLE 8.2
Queue Statistics for the Two Process Alternatives with One or Two Underwriting Teams

Statistic	One Team	Two Teams
Average queue length	3.219	0.156
Average waiting time in hours	3.183	0.155
Maximum queue length	24	8
Maximum waiting time in hours	22.994	4.215

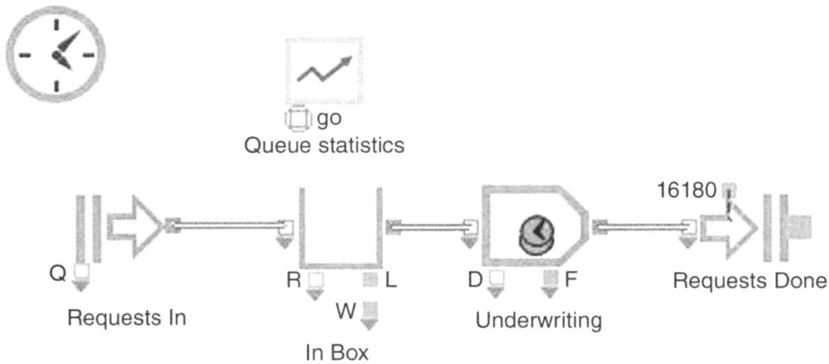

FIGURE 8.25
Underwriting process with two underwriting teams modeled with one Activity block.

Adding a team to the underwriting process significantly improves key measures of customer service, such as the maximum waiting time at the In Box block. Therefore, from the customers' point of view, adding a second underwriting team is undoubtedly beneficial. However, from a management perspective, the utilization of the teams should be considered before making a final decision about whether, or not, to add a team. The final average utilization is 50.1 percent for Team 1 and 31.6 percent for Team 2. The reason for the higher utilization of Team 1 is that if both teams are available, ExtendSim prioritizes sending requests to Team 1 because this block was the first to be connected to the Queue block when the model was built.

Running the alternative model with a single activity block as shown in Figure 8.25 results in corresponding queue statistics and an average use of 40.9 percent for the two teams.

8.7 Modeling Resources and Resource Pools

Most business processes require resources to perform activities. For example, a customer service representative is the resource used to perform the activities associated with assisting customers. In other words, resources are the entities that provide service to the items that enter a simulation model. In ExtendSim, resources can be modeled implicitly using an activity block, as we have seen so far, or explicitly using resource blocks. The explicit modeling of resources provides the flexibility of tying up resources for several activities. For example, suppose a nurse must accompany a patient through several activities in a hospital admissions process. In this case, the same nurse (resource) performs several activities and limits the capacity of the process.

To further explain and illustrate how resources can be modeled in ExtendSim, consider the aforementioned underwriting process and suppose that before the underwriting activity, a review activity takes place in which 80 percent of the requests are approved and 20 percent are rejected. There are two underwriting teams working independently of one another. Each team also performs the review activity. The review time is assumed to be uniformly distributed between 15 and 30 minutes. If the request is approved, the team proceeds with the underwriting activity. If the request is rejected, then the team is free to work on the next request. To modify the current model to deal with this situation, the Queue block in combination with a Resource Pool and two Resource Pool Release blocks from the Resources submenu of the Item library can be used. The latter blocks are depicted in Figure 8.26 and further explained below.

- The Resource Pool block holds a specified number of capacity units. The available units limit the capacity of one or several sections of a business process. It may, for example, represent the number of workers or machines in a workshop, the number of tables in a restaurant, or the number of underwriting teams in our tutorial example. The Resource Pool block is associated with at least one Queue block, where it is specified in the Queue tab of the dialog window that the queue is a "resource pool queue." In the table that appears, one has to specify which resource pool to take capacity from and how many units of capacity one item in the queue will need to be allowed to leave the queue. When the required units of capacity are available in the designated resource pool, the first item in the queue is allowed to leave, and the specified units of capacity are temporarily subtracted from the resource pool.
- The Resource Pool Release block releases a specified number of resource units from items passing through the block and returns these units of capacity to the specified Resource Pool block. For a functioning and stable model, it is important that all resource units assigned to an item are released and returned to the Resource Pool from where they originated before the item exits the model.

The ExtendSim model of the underwriting process with initial review and two underwriting teams is shown in Figure 8.27. As the restricted team capacity is modeled using the Resource Pool block, there should not be any

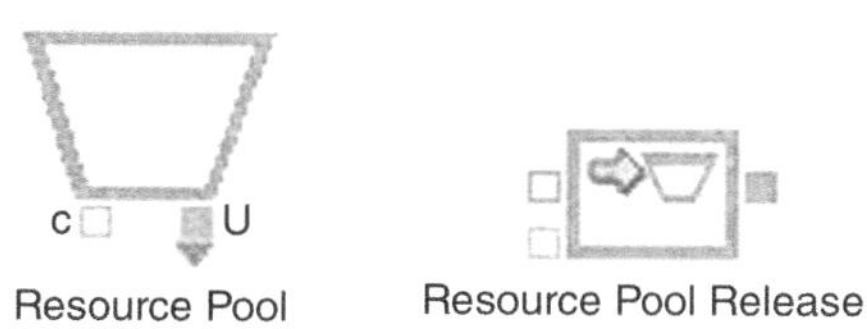

FIGURE 8.26
Resource Pool and Resource Pool Release blocks from the Item library.

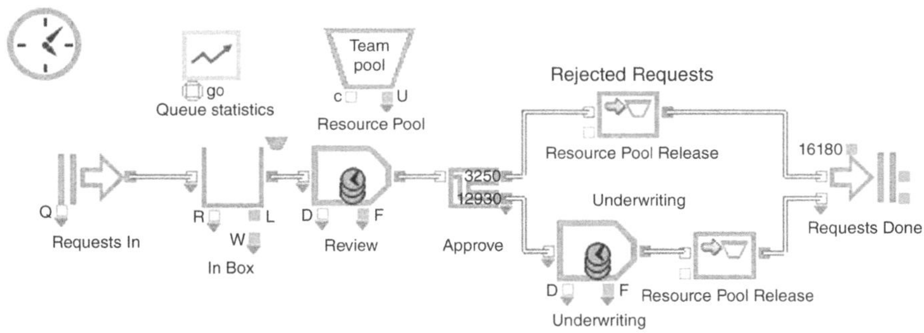

FIGURE 8.27
Underwriting process with an initial review activity and a resource pool.

additional limitations in the number of items allowed in the Activity blocks labeled "Review" and "Underwriting"; the option "Maximum items in activity" should therefore be set to infinity in these blocks. In the dialog window of the Select Item Out block named "Approve," choose the Option tab and set "Select output based on:" *random*. Thereafter, specify in the table the probability 0.2 for the top output connector going to the Resource Pool block and 0.8 for the bottom output going to the Underwriting block.

After the model is run (with random seed 1000 for 400 weeks), the following results are obtained:

- The use of the underwriting teams is approximately 52 percent. This can be seen in the Results tab in the Resource Pool block labeled "Team Pool."
- No requests are in the In Box at the end of the run, and the maximum number of requests in the In Box throughout the simulation is 8. The average and maximum waiting times in the In Box are 0.298 and 5.2 hours, respectively.
- Approximately 20 percent of the requests are rejected (3250/16,180).

An alternative approach to modeling capacity restricted resource pools, such as in the considered underwriting process, is to use a Resource Item block in combination with Batch and Unbatch blocks from the Batching submenu of the Item library. These blocks are depicted in Figure 8.28 and briefly

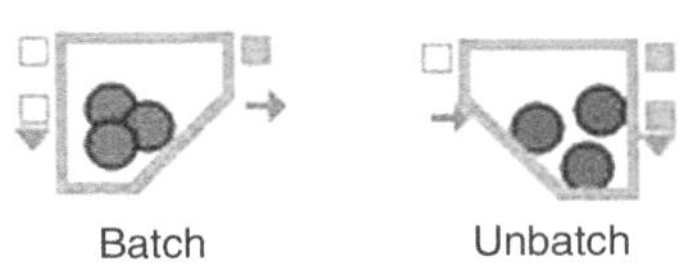

FIGURE 8.28
Batch and Unbatch blocks from the Item library.

described in the following. The advantage of using this modeling approach, which requires more blocks and connections than the modeling approach using the Resource Pool block described earlier, is that it allows the modeler to implement more advanced decision rules for routing of resources to different activities.

- The Batch block from the Batching submenu of the Item library is used to batch a chosen number of inputs into a single output. The number of required items from each input is specified in the dialog window of this block.
- The Unbatch block from the Batching submenu of the Item library is used to unbatch a single input to a chosen number of outputs. This block can be used to clone one item into several identical items, or it can be used to unbatch items that previously have been batched with a Batch block. For example, a Batch block could be used to batch a job with the worker who is going to perform the job. When the worker completes the job, the Unbatch block can be used to separate the worker from the job and return the worker to a Labor Pool. When using the Batch and Unbatch blocks in this way, it is prudent to check the Preserve Uniqueness box in both blocks. Preserve Uniqueness appears under the Unbatch tab in the dialog window of the Unbatch block and under the Options tab for the Batch block.

An ExtendSim model of the considered process can now be constructed as follows. Starting from the model in Figure 8.27, using the Resource Pool block, the following changes have to be made to arrive at the model shown in Figure 8.29.

1. Remove the Resource Pool block and the two Resource Pool Release blocks. Add a Batch block between the In Box block and the Review block. Also add two Unbatch blocks, one directly after the Underwriting block and the other between the top output of the Select Item Out block and the Exit block.
2. Connect the In Box block to the top input of the Batch block (named "Batch team"). Connect the output of the Resource Item block (now named "Team Pool") to the bottom input of the Batch block. Then, connect the output of the Batch block to the Activity block named "Review." Connect the blocks to make the model look like the one depicted in Figure 8.29. The model contains two named connections: Review Done and Underwriting Done. To make these connections, double-click where text is to be added. This opens a text box. Type the desired text, and when you have finished typing, click anywhere else in the model or press Enter. The text box can now be connected

to any block connector. Copy and paste the text box to establish the named connection.

3. In the dialog window of the Batch block, choose the menu option Batch, select the block behavior "Batch items into a single item," enter a quantity of one for the top and the bottom inputs, and check delay kit for both. If delay kit is checked, it means that an item is not pulled into the connector from the preceding block until the requested number of items is available at all input connectors. If the quantity is one at all inputs, they may all have delay kit checked (as in our model). If the quantity is larger than one, at least one input must have delay kit unchecked. Having delay kit checked for all inputs in the model means that the correct statistics are obtained from the Resource Item block and the Queue block. Finally, choose the Options tab and check the Preserve Uniqueness box.
4. In the dialog window of the Resource Item block labeled "Team Pool," set the initial number of items to 2.
5. In the dialog of the Unbatch blocks, choose the Unbatch menu tab and enter 1 in the Top and Bottom outputs. Also, check the Preserve Uniqueness box. This block separates the team from the request, and the team is sent back to the Labor Pool block. To avoid sending the team to the Exit block and the completed request to the Team Pool, it is important to unbatch the team from the bottom connector if it was batched with the request through the bottom connector of the Batch block.

Running the model in Figure 8.29 with the same random seed as the model in Figure 8.27 renders the same results.

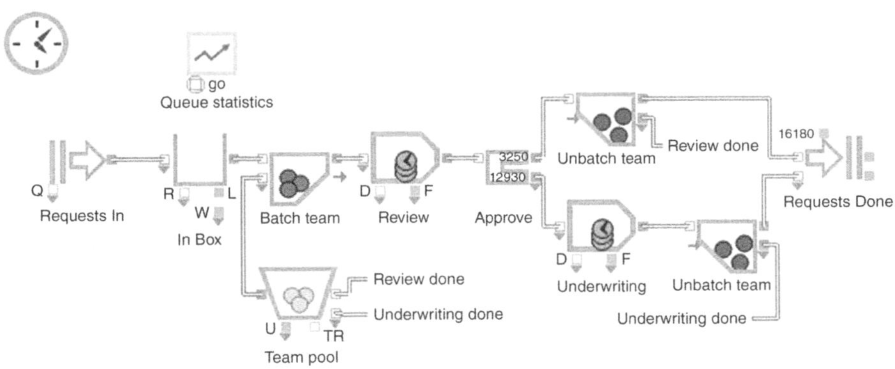

FIGURE 8.29
Underwriting process with initial review and a Resource Item block with batching.

8.8 Customizing the Animation

Model development and validation are key elements in computer simulations, but animation is an important additional key component. ExtendSim offers extensive opportunities for customizing the animation in a model. The Animate Value and Animate Item blocks from the Animation library show customized animation in response to model conditions. Based on choices in their dialogs, these blocks can display input values, show a level that changes between specified minimum and maximum values, display text or animate a shape when the input value exceeds a critical value, or change an item's animation icon. The blocks can be used for developing models with sophisticated animation features, including pictures and movies.

A simple way of customizing the animation in a simulation model consists of changing the appearance of the items in the model. Consider the model in Figure 8.29 of the underwriting process with a Team Pool. Turn on animation by selecting Run > Show 2D Animation and make sure that Animate Connection Lines also is selected in the Run menu. During the simulation run, ExtendSim's default item animation picture, a green circle, will flow along the connection lines between blocks. In the Animate tab of blocks that generate items, such as the Create block named "Requests In" in this model, the animation picture that will represent the items generated by this block can be chosen. In the Item Animation tab of the Requests In block, choose the paper picture to represent the requests as shown in Figure 8.30.

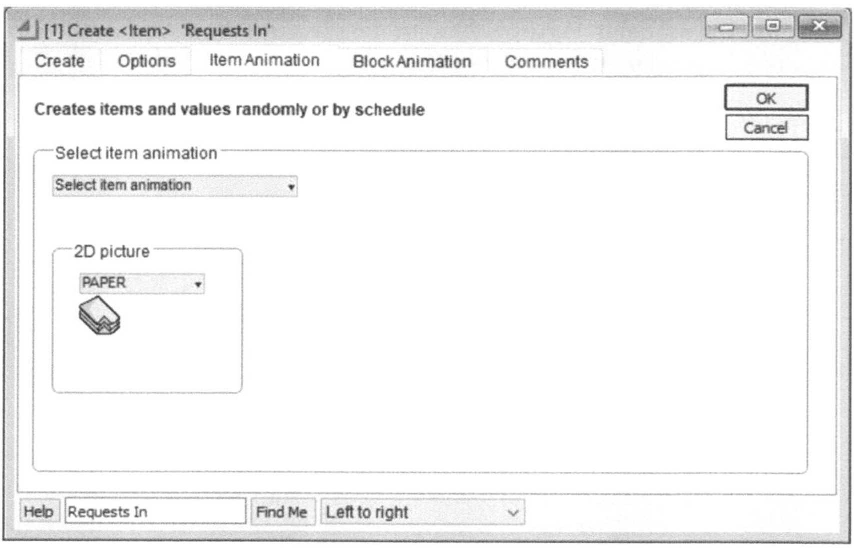

FIGURE 8.30
Item Animation tab of the Create block named "Requests In."

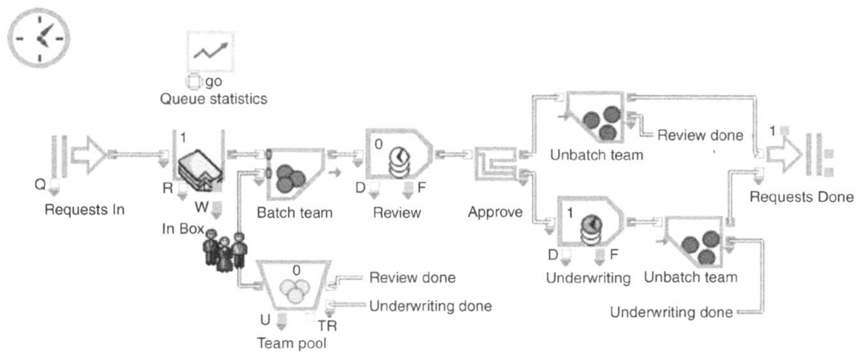

FIGURE 8.31
Underwriting model with paper and people animation of requests and teams.

The request generated in the Requests In block will now appear as paper flowing through the model. Make sure that in the Item Animation tab of all blocks that do not generate items, the option Do Not Change Animation Pictures is selected.

The Team Pool block provides people for the review and underwriting activities. The animation may accordingly be set in a similar way as for the Requests In block earlier. In the Item Animation tab of the Team Pool block (the Resource Item block), first choose the option "Change all Items to" and then choose the People picture in the 2D picture list below to represent the teams. Because of how attributes are merged in the batching activity, the teams will inherit the animation picture of the requests (paper picture) while the two items remain batched. The items recover their original animation after they are unbatched thanks to the Preserve Uniqueness option chosen. Figure 8.31 shows the model with the chosen animation pictures at a given instance of a simulation run.

8.9 Calculating Activity-Based Costs

Many companies use activity-based costing (ABC) as the foundation for designing their business processes. The ABC concept is based on the notion that every enterprise consists of resources, activities, and cost objects. Activities are defined by decomposing each business process into individual tasks. Then, the cost of all resources consumed by each activity and the cost of all activities consumed by each product or cost object are tracked (Nyamekye, 2000).

ABC is a method for identifying and tracking the operational costs associated with processing jobs. Typically, this approach focuses on some unit of output, such as a completed order or service, in an attempt to determine its total cost as precisely as possible. The total cost is based on fixed and variable costs of the inputs necessary to produce the specified output. ABC is used to identify, quantify, and analyze the various cost drivers (such as labor, materials, administrative overhead, and rework) and determine which ones are candidates for reduction.

When a simulation model is built, the outputs as well as the resources needed to create such outputs are fully identified. Adding ABC to the model entails entering the costing information into the appropriate block dialogs. Blocks that generate items (e.g., the Create block) or resources (e.g., the Resource Pool or Resource Item blocks) and blocks that process items (e.g., the Activity block) have tabs in their dialogs for specifying cost data. These tabs allow the modeler to enter variable cost per unit of time and fixed cost per item or use. After the cost information has been entered, ExtendSim keeps track of the cost automatically as the items enter the process, flow through the activities in their routing, and exit.

When considering ABC in a simulation model, every item is categorized as either a cost accumulator or a resource. Cost accumulators are the items (or jobs) being processed. Jobs accumulate cost as they wait, get processed, or use resources. The cost of a resource is used for calculating the activity-based costs that a job accumulates. For example, suppose the objective is to determine the cost of receiving crates at a warehouse. As each shipment arrives, a labor resource is required to unpack crates and stock the contents on the appropriate shelves. In this case, crates are being processed; therefore, they become the cost accumulators. A crate accumulates cost while waiting and while being processed by the labor resource. For example, if it takes an employee 30 minutes to unpack a particular crate at an hourly wage of $10, the accumulated labor cost for this crate is $5 ($10 × 30/60). The calculated cost is then added to the accumulated cost for this crate. More cost is added to the accumulated total as the crate flows through the receiving process. Note that the accumulated cost per item typically varies because of variations in processing times and waiting times.

Returning to our underwriting process example, ExtendSim's ABC feature can help in calculating the costs of reviewing and underwriting in the considered process. In this model, two things contribute to the total cost:

- The teams are paid $46.50 per hour. (Three team members each earn $15.50 per hour.)
- As part of the review activity, a report is generated that costs $25.

To show how the cost calculations can be implemented in ExtendSim, we first consider the model in Figure 8.27 with the Resource Pool block modeling the Team Pool.

In the Cost tab of the Team Pool block, check the option "Define Resource Pool costs" and fill in the team cost per hour ($46.50). Figure 8.32 displays the resulting Cost tab dialog window.

In the Cost tab of the Activity block named "Review," check the option "Define processing costs" and fill in the cost of the report (Cost per Item: $25) as shown in Figure 8.33.

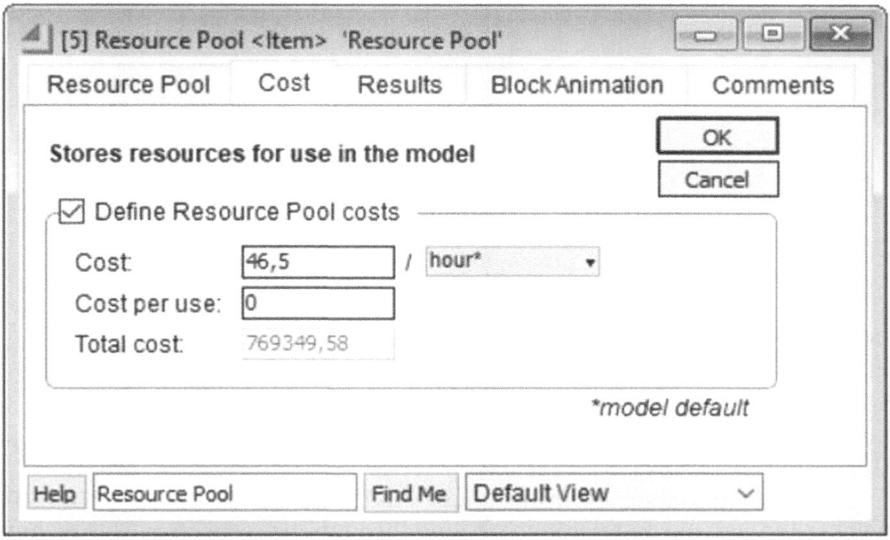

FIGURE 8.32
Cost tab of the Resource pool block named "Team pool."

When the simulation is run, the costs are tracked with each request. To collect the cost data, add two Cost by Item blocks from the Information submenu of the Item library and a Cost Stats block from the Report library to the model. The Cost Stats block can be placed anywhere in the model, and Cost

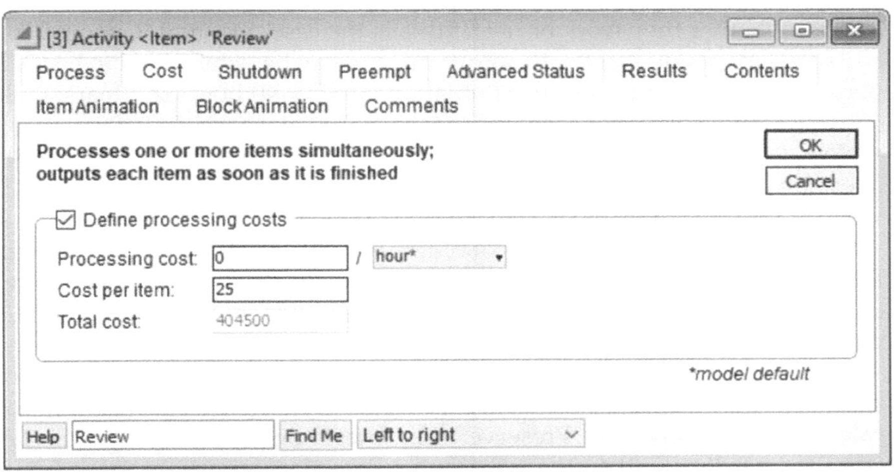

FIGURE 8.33
Cost tab of the Activity block named "Review."

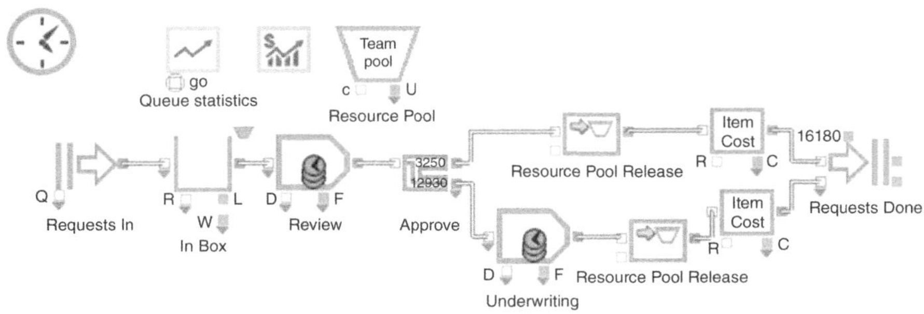

FIGURE 8.34
Underwriting process model with a Resource pool block and cost calculations.

by Item blocks should be placed after the Resource Pool Release blocks as shown in Figure 8.34.

After the simulation is run (random seed 1000 and Simulation time 400 weeks), the average cost per item passing through the respective Cost by Item blocks is found in the Cost tab of their dialog windows. By opening these blocks, one can see that the average cost per rejected request is approximately $42.40, and the average cost of a completed request is approximately $80.13. This information can provide the basis for setting the value for a possible application fee for this process. From the Cost Stats block, one can see that the total cost is $1,173,849.58 or $2934.62 per week ($1,173,849.58/400 weeks). This total cost consists of the cost for all review reports ($25 times 16,180 reports equals $404,500) and the Resource Pool costs for the two teams for the time that they are working ($46.5 × 16,000 hours × 2 teams × 51.7036% team use = $769,349.57). Clearly, one can argue that the hourly cost for the two teams accrues whether they work or not. If this is the case, the total Resource Pool costs are independent of the Resource use and can be determined to be $3720 per week without running the simulation ($46.5 × 40 hours × 2 teams). The total cost per week then becomes $4731.25 ($3720 plus $404,500/400).

Now, let us consider how to incorporate the same cost calculations in the alternative ExtendSim model of the process shown in Figures 8.29 and 8.31, where a Resource Item block is used for modeling the Team Pool.

In the Cost tab of the Team Pool block, check the option "Define Activity Based Costs," choose *resources* in the pop-up menu "Provides items that calculate costing as:", and fill in the cost per hour ($46.50) for the teams. Figure 8.35 displays the resulting Cost tab dialog window.

In the dialog of both Unbatch blocks, choose the Unbatch tab and "Select block behavior" *Release cost resources*, as shown in Figure 8.36. For the cost calculations to work, the "Preserve Uniqueness" option must be unchecked (see Figure 8.36). For the same reason, the Preserve Uniqueness option must also be unchecked in the Batch block (done in the Options tab in the Batch block dialog).

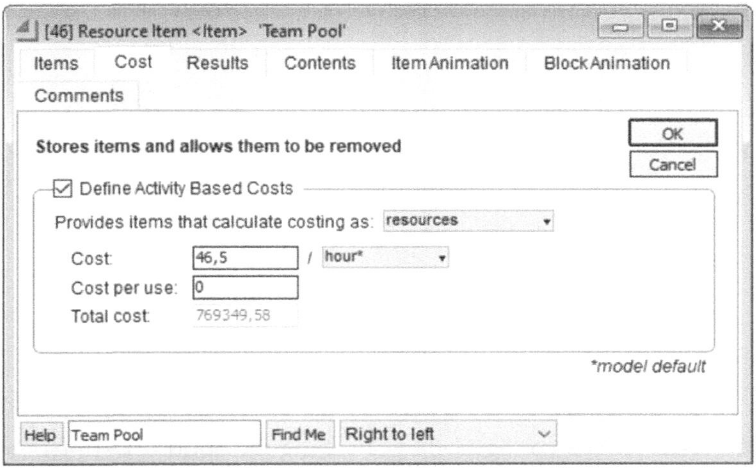

FIGURE 8.35
Cost tab of the Resource Item block named "Team Pool."

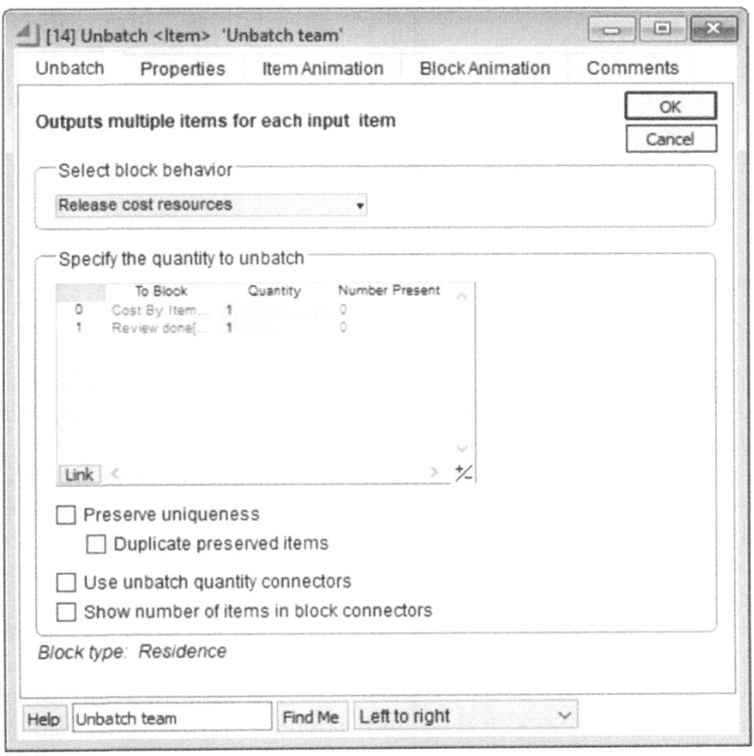

FIGURE 8.36
Unbatch block dialog window.

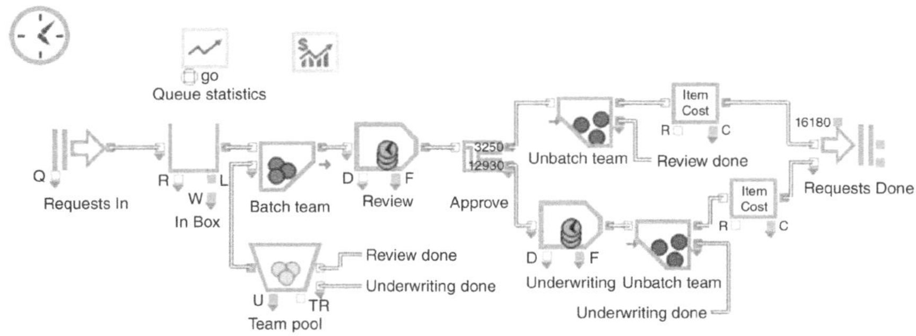

FIGURE 8.37
Underwriting process model with a Team Pool block and cost calculations.

The Review block is configured by checking the option "Define processing costs" in the Cost tab and filling in the cost of the report (Cost per Item: $25) as shown in Figure 8.33.

Finally, one Cost Stats block and two Cost by Item blocks are added to collect the desired cost statistics. The Cost by Item blocks should be placed after the Unbatch blocks as shown in Figure 8.37. The Cost Stats block can, of course, be placed anywhere in the model.

Running the simulation model in Figure 8.37 renders the exact same results as the model in Figure 8.34 using random seed 1000 and Simulation time 400 weeks.

8.10 Cycle Time Analysis

One of the most important measures of process performance is cycle time. In addition to total cycle time (the time required by an item to go from the beginning to the end of the process), the cycle times for different process segments or subprocesses are often of great importance to track. In other words, it is often valuable to know the cycle time of certain process segments in addition to knowing the cycle time for the entire process. ExtendSim provides simple ways of adding cycle time analysis to a simulation model and thereby determining any cycle times of interest.

To calculate the cycle time of individual items, ExtendSim uses the Information block in the Information submenu of the Item library, in combination with a Set block in the Properties submenu of the Item library and a Simulation Variable block in the Inputs submenu of the Value library. These three blocks are shown in Figure 8.38.

The Set block is placed at the beginning of the process segment for which the cycle time is to be measured, and the Information block is placed at the

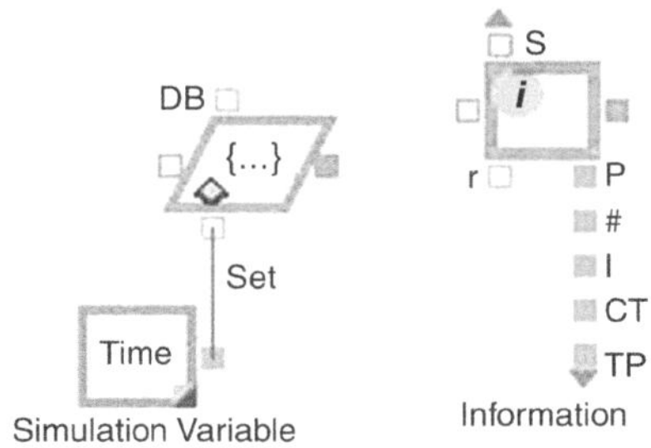

FIGURE 8.38
ExtendSim blocks for calculating cycle times.

end of this process segment. The items pass through the Set block, where a Timing attribute is given the value of the current simulation time. This is done by connecting the Simulation Variable block to the Set block through the input connector associated with the attribute name chosen by the modeler. Under the Options tab in the dialog window of the Simulation Variable block, the option "Current time" should be chosen as the system variable to output. At the end of the considered process segment, the items pass through the Information block, which (if appropriately configured) reads the Timing attribute and calculates the cycle time of each individual item. The configuration of the Information block is done by opening its dialog window, choosing the statistics tab, checking the "Calculate TBI and Cycle Time statistics" option, and specifying the name of the Timing attribute defined in the Set block earlier. Figure 8.39 shows how this modeling technique can be used for measuring the cycle time of customers visiting a travel agency modeled as a simple queuing system consisting of a single queue and two parallel servers.

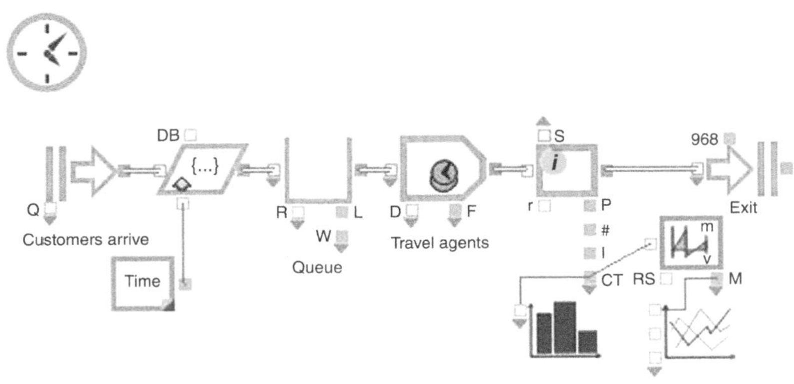

FIGURE 8.39
Cycle time analysis of a travel agency.

Customers arrive to the travel agency according to a Poisson process with a mean of 0.5 customers per minute. The arriving customers are served (in the order they arrive) by two travel agents, each with uniformly distributed service times between 1 and 5 minutes.

In Figure 8.39, the cycle time is measured from the moment a customer arrives to the Queue block until he or she has been serviced and leaves the Activity block labeled "Travel agents." Opening the Set block and the Information block, one can see that the name of the Timing attribute used in this model is Cycle Time, but this can be chosen freely by the modeler. The cycle time of each item passing through the Information block is available at the CT output connector at the bottom of the Information block.

A histogram of the observed cycle times is easily obtained by connecting a Histogram block to the CT connector. Figure 8.40 displays the result when running a simulation of 600 hours with random seed of 1000. To calculate the average cycle time, a Mean & Variance block is connected to the CT connector. To plot the average cycle time versus the simulation time, a Line Chart block is connected to the M output connector on the Mean & Variance block. The resulting plot is found in Figure 8.41.

Returning to the underwriting process with initial review, given in Figure 8.34, the described modeling technique can be used for analyzing the cycle time of approved requests as shown in Figure 8.42.

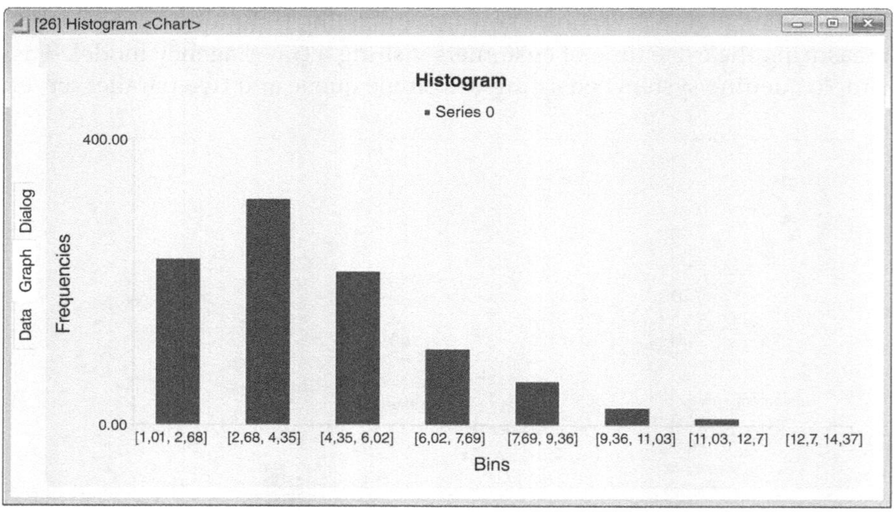

FIGURE 8.40
Histogram of observed cycle times for the travel agency.

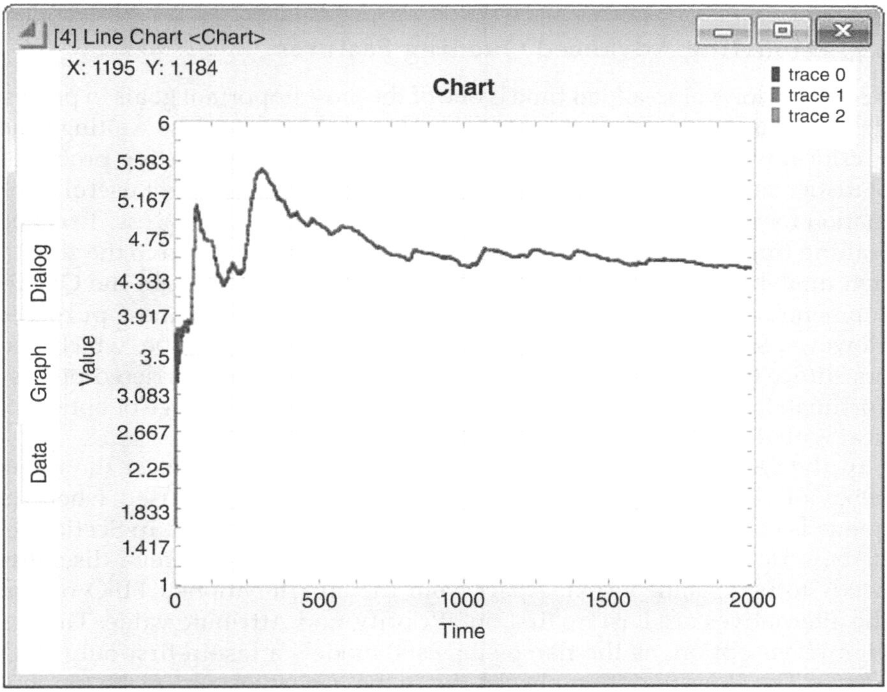

FIGURE 8.41
Average cycle time versus simulation time for the travel agency.

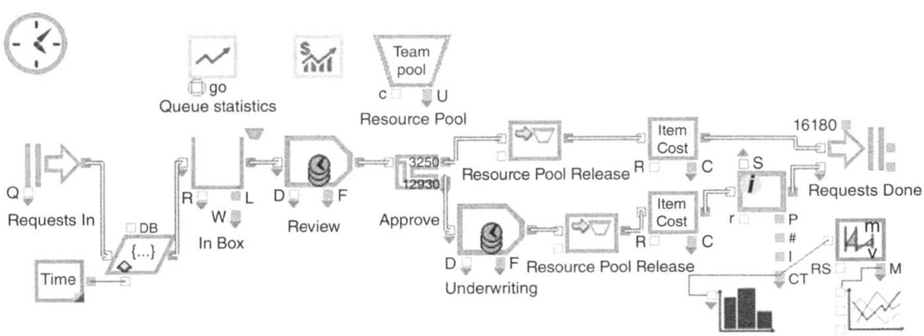

FIGURE 8.42
Cycle time analysis of approved requests in the model of the underwriting process with initial review.

8.11 Modeling Advanced Queuing Features

Reducing non-value-added time is one of the most important goals in process design. Quantifying the amount of unplanned delays (such as waiting time) is critical when designing new processes or redesigning existing processes. Statistics such as average and maximum waiting time represent useful information for making decisions with respect to changes in a process. To collect waiting time and queue length data, queues need to be added to the simulation model. As seen earlier in this chapter, the Queue block in the Queues submenu of the Item library in ExtendSim can be used for these purposes. However, so far, all queues have been of the basic FIFO type, which may not suffice to accurately model the queuing features of a considered process. Fortunately, ExtendSim and the Queue block offer a wide range of options to deal with different types of queues and queuing disciplines.

In the Queue tab of the Queue block dialog, one can select the queue behavior *sorted queue* or *resource pool queue*. The latter is used when the queue is combined with a Resource pool block as explained in Section 8.7. If the *sorted queue* option is chosen, the sort method or queue discipline needs to be specified next. Apart from the aforementioned FIFO option, the alternatives are Last in, first out, Priority, and Attribute value. The *Last in, first out* option, as the name suggests, models a last-in-first-out (LIFO) queue. The *Priority* option checks the items' priorities and picks the item with the highest priority (the smallest priority value) to be released first. If all the items in the queue have the same priority, then a FIFO discipline is used. The *Attribute value* option checks the value of a named attribute and sorts either from low to high values or from high to low values. Thus, the Priority and Attribute value options are quite similar but may be more or less convenient to use in a larger model.

The Options tab of the dialog window provides options to specify the maximum queue length (default is no limitation) and to Renege items after a specified time. The Reneging option can be used to specify how long an item will wait before it reneges (prematurely leaves). An item will wait in the queue until its renege time (the maximum amount of time the item is allowed to spend in the queue) has elapsed. When this happens, it will exit the Queue block through the lower (renege) output connector. The remainder of this chapter on advanced queuing features focuses on Blocking, Balking, Reneging, and Priority queues.

8.11.1 Blocking

Queues are used to avoid blocking. Blocking occurs when an item has finished processing but is prevented from leaving the block because the next activity is not ready to pick it up (or the next resource block or Queue block is full). Blocking can occur in serial processes where activities are preceded by queues. For example, consider two activities in a serial process, where Activity B follows Activity A. Suppose this process has two workers, one who performs Activity A and one who performs Activity B, and the workers don't have an In Box, or queue. If Activity A is completed while Activity B is still in process, the worker performing Activity A is blocked because that person cannot pass on the item until the other worker completes Activity B. Adding a queue (with sufficient capacity) to Activity B eliminates the blocking problems because completed items go from Activity A to the queue from which the worker performing Activity B picks them up.

8.11.2 Balking

In service operations, customers sometimes enter a facility, look at the long line, and immediately leave. As discussed in Chapter 6, this is called *balking*. One way to model balking in ExtendSim is to use a Select Item Out block from the Routing submenu of the Item library in combination with an Equation (I) block from the Properties submenu of the Item library. Figure 8.43 shows a simple model of a FIFO queue with balking in front of an automated teller machine (ATM). The processing times at the ATM are exponentially distributed with a mean of 1 minute. The model uses two Exit blocks to keep track of the number of served customers versus the number of balking customers.

Customers are generated in the Create block with exponentially distributed interarrival times (mean of 1 minute) and go to the Equation block, which checks the length of the queue. The length of the queue is passed from the L connector of the Queue block to the defined value input connector on the Equation block named QL. The current queue length is used to decide whether the customer will join the queue or balk and leave the system directly through the Exit block named "Balking customers." The routing of the customer to the queue or directly to the Exit block is done in the Select Item Out block, based on the signal sent from the Equation block. To enable the Select Item Out block to receive this signal, open the

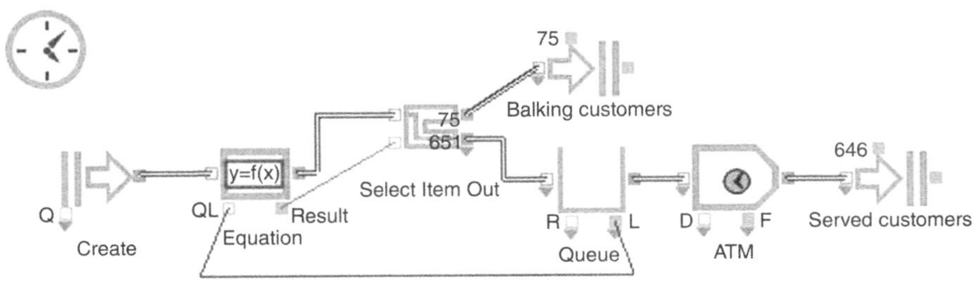

FIGURE 8.43
ExtendSim model of an ATM process with balking customers.

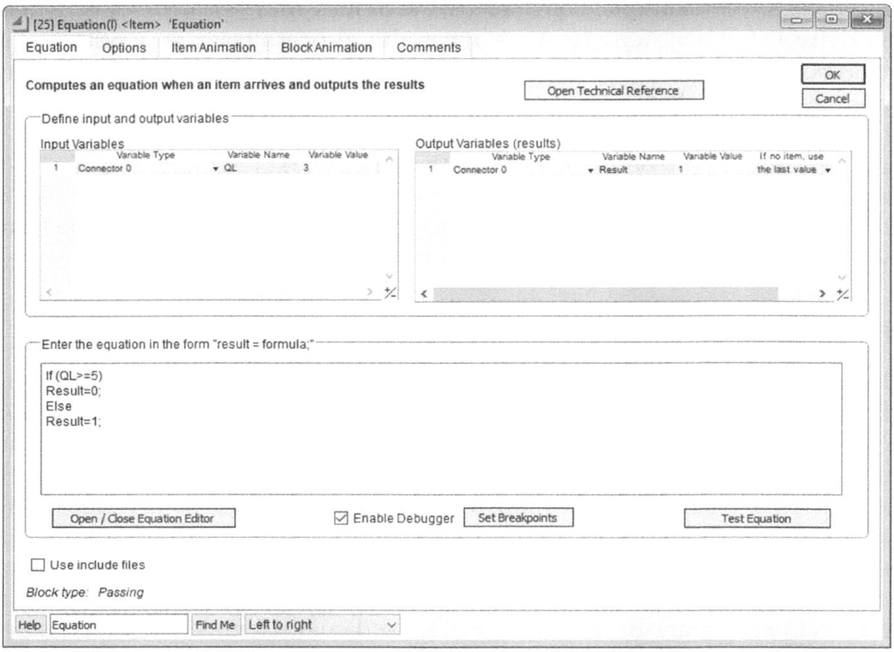

FIGURE 8.44
Dialog window of the Equation (I) block In the ATM model.

dialog window, click on the Options tab, and choose "Select output based on: select connector". In the current model, the customers balk if there are five or more customers in the queue. Figure 8.44 shows the dialog window of the Equation block, where the output variable *Result* is given the value 0 if the variable *QL* (which corresponds to the current queue length available through the value input connector) is greater than or equal to five. Otherwise, Result = 1.

The Result value is sent from the value output connector of the Equation block to the value input connector of the Select Item Out block, where a value of 0 means that the balking customer is routed through the top output connector directly to the Exit block named "Balking customers." A Result value of 1, on the other hand, means that the customer is routed through the bottom output connector of the Select Item Out block to the queue in front of the ATM. The customers served by the ATM leave the model through the Exit block named "Served customers." Running the simulation model for 12 hours with random seed 55,555 renders the results shown in Figure 8.43; 646 customers are served, and 75 customers balk because of long queues.

An alternative approach to model balking, without using the Equation block, is to use the connector priority option in the Select Item block. More precisely, with the model in Figure 8.43 as starting point, one can proceed as follows:

1. Remove the Equation block and connect the Select Item Out block directly to the Create block.
2. Set the maximum queue length to 5 in the Options tab of the Queue block.
3. Open the dialog of the Select Item Out block, and in the Options tab, choose "Select output based on: *connector priority*." Then in the following table, assign Priority 1 to the connector leading to the Queue block and Priority 2 to the Exit block labeled "Balking customers."

In the resulting model, the Select Item block always attempts to route customers through the bottom connector (with lowest priority value) to the Queue, but if this block is full (five customers waiting), customers are instead routed through the top connector to the Balking customer exit.

8.11.3 Reneging

As mentioned earlier, another important queuing phenomenon is reneging. Reneging occurs when an item that is already in the queue leaves before it is released for processing. An example of reneging is a caller to a service line hanging up after being put on hold for too long. Figure 8.45 shows a model that simulates this situation. Suppose the customer hangs up if the waiting time reaches 5 minutes.

The Create block in Figure 8.45 generates calls with exponentially distributed interarrival times with a mean of 2 minutes. It is assumed that the

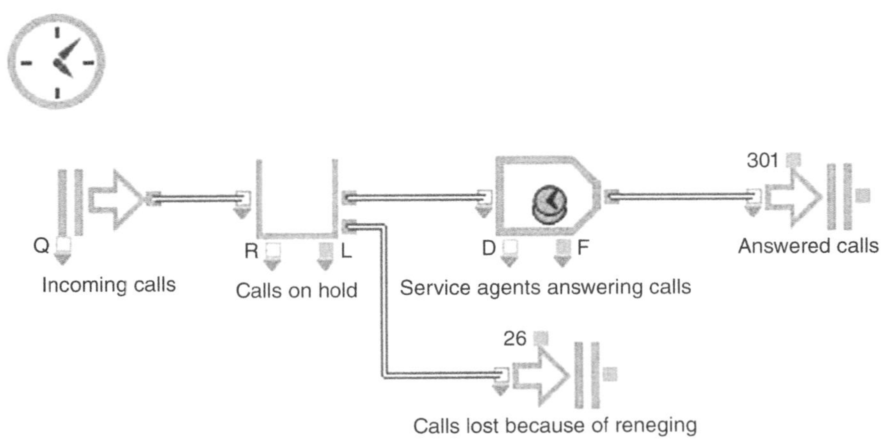

FIGURE 8.45
Model of a call center service line where customers renege after being on hold for a specified amount of time.

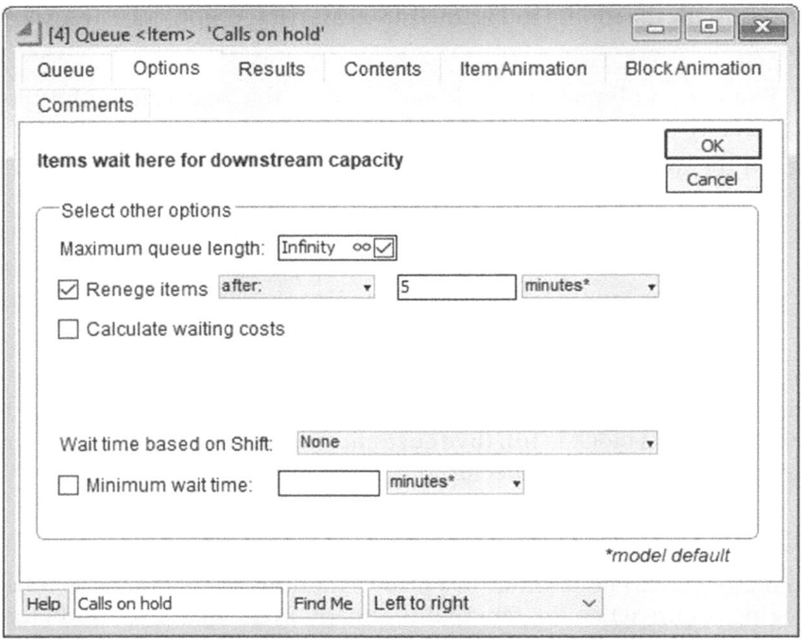

FIGURE 8.46
Queue block labeled "Calls on Hold" configured for reneging after 5 minutes in the block.

processing time of a call is uniformly distributed between 2 and 6 minutes and that there are two service agents answering calls. The Queue block (labeled "Calls on Hold") is configured for customers to renege after being on hold for 5 minutes. This is done by choosing the option "renege items after" in the Options tab of the block dialog and specifying the renege time to 5 minutes, as shown in Figure 8.46.

In this example (see Figure 8.45), a call waits on hold until one of the two representatives is available to answer the call. The Activity block (labeled "Service agents answering calls") is used to model the two customer service representatives. In the Process tab of the Activity block dialog, the Maximum number of items in activity is set to 2. The Queue block uses the waiting time of the calls on hold to decide whether to release a call to the Activity block through the upper output connector or to the Exit block (labeled "Calls lost because of reneging") through the lower output connector. From Figure 8.45, one can see that after the current simulation run (720 minutes with random seed 55,555), 301 calls have been answered, and 26 have been lost because of reneging customers.

As an alternative to reneging *after* a specified time in the queue, the option "Renege items *immediately*" can be used. In this case, reneging takes place as soon as there is a True signal (value >0.5) present at the R input value connector of the Queue block. This enables the modeling of dynamic reneging decisions that change during the simulation.

For convenience, the model in Figure 8.45 uses a separate Exit block to keep track of the number of lost calls. However, this information is also available in the Queue block, which counts the number of reneges and displays it in the Results tab. Because items that renege leave the Queue block through the output connector on the lower right, these items can be rerouted back to the original line or to other activities in the process.

8.11.4 Priorities and Priority Queues

Priorities are used to specify the relative importance of an item. When comparing two priority values, ExtendSim assigns top priority to the smallest value (including negative values). Priorities are useful when the processing of jobs does not have to follow a FIFO discipline. For example, priorities can be used to model a situation in which a worker examines the pending jobs and chooses the most urgent one to be processed next. In ExtendSim, priorities can be assigned to items using the Set block found in the Properties submenu of the Item library. Items passing through this block are assigned the priority value specified in the table in the Set properties tab of the block's dialog window. The procedure is to choose "_item priority" as the Property Name from the available list in the first column of the table and specify the associated priority value in the Value column as shown in Figure 8.47. Items can have only one priority but many attributes.

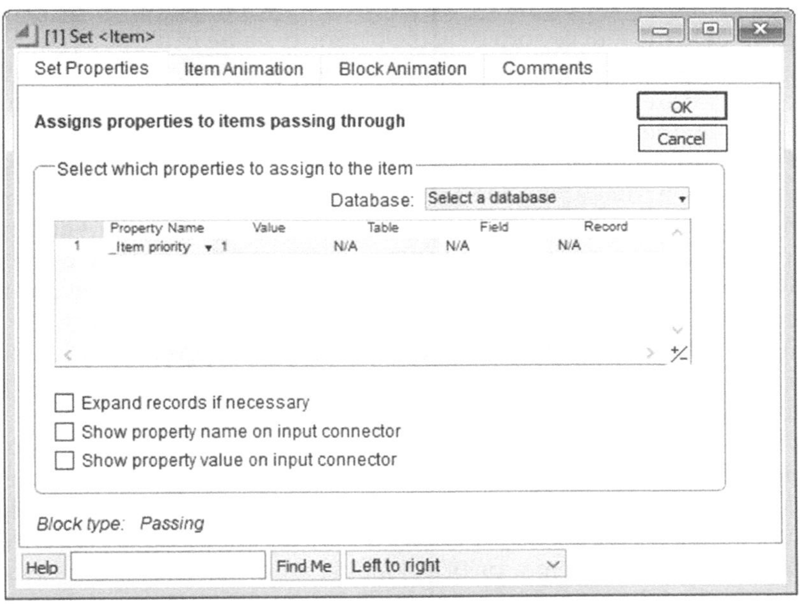

FIGURE 8.47
Dialog window of the Set block when assigning Priority 1 to all items passing through it.

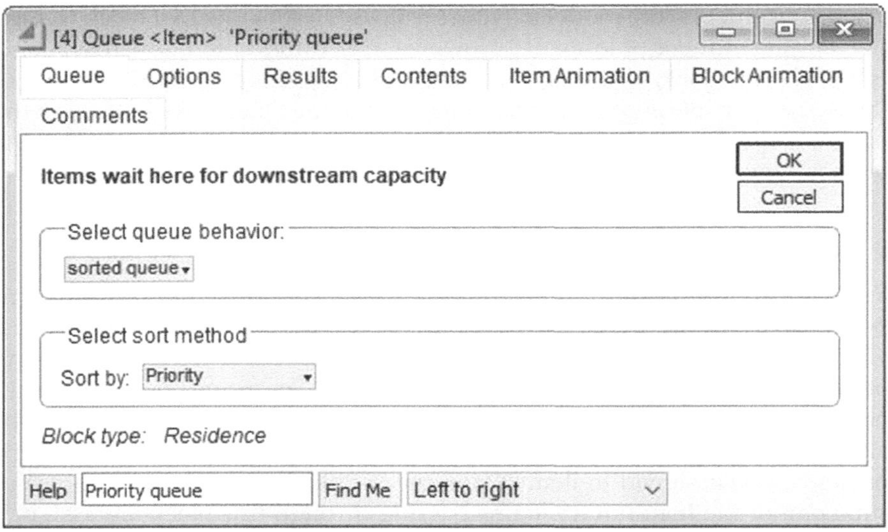

FIGURE 8.48
Dialog window of the Queue block when configured as a priority Queue.

To make the Queue block release items in order of priority, the options *sorted queue* and *Priority* need to be selected from the pop-up menus of the Queue tab in this block's dialog window, as shown in Figure 8.48. It is noteworthy that Items will be sorted by their priority value in a Queue block only if they have to wait there with other items. Moreover, if several items with the same priority are in the queue at the same time, they are sorted according to a FIFO discipline.

To illustrate the use of priority queues, consider the following situation. Patients arrive at a hospital admissions counter, but 20 percent of the time, they cannot be admitted because they need to fill out additional forms. After they have filled out these additional forms, they come back to the counter to complete the admissions process. Patients returning to the counter are given higher priority and can go to the front of the line. Figure 8.49 shows a simulation model of this admissions process. Arriving patients are generated in the Create block with exponentially distributed interarrival times with a mean of 7 minutes. The generated items then pass through the Set block, where they are assigned a priority value of 2. A priority value of 1 is used for patients who return to the priority queue at the counter after filling out the additional forms. This priority is assigned in the Set block labeled "Set priority 1" at the upper right corner of the model. The patients form a priority queue, modeled with a Queue block, in front of the admissions counter, modeled by the Activity block labeled "Admission form review."

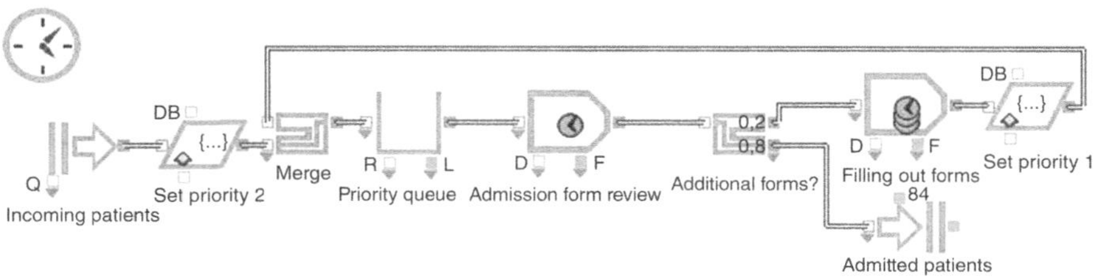

FIGURE 8.49
Model of admissions process with a priority queue.

A Select Item Out block, labeled "Additional Forms?", is used for simulating the percentage of patients needing to fill out the additional forms. In the Options tab of this block's dialog window, the *random* option should be chosen as selection condition. Furthermore, in the Select options table, a probability value of 0.2 should be specified for the top output connector (1) and 0.8 for the bottom output connector (2). The model assumes that a single clerk is performing the admission form review and that the processing time is uniformly distributed between 3 and 6 minutes. For patients who need to fill out additional forms, the time for this activity is assumed to be uniformly distributed between 5 and 10 minutes. After the additional forms are filled out (modeled by the Activity block labeled "Filling out forms"), the patients pass through the Set block, where the priority of the patients is changed to 1. They are then rerouted back to the priority queue, where they go to the front of the line. As mentioned earlier, ExtendSim uses a FIFO queuing discipline when all the items in a queue have identical priority values.

8.12 Modeling Routing in Multiple Paths and Parallel Paths

When modeling business processes, it is common to encounter situations where jobs come from different sources or follow different paths. For example, Figure 8.49 shows that the Priority queue at the admissions counter receives both newly arriving patients and patients who have completed additional paperwork. In other words, the Queue block, which simulates the queue, receives patients from two different sources: the Create block, which simulates the patients arriving to the hospital, and the Activity block simulating the patients filling out additional forms. The patients coming from these two sources are merged into one stream with the Select Item In block; however, the patients remain as individual items and retain their unique identity.

The Select Item In block in the Routing submenu of the Item library can merge items from many separate sources (the number of input connectors is easily changed by a simple drag and drop operation on the small arrow seen on the bottom input connector). Another example of merging items from different sources occurs when telephone orders and mail orders are directed to an order-entry department.

Chapter 5 introduced two types of routing: multiple paths and parallel activities. Next we will discuss how to simulate these routing conditions with ExtendSim.

8.12.1 Multiple Paths

Jobs do not usually arrive, join a queue, get completed in a single step, and leave. If this were true, every process could be simulated with a Create block (to generate arrivals), a Queue block (to simulate a queue), an Activity block (for the service activity), and an Exit block (for jobs leaving after the completion of a single activity). Real-world business processes call for routing of jobs for processing, checking, approval, and numerous other activities. The simulation models must be capable of routing jobs based on a probability value, logical and tactical decisions, or job characteristics.

Probabilistic routing occurs when a job follows a certain path a specified percentage of the time. For example, Figure 8.49 shows a model in which 20 percent of the time, hospital patients cannot be admitted because they need to fill out an additional form. A rework loop shares this characteristic. That is, after an inspection (or control) activity, jobs are sent back for rework with a specified probability. In the model in Figure 8.49, a Select Item Out block from the Routing submenu of the Item library was used to model a probabilistic routing with two paths. However, the same block can be used for probabilistic routing to any given number of paths by expanding the number of output connectors accordingly. This is done by a simple drag and drop operation on the downward arrow found at the bottom output connector on the right side of the block. Suppose a job follows one of three possible paths (labeled 1, 2, and 3) with probabilities 0.2, 0.3, and 0.5, as shown in Figure 8.50.

This situation is easily simulated by first generating jobs with a Create block and then adding a Select Item Out block to probabilistically route each job, as illustrated in Figure 8.51. Note that the Select Item Out block in this case is expanded to include three output connectors. As indicated earlier, this is done by positioning the marker on the downward arrow on the bottom output connector until a cross with arrows is shown, then pressing the left button on the mouse and pulling down until the desired number of output connectors is available.

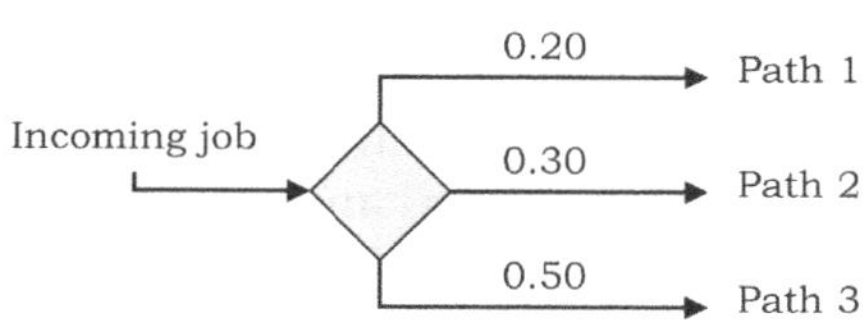

FIGURE 8.50
Example of probabilistic routing of an incoming job into three paths.

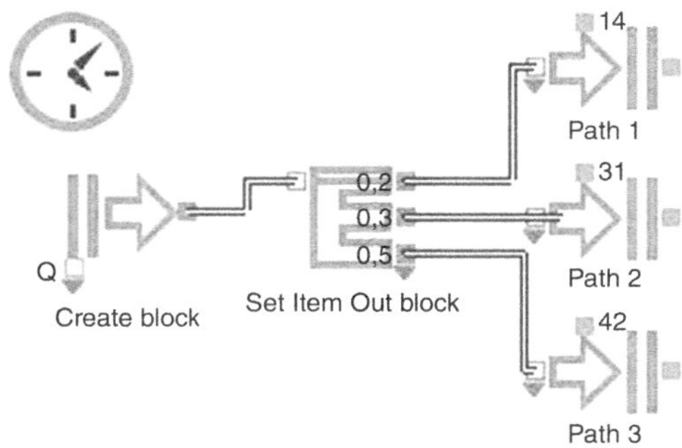

FIGURE 8.51
ExtendSim model with probabilistic routing.

To enter the desired probabilities for Paths 1, 2, and 3, open the Select Item Out block dialog window, go to the Options tab, and choose "Select output based on: *random*" as the selection conditions. The probabilities are then entered into the Select options table as shown in Figure 8.52.

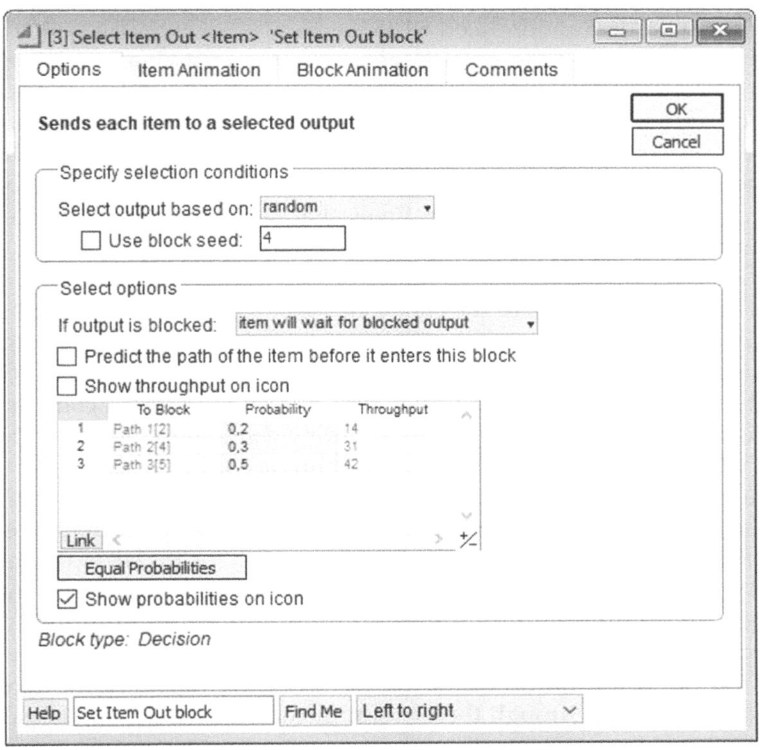

FIGURE 8.52
Select Item Out block dialog with probabilistic routing according to Figure 8.51.

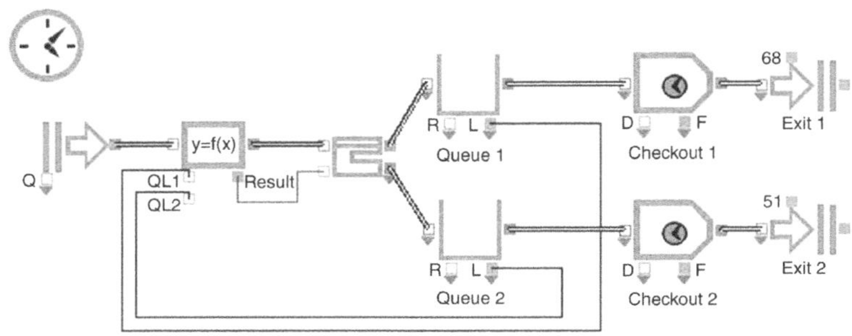

FIGURE 8.53
Illustration of tactical routing model with customers choosing the shortest line.

In addition to probabilistic routing, business process models often include tactical routing. This type of routing relates to the selection of paths based on a decision that typically depends on the state of the system. For example, suppose one wants to model the checkout lines at a supermarket where customers choose to join the shortest line. Figure 8.53 shows an ExtendSim model of this tactical routing situation (the results are based on a run length of 4 hours with the random seed 55,555).

The Create block in Figure 8.53 generates the customers arriving to the checkout counters with exponentially distributed interarrival times with a mean of 2 minutes. Each checkout station is modeled with an Activity block (labeled Checkout 1 and Checkout 2, respectively), and in front of each checkout, there is a queue. Each checkout station, which can serve at most one customer at a time, provides service times that are normally distributed with a mean of 3 minutes and a standard deviation of 1 minute. The two queues are modeled with Queue blocks (labeled Queue 1 and Queue 2, respectively). When a customer arrives to the checkout, he or she checks the lengths of the two queues and joins the shortest line. If the lines are of equal length, the customers choose to join Queue 1 because it is closest. An Equation (I) block from the Properties submenu of the Item library is used for comparing the lengths of Queue 1 and Queue 2 and sending a signal to the Select Item Out block, which routes the item to the shortest line accordingly. Figure 8.54 shows the dialog window of the Equation block, where the length of Queue 1, QL1, is compared with the length of Queue 2, QL2. If QL1 is smaller than or equal to QL2, the Result variable sent to the Select Item Out lock is set to 0; otherwise, Result = 1. The current lengths of the two queues are available on the value input connectors named QL1 and QL2 by connecting them to the L output connectors on the bottom of the two Queue blocks as shown in Figure 8.53.

Turning to the Select Item Out block, it is configured so that a Result value of zero routes the item through the top output connector to Queue 1, and otherwise, the item is routed through the bottom output to Queue 2. This configuration is done in the Options tab of the Select Item Out block's dialog window by choosing the options "Select output based on:" *select connector* and "Top output chosen by Select value:" 0 as shown in Figure 8.55.

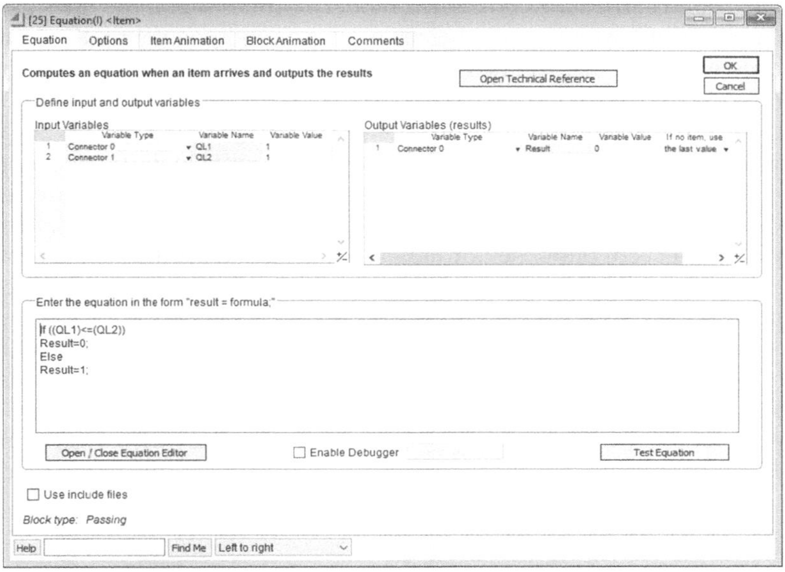

FIGURE 8.54
Equation (I) block dialog for the model in Figure 8.53.

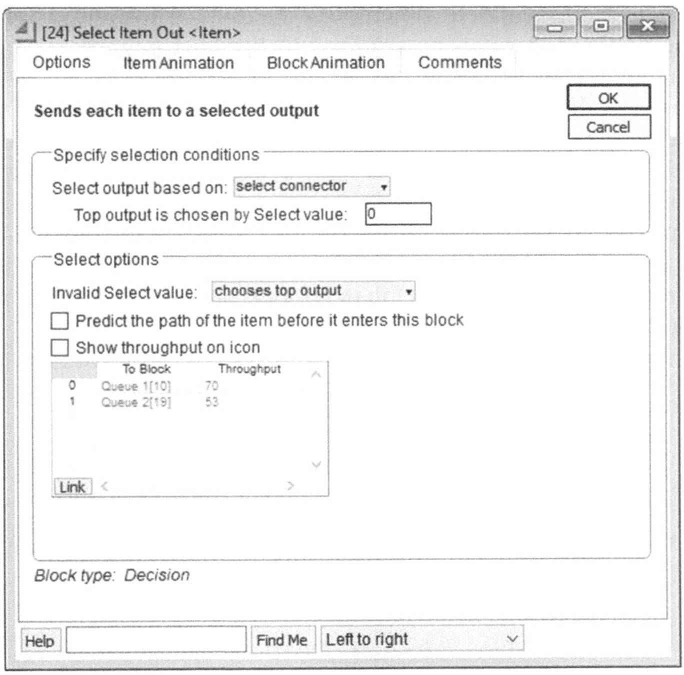

FIGURE 8.55
Select Item Out block dialog for the model in Figure 8.53.

As described earlier, the considered model assumes that if the lines have the same number of customers, the incoming customer joins the first line. This is why even when the cashiers work at the same speed, Checkout 1 ends up serving more customers than Checkout 2. (See the Exit blocks in Figure 8.53.) To model a situation where customers instead randomly choose one of the queues when they are of equal length requires some additional logic, but it is not hard to do. Another remark to be made is that the considered routing decision is only based on the queue lengths and does not consider customers that are currently being served. Suppose both queues are empty, but there is a customer being served in Checkout 1; the current model will then route an incoming customer to Checkout 1, where it will be first in line, instead of routing it to Checkout 2, where it would be served immediately. It is easy to modify the model to avoid this situation by basing the routing decision on a comparison of the total number of customers in Checkout 1 and Queue 1 with the total number of customers in Checkout 2 and Queue 2 instead of only considering the lengths of the queues. It is left to the reader to investigate this refined model.

8.12.2 Parallel Paths

Some business processes are designed in such a way that two or more activities are performed in parallel. For example, in an order fulfillment process, the activities associated with preparing an invoice can be performed while the order is being assembled. When both the invoice and the order assembly are complete, the order can be shipped to the customer. A simple process chart is depicted in Figure 8.56.

Modeling parallel paths in ExtendSim is easily done using the Batch and Unbatch blocks found in the Batching submenu of the Item library. This is illustrated in Figure 8.57, which shows an ExtendSim model of the order fulfillment process.

In this model, orders are generated in the Create block (labeled "Received orders"). Interarrival times are assumed to be exponentially distributed with a mean of 6 minutes. The Unbatch block separates each incoming item (order) into two copies. It then outputs them simultaneously, one through the top output connector and the other through the bottom output connector.

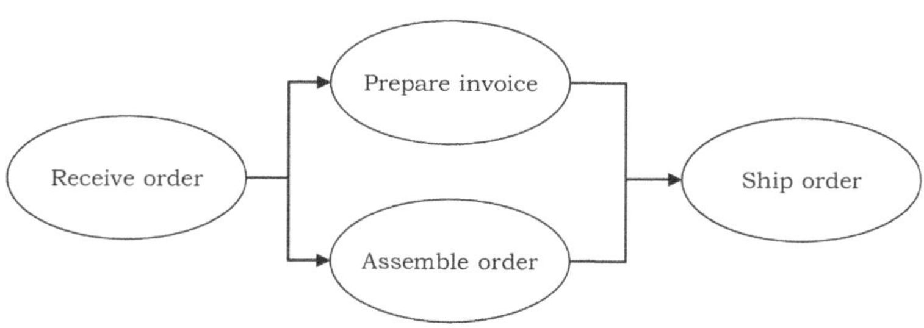

FIGURE 8.56
Parallel activities in an order fulfillment process.

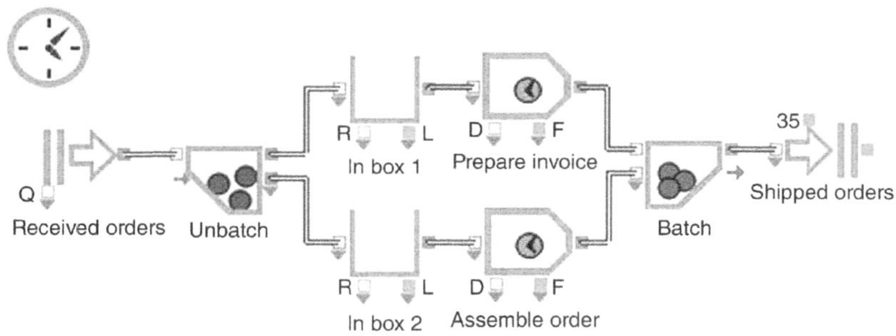

FIGURE 8.57
ExtendSim model of the order fulfillment process with two parallel paths.

Configuration of the Unbatch block is done in the Unbatch tab of the block's dialog. The block behavior is set to *Create multiple items* and the quantity to unbatch for each output connector is set to 1, as shown in Figure 8.58. In general, the Unbatch block allows the modeler to specify the number of item copies to be issued for each output connector and also to freely choose the number of connectors (greater than or equal to one).

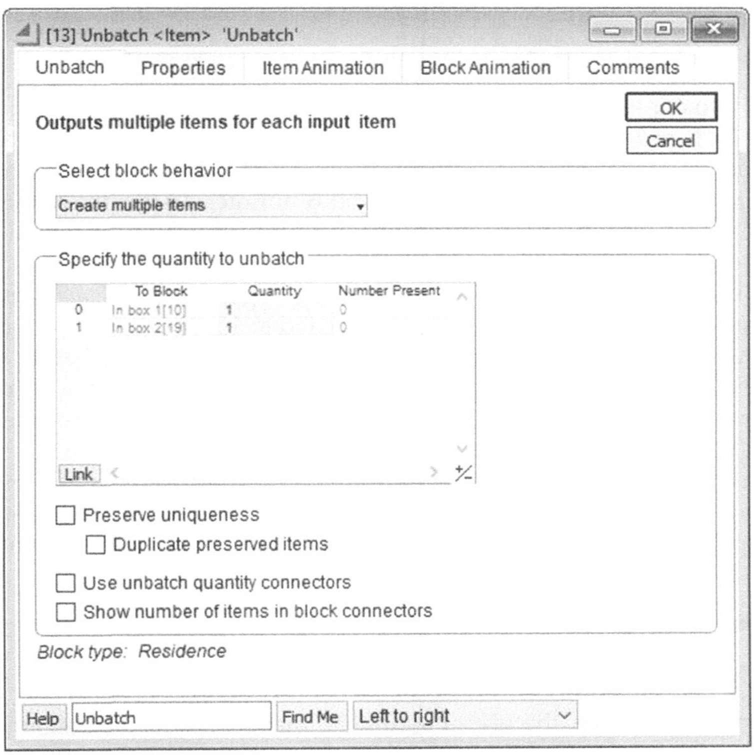

FIGURE 8.58
Dialog window of the Unbatch block in Figure 8.57.

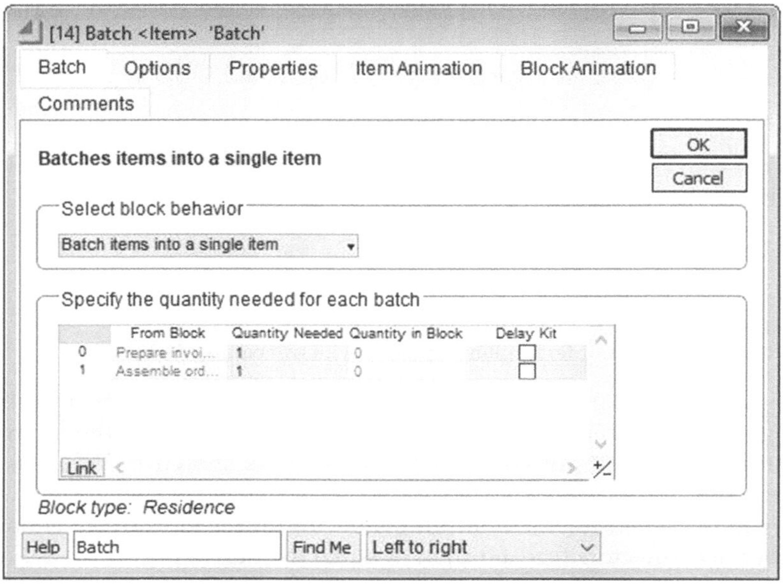

FIGURE 8.59
Dialog window of the Batch block in Figure 8.57.

The two copies of the order item issued from the Unbatch block proceed to In Box 1 and In Box 2 on the two parallel paths with invoice preparation (exponential processing time with a mean of 4 minutes) and order assembly (uniform processing time between 2 and 6 minutes). After the activities on each parallel path are completed, a Batch block batches the two items into a single order item, which is shipped. Configuration of the Batch block is done in the Batch tab of its dialog window by choosing the option *Batch items into a single item* and specifying the quantity needed to 1 for both input connectors, as shown in Figure 8.59. All required inputs must be available before the output item is released. It is noteworthy that the modeler determines the number of input connectors available on the Batch block as well as the number of items (>0) required at each connector to create a single output item.

Parallel paths can consist of more than one activity each. In the model in Figure 8.57, each parallel path consists of one activity, but the same modeling principles can be used to deal with multiple paths with multiple activities.

8.13 Model Documentation and Enhancements

Section 8.8 discussed the use of simple animation within ExtendSim. Other forms of documenting and enhancing a model are adding text, using named connections, adding controls such as a slider or a meter, and displaying results.

FIGURE 8.60
Clone Layer tool.

Text can be added with the Text tool. Also, the Text menu allows the size and font to be modified, and the Color tool allows the text color to be changed. Named connections are helpful because as the model grows, more connecting lines might intersect and make the model less readable. Named connections were introduced in the model in Figure 8.29. The idea is simple: a text box is created with the name of the connection, and then the box is duplicated (Edit > Duplicate); an output connector is connected to one copy of the text box, and the corresponding input connector is connected to the other copy of the text box.

Several blocks for interactive control can be found in the Model Control submenu of the Utilities library. For example, the Buttons block and Model Interface block both create pushbutton interfaces for the model. The Buttons block executes a specified equation every time a button is pushed. The Model Interface block can, for example, be used for running a single or multiple simulation runs by clicking buttons on the worksheet. More information about these and other blocks can be found in the ExtendSim User Reference (2024) or in the considered block's dialog window by clicking the Help button at its lower left corner.

This chapter has shown how to use Chart blocks such as the Histogram and the Line Chart blocks to graphically display results from other ExtendSim blocks using their appropriate output connectors. Another easy way of displaying results during the execution of the simulation is by cloning. Suppose one wants to display the average waiting time in a Queue block as it is updated during the simulation. Open the dialog of the Queue block and click on the Results tab. Then, click on the Clone Layer tool displayed in Figure 8.60, highlight the Average Wait text and value box, and drag them to the place in the model where they are to be displayed. The Clone Layer tool creates a copy of the chosen dialog boxes and allows them to be placed in the model to enhance documentation and animation.

8.14 Summary

This chapter introduced the main discrete-event simulation functionality of ExtendSim. A simple, one-stage process was used to illustrate how to build a simulation model using block diagrams. The basic concepts discussed in this chapter are sufficient to model and analyze fairly complicated business processes. However, the full functionality of ExtendSim goes beyond what is presented here and makes it possible to tackle almost any situation that might arise in the design of business processes.

Discussion Questions and Exercises

1. Consider a single-server queuing system for which the interarrival times are exponentially distributed. A customer who arrives and finds the server busy joins the end of a single queue. Service times of customers at the server are also exponentially distributed random variables. On completing service for a customer, the server chooses a customer from the queue (if any) in a FIFO manner.
   a. Simulate customer arrivals assuming that the mean interarrival time equals the mean service time (e.g., consider that both of these mean values are equal to 1 minute). Create a plot of the number of customers in the queue (y-axis) versus simulation time (x-axis). Is the system stable? (*Hint*: Run the simulation long enough [e.g., 10,000 minutes] to be able to determine whether or not the process is stable.)
   b. Consider now that the mean interarrival time is 1 minute, and the mean service time is 0.7 minute. Simulate customer arrivals for 5000 minutes and calculate: (i) the average waiting time in the queue, (ii) the maximum waiting time in the queue, (iii) the maximum queue length, (iv) the proportion of customers having a delay time in excess of 1 minute, and (v) the expected use of the server.
2. For the single-server queuing system in Exercise 8.1.b, suppose the queue has room for only three customers and that a customer arriving to find that the queue is full just goes away (balk). Simulate this process for 5000 minutes and estimate the same quantities as in Part (b) of Exercise 8.1 as well as the expected number of customers who balk.
3. A service facility consists of two servers in series (tandem), each with its own FIFO queue (see Figure 8.61). A customer completing service at Server 1 proceeds to Server 2, and a customer completing service at Server 2 leaves the facility. Assume that the interarrival times of customers to Server 1 are exponentially distributed with a mean of 1 minute. Service times of customers at Server 1 are exponentially distributed with a mean of 0.7 minute, and at Server 2, they are exponentially distributed with a mean of 0.9 minute.
   a. Run the simulation for 1000 minutes and estimate for each server the expected average waiting time in the queue for a customer and the expected use.

b. Suppose that there is a travel time from the exit of Server 1 to the arrival to Queue 2 (or Server 2). Assume that this travel time is distributed uniformly between 0 and 2 minutes. Modify the simulation model and run it again to obtain the same performance measures as in Part (a). (*Hint*: You can add an Activity block to simulate this travel time. A uniform distribution can be used to simulate the time.)

c. Suppose that no queuing is allowed for Server 2. That is, if a customer completing service at Server 1 sees that Server 2 is idle, the customer proceeds directly to Server 2, as before. However, a customer completing service at Server 1 when Server 2 is busy with another customer must stay at Server 1 until Server 2 gets done; this is called *blocking*. When a customer is blocked from entering Server 2, the customer receives no additional service from Server 1 but prevents Server 1 from taking the first customer, if any, from Queue 1. Furthermore, new customers may arrive to Queue 1 during a period of blocking. Modify the simulation model and rerun it to determine the same performance measures as in Part (a).

FIGURE 8.61
Process with two servers in series.

4. A bank with five tellers opens its doors at 9 a.m. and closes its doors at 5 p.m., but it operates until all customers in the bank by 5 p.m. have been served. Assume that the interarrival times of customers are exponentially distributed with a mean of 1 minute and that the service times of customers are exponentially distributed with a mean of 4.5 minutes. In the current configuration, each teller has a separate queue (see Figure 8.62). An arriving customer joins the shortest queue, choosing the shortest queue furthest to the left in case of ties.

   The bank's management team are concerned with operating costs as well as the quality of service currently being provided to customers, and they are thinking about changing the system to a single queue. In the proposed system, all arriving customers would join a single queue. The first customer in the queue goes to the first available teller. Simulate 5 days of operation of the current and proposed systems and compare their expected performance.

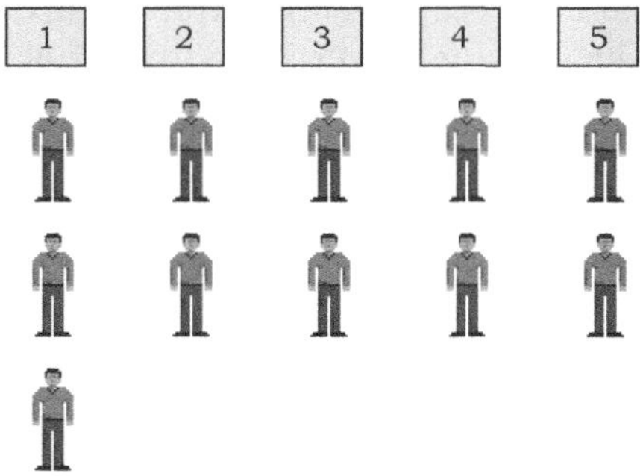

FIGURE 8.62
Teller configuration (multiple queues).

5. *Airline Ticket Counter*—At an airline ticket counter, the current practice is to allow queues to form before each ticket agent. Time between arrivals to the agents is exponentially distributed with a mean of 5 minutes. Customers join the shortest queue at the time of their arrival. The service time for the ticket agents is uniformly distributed between 2 and 10 minutes.
   a. Develop an ExtendSim model to determine the minimum number of agents that will result in an average waiting time of 5 minutes or less.
   b. The airline has decided to change the procedure involved in processing customers by the ticket agents. A single line is formed, and customers are routed to the ticket agent who becomes free next. Modify the simulation model in Part (a) to simulate the process change. Determine the number of agents needed to achieve an average waiting time of 5 minutes or less.
   c. Compare the systems in Parts (a) and (b) in terms of the number of agents needed to achieve a maximum waiting time of 5 minutes.
   d. It has been found that a subset of the customers purchasing tickets is taking a long period of time. By separating ticket holders from non-ticket holders, management believes that improvements can be made in the processing of customers. The time needed to check in a person is uniformly distributed between 2 and 4 minutes. The time to purchase a ticket is uniformly distributed between 12 and 18 minutes. Assume that 15 percent of the customers will purchase tickets, and develop a model to simulate this situation. As before, the time between all arrivals is exponentially distributed with a mean of 5 minutes. Suggest staffing levels for both counters, assuming that the average waiting time should not exceed 5 minutes.

6. *Inventory System*—A large discount store is planning to install a system to control the inventory of a particular video game system. The time between demands for a video game system is exponentially distributed with a mean time of 0.2 weeks. In the case where customers demand the system when it is not in stock, 80 percent will go to another nearby discount store to find it, thereby representing lost sales; the other 20 percent will backorder the video system and wait for the arrival of the next shipment. The store employs a periodic review–reorder point system whereby the inventory status is reviewed every 4 weeks to decide whether an order should be placed (ordering can only take place at these review instances). The policy is to place an order if the inventory position, consisting of the units in stock plus the units on order minus the units on backorder, is found to be less than or equal to the reorder point of 18 video systems. Moreover, the order size should be such that it raises the inventory position to an order-up-to level of 72 video systems. Thus, if the inventory position is $x \leq 18$, an order for $72 - x$ units is placed. The procurement lead time (the time from the placement of an order to its receipt) is constant and requires 3 weeks. Develop an ExtendSim model to simulate the inventory policies for a period of 6 years (312 weeks) to obtain statistics on (a) the number of video systems in stock, (b) the inventory position, (c) the safety stock (the number of video systems in stock at the time of receiving a new order), and (d) the number of lost sales. Investigate improvements on this inventory policy by changing the reorder point, the order-up-to level, and the time between reviews.
7. *Bank Tellers*—Consider a banking system involving two inside tellers and two drive-in tellers. Arrivals to the banking system are either for the drive-in tellers or for the inside tellers. The time between arrivals to the drive-in tellers is exponentially distributed with a mean of 1 minute. The drive-in tellers have limited waiting space. Queuing space is available for only three cars waiting for the first teller and four cars waiting for the second teller. The service time of the first drive-in teller is normally distributed with a mean of 2 minutes and a standard deviation of 0.25. The service time of the second drive-in teller is uniformly distributed between 1 and 3 minutes. If a car arrives when the queues of both drive-in tellers are full, the customer balks and seeks service from one of the inside bank tellers. However, the inside bank system opens 1 hour after the drive-in bank, and it takes between 0.5 and 1 minute to park and walk inside the bank. Customers who seek the services of the inside tellers directly arrive through a different process, with the time between arrivals exponentially distributed with a mean of 1.5 minutes. However, they join the same queue as the balkers from the drive-in portion. A single queue

is used for both inside tellers. A maximum of seven customers can wait in this single queue. Customers arriving when seven are in the inside queue leave. The service times for the two inside tellers are triangularly distributed between 1 and 4 minutes with a mode of 3 minutes. Simulate the operation of the bank for an 8-hour period (7 hours for the inside tellers). Assess the performance of the current system.

8. *Grocery Store*—You are hired by Safeway to help them build a number of simulation models to better understand the customer flows and queuing processes in a grocery store setting. The pilot project at hand focuses on an off-peak setting where at most two checkouts are open. To better understand the necessary level of detail and the complexities involved, Safeway wants a whole battery of increasingly more realistic and complex models. For each model, Safeway wants to keep track of (i.e., plot) the average cycle time, queue length, and waiting time in the queue. To understand the variability, they also want to see the standard deviation of these three metrics. In addition, they would like to track the maximum waiting time and the maximum number of customers in line. Furthermore, to better understand the system dynamics, plots of the actual queue lengths over time are required features of the model. The off-peak setting is valid for about 4 hours, so it is reasonable to run the simulation for 240 minutes. Furthermore, to facilitate an easier first-cut comparison between the models, a fixed random seed is recommended. Because Safeway plans to use these different models later, it is important that each model sheet has a limit of one model.
   a. In the first model, your only interest is in the queues building up at the checkout counters. Empirical investigation has indicated that it is reasonable to model the arrival process (to the checkout counters) as a Poisson process with a constant arrival intensity of three customers per minute. The service time in a checkout station is on average 30 seconds per customer and will, in this initial model, be considered constant. Inspired by the successes of a local bank, Safeway wants to model a situation with one single line to both checkout counters. As soon as a checkout is available, the first person in the queue will go to this counter. After the customers have paid for their goods, they immediately leave the store.
   b. On closer investigation, it is clear that the service time is not constant but rather normally distributed with mean = 30 seconds and standard deviation = 10 seconds. What is the effect of the additional variability compared with the results in Part (a)?

c. To be able to analyze the effect of different queuing configurations, Safeway wants a model in which each checkout counter has its own queue. When a customer arrives to the checkout point, he or she will choose the shortest line. The customer is not allowed to switch queues after making the initial choice. Can you see any differences in system performance compared with the results in Part (b)?

d. To make the model more realistic, Safeway also wants to include the time customers spend in the store walking around and picking up their groceries. Empirical investigation has shown that there are basically two types of customers, and they need to be treated somewhat differently:

   Type 1: The light shopper who buys only a few items (fewer than 15)

   - Represents about 60 percent of the arriving customers.
   - The shopping time follows a triangular distribution with a most likely value of 5 minutes, a minimum value of 2 minutes, and a maximum value of 8 minutes.
   - The service times for these customers at the checkout counter are exponentially distributed with a mean of 15 seconds.

   Type 2: The heavy shopper who buys several items (more than 15)

   - Represents about 40 percent of the arriving customers.
   - The shopping time is triangularly distributed with a most likely value of 10 minutes, a minimum value of 5 minutes, and a maximum value of 15 minutes.
   - The service times for these customers at the checkout counter are exponentially distributed with a mean of 52 seconds.

   The total arrival process to the store is still a Poisson process with a mean of three customers per minute. As for the queue configuration, Safeway feels that the setup in Part (b) with one line for each checkout is better for psychological reasons; one long line might deter customers from entering the store.

   Modify the simulation model to incorporate the described elements and make it more realistic. Analyze the performance of the current process using the performance measures discussed previously as well as the following additional measures:

   - The time spent shopping (average and standard deviation)
   - The number of customers (average and standard deviation)
   - The separate cycle times for heavy and light shoppers (average and standard deviation)

e. To improve the service for the light shoppers, Safeway is thinking about dedicating one of the checkout counters to this customer group. In other words, only light shoppers are allowed to use Checkout 1. The other checkout (Checkout 2) will handle both heavy and light shoppers. However, empirical interviews indicate that no light shopper will choose the regular lane unless at least two more shoppers are waiting in line at the express lane. How does this design change affect the cycle times for the two customer groups and for the average customer?

9. *Measuring Cycle Times of Different Types of Jobs*—Three types of jobs arrive to a process at a rate of four jobs per hour. The interarrival times are exponentially distributed. On average, 40 percent of the jobs are Type I, 35 percent are Type II, and 25 percent are Type III. It can be assumed that all of the processing times are exponentially distributed; however, the mean times depend on the job type. Table 8.3 shows the mean processing time for the activities associated with each job type. (All times are given in minutes.)

TABLE 8.3

Mean Processing Times in Minutes for Three Different Job Types

	Job Type		
Activity	**I**	**II**	**III**
A	15	20	35
B	34		
C	12		
D		9	14
E		45	24
F	25	18	22
G	14	27	12

A flowchart of the process is depicted in Figure 8.63. Note that Type I jobs are routed through Activities B and C, and Types II and III are routed through Activities D and E.

The percentage of rework also depends on the job type. About 5 percent of the Type I jobs are reworked, about 4 percent of the Type II jobs are reworked, and about 6 percent of the Type III jobs are reworked. In terms of resources, the process has three types of workers. Five superworkers can handle any job; therefore, managers employ them to work on Activities A, F, or G. Three Type I specialists can handle only Type I jobs; therefore, management employs them to work on Activities B and C. Finally, three Type II and III specialists can handle only Type II and Type III jobs; therefore, management employs them to work on Activities D and E.

a. Develop a simulation model for this process.
b. Run the simulation for 250 hours to generate approximately 1000 job arrivals and collect data on cycle times for each job type separately.
c. Assess the performance of the process based on your analysis in Part (b).

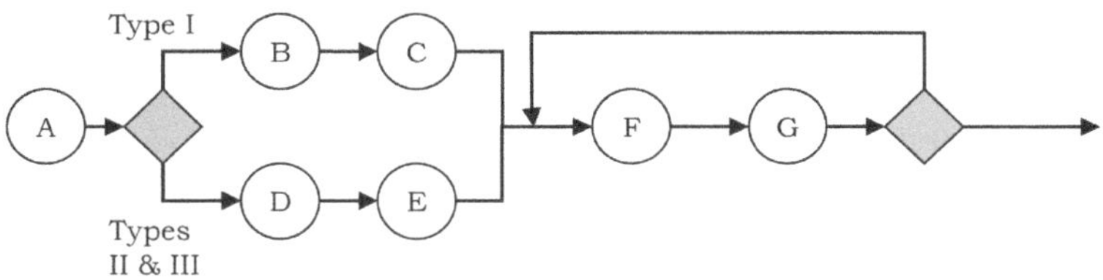

FIGURE 8.63
Measuring cycle times of different types of jobs.

10. *Investigating the Effect of Pooling Resources*—Three types of jobs arrive to a process at a rate that randomly varies between two and five jobs per hour; that is, the interarrival times are governed by a uniform distribution with mean equal to three and a half jobs per hour.

The process is currently configured in such a way that each arrival is sent to a team of specialists according to the type of job. Each team consists of three members. A team performs three activities to complete each job, and any team member can work on any of the three activities (A, B, or C). Figure 8.64 shows a flowchart of the process.

The arrivals are not equally likely for each job type. Typically, 30 percent of the jobs are Type 1, 35 percent are Type 2, and 35 percent are Type 3. Processing times (in minutes) associated with the activities depend on the job type and are summarized in Table 8.4.

TABLE 8.4
Probability Distributions for Processing Times (in Minutes)

		Job Type		
Activity	**Distribution**	**1**	**2**	**3**
A	Exponential	28	20	40
B	Normal	(34, 5)	(38, 7)	(21, 3)
C	Lognormal	(19, 3.5)	(28, 2.3)	(17, 3.1)

(Mean, Standard Deviation)

Management is concerned with the current performance of the process and is considering cross-training the teams so that team members can work on any job type.

a. Develop a simulation model of this process that can be used to model 1 day of operation; that is, one run of the model will consist of 8 hours of operation.

b. Use your model to assess performance with the following measures: cycle times, resource use, and overtime (the time needed to finish all jobs after the process closes). Simulate 10 days of operation to collect data for your analysis.

a. Assume that the teams are cross-trained and that all nine people can perform any of the activities in the process. Compare this design with the original process using the measures in Part (b).

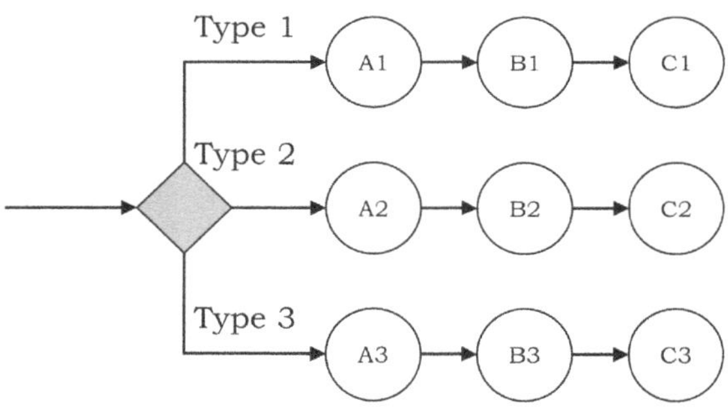

FIGURE 8.64
Investigating the effect of pooling resources.

11. *Investigating the Effect of Rework Rates*—A proposed business process consists of five serial workstations. One caseworker is positioned in each workstation. The processing times at each workstation are exponentially distributed with mean values of 11, 10, 11, 11, and 12 minutes, respectively. The interarrival times are uniformly distributed between 13 and 15 minutes. The workstations have an unlimited queuing capacity, and it is assumed that the downstream transfer time is negligible. The unique aspect of this process is that between Workstations 2 and 5, it is possible that the job will need to be reprocessed by the workstation immediately preceding the current one. For example, after Workstation 3, it is possible that the job will be sent back to the queue in Workstation 2. When this occurs, the transfer requires 3 minutes. The probability

of rework remains the same regardless of the number of times a job is sent back for reprocessing. The rework probability is considered to be the same for all workstations and is currently estimated to be between 5 and 10 percent.

a. Develop a simulation model of this process that is capable of running 10,000 minutes of operation.
b. Run 10,000 minutes of operation for rework probabilities of 5, 6, 7, 8, 9, and 10 percent. Collect cycle time data for each run.
c. Construct a plot of the average cycle time (y-axis) versus the rework probability (x-axis). Also, construct a plot of the maximum cycle time versus the rework probability.
d. Assume that the processing times are constant. Rerun the simulation six times with the rework probabilities in Part (b).
e. Construct plots of average cycle time and maximum cycle time versus the rework probability. Compare these plots with the ones obtained in Part (c).

12. *Assessing Process Performance*—The process of insuring a property consists of four main activities: review and distribution, underwriting, rating, and policy writing. Four clerks, three underwriting teams, eight raters, and five writers perform these activities in sequence. The time to perform each activity is exponentially distributed with an average of 40, 30, 70, and 55 minutes, respectively. On the average, a total of 40 requests per day are received. Interarrival times are exponentially distributed. A flowchart of the process is depicted in Figure 8.65.
    a. Develop a simulation model of this process. The model should simulate 10 days of operation. Assume that work in process at the end of each day becomes the beginning work in process for the next day.
    b. Add data collection to calculate the following measures: resource use, waiting time, length of the queues, work in process at the end of each day, and average daily throughput (given in requests per day).
    c. Assess the performance of the process with the data collected in Part (b).

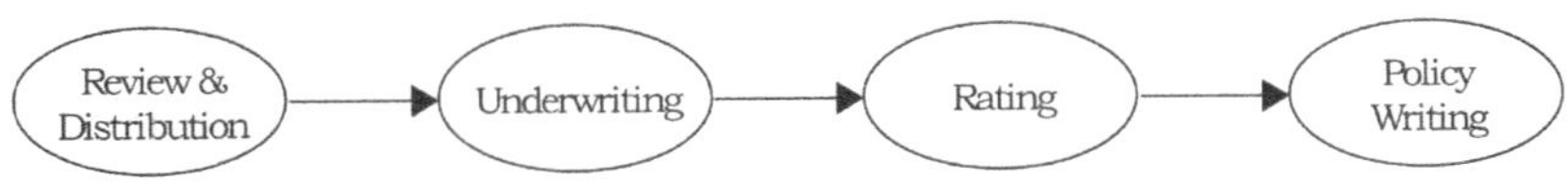

FIGURE 8.65
Assessing process performance.

13. *Variable Resource Availability*—Travelers arrive at the main entrance door of an airline terminal according to an exponential interarrival time distribution with a mean of 1.6 minutes. The travel time from the entrance to the check-in is distributed uniformly between 2 and 3 minutes. At the check-in counter, travelers wait in a single line until one of five agents is available to serve them. The check-in time follows a normal distribution with a mean of 7 minutes and a standard deviation of 2 minutes. On completion of their check-in, travelers walk to their gates.
    a. Create a simulation model of the check-in process.
    b. Run the simulation for 16 hours and collect cycle time and cycle time efficiency for each traveler. Plot a frequency distribution of these values. Create a line plot of the cycle time versus the clock and the cycle time efficiency versus the clock. Also, analyze queue information.
    c. Assume that the 16 hours are divided into two 8-hour shifts. Agent breaks are staggered, starting at 90 minutes into each shift. Each agent is given one 15-minute break. Agent lunch breaks (30 minutes) are also staggered, starting 3.5 hours into each shift. Compare the results of this model with the results without agent breaks. Use cycle time and queue data to make the comparison.
14. *Multiple Queues*—The order fulfillment process of an entrepreneurial catalog business operates as follows. Orders arrive with exponential interarrival times with a mean of 10 minutes. A single clerk accepts and checks the orders and processes payment. These activities require a random time that is uniformly distributed between 8 and 10 minutes. On completion of these activities, orders are assigned randomly to one of two inventory workers, who retrieve the orders from the warehouse. The time to retrieve an order randomly varies between 16 and 20 minutes. The inventory workers retrieve only their assigned orders.
    a. Develop a simulation model that is capable of running 5000 minutes of this process.
    b. Assess the performance of the process using measures such as cycle time, queue statistics, and resource use.
    c. A manager points out that the assignment of orders should not be made at random. Instead, the manager suggests that the next order should be assigned to the inventory worker with the shortest queue. He also suggests breaking ties arbitrarily. Follow these recommendations and change your model accordingly. Compare the performance of this process with the original.
    d. A bright young "reengineer" recommends that the company should eliminate the assignment of an order to a specific person and allow both inventory workers to select their next order from a single queue. Modify your model to simulate this suggestion. Compare the performance with the previous two configurations.

15. *Priority Queues*—A business process consists of six activities, as shown in the flowchart of Figure 8.66. The activity times (in minutes) are normally distributed with mean values of 15, 10, 8, 8, 13, and 7, respectively, for Activities A through F. Similarly, the standard deviations are 3, 2.5, 2, 3, 3.5, and 2.

    Jobs arrive every 10 minutes on the average, with actual interarrival times following an exponential distribution. Caseworkers process the jobs. Two caseworkers can perform Activities A and B. Five caseworkers can perform Activities C, D, E, and F. Activity C requires two caseworkers per job, and the other activities require only one caseworker per job. When jobs arrive, they are assigned a due time, which is calculated as follows:

    Due time = arrival time + random number between 30 and 50.

    a. Develop a simulation model for this process.
    b. Simulate the arrival of 200 jobs (i.e., simulate the process for approximately 2000 minutes) and collect data on the tardiness for each completed job. (*Hint*: Consider using the Max & Min block from the Math submenu of the Value library to calculate tardiness. This block calculates the maximum and minimum values of up to five inputs.)
    c. Change the queues so that jobs with the earliest completion time are given priority. Compare the process performance with the process that uses a FIFO queuing discipline. (Note that Priority queues can be used to model a priority queuing discipline. Because the item with the smallest priority value is processed first, consider assigning the due date as priority value.)

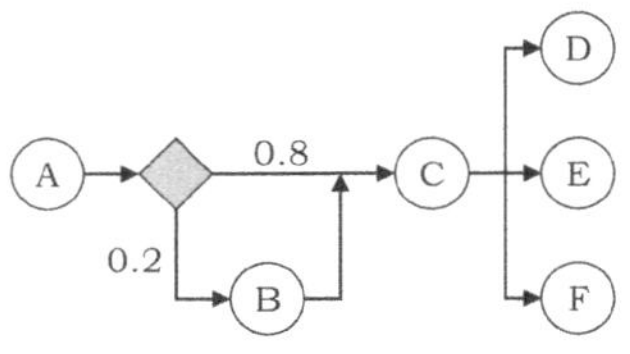

FIGURE 8.66
Priority queues.

16. Figure 8.67 shows the flowchart of a business process that receives an average of one job every 10 minutes with a standard deviation of 2.5 minutes. The actual interarrival times approximately follow a normal distribution. The processing times are exponentially

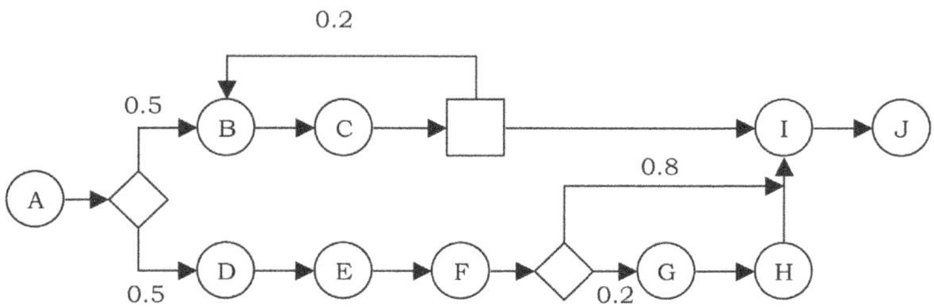

FIGURE 8.67
Flowchart for Exercise 16.

distributed with the mean values (in minutes) shown in Table 8.5. Three teams work in this process. The teams are assigned to activities as shown in Table 8.6.

a. Create a simulation model of this process.
b. Simulate the process for 10 working days, where each working day consists of 8 hours.
c. Assess the performance of the process considering waiting times, the use of the teams, and the work in process at the end of the 10-day period.
d. Add cycle time data collection and evaluate process performance considering the distribution of cycle times.

TABLE 8.5
Mean Processing Times (in Minutes) for Exercise 8.16

	A	B	C	D	E	F	G	H	I	J
Mean time	6	3	10	2	6	4	5	10	9	2

TABLE 8.6
Activity Assignments for the Processing Times (in Minutes) in Exercise 8.16

Team	Activities	Number of Team Members
1	A, B, and C	2
2	D, E, and F	2
3	G, H, I, and J	2

Notes

1. ExtendSim is a trademark of ANDRITZ Inc, 13560 Morris Road, Suite 1250, Alpharetta, GA 30004 USA 770.640.2500, Info.ExtendSim@Andritz.com, www.ExtendSim.com. The descriptions in Chapters 8 through 10 are based on ExtendSim 10 for Windows.
2. The use of optimization in the context of discrete-event simulation is discussed in Chapter 10.
3. This queuing discipline is also known as *first-come-first-served* (FCFS). The FIFO notation is used in ExtendSim.
4. The notation Menu > Menu Item will be used to indicate the menu item to select. For instance, File > New Model indicates to click on the File menu and then choose the New Model item.

References

ExtendSim User Reference. 2024. www.ExtendSim.com, ANDRITZ Inc., 13560 Morris Road, Suite 1250, Alpharetta, GA 30004 USA 770.640.2500, Info.ExtendSim@Andritz.com.

Nyamekye, K. 2000. New tool for business process re-engineering. *IIE Solutions* 32(3): 36–41.

9

Input and Output Data Analysis

Data analysis is a key element of computer simulation. Without analysis of input data, a simulation model cannot be built and validated properly. Likewise, without appropriate analysis of simulation output data, valid conclusions cannot be drawn, and sound recommendations cannot be made. In other words, without input data analysis, there is no simulation model, and without output data analysis, a simulation model is more or less worthless.

Because business processes are rarely deterministic, the use of statistics is important for the analysis of input data. Factors such as arrival rates and processing times affect the performance of a process, and they are typically random (stochastic) in nature. For instance, the time elapsed between the arrival of one job and the next generally follows a nondeterministic pattern that needs to be studied and understood to build a simulation model that accurately represents the real process. In addition, building a simulation model of a process entails the re-creation of the random elements in the real process. Therefore, three main activities are associated with the input data necessary for building a valid simulation model:

1. Analysis of input data
2. Random number generation
3. Generation of random variates

The first part of this chapter is devoted to these three activities. Input data are analyzed to uncover patterns such as those associated with interarrival times or processing times. After the analysis is performed, the patterns are mimicked using a stream of random numbers. The stream of random numbers is generated using procedures that are based on starting the sequence in a so-called *seed value*. Because the patterns observed in the real process may be diverse (e.g., a pattern of interarrival times might be better approximated with an exponential distribution, and a pattern of processing times might follow a uniform distribution), some random numbers need to be transformed in such a way that the observed patterns can be approximated during the execution of the simulation model. This transformation is referred to as the *generation of random variates*.

The second part of the chapter is devoted to the analysis of output data. One of the main goals of this analysis is to determine the characteristics

DOI: 10.1201/9781032617237-9

of key measures of performance such as cycle time, work in process (WIP), throughput, and costs for given input conditions such as changes in resource availability or demand volume. This can be used either to understand the behavior of an existing process or to predict the behavior of a proposed process. Another important goal associated with the analysis of output data is to be able to compare the results from several simulated scenarios and determine the conditions under which the process is expected to perform most efficiently.

The last part of this chapter illustrates how ExtendSim can be used for the modeling and analysis of two process design cases. These cases integrate the input and output data analysis with the modeling features of ExtendSim described in Chapter 8. They also illustrate how simulation analysis can provide ideas for process improvement and for comparing the performance of alternative designs.

9.1 Dealing with Randomness

One option for mimicking the randomness of a real process is to collect a sufficient amount of data from the process and then use the historical data as the input values for the simulation model. For example, in a simulation of a call center, it might be possible to record call data for several days of operation and then use these data to run the simulation model. Relevant data in this case might be the time between the arrival of calls, the pattern of call routings within the center, and the time to complete a call.

However, this approach has several shortcomings. First, field data (also called *real data*) typically are limited in quantity because, in most processes, the collection of data is expensive. Therefore, the length of the simulation would be limited to the amount of available data. Second, field data are not available for processes that are not currently in operation. However, an important role of simulation is to aid in the design of processes in situations where an existing process is not available. Field data can be used on a proposed redesign of an existing process, but it is likely that the problem of not having enough data to draw meaningful conclusions would be encountered. Third, the lack of several scenarios represented by field data limits the possibility of performing sensitivity analysis to disclose the behavior of the process under a variety of conditions. Finally, real data may not include extreme values that might exist but that do not appear in the data set collected. Probability distribution functions include these extreme values in the tails.

Although field data are not useful for gaining a full understanding of the behavior of a process, they are useful for model validation. Validating a model entails checking that the model behaves like the actual process when both are subject to the same data patterns. In the call center example, one should expect that given all relevant data patterns (e.g., the actual interarrival times for the calls during a given period of time), the model would produce the same values for the performance measures (e.g., average waiting time) as observed in the actual call center. Model validation is necessary before the model can be used for analysis and prediction. A valid model can also be used for attaining buy-in from the process stakeholders; that is, if the stakeholders are convinced that the model represents the actual process, they are more likely to accept recommendations that are drawn from testing design alternatives using the computer simulation.

Due to these limitations, simulation models are typically built to run with artificially generated data. These artificial data are generated according to a set of specifications to imitate the pattern observed in the real data; that is, the specifications are such that the characteristics of the artificially generated data should reflect those of the real data. The procedure for creating representative artificial data is as follows:

1. A sufficient amount of field data is collected to serve as a representative sample of the population. The collection of the data must observe the rules for sampling a population so that the resulting sample is statistically valid.
2. Statistical analysis of the sample is performed to characterize a probability distribution that adequately describes the population from which the sample was taken. The analysis must be such that the probability distribution function is identified, and the appropriate values of the associated parameters are determined. In other words, the population must be fully characterized by a probability distribution after the analysis is completed.
3. A mechanism is devised to generate an unlimited number of random variates from the probability distribution identified in Step 2.

The analyst performs Step 1 by direct observation of the process or by gathering data using historical records, including those stored in databases.

The analysis in Step 2 to determine a probability distribution that adequately characterizes the field data may be divided into three phases:

Phase 1: Identify an appropriate distribution family. This may be done by plotting histograms of the field data and comparing them graphically ("eyeballing") with shapes of well-known distributions, for example, the normal and exponential distributions.

Phase 2: Based on the chosen distribution family, estimate the distribution parameters from the field data (e.g., the mean and standard deviation for the normal distribution) to arrive at a hypothesis for a specific distribution that is appropriate.

Phase 3: Perform a statistical goodness-of-fit test to investigate whether the hypothesis may be rejected. In this case, the procedure restarts in Phase 1 and continues until a distribution hypothesis that cannot be rejected is found. If no well-known distribution family is appropriate, an empirical distribution may be used.

Manually performing the described analysis for a simulation model with large data sets and many different processing steps and arrival processes can be very time-consuming. Fortunately, there exist many software packages that provide distribution-fitting tools to perform Step 2 in a more or less automated fashion. These tools compare the characteristics of the sample data with theoretical probability distributions. The analyst can choose the probability distribution based on how well each of the tested distributions fits the empirical distribution drawn from the sample data. ExtendSim includes such a built-in tool for fitting distributions to filed data, called Stat::Fit. This software can be launched from the Run tab in the main ExtendSim window by choosing the option "launch StatFit." Alternatively, Stat::Fit can be opened from a Random Number block from the Inputs submenu of the Value library. Figure 9.1 shows the distribution-fitting tab of the dialog for this block. The use of Stat::Fit is explained and illustrated in Section 9.2.2. Examples of other popular distribution-fitting software are BestFit and ExpertFit.

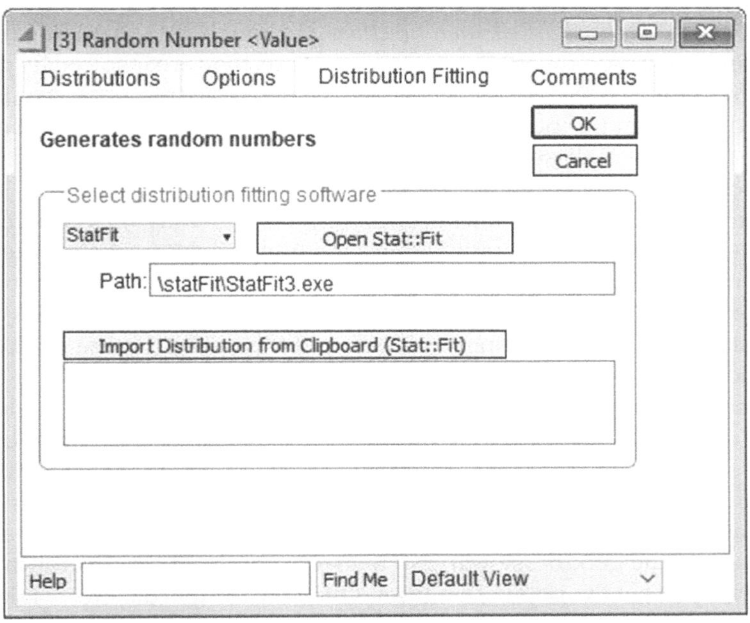

FIGURE 9.1
Distribution-fitting tab in the random number block of ExtendSim.

Finally, most simulation software packages (including ExtendSim) provide a tool for performing the third step of generating an unlimited number of random variates. These tools are designed to give users the ability to choose the generation of random variates from an extensive catalog of theoretical probability distributions (such as exponential, normal, gamma, and so on).

9.2 Characterizing Probability Distributions of Field Data

A probability mass function (PMF)—or if the data values are continuous, a probability density function (PDF)—is a model to represent the random patterns in field data. A PMF (sometimes referred to as a *probability distribution function*) is a mathematical function that assigns a probability value to a given value or range of values for a random variable. The processing time, for instance, is a continuous random variable in most business processes.

Researchers have empirically determined that many random phenomena in practice can be characterized using a fairly small number of probability mass and density functions. This small number of PMFs and PDFs seems to be sufficient to represent the most commonly observed patterns occurring in systems with randomness. This observation has encouraged mathematicians to create a set of PMF and PDF structures that can be used to represent each of the most commonly observed random phenomena. The structures are referred to as *theoretical* or *standardized distribution functions*. Although these standardized distributions are referred to as theoretical in nature, they are inspired by patterns observed in real systems. Some of the best-known distributions are uniform, triangular, normal, exponential, binomial, Poisson, Erlang, Weibull, beta, and gamma. The use of these distributions is well established, and guidelines exist to match a given distribution to specific applications.

For the simulation of business processes, it is generally the randomness of processing times and interarrival times that is of key interest. Because time is continuous, it is theoretical PDFs that are of primary importance for business process simulations. Therefore, in the following, the focus is placed on data analysis of continuous random variables and their PDFs. Figure 9.2 shows the shape of some well-known theoretical PDFs.

To facilitate the generation of random variates, it is necessary to identify a PDF that can represent the random patterns associated with a real business process appropriately. It is also desirable to employ a known theoretical PDF, if one can be identified as a good approximation of the field data. At this stage of the analysis, several PDFs must be considered while applying statistical methods (such as the goodness-of-fit test) to assess how well the theoretical distributions represent the population from which the sample data were taken.

As indicated earlier, a simple method for identifying a suitable theoretical probability distribution function is a graphical analysis of the data using a frequency chart or histogram. Histograms are created by dividing the range

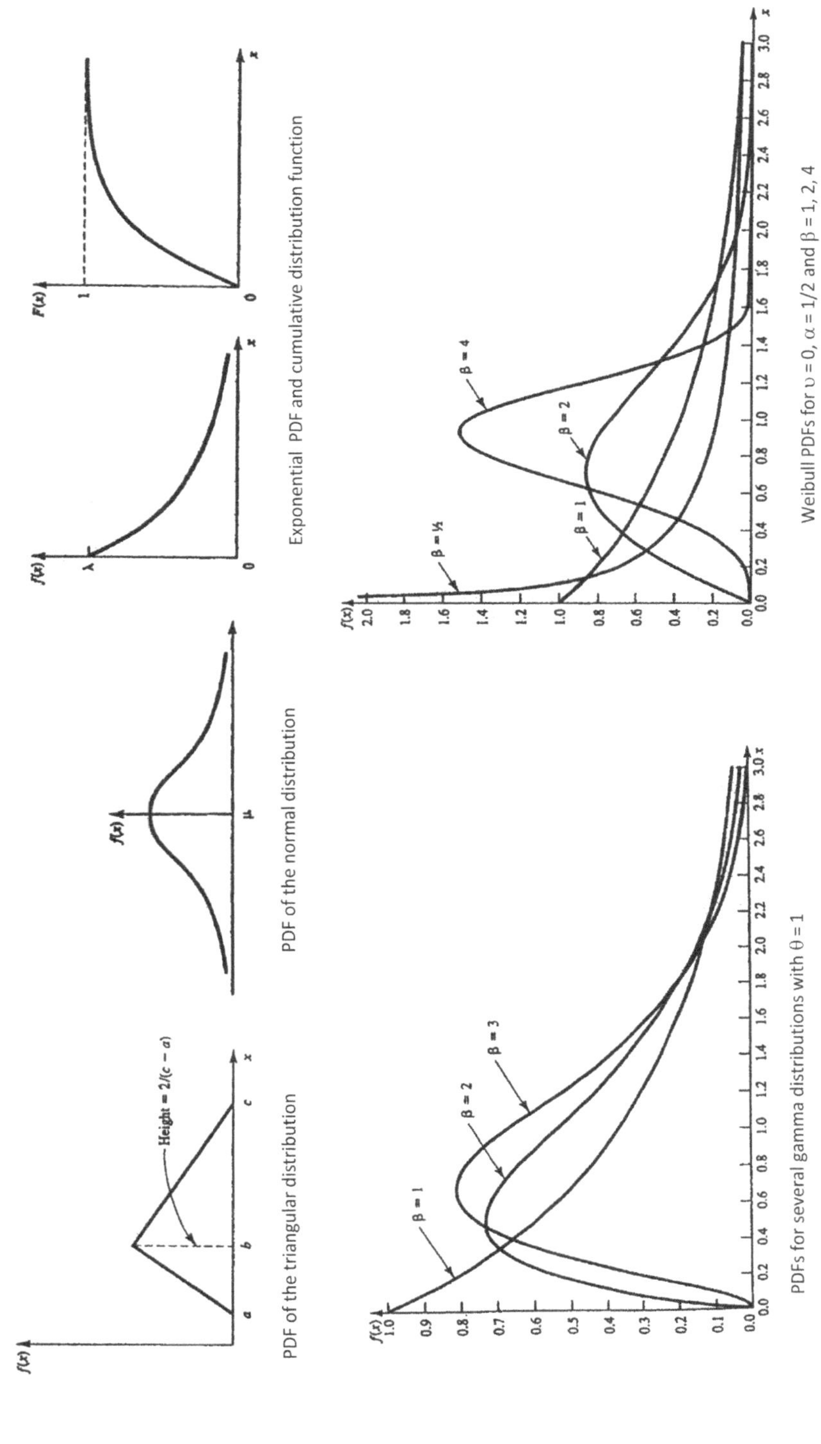

FIGURE 9.2
Shapes of some well-known probability density functions.

of the sample data into a number of bins or intervals (also called *cells*) of equal width. No precise formula exists to calculate the best number of bins for each situation, but larger samples allow more bins. If there are many bins with very few or no observations in them, this suggests that the number of bins should be reduced. It has been suggested that the square root of the number of observations in the sample may be a good value for the number of bins. Another rule of thumb sometimes used is that the number of bins should not exceed 15. No matter how many bins are used, they must be defined in such a way that they cover all possible outcomes, and no overlap between them exists. For example, the following three ranges are valid non-overlapping bins for building a histogram with sample data of a random variable x:

$$0 \leq x < 10$$
$$10 \leq x < 20$$
$$20 \leq x < 30$$

After the bins have been defined, frequencies are calculated by counting the number of sample values that fall into each range. Therefore, the frequency value represents the number of occurrences of the random variable values within the range defined by the lower and upper bounds of each bin. In addition to the absolute frequency values, relative frequencies can be calculated by dividing the absolute frequency of each bin by the total number of observations in the sample. The relative frequency of a bin can be considered an estimate of the probability that the random variable takes on a value that falls within the bounds that define the bin. This estimation process makes histograms approach the shape of the underlying distribution for the population. Because the number of bins and their corresponding width affect the shape, the selection of these values is critical to the goal of identifying a suitable PDF for the random variable of interest. The shape of the histogram is then compared with the shape of theoretical distribution functions (such as those shown in Figure 9.2), and a list of candidate PDFs is constructed to perform the goodness-of-fit tests. The following example shows the construction of a histogram using Microsoft Excel.

Example 9.1

Suppose the 60 data points in Table 9.1 represent the time in minutes between arrivals of claims to an insurance agent.

The characterization of the data set starts with the following statistics:

- Minimum = 6
- Maximum = 411
- Sample mean = 104.55
- Sample standard deviation = 93.32

Setting the bin width at 60 minutes, we can use Excel's Data Analysis tool to build a histogram. The resulting histogram consists of seven bins with the ranges and frequencies given in Table 9.2. Note that in Excel's Data Analysis tool, it is necessary to provide the Bin Range, which

TABLE 9.1

Interarrival Times

13	294	134	16	107	87
242	164	82	32	84	411
204	83	89	77	23	16
21	55	27	8	34	71
6	315	280	166	323	61
172	18	118	52	135	52
29	72	95	150	219	58
7	153	72	299	66	221
22	70	108	115	85	15
53	114	55	25	38	60

TABLE 9.2

Frequency Distribution for Interarrival Times

Bin	Frequency
0–59	23
60–119	20
120–179	7
180–239	3
240–299	4
300–359	2
360–419	1

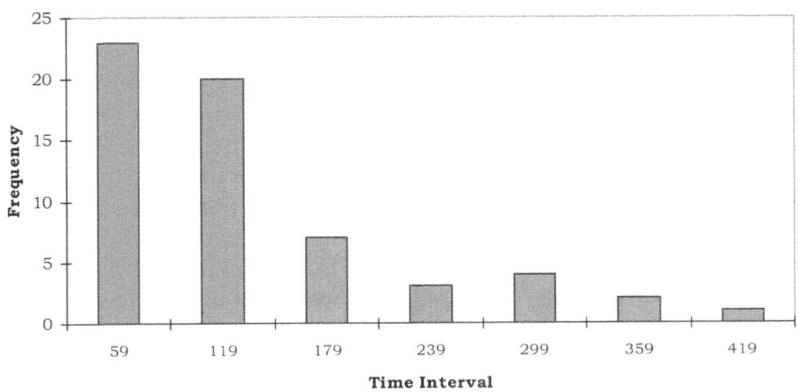

FIGURE 9.3
Histogram of sample interarrival times.

corresponds to the upper bounds of the bins in Table 9.2; that is, the Bin Range is a column of the spreadsheet that contains the upper bound values for each bin in the histogram (e.g., 59, 119, 179, and so on). After the data and the bin ranges have been specified, the histogram (or frequency table) in Figure 9.3 is obtained. ■

The next step of the analysis consists of testing the goodness of fit for the probability distributions that are candidates for approximating the real distribution.

9.2.1 Goodness-of-Fit Tests

Before describing the goodness-of-fit tests, it is important to note that these tests cannot prove the hypothesis that the sample data follow a particular theoretical distribution. In other words, the tests cannot be used to conclude, for instance, that the sample data in Example 9.1 are data points from an exponential distribution with a mean of 100 minutes. What the tests can do is rule out some candidate distributions and possibly conclude that one or more distributions are a good fit for the sample data.

The tested hypothesis is that a given number of data points are independent samples from a specified theoretical probability distribution. If the hypothesis is rejected, then it can be concluded that the distribution in question is not a good fit for the sample data. If the hypothesis cannot be rejected, this suggests that the candidate theoretical distribution is a good fit for the sample data. Failure to reject the hypothesis does not mean that it is true; several candidate theoretical distribution functions might be considered a good fit

The tests are based on detecting differences between the pattern of the sample data and the pattern of the candidate theoretical distribution. When the sample is small, the tests are capable of detecting only large differences, making them unreliable. When the sample is large, the tests tend to be sensitive to small differences between the sample data and the theoretical distribution. In this case, the tests might recommend rejecting the hypothesis of goodness of fit even if the fit could be considered close by other means, such as a visual examination of the frequency histogram. There exist many different goodness-of-fit tests developed for specific conditions. Two of the most popular and well-known general tests are the Chi-square test and the Kolmogorov–Smirnov test. These tests are described in detail and illustrated with numerical examples in Appendix 9A.1. Next follows an explanation of how distribution fitting may be done with the Stat::Fit software included in ExtendSim.

9.2.2 Using Stat::Fit for Distribution Fitting

Stat::Fit is a distribution-fitting package that is bundled with ExtendSim and that can be used to model empirical data with a probability distribution function. It can be launched from the Run tab in the main ExtendSim window by choosing "launch StatFit." Alternatively, it can be opened from the dialog window of a Random Number block of the Inputs submenu of the Value library. The Open Stat::Fit button is located under the Distribution-Fitting tab (see Figure 9.1).

The main Stat::Fit window is shown in Figure 9.4. Data files can be imported into the software with the File > Open… function. Alternatively, the data can be input by typing values or pasting copied values in the empty table that appears when Stat::Fit is called from ExtendSim. The data shown in Figure 9.4 correspond to the interarrival times of Table 9.1.

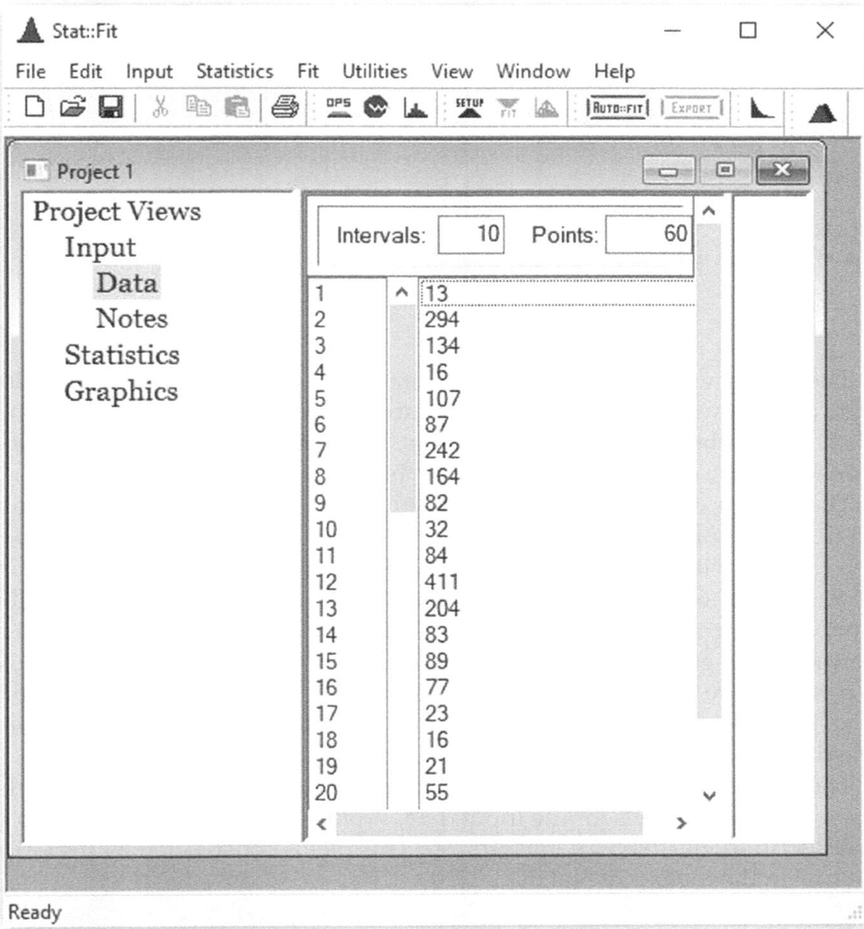

FIGURE 9.4
Stat::Fit window showing Table 9.1 data.

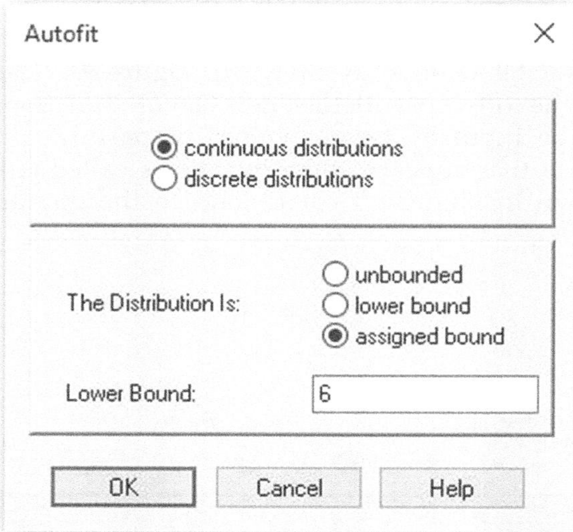

FIGURE 9.5
Auto Fit dialog window.

The fastest way of fitting a distribution is to select the Auto Fit function, which is shown in the toolbar of Figure 9.4. After the Auto Fit button is clicked, the dialog in Figure 9.5 appears. In this case, a continuous distribution with a lower bound of 6 has been chosen for the Auto Fit.

The results of the Auto Fit are shown in Figure 9.6. Stat::Fit tried to fit 25 different continuous distributions; the Weibull distribution was ranked the highest with rank 100, followed by the Gamma distribution with rank 96.9, Pearson 6 distribution with rank 87.4, and the exponential distribution with rank 70. There are also several other distributions that could be considered a good fit with the sample data. However, there are also quite a few that are rejected by the test, including the triangular and uniform distributions. The rejected distributions should not be used for describing the data material.

The fitted distribution can be plotted to provide visual confirmation that these are good models for the input data. A plot of the input data and two fitted distributions is shown in Figure 9.7. This plot is constructed by the Graph Fit tool in the toolbar of Stat::Fit.

Instead of using the Auto Fit function, it is possible to use Stat::Fit to test the goodness of fit of a particular distribution, just as is done by hand in Section 9A.1. To do this, open the Fit > Setup… dialog, then select the distribution(s) to be tested in the Distributions tab and the tests to be performed in the Calculations tab. By choosing the exponential distribution in autofit of distributions

distribution	rank	acceptance
Weibull(6, 1.06, 102)	100	do not reject
Gamma(6, 1.11, 88.5)	96.9	do not reject
Pearson 6(6, 1.95e+003, 1.13, 23.5)	87.4	do not reject
Exponential(6, 98.5)	70.9	do not reject
Loglogistic(6, 1.51, 66)	53.2	do not reject
Beta(6, 464, 0.839, 2.87)	40	do not reject
Lognormal(6, 4.08, 1.21)	9.03	do not reject
Inverse Weibull(6, 0.673, 0.0327)	0.0754	reject
Power Function(6, 445, 0.499)	0.00423	reject
Pearson 5(6, 0.575, 11.5)	0.000246	reject
Triangular(5, 423, 5)	3.65e-005	reject
Rayleigh(6, 96.4)	0	reject
Pareto(6, 0.412)	0	reject
Inverse Gaussian(6, 25.1, 98.5)	0	reject
Uniform(6, 411)	0	reject
Chi Squared(6, 60)	0	reject
Erlang	no fit	reject
Johnson SB	no fit	reject

FIGURE 9.6
Auto Fit results.

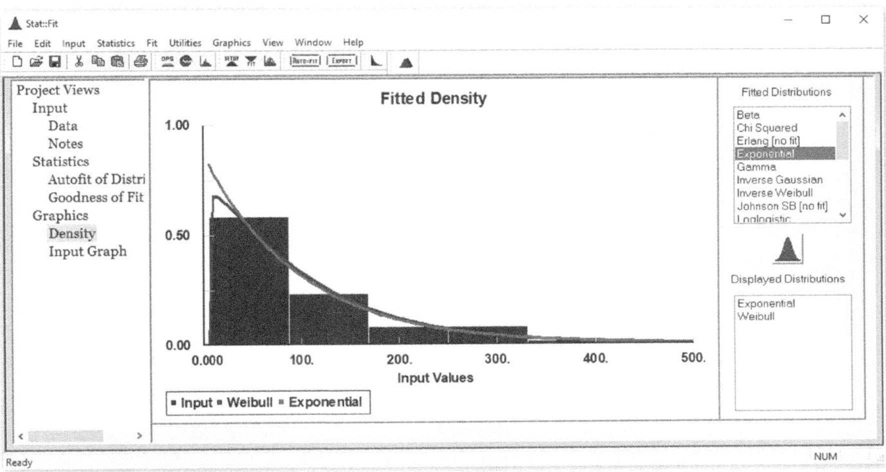

FIGURE 9.7
Plot of the input data and two fitted probability density functions.

the Distributions tab and the Chi-square and KS (Kolmogorov–Smirnov) tests in the Calculations tab, the results shown in Figure 9.8 are obtained after choosing Fit > Goodness of Fit (alternatively clicking the Fit button in the Stat::Fit toolbar). Note that the Chi-square value calculated by Stat::Fit matches the one calculated by hand in Table 9.22.

The results of the distribution fitting performed with Stat::Fit can be exported to ExtendSim. To do this, simply click on the Export button in the toolbar. Pick ExtendSim under the application window and clipboard under output and then click OK. Go back to ExtendSim and the Random Number block from which Stat::Fit was launched. Choose the Distribution fitting tab and click on "Import Distribution from Clipboard". The fitted distribution will then appear as the specified distribution for this block.

goodness of fit

data points	60
estimates	maximum likelihood estimates
accuracy of fit	0.0003
level of significance	0.05

summary

distribution	Chi Squared	Kolmogorov Smirnov
Exponential	2.5 (4)	0.0896

detail

Exponential		
minimum =	6 [fixed]	
beta =	98.55	
Chi Squared		
total classes		5
interval type		equal probable
net bins		5
chi**2		2.5
degrees of freedom		4
alpha		0.05
chi**2(4,0.05)		9.49
p-value		0.645
result		DO NOT REJECT
Kolmogorov-Smirnov		
data points		60
ks stat		0.0896
alpha		0.05
ks stat(60,0.05)		0.172
p-value		0.687
result		DO NOT REJECT

FIGURE 9.8
Goodness-of-fit test results obtained from Stat::Fit.

9.2.3 Choosing a Distribution in the Absence of Sample Data

When using simulation for process design, it is possible (and even likely) to be in a situation where field data is not available. When this situation arises, it is necessary to take advantage of the expert knowledge of the people involved in the process under consideration. For example, the analyst could ask a clerk to estimate the length of a given activity in the process. The clerk could say, for instance, that the activity requires "anywhere between 5 and 10 minutes." The analyst can translate this estimate into the use of a uniform (or rectangular) distribution with parameters 5 and 10 for the simulation model.

Sometimes, process workers are able to be more precise about the duration of a given activity. For instance, a caseworker might say: "This activity requires between 10 and 30 minutes, but most of the time it can be completed in 25 minutes." This statement can be used to model the activity time as a triangular distribution with parameters (10, 25, 30), where $a = 10$ is the minimum, $b = 30$ is the maximum, and $c = 25$ is the mode. The additional piece of information (the most likely value of 25 minutes) allows the use of a probability distribution that reduces the variability in the simulation model when compared with the uniform distribution with parameters 10 and 30.

If the analyst is able to obtain estimates representing the minimum (a), the maximum (b), the mean (μ), and the most likely value (c), a generalization of the triangular distribution also can be used. This generalization is known as the *beta distribution*. The beta distribution is determined by two shape parameters, α and β. Given these parameters, the corresponding mean, μ, and mode, c, are obtained from the following expressions:

$$\mu = a + \frac{\alpha(b-a)}{\alpha+\beta} \text{ and } c = a + \frac{(\alpha-1)(b-a)}{\alpha+\beta-2}$$

If an estimate $\bar{x}$ of the true mean, μ, is available along with an estimate of the mode, c, and estimates for the minimum, a, and maximum, b, the following relationships can be used to obtain the values of the α and β parameters for the beta distribution:

$$\alpha = \frac{(\bar{x}-a)(2c-a-b)}{(c-\bar{x})(b-a)}$$

$$\beta = \frac{\alpha(b-\bar{x})}{(\bar{x}-a)}$$

The beta distribution has some interesting properties. For example, the beta distribution with parameters (1, 1) corresponds to a uniform distribution with parameters (0, 1). A beta distributed random variable X in the interval from 0 to 1 can be rescaled and relocated to obtain a beta distributed random variable Y in the interval between A and B with the transformation $Y = A + (B - A)X$.

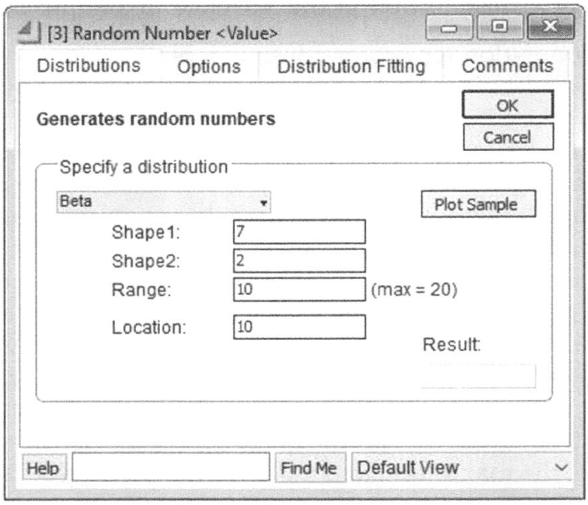

FIGURE 9.9
Beta distribution modeled with the random number block.

Also, the beta distribution with parameters (1, 2) is a left triangle, and the beta distribution with parameters (2, 1) is a right triangle. The selection of the parameter values for the beta distribution can change the shape of the distribution drastically. For example, when $\bar{x}$ is greater than c, the parameter values α and β calculated as shown previously result in a beta distribution that is skewed to the right. Otherwise, the resulting beta distribution is skewed to the left. A beta distributed random variable can be modeled in ExtendSim with a Random Number block from the Inputs submenu of the Value library. Figure 9.9 shows the dialog screen for the Random Number block when the beta distribution is chosen from the Distribution menu. The Shape 1 parameter is α, the Shape 2 parameter is β, the Location parameter corresponds to A, and the Range parameter equals $B - A$, where B and A are the maximum and minimum values as previously specified.

Example 9.2

Suppose an analyst would like to use the beta distribution to simulate the processing time of an activity. No field data are available for a goodness-of-fit test, but the analyst was told that the most likely processing time is 18.6 minutes, with a minimum of 10 minutes and a maximum of 20 minutes. It is also known that the average time is 17.8 minutes. With this information, the analyst can estimate the value of the shape parameters as follows.

$$\alpha = \frac{(17.8-10)(2\times 18.6-10-20)}{(18.6-17.8)10} = 7.02 \approx 7$$

$$\beta = \frac{7.02(20-17.8)}{(17.8-10)} = 1.98 \approx 2$$

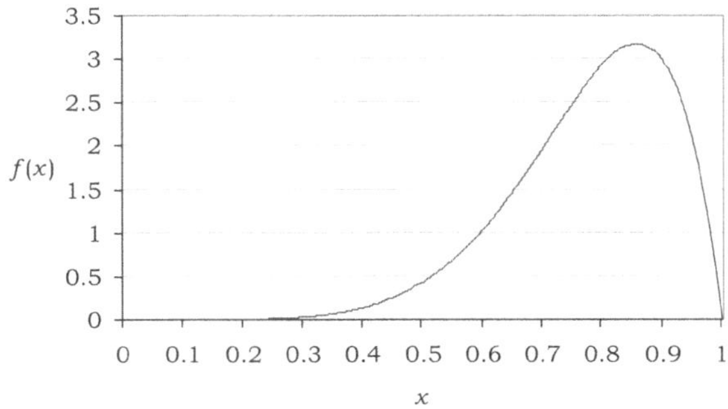

FIGURE 9.10
Beta distribution with $\alpha = 7$, $\beta = 2$, $A = 0$, and $B = 1$.

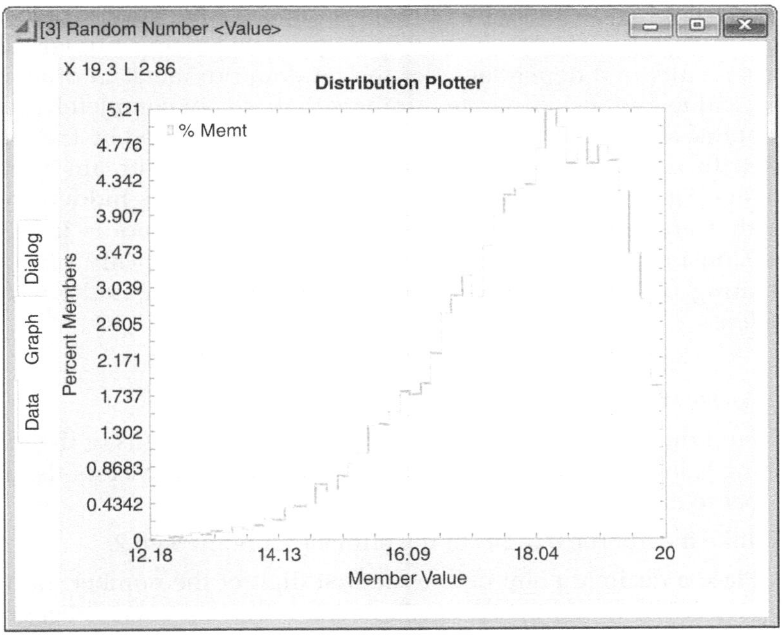

FIGURE 9.11
Beta distribution with $\alpha = 7$, $\beta = 2$, $A = 10$, and $B = 20$.

The shape of the beta distribution with $\alpha = 7$ and $\beta = 2$ is shown in Figure 9.10. Note that the x values in Figure 9.10 range from 0 to 1. To obtain processing times in the range from 10 to 20 minutes, simply use the transformation $10 + (20 - 10)x$. Figure 9.11 shows a sample plot of this distribution obtained from the Random Number block configured as in Figure 9.9 with Location = 10 and Range = 10. ■

9.3 Random Number Generators

Random numbers are needed to create computer simulations of business processes. The generation of random numbers is an important element in the development of simulation software such as ExtendSim. The random number generators are usually hidden from the user of the simulation software, so it might seem that these numbers are generated magically and that there should be no concern regarding how they were obtained. However, random number generation is one of the most fundamental elements in computer-based simulation techniques. For the conceptual understanding of this type of simulation method, it is therefore imperative to address a few basic issues related to random number generators.

Computer software uses a numerical technique called *pseudo-random number generation* to generate random numbers. The word *pseudo-random* indicates that the stream of numbers generated using numerical techniques results in a dependency of the random numbers. In other words, numerical techniques generate numbers that are not completely independent of one another. Because the numbers are interlinked by the formulas used with a numerical technique, the quality of the stream of random numbers depends on how well the interdependency is hidden from the tests that are used to show that the stream of numbers is indeed random. Consider, for example, a simple but poor numerical technique for generating random numbers. The technique is known as the *mid-square procedure*.

1. Select an n-digit integer, called the *seed*.
2. Find the square of the number. If the number of digits of the resulting value is less than $2n$, append leading zeros to the left of the number to make it $2n$ digits long.
3. Take the middle n digits of the number found in Step 2.
4. Place a decimal point before the first digit of the number found in Step 3. The resulting fractional number is the random number generated with this method.
5. Use the number found in Step 3 to repeat the process from Step 2.

Example 9.3

Use the mid-square procedure to create a stream of five four-digit random numbers in the range from 0 to 1 starting with the seed value 4151. The stream of random numbers generated from the given seed is shown in Table 9.3.

TABLE 9.3

Random Number Generation with Mid-Square Procedure

No.	Square of Current Number	Random Number
1	$(4151)^2 = 17230801$	0.2308
2	$(2308)^2 = 05326864$	0.3268
3	$(3268)^2 = 10679824$	0.6798
4	$(6798)^2 = 46212804$	0.2128
5	$(2128)^2 = 04528384$	0.5283

■

A problem with the mid-square procedure is that it can fail to generate random numbers if the seed is such that the square of the current number creates a pattern of a few numbers that repeat indefinitely. For example, if the seed 1100 is used, the method generates the sequence 0.2100, 0.4100, 0.8100, and 0.6100 and repeats this sequence indefinitely. The seed value of 5500 always generates the same "random" number—0.2500. Try it!

Random number generators in simulation software are based on a more sophisticated set of relationships that provide a long stream of random numbers before the sequence of values repeats itself. One popular procedure is the so-called *linear congruential method*. The mathematical relationship used to generate pseudo-random numbers according to this method is

$$Z_i = (aZ_{i-1} + c) \bmod m$$

$$r_i = Z_i / m$$

The Z values are integer values with Z_0 being the seed. The r values are the random numbers in the range from 0 to 1. The mod operator indicates that Z_i is the remainder of the division of the quantity between parentheses by the value of m. The parameter values a, c, and m as well as the seed are nonnegative integers satisfying $0 < m$, $a < m$, $c < m$, and $Z_0 < m$.

Example 9.4

Create a stream of five four-digit random numbers in the range from 0 to 1 using the linear congruential method with parameters $a = 23$, $c = 7$, $m = 34$, and a seed of 20.

$Z_1 = (20 \times 23 + 7) \bmod 34 = 25$	$r_1 = 20/34 = 0.7353$
$Z_2 = (25 \times 23 + 7) \bmod 34 = 4$	$r_2 = 20/34 = 0.1176$
$Z_3 = (4 \times 23 + 7) \bmod 34 = 31$	$r_3 = 20/34 = 0.9118$
$Z_4 = (31 \times 23 + 7) \bmod 34 = 6$	$r_4 = 20/34 = 0.1765$
$Z_5 = (6 \times 23 + 7) \bmod 34 = 9$	$r_5 = 20/34 = 0.2647$

■

The linear congruential method may degenerate if the parameter values are not chosen carefully. For example, if the parameter m is changed to 24 in Example 9.4, the sequence of random numbers degenerates in such a way that 0.8333 and 0.4583 repeat indefinitely.

Numerical methods for generating pseudo-random numbers are subjected to a series of tests that measure their quality. High-quality generators are expected to create random numbers that are uniform and independent. Goodness-of-fit tests can be used to check the uniformity. The runs test, discussed below, may be used to check the independency.

9.3.1 The Runs Test

The runs test is used to detect dependencies among values in a data set represented as a sequence of numbers. A run is a succession of an increasing or decreasing pattern in the sequence. The length of the run is given by the number of data points within one pattern before the pattern changes. Typically, the plus (+) sign is used to indicate a pattern of increasing numbers, and a minus (–) sign indicates a pattern of decreasing numbers. For example, the sequence 1, 7, 8, 6, 5, 3, 4, 10, 12, 15 has three runs:

$$+ + - - - + + + +$$

First is an increasing pattern that changes when the sequence goes from 8 to 6. The decreasing pattern continues until the sequence goes from 3 to 4, from which the increasing pattern continues until the end of the sequence. Note that any sequence of n numbers cannot have more than $n - 1$ runs or less than 1 run.

The runs test is based on comparing the expected number of runs in a true random sequence and the number of runs observed in the data set. If R is the number of runs in a true random sequence of n numbers, then the expected value and the standard deviation of R have been shown to be

$$\mu_R = \frac{2n-1}{3} \quad \text{and} \quad \sigma_R = \sqrt{\frac{16n-29}{90}}$$

Also, it has been shown that the distribution of R is normal for $n > 20$. The runs test has the following form:

(Null hypothesis) H_0: The sequence of numbers is independent.

(Alternative hypothesis) H_A: The sequence of numbers is not independent.

The statistic used to test this hypothesis is

$$Z = \frac{R - \mu_R}{\sigma_R}$$

where R is the number of runs in the data set with μ_R and σ_R computed as shown earlier. The null hypothesis cannot be rejected at the α level of significance if the following relationship holds:

$$-Z_{\alpha/2} \leq Z \leq Z_{\alpha/2}$$

In other words, if the Z statistic falls outside the range specified by the level of significance, the hypothesis that the numbers in the sequence are independent is rejected.

Example 9.5

Consider the following sequence of 22 numbers. Perform the runs test to determine, at a 95 percent confidence level, whether the numbers in the sequence (ordered by rows) may be considered independent.

The following represents the runs in the sequence:

30	34	21	89	81	8	94	67	58	19	18
22	37	33	88	56	67	77	36	27	67	90

$$+-+--+----++-+-++--++$$

Then, the number of runs R, the expected number of runs μ_R, and the standard deviation of the number of runs σ_R are

$$R = 13$$

$$\mu_R = \frac{2 \times 22 - 1}{3} = 14.33$$

$$\sigma_R = \sqrt{\frac{16 \times 22 - 29}{90}} = 1.89$$

The Z statistic is calculated as follows:

$$Z = \frac{13 - 14.33}{1.89} = -0.71$$

The critical value can be found from the normal distribution tables in Appendix 9A.4 or with the following Microsoft Excel function:

$$Z_{\alpha/2} = \text{NORM.S.INV}(0.975) = 1.96$$

Because Z is within the interval defined by $-Z_{\alpha/2}$ and $Z_{\alpha/2}$, the null hypothesis, which states that the numbers in the sequence are independent, cannot be rejected. ■

9.4 Generation of Random Variates

The previous section showed two methods for generating random numbers uniformly distributed between 0 and 1. However, in most simulation models, it is necessary to generate random numbers from probability distributions other than the uniform (or rectangular) distribution. These numbers are referred to as *random variates*. A general technique to generate a set of random variates from a specified distribution, based on a set of uniformly distributed random numbers, is the so-called *inverse transformation technique*. This technique takes advantage of the fact that cumulative distribution functions (CDFs) are defined in the range between 0 and 1. In addition, the CDF values corresponding to the values of the random variable are uniformly distributed, regardless of the probability distribution of the random variable. Mathematically, this is expressed as follows:

$$F(x) = r$$

where

$F(x)$ is the cumulative distribution function of the random variable x

r is a random number uniformly distributed between 0 and 1

If the CDF is known, then random variates of the random variable x can be found with the following inverse transformation:

$$x = F^{-1}(r)$$

Example 9.6

Use the uniform random numbers generated in Example 9.4 to generate variates of an exponentially distributed random variable with a mean of 10 minutes. The exponential cumulative distribution function has the following form:

$$F(x) = 1 - e^{-x}$$

By applying the inverse transformation, the following is obtained:

$$F(x) = 1 - e^{-\mu x}$$

$$-\mu x = \ln(1 - r)$$

$$x = -\frac{1}{\mu}\ln(1 - r) = -\frac{1}{\mu}\ln(r)$$

Note that if r is a random variable uniformly distributed between 0 and 1, then $1 - r$ is also a random variable uniformly distributed between 0 and 1. Therefore, random variates of the exponential distribution with mean value of 10 can be generated as follows.

$r_1 = 0.7353$	$x_1 = -10 \times \ln(0.7353) = 3.075$
$r_2 = 0.1176$	$x_2 = -10 \times \ln(0.1176) = 21.405$
$r_3 = 0.9118$	$x_3 = -10 \times \ln(0.9118) = 0.923$
$r_4 = 0.1765$	$x_4 = -10 \times \ln(0.1765) = 17.344$
$r_5 = 0.2647$	$x_5 = -10 \times \ln(0.2647) = 13.292$

■

Although the inverse transformation technique is a good method for generating random variates, there are some distributions for which the inverse of the CDF does not have a closed analytical form. For example, random variates of the normal distribution can for that reason not be obtained with a direct application of the described inverse transformation technique. However, the technique can in these situations be combined with numerical search methods and other approximations beyond the scope of this book.

Fortunately, all simulation software provides an automated way of generating random variates of the most popular theoretical distributions. In ExtendSim, the Random Number block from the Inputs submenu of the Value library is used for the purpose of generating random variates from a catalog of well-known probability distribution functions. As shown in Figure 9.9, one simply selects the probability distribution from the drop-down menu and then enters the appropriate parameter values.

Occasionally, it is necessary to simulate processes for which some of the uncertainty is due to random variables with discrete probability distributions of an arbitrary form. For example, the transportation time from a distribution center to a customer could be specified as being 2 days 20 percent of the time, 3 days 50 percent of the time, and 4 days 30 percent of the time. This distribution of transportation times is clearly an arbitrary discrete probability distribution function. Table 9.4 shows the CDF associated with the transportation times.

TABLE 9.4

Arbitrary Discrete PDF and CDF

Number of days x	Probability $P(x)$	Cumulative Probability $P(x)$
2	0.20	0.20
3	0.50	0.70
4	0.30	1.00

TABLE 9.5

Inverse Transformation for an Arbitrary Discrete PDF

Uniform Random Number Range	Random Variate
$0.00 \le r < 0.20$	2
$0.20 \le r < 0.70$	3
$0.70 \le r < 1.00$	4

The inverse transformation technique can be used to generate random variates of an arbitrary discrete distribution function. Once again, a uniformly distributed random variable r is used. The transformation associated with the probability distribution in Table 9.4 is given in Table 9.5.

Example 9.7

Use the uniform random numbers generated in Example 9.4 to generate variates of the arbitrary discrete probability distribution function in Table 9.4. When the inverse transformation in Table 9.5 is applied to the uniform random numbers generated in Example 9.4, the following random variates are generated.

$r_1 = 0.7353$	$x_1 = 4$
$r_2 = 0.1176$	$x_2 = 2$
$r_3 = 0.9118$	$x_3 = 4$
$r_4 = 0.1765$	$x_4 = 2$
$r_5 = 0.2647$	$x_5 = 3$

■

ExtendSim provides an easy way of generating random variates from arbitrary discrete probability distribution functions. The Random Number block from the Inputs submenu of the Value library provides this functionality. Figure 9.12 shows the dialog of the Random Number block when configured to generate random variates according to the discrete distribution in Table 9.4. The Empirical Table is chosen from the Distribution menu, and the Empirical Values option is set to Discrete.

The Random Number block also provides the functionality to generate random variates from an arbitrary continuous distribution. Suppose that instead of the discrete probability distribution in Table 9.4, one would like to use a continuous probability distribution of the following form:

$$f(x) = 0.2 \quad 1 \le x < 2$$
$$f(x) = 0.5 \quad 2 \le x < 3$$
$$f(x) = 0.3 \quad 3 \le x < 4$$

The inverse transformation technique could be applied using integral calculus to find the inverse of the CDF. In ExtendSim, however, this arbitrary continuous distribution can be modeled easily with the Random Number

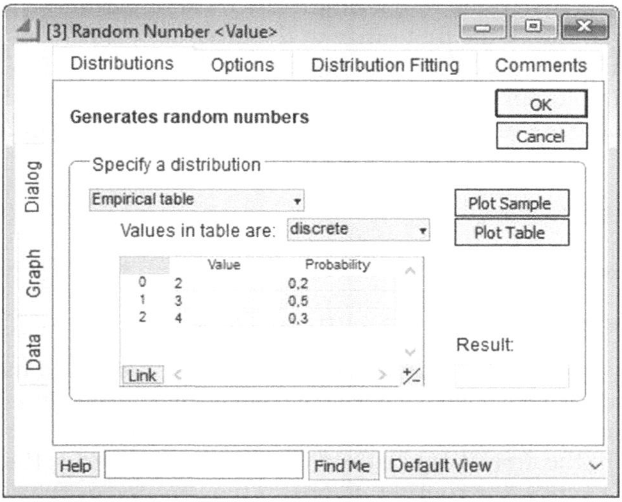

FIGURE 9.12
An arbitrary discrete PDF modeled with ExtendSim.

block. The Empirical Table is chosen from the Distribution menu and then the Stepped option is used for the Empirical values. Figure 9.13 shows the dialog for this ExtendSim block along with the corresponding plot of the arbitrary continuous distribution defined earlier.

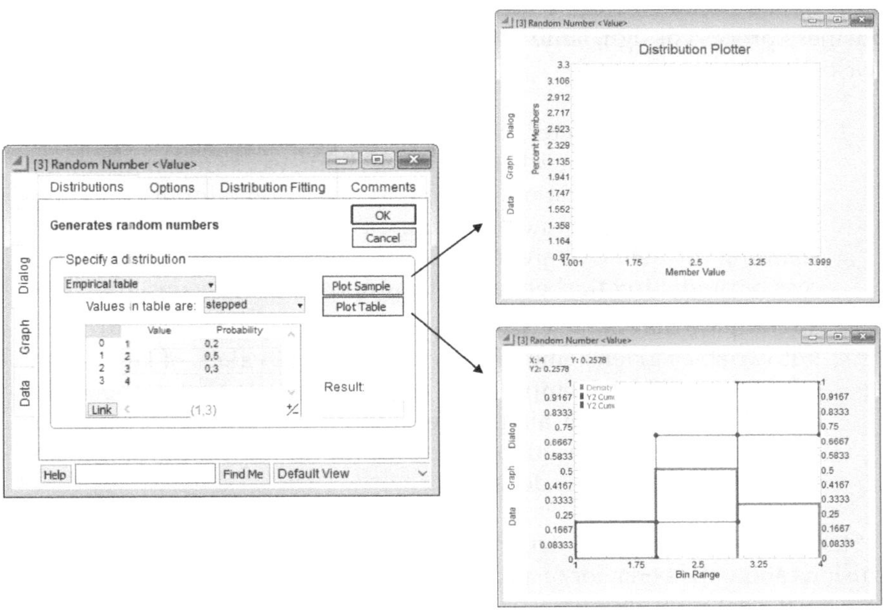

FIGURE 9.13
An arbitrary continuous PDF modeled with ExtendSim.

Note that the first value in the Empirical Table of Figure 9.13 corresponds to the lower bound of the first range in the arbitrary continuous distribution (which is 1 in this case). The bounds of the subsequent ranges are given in order until the upper bound of the last range is reached (which is 4 in this case). Also, note that the last probability value is left blank to indicate the end of the PDF.

9.5 Analysis of Simulation Output Data

The previous sections have addressed issues associated with modeling process uncertainty using discrete-event simulation. The stochastic behavior of the output of a process is due to the randomness of its inputs (e.g., the uncertainty in the arrival of jobs) and internal elements (e.g., the uncertainty of processing times, the routing of jobs in the process, or the availability of resources). Hence, the output of a simulation model should be treated as random (or stochastic) variables. For example, the waiting time of jobs is a random variable that depends on the inputs to the simulation model as well as the model's internal elements. Examples of other outputs from process simulations that are random variables include cycle times, throughput, WIP, utilization, and costs.

The goal of this section is to discuss some basic concepts of statistical analysis of data obtained from running a process simulation. In the context of business process design, simulation studies are performed for the following reasons:

1. To estimate the characteristics (e.g., mean and standard deviation) of output variables (e.g., cycle time, throughput, work-in-process inventories, resource utilization, and costs) given some input conditions and values of key parameter settings. This estimation aids in understanding the behavior and performance of an existing business process or predicting the behavior of a proposed process design.
2. To compare the characteristics (e.g., minimum and maximum) of output variables given some input conditions and values of key parameter settings. These comparisons help the analysts to choose the best design out of a set of alternative process configurations. Also, the comparisons can be used to determine the best operating conditions for a proposed process design.

Statistical analysis of simulation output data is necessary to draw valid conclusions about the behavior of a process. When performing statistical analysis, one must deal with the issues related to sampling and sample sizes. These issues cannot be ignored because valid conclusions cannot be drawn from a single simulation run of arbitrary length. To illustrate the danger of ignoring proper statistical sampling techniques, consider the following example.

Example 9.8

A simulation model has been built to estimate the number of insurance policies that are in process at the end of a working day in an underwriting department. After running the simulation model one time, the end-of-day WIP is estimated to be 20 policies. However, making decisions based on this estimate would be a terrible mistake. First of all, this estimate is based on a single run that represents a day of operation starting with an empty process. In other words, the WIP is zero at the beginning of the simulation. If 30 days of operation are simulated by running the model 30 times, the wrong assumption would be made that the WIP is always zero at the beginning of each day. It would be assumed that the policies that were in process at the end of one day would disappear by the beginning of the next business day. What needs to be done is to run the simulation model for $30 \times 8 = 240$ hours and record the WIP at the end of each 8-hour period. After running this simulation, one would observe that the WIP at the end of the first day is 20, but the WIP at the end of the 30th day is 154 insurance policies. This might indicate that the process is not stable and that the WIP will keep increasing with the number of simulated days. This is confirmed by a simulation run of 300 days that results in a final WIP of 1487 insurance policies. ■

The previous example illustrates the importance of choosing an appropriate value for the simulation run length. Clearly, in this example, simulating 10 days of operation was not enough to determine whether or not the current process configuration is indeed stable. The example also addresses the difference between a terminating and a nonterminating process. Figure 9.14 shows the types of statistical analysis associated with simulation models for terminating and nonterminating processes.

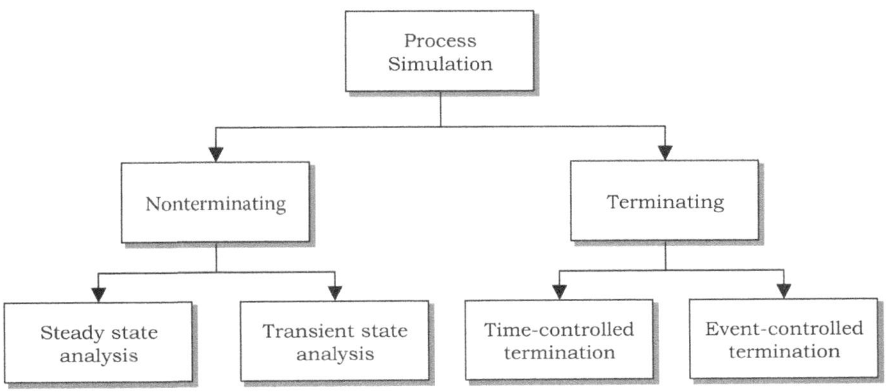

FIGURE 9.14
Analysis of output data according to simulation characteristics.

The following two subsections comment on the differences between the analyses that are appropriate for each of these simulation types.

9.5.1 Nonterminating Processes

A nonterminating process does not end naturally within a practical time horizon. Many business processes are nonterminating. For example, order fulfillment, credit approval, and policy underwriting are nonterminating processes. These processes are correctly viewed as nonterminating because the ending condition one day is the starting condition of the next day. Other systems, such as traffic and telecommunications networks, are clearly nonterminating because their operation is not divided into discrete units of time (such as working days). Most nonterminating processes eventually reach a steady state—a condition in which the process state distribution no longer changes with time and is no longer affected by the initial state in which the simulation was started. Example 9.8 is an illustration of a nonterminating process that doesn't reach a steady state. Typically, nonterminating processes go through a transient period before reaching a steady state. An output variable, such as the utilization of a resource, exhibits a transient period at the beginning of a simulation before reaching a steady state that represents the long-term condition for the given resource.

Determining the end of a transient period is an essential part of studying the steady-state behavior of a process. The data collected during a transient period belong to a different statistical population than data collected during a steady-state period. Mixing data from a transient period with data from the steady state period results in unreliable estimates of key output statistics. Therefore, the steady-state analysis should be performed with data collected after the end of the transient period.

Some statistical techniques have been used to determine the end of the transient period. For example, the runs test has been applied for this purpose. A simple way of finding the transition between the transient period and a steady state is to examine a line plot of the output variable of interest. Figure 9.15 shows the cycle time values from a simulation of a given process. The figure also shows the cumulative average cycle time as the simulation progresses. Note that the running average has a smoother pattern of change, which makes it easier to identify underlying changes in the process behavior.

In the line graph depicted in Figure 9.15, a change in the cycle times can be identified at Simulation Time 15. After detecting the length of the transient period using a pilot run, additional runs can be made with instructions to the software to start data collection after the so-called *warm-up period* has ended. In ExtendSim, this can be done using a Clear Statistics block from the Statistics submenu of the Value library. In the Options tab of this block's dialog window, one can select the types of blocks where the statistics calculations should be cleared and the time at which this should be done. The Clear Statistics block can be placed anywhere in the model and clears the statistics calculations in all specified block types without any connection lines to be drawn. Figure 9.16 shows the dialog when the statistics calculations in all listed block types are

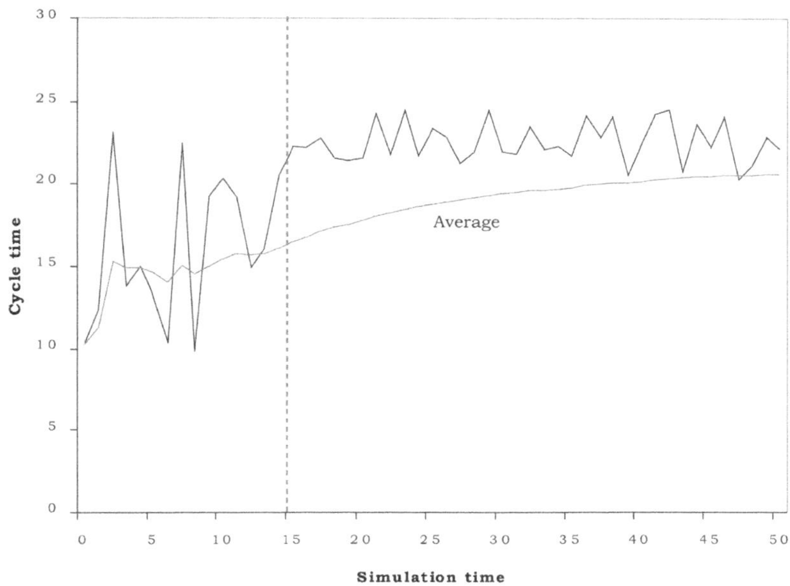

FIGURE 9.15
Line plot of current and average cycle times for a simulation run

cleared after 15 time units. An alternative to specifying the clearing time in the Options tab is to send an item or a value larger than 0.5 through the clear input connector at the bottom of the block. For example, one can connect a Create block and configure it to generate an item when the statistics should be cleared.

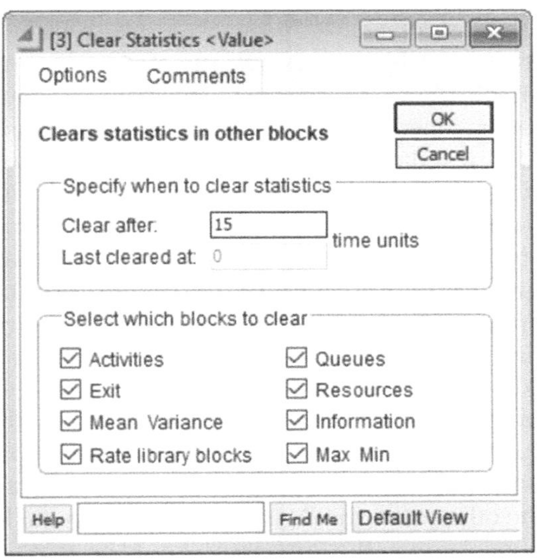

FIGURE 9.16
Clearing Statistics after a warm-up period with the Clear Statistics block.

The initial conditions of the process generally affect the steady-state analysis. However, the effect of the initial conditions decreases with the length of the simulation. For the simulation of an existing process, it is possible to use average values to create an initial condition for the simulation. For example, average values can be used to initialize the queues in a process in such a way that when the simulation starts, the process is not empty.

9.5.2 Terminating Processes

A terminating process typically starts from an empty state and ends at an empty state. The termination of the process happens after a certain amount of time has elapsed. For example, a bank office servicing walk-in customers opens in the morning at an empty state and closes after 8 hours of operation at an empty state.

The state of a terminating process is empty when it closes. Terminating processes might or might not reach a steady state. Many terminating processes might not even have a steady state. To find out whether or not a terminating process has a steady state, a simulation model of the process can be executed beyond the natural termination of the process. For example, the bank office process mentioned previously could be simulated for 100 continuous hours to determine whether or not the process reaches a steady state. The steady-state behavior of a terminating process can be used for estimating critical measures of performance and for making decisions about the process configuration.

The estimation of performance measures for terminating processes is generally achieved with statistical analysis of multiple independent runs. In this case, the length of each run is determined by the natural termination of the process. The number of runs is set to collect enough data to perform statistical analysis that allows the analyst to draw valid conclusions about the performance of the process.

The amount of data available after a single run of a terminating process depends on the output variable under consideration. For example, suppose a simulation model of the aforementioned bank office process is run once and that 100 customers are serviced during a day. At the end of the run, a data set consisting of 100 observations is available for output variables such as cycle time and cycle time efficiency (because these values can be recorded for each customer). On the other hand, there will be only a single observation of output variables such as daily throughput rate and overtime. In this case, a meaningful sample is taken, executing the simulation for a number of independent runs. Each run starts with the same initial conditions, but the runs use different streams of random numbers. The Setup tab under Simulation Setup in the Run menu of ExtendSim allows the number of runs for a single execution of the simulation model to be set. When more than one run is specified, the execution does not stop until all runs are completed. To assure that each run is based on a different set of random numbers, independent of the other runs, the *Random seed* choice in the Random Numbers tab under Simulation Setup in the Run menu should be zero or left blank.

This method results in samples of random variables for which only one observation can be obtained per run. The samples consist of independent observations of the random variable. In addition, it is valid to assume that the observations are identically distributed. These properties make the application of traditional statistical methods for estimation and hypothesis testing a robust approach to analyze output simulation data.

Example 9.9

A project consists of seven activities with stochastic duration and precedence relationships according to Table 9.6. An activity can be started only after all preceding activities are completed. An important aspect of successful project management is the ability to estimate the duration of the project. When the durations of the activities in the project are uncertain, project managers often rely on a methodology called PERT (project evaluation and review technique) to obtain an estimate of the expected completion time and its associated standard deviation. PERT calculations assume that the activity times are distributed according to a beta distribution. A minimum, a maximum, and a most likely value for the duration of each activity are used to perform the analysis. The advantage of using simulation in this context is that one does not have to assume that the activity times are beta distributed. An ExtendSim model of this project is shown in Figure 9.17. The completion time (or cycle time) of the project is calculated using an Information block in combination with a Set block and a Simulation Variable block as described in Chapter 8.

TABLE 9.6

Project Activities with Random Duration

Activity	Duration (Days)	Immediate Predecessor
A	Uniform(2, 6)	None
B	Uniform(2, 17)	None
C	Exponential(7)	A and B
D	Normal(8, 2)	B
E	Normal(4, 1)	C and D
F	Beta(7, 2) with Min = 3, Max = 9	D
G	Exponential(5)	E and F

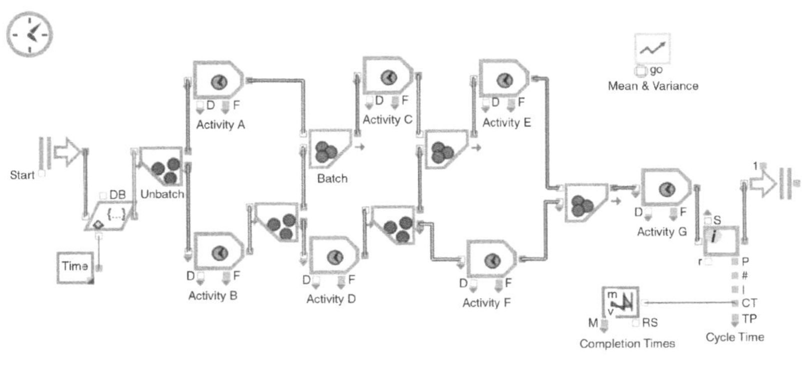

FIGURE 9.17
ExtendSim model of a project with seven activities.

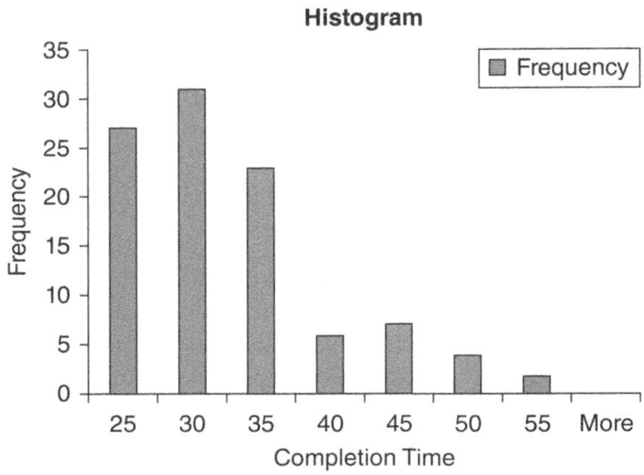

FIGURE 9.18
Histogram over completion times.

To estimate the project completion time (or project cycle time), the model in Figure 9.17 is run 100 times. At the beginning of each run, the Create block generates a single item representing the start of the project. The simulation time of each run needs to be long enough for the project to be completed. When running the model in Figure 9.17, a run time of 200 time units is used. The completion times for the 100 runs are collected using a Statistics block in combination with a Mean & Variance block. After the 100 runs are completed, the 100 project completion times are found in the Statistics block. It is easy to cut and paste this data into Excel for further analysis or to compute confidence intervals directly in the Statistics block. A histogram over the sample data (constructed by importing the output file to Excel) is displayed in Figure 9.18. This histogram shows that more than 80 percent of the time, the project is completed within 35 days. The expected completion time is 30.2 days with a standard deviation of 7.4 days. In addition to calculating simple statistics, the sample data can be used to estimate confidence intervals, to calculate an appropriate sample size for a specified interval precision, and to test hypotheses, as described in the following sections. ■

9.5.3 Confidence Intervals

Statistical estimation of characteristics (e.g., the mean) of a population is typically done in two ways: point estimates and confidence intervals. Point estimates are obtained by calculating statistics from a sample of the population. For example, the sample mean is a point estimate of the population mean, which is generally unknown. The average waiting time in a queue calculated from simulation output is a point estimate of the unknown mean of the population of all waiting times for the queue. Confidence intervals expand this analysis by expressing the estimates as intervals instead of single values. The intervals might contain the true value of the parameter; however, a so-called *confidence level* represents the accuracy of the estimation. If two intervals have the same confidence level, the narrower of the two is preferred. Similarly, if two intervals have the same width, the one with a higher confidence level is preferred. Typically, the analyst chooses a desired confidence level and constructs an appropriate interval according to this selection.

The sample size and the standard deviation are the main factors that affect the width of a confidence interval. For a given confidence level, the following relationships hold:

- The larger the sample size, the narrower the confidence interval.
- The smaller the standard deviation, the narrower the confidence interval.

Generally, the first relationship is used to approach the estimation problem from the viewpoint of determining a sample size that will satisfy a set of requirements for the confidence interval. In other words, the analyst can specify a level of confidence and a desired width for the confidence interval first and then use this to determine an appropriate sample size. In a discrete-event simulation context, this usually corresponds to the number of runs of a certain length.

9.5.3.1 Confidence Interval for a Population Mean

Mean values are often the focus of statistical analysis of simulation output. Average cycle time, average waiting time, average length of a queue, and average resource utilization are just a few examples of average values that are used to measure process performance. The most widely reported statistics associated with simulation output are the mean and the standard deviation. Suppose the values of a random output variable X are represented by x_1, x_2, ..., x_n (where n is the number of observations in the sample). The mean and the variance of X are estimated as follows:

$$\bar{x} = \frac{x_1 + x_2 + \ldots + x_n}{n}$$

$$s^2 = \frac{\sum_{i=1}^{n} (x_i - \bar{x})^2}{n-1}$$

If the random variables are independent and identically distributed, the sample mean $\bar{x}$ follows a normal distribution for a sufficiently large sample size (an often used rule of thumb is 30 observations or more). This is a result from the central limit theorem, which states that for a sufficiently large sample size, the distribution of the sample mean follows a normal distribution regardless of the distribution of the individual observations, as long as they are independent and identically distributed. The random variable Z is defined as follows:

$$Z = \frac{\bar{x} - \mu}{\sigma_{\bar{x}}}$$

This random variable follows a standard normal distribution, meaning a normal distribution with a mean of 0 and a standard deviation of 1. The population mean is represented by μ, and $\sigma_{\bar{x}} = \sigma / \sqrt{n}$ is the standard deviation of distribution of the sample mean. In this case, μ and σ are the true but unknown parameters for the population of X. The mean value μ is the parameter that is to be estimated. If the population standard deviation σ is not known (as is typically the case), then it is possible to use the sample standard deviation s as a reasonable approximation as long as the sample size is large (at least 30).

At a confidence level of $1 - \alpha$, a symmetrical confidence interval can be constructed from the following probability statement:

$$P\left\{-Z_{\alpha/2} \leq Z \leq Z_{\alpha/2}\right\} = 1 - \alpha$$

Using the definition of Z, it is possible to substitute this into the probability statement and solve for μ to obtain the confidence interval:

$$\bar{x} - \frac{\left(Z_{\alpha/2}\right)\sigma}{\sqrt{n}} \leq \mu \leq \bar{x} + \frac{\left(Z_{\alpha/2}\right)\sigma}{\sqrt{n}}$$

When the sample size contains fewer than 30 observations, the use of the normal distribution, as implied by the central limit theorem, can be questioned. Instead of the normal distribution, the Student t distribution can be employed in the following way:

$$\bar{x} - \frac{\left(t_{n-1,\alpha/2}\right)s}{\sqrt{n}} \leq \mu \leq \bar{x} + \frac{\left(t_{n-1,\alpha/2}\right)s}{\sqrt{n}}$$

This confidence interval, based on the Student t distribution, is exact only if the observed x values are normally distributed. However, it is often used as an approximation when this condition is not satisfied. If we compare the critical values for the standard normal distribution with those obtained with the Student t distribution for a given confidence level, $1 - \alpha$, then $t_{n-1,\alpha/2}$ is strictly greater than $Z_{\alpha/2}$. Consequently, using the Student t distribution produces a wider confidence interval. This is why the Student

t distribution produces confidence intervals that are closer to the desired confidence level $1 - \alpha$ when the sample size n is small. As the sample size becomes larger, $t_{n-1,\alpha/2}$ approaches $Z_{\alpha/2}$. The critical values for the standard normal distribution and the Student t distribution can be found in Appendix 9A.4. Alternatively, these values also can be found using the following Microsoft Excel functions:

$$Z_{\alpha/2} = NORM.S.INV(1-\alpha/2)$$

$$t_{n-1,\alpha/2} = T.INV(1-\alpha/2, n-1)$$

Example 9.10

Table 9.7 shows 50 observations of the cycle time for a given business process. Construct a 95 percent confidence interval to estimate the true but unknown value of the mean cycle time for this process.

TABLE 9.7

Sample Cycle Times

6.8	6.3	6.5	5.1	5.3	8.9	5.9	8.2	10.0	8.7
8.8	9.7	9.6	9.5	9.6	7.4	7.5	6.4	5.2	6.5
7.5	5.9	5.3	9.3	9.2	7.1	9.2	7.2	7.0	6.0
5.5	6.3	8.0	7.8	5.5	7.4	8.0	10.0	5.4	8.4
5.3	9.6	8.6	7.5	9.3	6.6	6.1	8.9	5.0	6.1

The point estimates for the population mean and standard deviation are $\bar{x} = 7.418$ and $s = 1.565$.

The Z value for a 95 percent confidence level is $Z_{0.975} = 1.96$. Then, the 95 percent confidence interval for the population mean is

$$7.418 - \frac{(1.96)1.565}{\sqrt{50}} \le \mu \le 7.418 + \frac{(1.96)1.565}{\sqrt{50}}$$

$$6.98 \le \mu \le 7.85$$

■

In ExtendSim, the Statistics block and the Mean & Variance block from the Statistics submenu of the Value library and the Cost Stats block from the Information submenu of the Item library automatically construct confidence intervals. These blocks are described in Section 8.4. For illustration, Figure 9.19 shows the dialog of the Statistics block in the model of the underwriting process in Figure 8.8 after 15 runs. The Statistics block in this model is configured to collect results from queues, which in this model means the Queue block labeled "In Box."

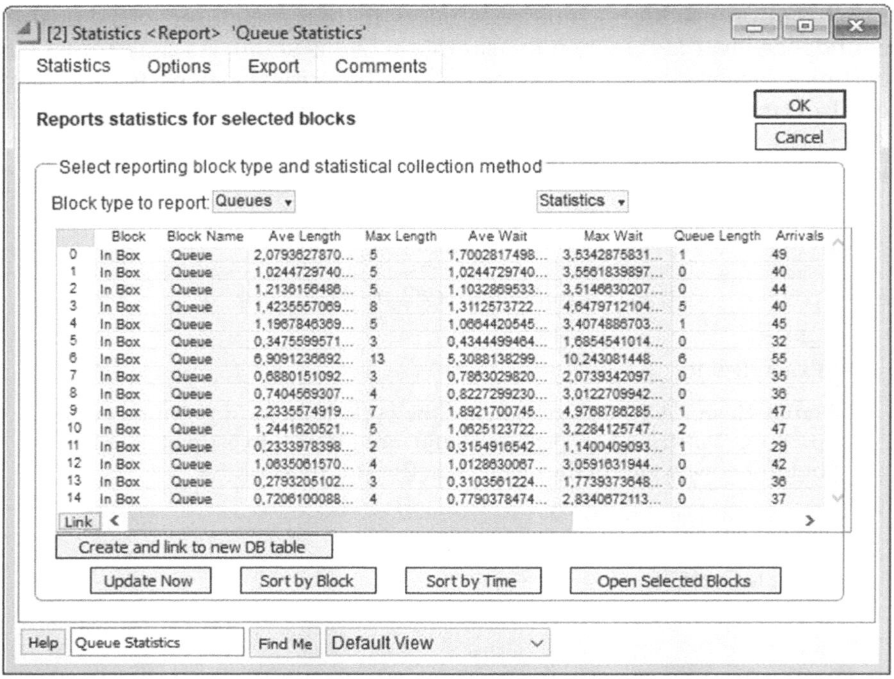

	Block	Block Name	Ave Length	Max Length	Ave Wait	Max Wait	Queue Length	Arrivals
0	In Box	Queue	2,0793627870...	5	1,7002817498...	3,5342875831...	1	49
1	In Box	Queue	1,0244729740...	5	1,0244729740...	3,5561839897...	0	40
2	In Box	Queue	1,2136156486...	5	1,1032869533...	3,5146630207...	0	44
3	In Box	Queue	1,4235557069...	8	1,3112573722...	4,6479712104...	5	40
4	In Box	Queue	1,1967846369...	5	1,0664420545...	3,4074886703...	1	45
5	In Box	Queue	0,3475599571...	3	0,4344499464...	1,6854541014...	0	32
6	In Box	Queue	6,9091236692...	13	5,3088138299...	10,243081448...	6	55
7	In Box	Queue	0,6880151092...	3	0,7863029820...	2,0739342097...	0	35
8	In Box	Queue	0,7404569307...	4	0,8227299230...	3,0122709942...	0	36
9	In Box	Queue	2,2335574919...	7	1,8921700745...	4,9768786285...	1	47
10	In Box	Queue	1,2441620521...	5	1,0625123722...	3,2284125747...	2	47
11	In Box	Queue	0,2333978398...	2	0,3154916542...	1,1400409093...	1	29
12	In Box	Queue	1,0635061570...	4	1,0128630067...	3,0591631944...	0	42
13	In Box	Queue	0,2793205102...	3	0,3103561224...	1,7739373648...	0	36
14	In Box	Queue	0,7206100088...	4	0,7790378474...	2,8340672113...	0	37

FIGURE 9.19
Statistics block with queue statistics from the In Box block of the underwriting model in Figure 8.8 after 15 runs.

Figure 9.19 shows the recorded statistics from the In Box block at the end of each run. Because these data come from 15 independent runs, they can be used to set confidence intervals. The result of choosing the option *confidence interval (compressed)* instead of *statistics* in the box directly above the table to the right is shown in Figure 9.20. The default is a 95 percent confidence interval, but it is easy to change this to, for example, 99 percent in the box that appears when the confidence interval option is chosen. Using the data in Figure 9.19, one can verify that ExtendSim constructs the confidence intervals using the equation with the Student *t* distribution as described earlier.

In simulating business processes, many interesting output variables are often proportions, such as proportions of customers serviced on time, percentage of jobs being reworked, etc. Constructing confidence intervals for population proportions can therefore be of great interest for the business process analyst. How this is done is further explained in Appendix 9A.2.

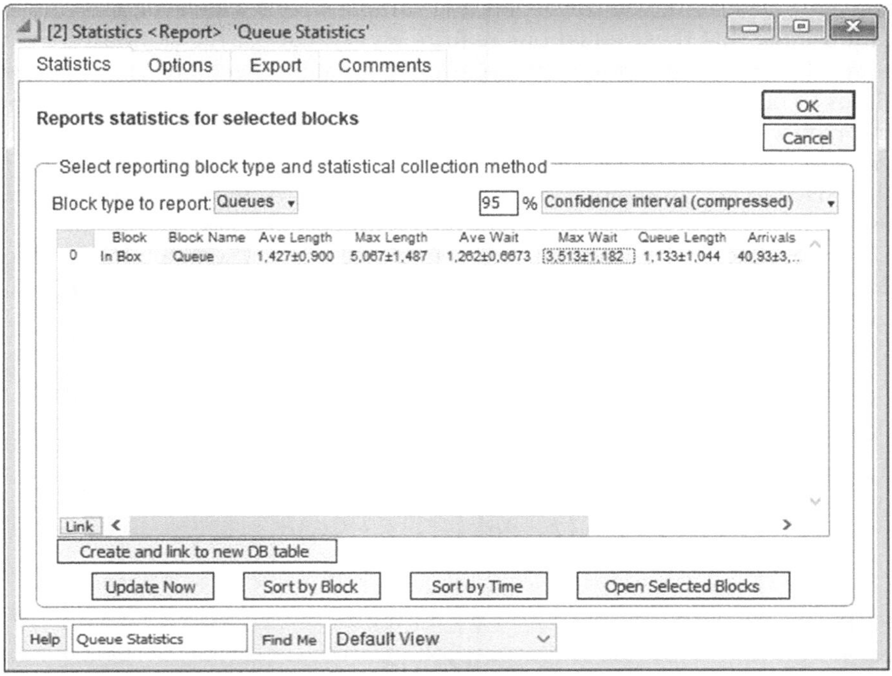

FIGURE 9.20
Statistics block in Figure 9.19 with calculated 95 percent confidence intervals after 15 runs.

9.5.4 Sample Size Calculation

The previous section assumed that the sample size was given. In practice, however, one of the key issues in the statistical analysis of simulation output is the determination of an appropriate sample size. The sample size is related to either the length of the simulation or the number of simulation runs, depending on the model. The theory of statistics includes what is known as the *law of large numbers*, which refers to the property that statistics calculated using large samples from independent random variables are nearly exact estimates of the population parameters. Today's simulation software coupled with the power of state-of-the-art computer hardware makes it possible to perform many lengthy simulation runs within a reasonable amount of time. Nevertheless, the question of what can be considered an appropriate sample size still remains. The answer depends on the variation of the random variable of interest.

An important issue in determining an appropriate sample size is the desired precision level of the calculated estimates. The analyst must set a desired width of the confidence interval and also specify a level of confidence. With this in place, the equations derived in the previous section can be solved to determine the corresponding sample size, n. However, it is important to note that these equations require an estimate of the standard deviation. This means that the simulation model usually needs to be run a few times to obtain such an estimate before a sample size can be calculated. So in practice, a provisional sample size is used for an initial run that can yield enough data to estimate the standard deviation. The resulting estimate is then used to derive an appropriate sample size, which in turn yields the necessary data for constructing the desired confidence interval.

Suppose the goal is to construct a confidence interval with a total width of $2d$, where d is half of the interval's width. The confidence interval for a population mean can be expressed as follows:

$$\bar{x} - d \leq \mu \leq \bar{x} + d$$

Comparing this inequality to the expression for the confidence interval for a normally distributed mean μ derived in the previous section,

$$\bar{x} - \frac{(Z_{\alpha/2})\sigma}{\sqrt{n}} \leq \mu \leq \bar{x} + \frac{(Z_{\alpha/2})\sigma}{\sqrt{n}}$$

We can conclude that

$$d = \frac{(Z_{\alpha/2})\sigma}{\sqrt{n}}$$

Because the true standard deviation is generally not known, the sample standard deviation s can be used to estimate it. The following equation is used to solve for the sample size:

$$n = \left(\frac{(Z_{\alpha/2})s}{d} \right)^2$$

The resulting sample size value guarantees a $(1 - \alpha)$ percent confidence interval of width $2d$ as long as the sample standard deviation, s, represents a good estimate of the population standard deviation, σ.

Example 9.11

Consider once again the cycle time data in Table 9.7. The confidence interval constructed in Example 9.10 with these data has a total width of 0.868 (i.e., $2d = 0.868$ or $d = 0.434$). Suppose one wants to construct a confidence interval with the same confidence level of 95 percent but with a total width of 0.6 (i.e., $2d = 0.6$ or $d = 0.3$). Calculate the sample size needed to achieve the desired precision level.

$$n = \left(\frac{(1.96)1.565}{0.3} \right)^2 \approx 104.54 \approx 105$$

■

A similar calculation can be made to find the sample size necessary to construct a confidence interval of a given precision for a population proportion, p. Again, assume that the analyst desires a $(1 - \alpha)$ percent confidence interval with a total width of $2d$. From Appendix 9A.2, the confidence interval can be expressed as follows:

$$\bar{p} - d \le p \le \bar{p} + d$$

where

$$d = Z_{\alpha/2}\sqrt{\bar{p}(1-\bar{p})/n}$$

Solving for the sample size n, one obtains the following:

$$n = \frac{(Z_{\alpha/2})^2\,\bar{p}(1-\bar{p})}{d^2}$$

Note that in this calculation, it is also necessary to execute a preliminary simulation run to be able to estimate the true proportion p with the sample proportion $\bar{p}$. Because an estimated value is being used to calculate the sample size, the accuracy of the calculation depends on the accuracy of the estimate obtained after the preliminary run. A more conservative approach is to use the maximum standard deviation value instead of the one calculated with the available $\bar{p}$. The maximum standard deviation of 0.25 occurs when $p = 0.5$. Therefore, a conservative and accurate sample size calculation for constructing a confidence interval on a proportion is

$$n = \frac{(Z_{\alpha/2})^2}{4d^2}$$

The advantage of using this equation is that it does not depend on any population parameters; therefore, the sample size can be calculated without performing a preliminary simulation run.

Example 9.12

Consider Example 9.16 in Appendix 9A. Calculate the sample size needed to construct a 99 percent confidence interval with a total width of 0.2. Use the formula that considers maximum standard deviation of the population.

$$n = \frac{(2.576)^2}{4(0.1)^2} \approx 165.89 \approx 166$$

■

9.5.5 Comparing Output Variables for Different Process Designs

The previous sections addressed several issues associated with statistical estimation. However, in business process design projects, simulation is often used as a tool for comparing the performance of different design alternatives. It is then important not only to estimate the expected value of an output variable but also to investigate whether this value is equal to a critical benchmark or target value, or whether there are statistically significant differences between the mean performance measures of two process design alternatives. Because the output variables in general are random variables, it is important that such comparative analysis takes the uncertainty of the estimation into account. A statistically correct way of doing this is to use hypothesis testing in some form. The statistical background for setting up and performing hypothesis testing is the same as for establishing confidence intervals. In particular, it is assumed that the sample means are normally distributed. A thorough explanation of hypothesis testing and how it may be used for the comparative analysis of simulation output variables is provided in Appendix 9A.3. This important material is deferred to the appendix because it is expected that many readers are familiar with hypothesis testing from their basic courses in statistics.

A simple pragmatic approach for comparing performance measures of alternative business process designs evaluated by simulation is to use a criterion of non-overlapping confidence intervals. This method is crude in comparison to conventional hypothesis testing, but it offers a simple, statistically sound way of identifying large differences between estimated performance measures. It may be viewed as a screening method to identify large differences where further hypothesis testing is unnecessary. Its usefulness in comparative simulation analysis of business processes is based on two observations:

1. Business process design projects are usually focused on identifying new designs with large improvements in process performance. Implementing a new process design is generally costly and a risky endeavor on which one does not embark unless there is a large potential reward in terms of improved process performance. Identifying large differences between estimated output variables, taking their uncertainty into account, is therefore often sufficient.
2. ExtendSim and other simulation software automatically generates confidence intervals for estimated average performance measures. This means that the criterion of non-overlapping confidence intervals is very easy to use and does not require any further statistical analysis in other software packages.

To explain the use of the criterion of non-overlapping confidence intervals, consider a business process, where the objective is to improve performance by reducing the cycle time. Simulating the current process design over multiple runs with ExtendSim renders the estimated expected cycle time, y_1, and the associated 99% confidence interval for the true expected cycle time μ_1: $y_1 - d_1 \leq \mu_1 \leq y_1 + d_1$.

Similarly, simulating an alternative process design in ExtendSim renders the estimated expected cycle time, y_2, with the associated 99% confidence interval $y_2 - d_2 \leq \mu_2 \leq y_2 + d_2$ for its true expected cycle time μ_2.

If $y_2 + d_2 \leq y_1 - d_1$, the confidence intervals are non-overlapping, and the entire confidence interval for μ_2 is below the confidence interval for μ_1. This means that there is a very low probability (less than 0.25 percent) that $\mu_1 < \mu_2$. Thus, one may conclude that μ_2 is smaller than μ_1 under the significance criteria of non-overlapping 99 percent confidence intervals.

Conversely, if $y_1 + d_1 \leq y_2 - d_2$, the confidence intervals are again non-overlapping, but now with the entire confidence interval for μ_1 below the interval for μ_2. This means that there is a very low probability (less than 0.25 percent) that $\mu_2 < \mu_1$, and one can conclude that the cycle time of the current process, μ_1, is shorter than for the alternative design, μ_2, under the significance criteria of non-overlapping 99 percent confidence intervals.

Finally, if $y_2 + d_2 > y_1 - d_1$ and $y_2 - d_2 < y_1 + d_1$, OR if $y_1 + d_1 > y_2 - d_2 <$ and $y_1 - d_1 < y_2 + d_2$, the confidence intervals are overlapping. In these cases, the significance criterion of non-overlapping 99 percent confidence intervals cannot be used to conclude that there is a difference between μ_1 and μ_2 (the hypothesis that $\mu_1 = \mu_2$ cannot be rejected using the criteria of non-overlapping 99 percent confidence intervals). If this happens, either the process analyst concludes that the new process design does not offer a large enough performance improvement to be of interest or she turns to conventional hypothesis testing. The objective is then to investigate whether a more precise analysis can assert that there exists a difference in expected cycle times that is statistically significant on some significance level α that is deemed small enough. A third alternative is, of course, to run more simulations to reduce the uncertainty of the estimates and the width of the confidence intervals and thereby detect smaller differences between μ_1 and μ_2.

Example 9.13

Consider the underwriting process described in Chapter 8 and assume as in Section 8.5 that the process contains a single underwriting team having exponentially distributed service times with a mean of 0.8 hours. The interarrival times of customer requests are exponentially distributed with a mean of 1 hour. A key performance measure for the process is the cycle time for handling the requests. Figure 9.21 shows an ExtendSim model of the process where cycle time statistics are collected over 30 runs simulated for 1600 hours each. Figure 9.22 shows the dialog window of the Statistics block labeled *Cycle Time Statistics*, where the 99 percent confidence interval for the mean cycle time CT_1 is determined to be 3.698 ± 0.3157 hours;

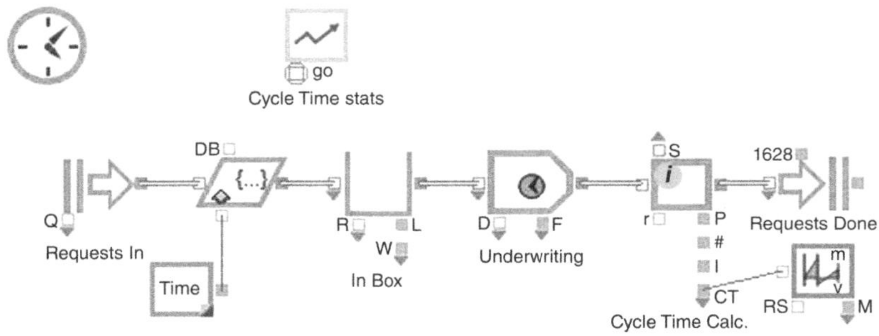

FIGURE 9.21
ExtendSim model of the underwriting process with a single underwriting team after 30 runs.

that is, $3.38 \leq CT_1 \leq 4.01$. To reduce the cycle time, an alternative process design with two underwriting teams working in parallel has been proposed. Figure 9.23 shows an ExtendSim model of this alternative process after 30 runs. Opening the dialog of the Statistics block, one finds that the 99 percent confidence interval for the mean cycle time, CT_2, is 0.9535 ± 0.020 or, equivalently, $0.934 \leq CT_2 \leq 0.974$. Comparing

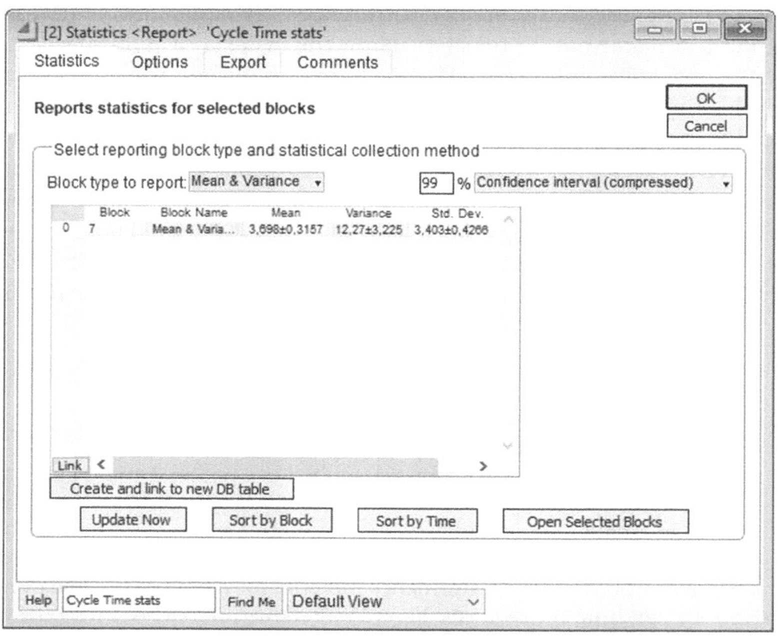

FIGURE 9.22
Dialog window of the cycle time statistics block with computed confidence intervals.

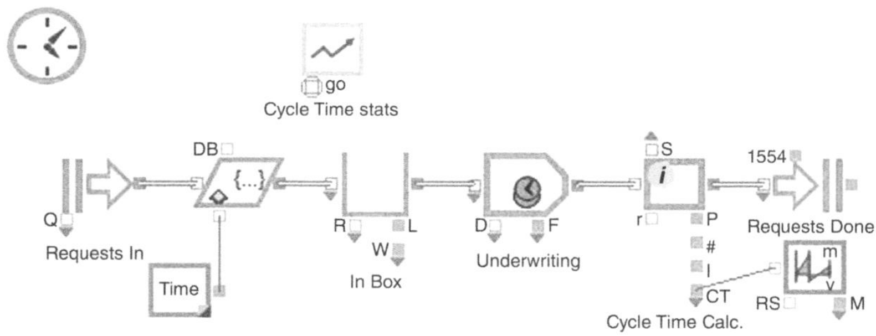

FIGURE 9.23
ExtendSim model of the underwriting process with two underwriting teams after 30 runs.

the confidence intervals for the two process designs, we can see that they are non-overlapping and that $CT_2 \leq 0.974 < 3.38 \leq CT_1$. Hence, the conclusion is that under the significance criterion of non-overlapping 99 percent confidence intervals, the expected cycle time for the process with two underwriting teams is shorter than for the current process with a single team ($CT_2 < CT_1$). ■

9.6 Modeling and Analysis of Process Design Cases

In the following, a few examples of business process design cases are considered. The objective is to illustrate how small cases can be modeled and analyzed by the use of simulation software packages such as ExtendSim. The analysis requires a synthesis of the material presented in Chapters 8 and 9 up to this point and assumes that the reader is familiar with this content.

9.6.1 Process Design of a Call Center for Software Support[1]

A software support organization currently provides help to licensed users over the phone and by e-mail. The manager wants to use simulation to explain why the productivity of the group is lower than he thinks it should be. His goal is to redesign the support processes so that the time to completion for problem resolution is reduced. The facilitator working with the manager would like to use this goal to define parameters by which the process can be measured. The facilitator knows that establishing a goal early in a modeling effort will allow her to determine the attributes of the process that will have to be investigated. In addition, it will help in the task of interviewing process participants.

Before interviewing the process participants, the facilitator asked for any documentation about the process that was available. The documentation of the process and the flowcharts treat phone and e-mail support as separate

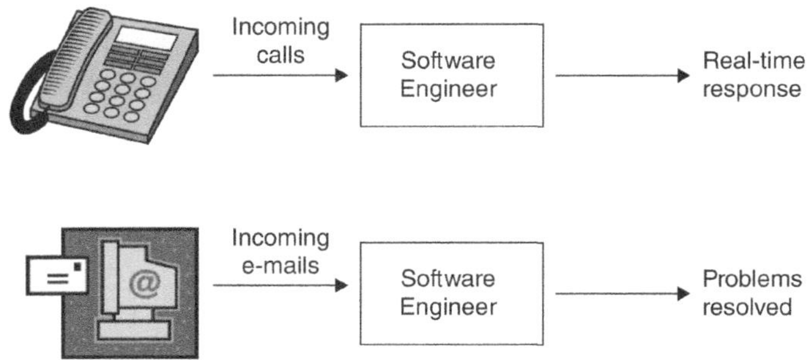

FIGURE 9.24
Documented software support process.

processes (see Figure 9.24). However, the reality is that in addition to reviewing and solving problems that have been submitted in writing, support personnel also have to answer phone calls. In fact, answering the telephone and providing real-time responses was given priority over providing responses via e-mail. The support personnel have suggested to management that one person should handle all the phone calls, and the others should handle problems submitted via e-mail. The request has been ignored because management has concluded that (1) there is time to perform both activities and (2) with three support personnel, if one only handles phone calls, a 33 percent reduction in problem-solving productivity will result.

In an effort to better understand the interaction between the e-mail software support system and the phone support process, the facilitator has collected data and made the following assumptions:

1. Requests for software support arrive at a rate of about 18 per hour (with interarrival times governed by an exponential distribution). Two-thirds of the requests are e-mails, and one-third are phone calls.
2. E-mails require an average of 12 minutes each to resolve. It can be assumed that the actual time varies according to a normal distribution with a mean of 12 minutes and a standard deviation of 2 minutes.
3. The majority of the phone calls require only 8 minutes to resolve. Specifically, it can be assumed that the time to serve a customer call follows a discrete probability distribution where 50 percent of the customers require 8 minutes, 20 percent require 12 minutes, 20 percent require 17 minutes, and 10 percent require 20 minutes.

Given this information, the manager of the group has concluded that he is correct in his assumption that with a little extra effort, support personnel can handle e-mails as well as phone calls. His logic is the following: 96 e-mails (12 e-mails/hour × 8 hours) at 12 minutes each require 1152 minutes, and 48 phone calls (6 calls/hour × 8 hours) at an average of 8 minutes each require 384 minutes. That totals 1536 minutes, or 3 working days of 8 hours and 32 minutes. He reasons that his personnel are professionals and will work the required extra 32 minutes.

9.6.1.1 Modeling, Analysis, and Recommendations

First, one should take a closer look at the manager's reasoning. The manager has based his calculations on the assumption that it takes an average of 8 minutes to resolve problems reported by phone. Although half (50 percent) of the calls require 8 minutes, the average call requires 11.8 minutes ($8 \times 0.5 + 12 \times 0.2 + 17 \times 0.2 + 20 \times 0.1$). Furthermore, the probability is fairly large (30 percent) that a phone call will require 17 or more minutes, which more than doubles what the manager is using for his calculations. He is also ignoring the variability of the arrivals of requests for technical support.

The manager is concerned with the group's productivity, which he most likely defines as the ratio of resolved problems to total requests. In addition to this, he would like to reduce the time needed to resolve software problems. Controlling the time to resolve a problem might be difficult, so the reduction in cycle time must come from reductions in waiting time. This is particularly important when handling phone calls because customers are generally irritated when they have to spend a long time on hold.

The process currently in use can be analyzed by modeling the work of one of the support people. To do this, adjust the arrival rates of e-mails and phone calls to represent what a single support person observes. According to the aforementioned arrival rates, the interarrival times are 5 and 10 minutes for e-mails and phone calls, respectively. If three support people are working, each of them experiences interarrival times of 15 and 30 minutes for e-mails and phone calls, respectively.

Figure 9.25 shows an ExtendSim model of the actual process for one of the support persons. Two Create blocks are used to model the arrivals

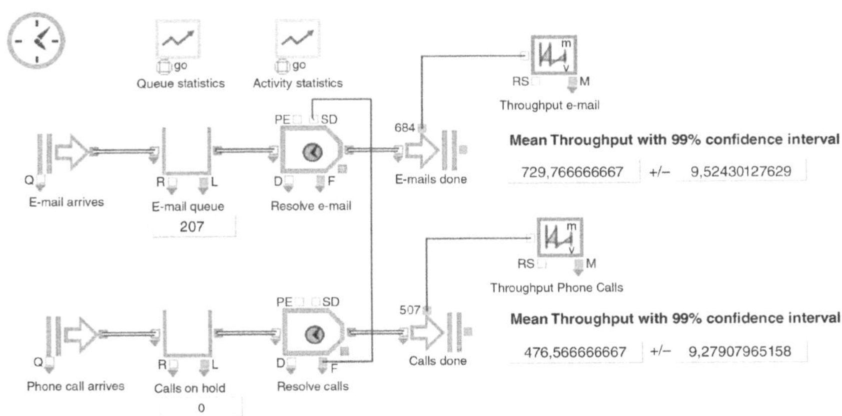

FIGURE 9.25
Simulation model of the actual process for one of the support persons.

of e-mails and phone calls separately. E-mails are set to arrive one every 15 minutes, and phone calls are set to arrive one every 30 minutes. Both interarrival times are set to follow exponential distributions in the Create blocks.

After an e-mail or phone call is generated, they are sent to two different Queue blocks labeled "E-mail queue" and "Calls on hold," respectively. Both these queues are set to use the first-in-first-out (FIFO) queuing discipline.

The processing times for resolving e-mail questions and phone calls are specified as delay times in the two Activity blocks labeled "Resolve e-mail" and "Resolve calls," respectively. The blocks are configured to allow at most one item in the block. The delay for e-mails is set to be drawn from a normal distribution with a mean of 12 minutes and a standard deviation of 2 minutes. The delay time for phone calls is set to be drawn from an empirical discrete distribution with values as specified previously.

To model that phone calls have priority over e-mail questions and are answered first, the Resolve e-mail block is shut down whenever there is a phone call to answer. This is achieved by choosing *Enable shutdown* in the Shutdown tab of this block's dialog. Also, choose the options: (i) SD (shutdown) input is from: *value connection,* (ii) when the signal is received at SD input, shutdown: *entire block,* and (iii) when activity shuts down: keep items, resume process after shutdown. The signal to shut down the block is taken from the F output connector (the L connector would work just as well) of the Resolve calls block, which has the value 1 whenever a phone call is answered and the block is full.

The model in Figure 9.25 shows the results after 30 runs of 240 hours each; the mean throughputs with 99 percent confidence intervals are computed in the Mean & Variance blocks labeled "Throughput e-mail" and "Throughput Phone Calls," respectively. The results (729.77 ± 9.52 and 476.57 ± 9.28) are made visible on the model sheet by using the Clone tool. The throughputs of the last simulation run of 240 hours (684 e-mails and 507 phone calls) are shown at the Exit blocks labeled "E-mails done" and "Calls done," respectively. Similarly, below the Queue blocks "E-mail queue" and "Calls on hold," the current numbers in these queues are displayed by use of the Clone tool. There are 207 unresolved e-mail questions but no phone calls on hold. It should be noted that each simulation run of 240 hours corresponds to 30 days of 8 hours each. E-mail questions are retained in the queue from one day to the next, while in the real system, phone calls cannot be put on hold from one day to the next. Thus, running the model for 240 hours is only appropriate if the process manages to clear the phone calls quickly enough for the number of calls on hold at the end of each day to be negligible. The results for the last run found in the Calls on hold block support this claim. A further investigation and validation of this assumption is a good exercise.

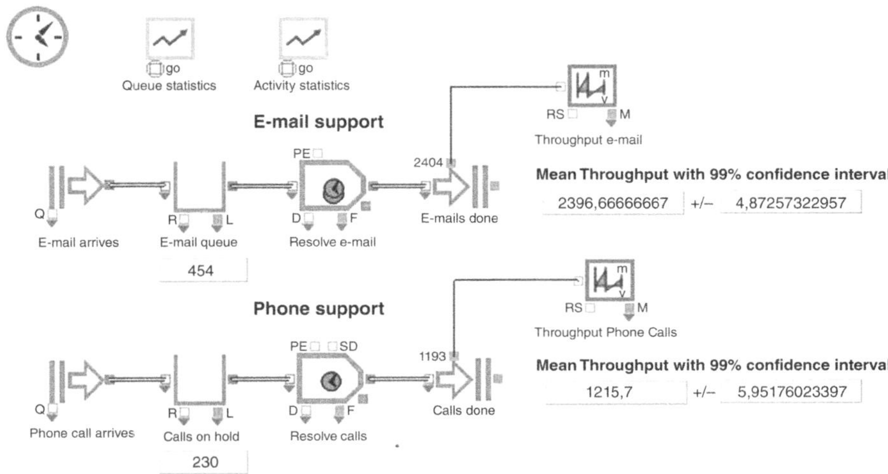

FIGURE 9.26
Simulation model of the documented process.

Figure 9.26 shows an ExtendSim model of the documented process, which is also the implementation proposed by the support personnel. In this model, requests are generated and routed independently. Two support persons only answer e-mails, and one only answers phone calls. The interarrival times for e-mail requests are exponentially distributed with a mean of 5 minutes. The time between phone calls is exponentially distributed with a mean of 10 minutes. The e-mail queue and the queue of calls on hold are served according to FIFO rules. The Activity block labeled "Resolve e-mail" is configured to allow at most two items in the block. Analogously, the Activity block labeled "Resolve calls" is set to a maximum of one item.

Table 9.8 summarizes the performance of both systems with three software engineers (one to answer phone calls exclusively in the documented process) when simulating 30 runs of 30 days of operation (240 hours of operation). Times are given in minutes, and all results are shown with 99 percent confidence intervals determined by the Statistics block labeled "Queue statistics."

TABLE 9.8
Performance Comparison with Three Engineers

		Documented Process		Actual Process	
Measures of Performance		**Phone Calls**	**Emails**	**Phone Calls**	**Emails**
Waiting Time (in minutes)	Average	1129.2 ± 101.3	1236.0 ± 63.9	4.2 ± 0.4	1749.6 ± 127.3
	Max.	2259.0 ± 182.0	2437.8 ± 124.0	45.5 ± 6.9	3490.0 ± 193.6
Utilization	Average	0.999 ± 0.001	0.999 ± 0.001	0.998 ± 0.002	

TABLE 9.9

Performance Comparison with Five Engineers

Measures of Performance		Documented Process		Actual Process	
		Phone Calls	Emails	Phone Calls	E-mails
Waiting Time (in minutes)	Average	3.7 ± 0.3	7.3 ± 0.6	2.0 ± 0.2	20.2 ± 1.9
	Max.	44.1 ± 5.4	57.3 ± 6.2	31.1 ± 4.0	137.3 ± 16.5
Utilization	Average	0.589 ± 0.009	0.805 ± 0.006	0.643 ± 0.01	

It is clear that neither the documented process nor the actual process is able to handle the volume of requests with three people. The utilization values show that the volume of work is unreasonably large considering the current level of resources. The average waiting time is somewhat reasonable for phone calls arriving to the actual process due to the priority system that is in place. When a dedicated server is used for phone call requests, the average waiting time explodes to more than 18 hours.

Based on the results in Table 9.8, one may conclude that both processes need additional staff. After both processes were simulated with five people (two dedicated to phone calls in the documented process), the summary statistics in Table 9.9 were found for 30 runs of 240 hours each. All results are shown with 99 percent confidence intervals, and times are given in minutes. The actual process model is modified by using a mean interarrival time of 25 minutes for e-mails and 50 minutes for phone calls.

The results in Table 9.9 indicate that both processes are now stable. There is no queue of e-mails that keep accumulating over time causing extremely long waiting times. Focusing on the average and maximum waiting times of the phone calls, one can conclude that the actual process performs better than the documented process using the significance criterion of non-overlapping 99 percent confidence intervals. Under the same criterion, the documented process works better considering all the waiting time values for resolving e-mail requests. The manager can now use this information to make a decision regarding the final configuration of the process, and the support personnel can use this information to show that they need additional help. There are clearly higher salary costs associated with employing five engineers instead of three, so quantifying the trade-off in reduced waiting time is key for making an informed decision.

9.6.2 Design of a Hospital Admissions Process

The considered hospital admissions process deals with routing several types of patients and managing several types of resources.[2] The main performance measure of interest is cycle time. Three types of patients are processed by the admissions function, as indicated in Table 9.10. Service times in the admissions office vary according to patient type as given in Table 9.10. All times are

TABLE 9.10

Data for Three Types of Patients with Service Time in Minutes

Type	Description	Service Time
1	Patients who seek admission and have previously completed their preadmission forms and tests	Normal(15, 1.5)
2	Patients who seek admission but have not completed preadmission	Normal(40, 4)
3	Patients who are only coming in for preadmission testing and information gathering	Normal(30, 3)

expressed in minutes. On arrival to admitting, a person waits in line if the two admissions officers are busy. When idle, an admissions officer selects a patient who is to be admitted before those who are only to be preadmitted. From those who are being admitted (Types 1 and 2), Type 1 patients are given higher priority.

Type 1 Process: After filling out various forms in the admitting office, Type 1 patients are taken to their floors by an orderly. Three orderlies are available to escort patients to the nursing units. Patients are not allowed to go to their floor on their own as a matter of policy. If all the orderlies are busy, patients wait in the lobby. After patients have been escorted to a floor, they are considered to be beyond the admitting process. The travel time between the admitting desk and a floor is uniformly distributed between 3 and 8 minutes. There is an 80 percent probability that the orderly and the patient have to wait 10 minutes at the nursing unit for the arrival of the paperwork from the admitting desk. It takes 3 minutes for the orderly to return to the admitting room.

Type 2 Process: After finishing the paperwork at the admitting office, patients walk to the laboratory for blood and urine tests. These patients are ambulatory and, as a result, require no escorts. After arriving at the lab, they wait in line at the registration desk, where one person is working. The service time at the registration desk follows a gamma distribution with a scale parameter of 2.5, a shape parameter of 1.6, and a location of 1. This service time includes copying information from the admission forms onto lab forms. The lab technicians use the lab forms to perform the indicated tests. After registration, patients go to the waiting room outside the lab and wait until they are called by one of the two lab technicians. The time spent drawing a lab specimen follows an Erlang distribution with a mean of 5 minutes, a k value of 2, and a location of 1. After the samples are drawn, patients walk back to the admitting office. On return to the admitting office, they are processed as regular Type 1 patients. The travel time between the admitting office and the lab is uniformly distributed between 2 and 5 minutes.

Type 3 Process: These patients follow the same procedure as Type 2 patients. The registration desk as well as the lab technicians services Type 2 and Type 3 patients in the order they arrive, disregarding any priorities. After the samples are drawn, the Type 3 patients leave the hospital.

Arrivals and Office Hours: The time between arrivals to the admitting office is exponentially distributed with a mean of 15 minutes. Before 10 a.m., the probability of a Type 1 arrival is 90 percent, and the probability of a Type 2 arrival is 10 percent. No preadmissions (Type 3) are scheduled until 10 a.m. because of the heavy morning workload in the lab. After 10 a.m., the probability of a Type 1 arrival is 50 percent, and the probabilities are 10 percent and 40 percent for Type 2 and 3 arrivals, respectively. The admitting office is open from 7:00 a.m. until 6:00 p.m. At 4:00 p.m., incoming admissions are sent to the outpatient desk for processing. However, Type 2 patients returning from the lab are accepted until 6:00 p.m., which is when both admitting officers go home and the office is closed. Experience has shown that not allowing new patients into the system after 4:00 p.m. means that the process is cleared of patients by 6:00 p.m. A graphical representation of the admitting process is shown in Figure 9.27. Travel times are indicated as well as waiting lines. All queues are infinite with FIFO disciplines except where noted. Activity and travel times are given in minutes.

Figure 9.28 shows an ExtendSim model of the current process with cycle time and throughput statistics computed for the three patient types after simulating 30 days of operation (30 runs of 11 hours each) The model uses four hierarchical blocks labeled Arrivals, Admissions, Getting Rooms, and Lab Tests. Hierarchical blocks contain process segments and make simulation models more readable; see ExtendSim User Reference (2024).

To make a hierarchical block, highlight the desired process segment and then choose Model > Make Selection Hierarchical and enter a name for the block when prompted. To assign item connectors through which items can be sent in and out of the block, mark it by a single left click with the mouse and then choose Model > Open Hierarchical Block Structure.

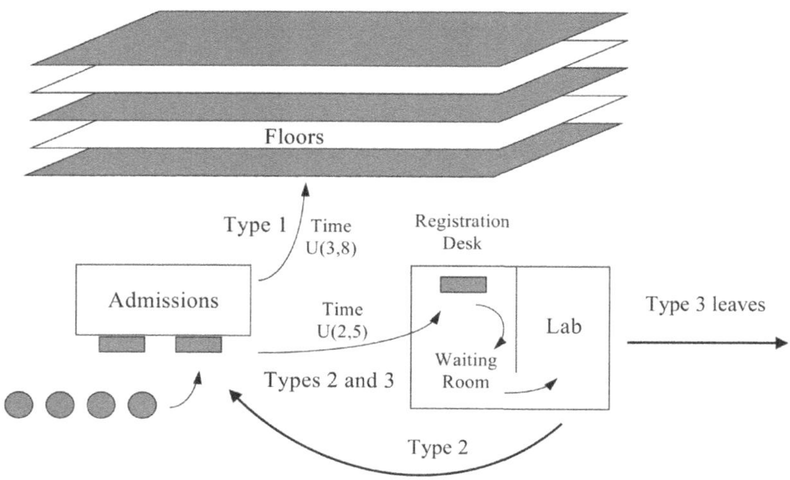

FIGURE 9.27
Schematic representation of the hospital admissions process.

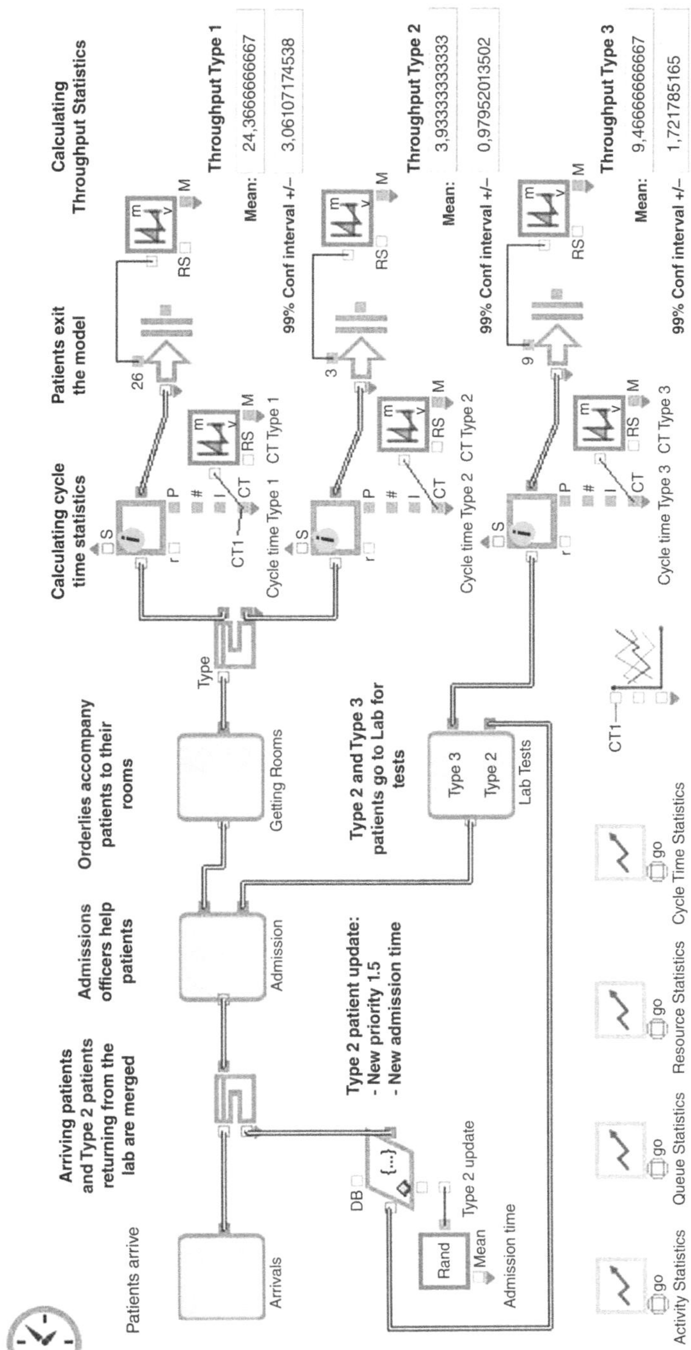

FIGURE 9.28
ExtendSim model of the current admissions process.

A hierarchical structure window for the block then opens up. Figure 9.29 shows the hierarchical structure window for the Arrivals block in Figure 9.28. The worksheet tab in the vertical menu bar to the left contains the process segment included in the hierarchical block, while the Icon tab displays the block's icon. Item and Value connectors can be added to the block's icon from the Icon toolbar activated under the Tools tab in the menu bar of the main ExtendSim window. When the Icon option is checked, the Icon toolbar is displayed on the right side of the ExtendSim Window. To add an Item connector (the procedure is the same for a Value connector), first make sure the graphics marker is chosen in the horizontal toolbar of the ExtendSim window, then left click on the corresponding connector type in the Icon toolbar, move the marker to the icon window of the hierarchical block, and left click again to place it where it should be. The item connector then appears on the icon and it can easily be moved around with the graphics marker. At the same time, *Con0In* appears as a small icon in the Connectors list to the left of the Icon window. If the procedure is repeated, the second Item connector is labeled *Con1In* and so on.

All item (and value) connectors are initially input connectors, but an input connector is easily made into an output connector just by changing the name in the connector list to the left to, for example, *Con1Out*. This is done by double-clicking on the text in the list. In the Worksheet window, the considered process segment can now be connected to the created input and output connector icons by drawing connection lines as usual. When all connections

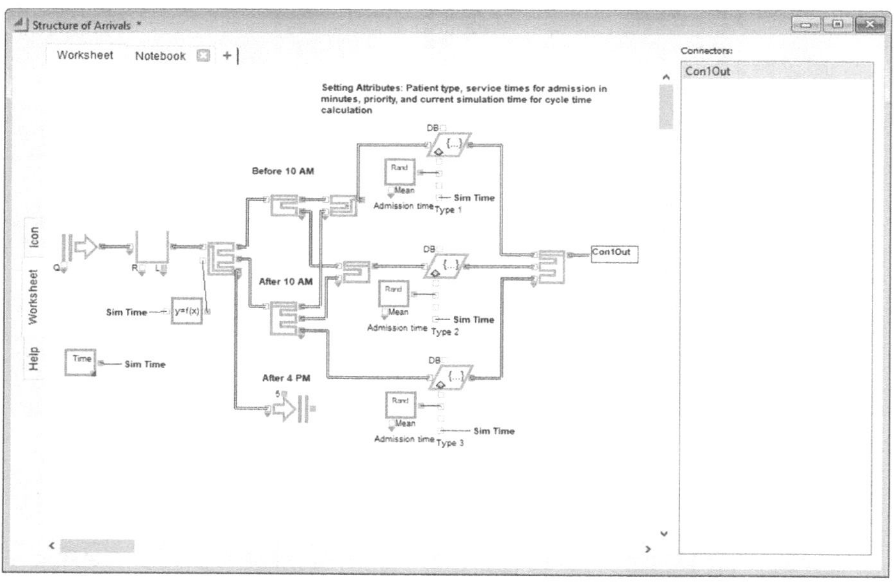

FIGURE 9.29
Hierarchical structure window for the hierarchical block labeled "Arrivals" in Figure 9.28.

are completed, the hierarchical structure window should be closed. When prompted, the option *Save this block only* should be chosen.

The first block in the ExtendSim model of the current process in Figure 9.28 is the hierarchical block labeled "Arrivals." Figure 9.29 shows that it contains a Create block and ancillary blocks that model the arrivals of the three different types of patients. The probability that an arriving patient is of Type 1, 2, or 3 changes depending on the time of the day. Therefore, the model uses the Simulation Variable block (set at the current time), together with an Equation block and a Select Item Out block, to switch between the three different patient type distributions before 10 a.m., after 10 a.m., and after 4:00 p.m. After 4:00 p.m., all arriving patients are turned away and referred to the outpatient desk. Thus, these patients are directly routed to an Exit block. For the other two time periods (before and after 10 a.m.), two Select Item Out blocks with probabilistic routing are used for dividing arriving patients into Types 1, 2, and 3. After the arriving customers are separated into the three different patient types according to the given probabilities, attributes are assigned using three Set blocks, one for each patient type. In these Set blocks, each item is assigned a priority (1, 2, or 3) and three attributes. The attributes are named Type (specifies the patient type 1, 2, or 3), Cycle Time, and Admissions time. The Cycle Time attribute is assigned the current time as a value in the Set block. Later, this attribute will be used for determining the cycle times of the three patient types. The Admissions time attribute specifies the delay time in the admissions activity. Values are generated by the Random Number blocks, in which the probability distribution for admissions processing of each patient type is specified. After the priority and attributes are assigned, the items are sent out from the hierarchical block via the Con1Out output connector.

Continuing to follow the process model in Figure 9.28, the arriving patients generated in the Arrivals block are merged with Type 2 patients returning from the lab. The combined flows of patients are then sent into the hierarchical block labeled "Admission." This block models the activities associated with the admitting officers, as shown in Figure 9.30.

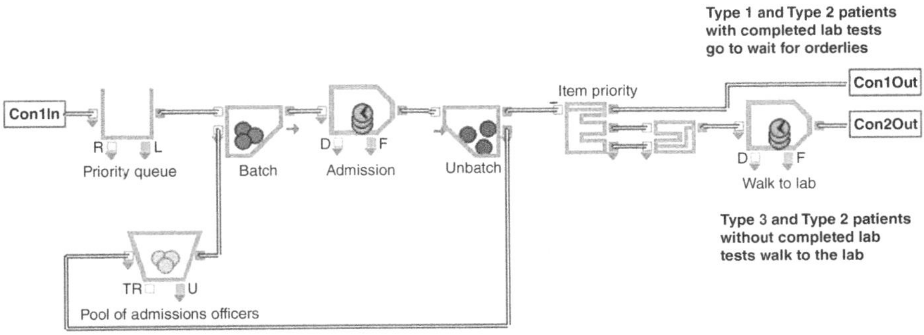

FIGURE 9.30
Process segment contained in the hierarchical Admissions block in Figure 9.28.

Patients form a line until an admitting officer becomes available. The Queue block labeled "Priority queue" is set to follow a Priority queue discipline, where the highest priority is 1 and the lowest is 3. A Resource Item block is used to model the pool of available admissions officers. When both an officer and a patient are available, the two are batched and sent into the Activity block labeled "Admission." This block is configured to delay the batched items for the time given by the attribute Admissions time specified in the Arrivals block (drawn from different distributions depending on the patient type).

After the admission activity is completed, the admissions officer and the patient are unbatched from one another, and the former returns to the pool of available admissions officers. The patient either proceeds to the top output connector, Con1Out (if it is of Type 1 or Type 2 with completed lab tests), or to the bottom output connector, ConOut2 (if it is of Type 3 or Type 2 without completed lab tests). The routing is done with a Select Item Out block based on the items' priority values. Type 1 patients have priority 1, and Type 2 patients with completed lab tests have priority 1.5. These items are routed through the top connector of the Select Item Out block. Type 3 patients have priority 3, and Type 2 patients who have yet to get their lab work done have priority 2. These items are routed through the two lower connectors. They then proceed to the lab as modeled by the Activity block labeled "Walk to lab."

From the process model in Figure 9.28, it is clear that the patients who have walked to the lab are routed through the bottom output connector of the Admissions block to the hierarchical block labeled "Lab Tests." The patients with completed lab work are routed through the top connector to the hierarchical block labeled "Getting Rooms." The hierarchical block Lab Tests consists of registration and laboratory activities, as shown in Figure 9.31. Patients form a line at the registration desk and wait for the registration clerk to complete the registration form. After registration, the patients go to the waiting room at the lab and wait for a lab technician to do the tests. After the tests are completed, Type 3 patients leave the process, and Type 2 patients return to admissions. When Type 2 patients return to admissions, the model uses an Activity block (with infinite capacity) to simulate the

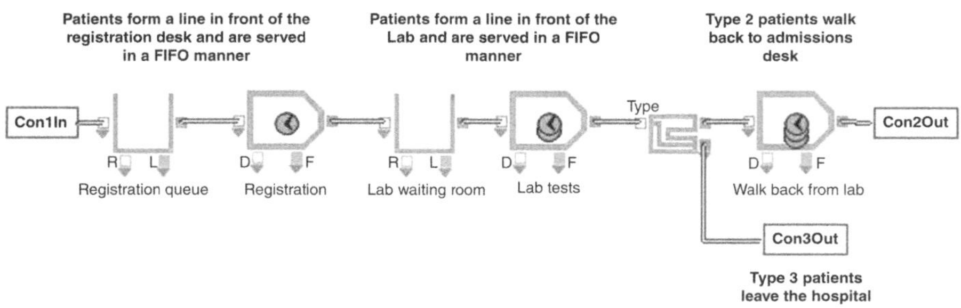

FIGURE 9.31
Process segment contained in the hierarchical block Lab Tests in Figure 9.28.

time to walk back. The main model in Figure 9.28 shows that after the Type 2 patients leave the Lab Tests block, they are routed back to the Admissions block. However, before being merged with new arrivals, their priority is changed from 2 to 1.5, and the attribute Admission time is updated with a new value drawn from the same distribution as Type 1 patients. The purpose of this change is to identify that these Type 2 patients have already completed the lab tests and are ready to be admitted in the same way as Type 1 patients.

The final hierarchical block, labeled "Getting Rooms," is shown in Figure 9.32. It consists of the activities related to orderlies taking patients to their rooms. The batching of a patient from the initial Queue block with an orderly from the Resource Item block labeled "Pool of orderlies" is done in the same way as the batching of a patient with an admission officer described earlier. The batching models that the patient and the orderly walk together to the room. When reaching the nursing station on the floor where the room is located, they might need to wait for paperwork to arrive. After the paperwork is cleared, the patient and the orderly part ways (modeled with an Unbatch block), and the patient enters his or her room. The orderlies walk back to admissions and join the pool of available orderlies. The time to walk back is modeled by an Activity block with infinite capacity labeled "Orderlies walk back." Patients leave this process segment through the connector named Con1Out.

Before the patients of Types 1, 2, and 3 leave the model in Figure 9.28 through the final Exit blocks, Cycle Time statistics are calculated. To do this, the model uses three Information blocks (labeled "Cycle time Type 1, 2, and 3," respectively), with connected Mean & Variance blocks (labeled "CT Type 1, 2, and 3," respectively), and a Statistics block labeled "Cycle Time Statistics." At the end of the model to the right, throughput statistics are calculated by the Mean & Variance blocks connected to the Exit blocks. The resulting mean and 99 percent confidence intervals are made visible in the main model by cloning them from the Mean & Variance Results tab.

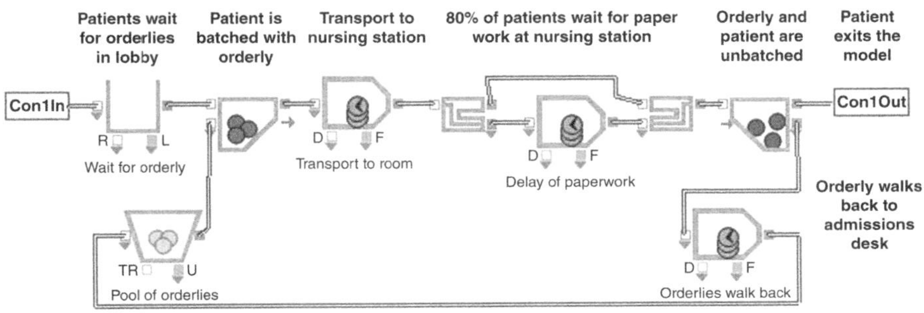

FIGURE 9.32
Process segment in the hierarchical block Getting Rooms in Figure 9.28.

TABLE 9.11

Estimated Mean Cycle Time and Daily Throughput with 99 Percent Confidence Intervals for the Current Process

Patient Category	Cycle Time (Minutes)	Daily Throughput (Number of Patients)
Type 1	34.04 ± 1.57	24.37 ± 3.06
Type 2	103.6 ± 11.86	3.93 ± 0.98
Type 3	79.51 ± 17.82	9.47 ± 1.72

Table 9.11 summarizes the cycle time and daily throughput statistics for the current hospital admissions process after 30 days of operation. The table shows the mean values with 99 percent confidence intervals computed by the Statistics blocks and the Mean & Variance blocks in ExtendSim. Each day, 11 hours of operation is simulated (opening hours are from 7 a.m. to 6 p.m.) All cycle time values are given in minutes.

Table 9.11 indicates that all patients experience fairly long cycle times. On average, a Type 2 patient seeking admission will require around 100 minutes to complete the process. If the patient has been preadmitted (Type 1), the average time is just above 30 minutes. The cycle time values associated with Type 3 patients only seeking preadmission are also quite long (about 80 minutes). Type 3 patients experience the largest variation in their cycle times, as indicated by the width of the confidence interval. To complement the analysis of the mean cycle times in Table 9.11, Figure 9.33 shows a frequency distribution of the cycle times for Type 1 patients based on 10 days of operation. Figure 9.33 reveals that about two-thirds of the Type 1 patients (about 67 percent) experience cycle times longer than 30 minutes. It is assumed that the hospital would like to minimize the number of Type 1 patients who experience long cycle times because this can discourage future patients from following the preadmission process. In other words, a patient would prefer to seek admission directly as a Type 2 patient instead of going through preadmission first as a Type 3 patient and then returning to be admitted as a Type 1 patient.

It might be possible to decrease the cycle time by adding more staff to the admissions desk because the admitting officers have the highest average utilization of 63.4 ± 6.2 percent (99 percent confidence interval). However, it seems more promising to explore the potential for improvement by redesigning the process instead of taking the simple approach of adding more staff.

The process design project can begin with the formulation of a vision statement. A vision statement for this process might include a reference to improving service as measured by the cycle time that each patient type experiences; for instance:

> The hospital admissions process will preadmit patients in an average of less than 35 minutes and never more than 45 minutes and admit patients who have been previously preadmitted in an average of less than 15 minutes and never more than 30 minutes.

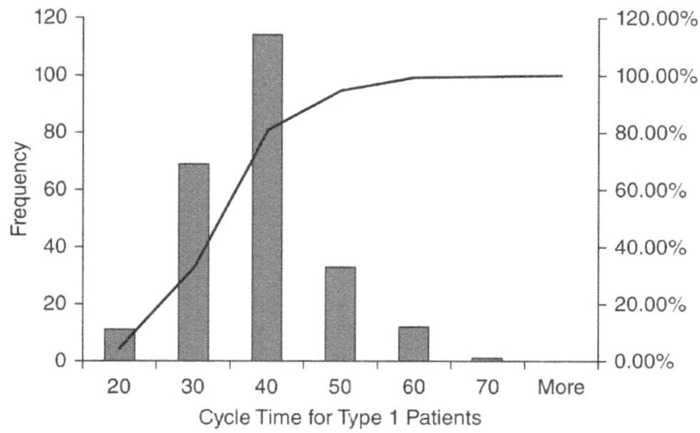

FIGURE 9.33
Histogram of cycle times for Type 1 patients.

The case for action in this situation might focus on the need to improve customer service in the highly competitive health care business. The following redesign opportunities are identified:

- All information about a patient can be captured at the admissions desk and entered into the hospital's electronic system. It is then seamlessly available from all computer terminals in the hospital.
- All paperwork and information gathering can take place during preadmission.

Therefore, the service time for Type 1 patients in the admissions desk can be minimized.

- The registration desk at the lab is unnecessary.
- It is unnecessary for patients to wait for paperwork at the nurses' station.
- Type 2 patients could walk directly from the lab to the lobby to wait for an orderly instead of going back to the admissions desk.

A model of the redesigned process is depicted in Figure 9.34. The features of this model are as follows:

- The lab registration desk has been eliminated, and the person previously working there has been reassigned to the admissions desk. Patients (Types 2 and 3) are instructed to go directly to a lab technician for service. The lab technicians can access all the information about a patient from a computer terminal. The processing time at the lab remains unchanged.
- Preadmission now includes all the paperwork and information gathering necessary to admit a patient. This reduces the service time required by Type 1 patients. It is assumed that their new service time at

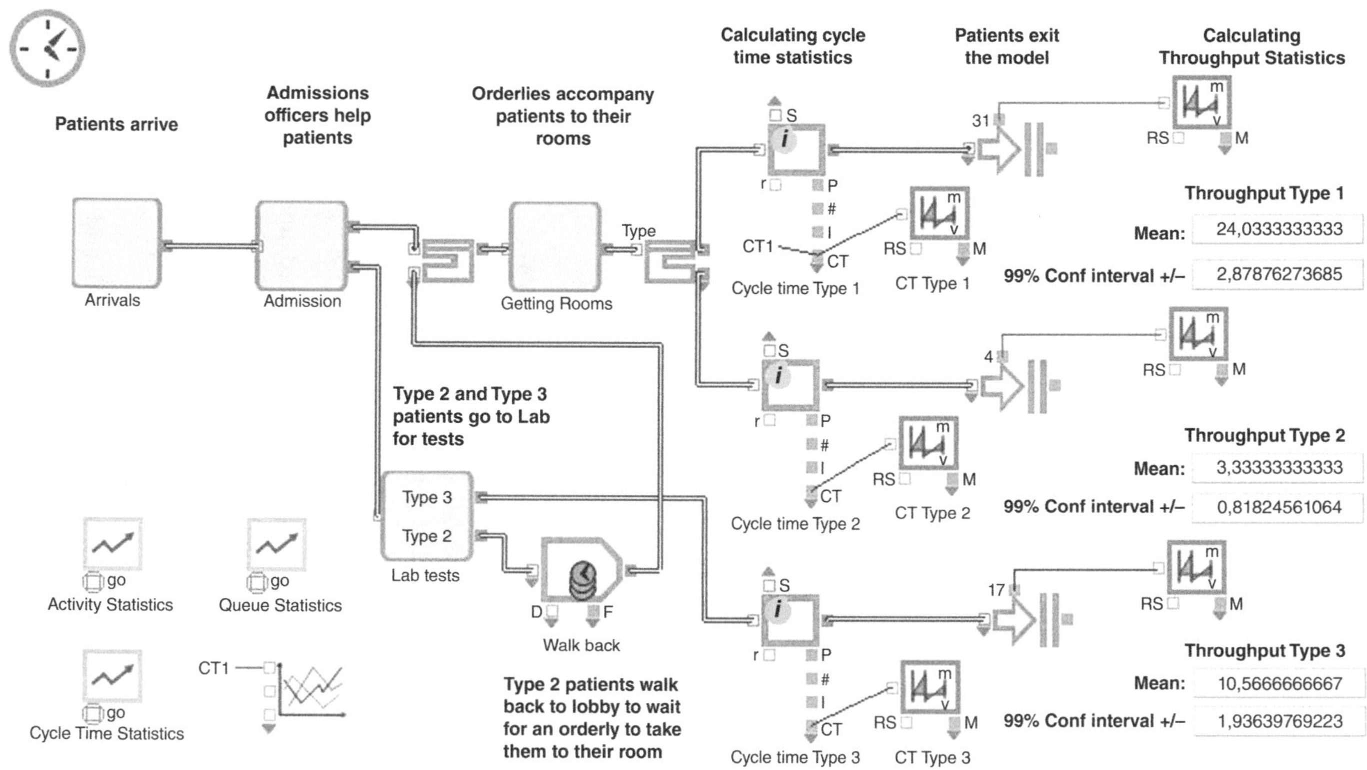

FIGURE 9.34
ExtendSim model of the redesigned admissions process.

TABLE 9.12

Estimated Mean Cycle Time and Daily Throughput with 99 Percent Confidence Intervals for the Redesigned Process

Patient Category	Cycle Time (Minutes)	Daily Throughput (Number of Patients)
Type 1	10.8 ± 0.3	24.0 ± 2.9
Type 2	50.5 ± 5.0	3.3 ± 0.8
Type 3	29.2 ± 2.9	10.6 ± 1.9

the admissions desk is normally distributed with a mean of 5 minutes and a standard deviation of 0.5 minutes. Also, the processing times for Type 2 and Type 3 patients are reduced due to more effective (electronic) data handling. Their new processing times at the admissions desk are assumed to be 10 minutes less than the original times.

- Type 2 patients are not required to go back to the admissions desk after completing the lab tests. Instead, they walk directly from the lab to the lobby, where they wait next to the admissions desk for an orderly to take them to their floor.
- Patients do not have to wait for paperwork at the nurses' station. All information is available through the local area network.

Table 9.12 summarizes the results of simulating 30 days of operation of the redesigned process. A comparison with the results for the current process in Table 9.11 clearly shows that the redesigned process produces shorter cycle times using the significance criterion of non-overlapping confidence intervals. Moreover, Table 9.12 shows that the redesigned process satisfies the goals for the average cycle times specified in the vision statement. Under the criterion of non-overlapping 99% confidence intervals, the average cycle time for Type 1 patients is less than 15 minutes, and for Type 3 patients, it is less than 35 minutes.

Table 9.13 shows cumulative frequency distributions for the cycle times of Type 1 and Type 3 patients estimated from 10 days of operation. As can be seen, no Type 1 patient experiences more than 14 minutes of cycle

TABLE 9.13

Cumulative Frequency Distributions of Cycle Times for Type 1 and Type 3 Patients in the Redesigned Process

Type 1 Patients			Type 3 Patients		
Cycle Time (Minutes)	Frequency	Cumulative (%)	Cycle Time (Minutes)	Frequency	Cumulative (%)
10	77	34.8	30	59	67.8
12	97	78.7	35	23	94.2
14	47	100.0	40	3	97.7
16	0	100.0	45	2	100.0
			50	0	100.0

time under the new design, which more than satisfies the goal of at most 30 minutes. Furthermore, no Type 3 patient experiences a cycle time longer than 45 minutes, which was the stated goal. As a matter of fact, fewer than 3 percent have cycle times above 40 minutes.

The conclusions from the simulation analysis are that the redesigned process produces much shorter cycle times than the existing process and that it achieves the goals for the cycle time performance articulated in the vision statement.

Although the redesigned process satisfies the stated cycle time goals, it is not necessarily an optimal design. A common managerial concern is the trade-off between customer service and the cost of installing new technology. One possible cost-saving strategy is to reduce the staff by operating with two admissions officers as in the original system. The cycle time statistics must be recomputed to figure out how close they are to the numerical goals specified in the vision statement. How simulation may be used to aid in optimizing business process performance is further discussed in Chapter 10.

9.7 Summary

This chapter presented some basic tools for statistical analysis of input and output simulation data. On the input side, it was shown how goodness-of-fit tests can be employed to determine suitable distributions for random variables such as interarrival times and processing times. The chapter also explored the mechanics involved in the generation of uniformly distributed random numbers and random variates from well-known theoretical distributions.

On the output side, the focus was placed on the problem of estimation and hypothesis testing. These statistical tools are necessary to be able to draw valid conclusions with respect to the performance of business processes.

The last part of the chapter focused on the analysis of two process design cases. These cases illustrate how input/output data analysis and simulation modeling in ExtendSim can be used for analyzing process performance and for comparing different design alternatives.

9.8 Training Cases

9.8.1 Case 1: Improving the X-Ray Process at County Hospital[3]

County Hospital wishes to improve the service level of its regular X-ray operation, which runs from 8 a.m. to 8 p.m. Patients have identified the total required time as their main concern with this process. Management, on the other hand, is concerned with the use of available resources. Management has created a process-improvement team to study this problem. The process might be redesigned as a result of the team's recommendations.

The team has defined the entry point to the X-ray process to be the instant a patient leaves the physician's office en route to the X-ray lab. The exit point has been defined as the instant at which the patient and the completed X-ray re-enter the physician's office.

Broadly speaking, two types of patients arrive at the X-ray process: emergency and nonemergency patients (priority levels 1 and 2, respectively). The emergency patients arrive according to a Poisson process with a mean of four patients per hour. The nonemergency patients are registered as they enter, and a sample of the arrival time data is provided in Table 9.14. Until now, no attempt has been made to further analyze this data, so there is no insight into what the arrival process looks like.

TABLE 9.14

Sample Arrival Times of Nonemergency Patients

Patient #	Time of Arrival (in Minutes from Time Zero)	Patient #	Time of Arrival (in Minutes from Time Zero)
1	6.30	31	197.89
2	10.13	32	205.50
3	17.07	33	215.42
4	17.09	34	219.95
5	23.94	35	223.50
6	26.06	36	233.33
7	27.65	37	234.89
8	29.21	38	239.20
9	41.65	39	244.29
10	44.69	40	247.29
11	49.79	41	249.90
12	60.07	42	250.25
13	70.34	43	256.34
14	70.73	44	257.90
15	74.32	45	268.97
16	84.59	46	276.82
17	91.77	47	280.43
18	95.78	48	281.94
19	98.20	49	293.23
20	117.24	50	293.57
21	122.85	51	299.79
22	130.58	52	303.75
23	137.46	53	306.58
24	139.76	54	308.13
25	142.52	55	314.06
26	150.70	56	322.84
27	151.95	57	326.51
28	154.74	58	338.21
29	157.48	59	339.91
30	193.25	60	365.79

The team has identified 12 activities in the current X-ray process (see Table 9.15), which are the same for all patient types. The only differences between patient categories are the activity times and their distributions, specified in Table 9.16.

The patient priority levels determine the service order of all the X-ray activities. Emergency patients (Priority 1) come first at the expense of nonemergency patients (Priority 2). However, after service is started, it will never be interrupted to benefit a high-priority patient.

TABLE 9.15

Activities in the Current X-ray Process

Activity	Description	Type
1	The patient leaves physician's office with instructions	Start of the X-ray process
2	The patient is taken to the lab by an orderly, on foot, in wheelchair, or lying in bed	Transportation
3	The patient is left in the waiting area outside the X-ray lab in anticipation of an X-ray technician	Waiting
4	An X-ray technician fills out a standard form based on information supplied by the physician and the patient (done outside the X-ray lab). The technician then leaves the patient, who queues up in front of the X-ray labs	Business value added
5	The patient enters the X-ray lab, undresses, and an X-ray technician takes the required X-rays (all done in the X-ray lab)	Value added
6	A dark room technician develops the X-rays. (Assume that the patient and the X-ray technician accompanies the X-rays)	Value added
7	The dark room technician and the X-ray technician check the X-rays for clarity. (Assume that the patient accompanies his/her X-rays)	Inspection
8	If X-rays are not clear, then the patient needs to go back to the waiting room in anticipation of repeating steps 5, 6, and 7. Historically, the probability of rejecting X-rays has been 25%. If the X-rays are acceptable, the patient proceeds to activity 9 while the X-rays are put in the outbox, where eventually the messenger service will pick them up	Decision
9	A patient waits for an orderly to take him/her back to the physician's office	Waiting
10	The patient is taken back to the physician's office by an orderly	Transportation
11	A messenger service transfers the X-rays to the physicians in batches of five jobs	Transportation
12	Patient and X-rays enter the physician's office together	End

TABLE 9.16

Activity Times for X-ray Process

Activity	Patient Type	Activity Time Distribution	Parameter Values (Minutes)
1	All types	Not Applicable	Not Applicable
2	Emergency patients	Uniform	Max = 9, Min = 5
	Nonemergency patients	Uniform	Max = 12, Min = 5
3	All types	Not Applicable	Not Applicable
4	All types	Uniform	Max = 6, Min = 4
5	Emergency patients	Normal	$\mu = 9, \sigma = 4$
	Nonemergency patients	Normal	$\mu = 11, \sigma = 4$
6	All types	Normal	$\mu = 12, \sigma = 5$
7	All types	Constant	Value = 2
8	All types	Constant	Value = 0
9	All types	Not Applicable	Not Applicable
10	Emergency patients	Uniform	Max = 9, Min = 5
	Nonemergency patients	Uniform	Max = 12, Min = 5
11	All types	Uniform	Max = 7, Min = 3
12	All types	Not Applicable	Not Applicable

The resource data for the X-ray process is specified in Table 9.17. The orderlies will always take one patient back from the X-ray lab when they have dropped one off. Assume that the transportation time back from the X-ray lab is the same as the transportation time to the X-ray lab. If no patient is ready to go back, the orderly will wait for 5 minutes; if no patient becomes available during this time, the orderly will return to the ward without a patient. The time for an orderly to walk back without a patient is always 5 minutes. The orderlies will never go and pick up a patient at the X-ray area without bringing another patient with them from the ward.

TABLE 9.17

Resource Data for X-ray Process

Resource	Activities	No. of Units Available
Orderlies	2 and 10	3
X-ray technician	4, 5, 6, and 7	3
X-ray lab	5	2
Dark room technician	6 and 7	2
Dark room	6	1

Part I: Analyzing the Current Process Design

1. Draw a flowchart of the current X-ray process.
2. Develop a simulation model of this process.
   - The model requires analysis of input data regarding the arrival process of nonemergency patients.
   - *Modeling Hint:* Build the model incrementally based on your flowchart. Do not try to put everything together at once and then test whether it works.
   - As a first check that everything works as it is supposed to, it is often useful to run a shorter simulation with animation. Use different symbols for different types of items, such as different types of labor and different types of jobs.
3. For a first-cut analysis, run a 1-day simulation with the random seed set at 100 using the correct activity time distributions. Look at the average cycle time, the throughput rate, the resource, the queue, and the activity statistics. What are the problems in this process?
4. Simulate 30 days of operation and compute the cycle time and daily throughput (average, standard deviation, and 95 percent confidence intervals). Also, compute the activity and resource utilization statistics and queue statistics with 95 percent confidence intervals. (Use the Statistics block in the Value library of ExtendSim.) Assume that any patients remaining in the system at the end of the day will be taken care of by the night shift. Every morning, the system is assumed to be empty. Are there any surprises when you compare these results with the ones from the first cut analysis in 3. above?
5. Assess the performance of the process using the values calculated in Question 4. Where is the bottleneck? Which are the problems for reducing the cycle time and increasing the throughput rate?

Part II: Suggest and Evaluate a New Process Design

6. Based on your insight about the current operations, identify a plausible way of reducing the average cycle time by redesigning the process. For example: What happens if the X-ray technician no longer has to stay with the patient while the X-rays are developed? What if the messenger batch size changes? Is the messenger service necessary? What if more personnel are hired? What type of personnel would be most useful?
7. Investigate the performance of the redesigned process in terms of the cycle time and the daily throughput. Also, look at the activity and resource utilization statistics and queue statistics with 95 percent confidence intervals as before. What are your conclusions? Is the new design significantly better than the old process with regard to the cycle time and throughput? Are there any obvious drawbacks?

9.8.2 Case 2: Process Modeling and Analysis in an Assembly Factory

The LeedSim factory is a traditional assembly facility working as a subcontractor to the telecommunications industry. Its main product is a specialized switchboard cabinet used in network base stations. The company has been successful on the sales side, and the orders are piling up. Unfortunately, the operations department has had some problems with reaching the desired (and necessary) productivity levels. Therefore, they have decided to seek your help to create a simulation model of the involved processes as a first step toward analyzing and improving the process design.

To find the right level of detail in the model description, they want to start with a simple model and then successively add more details until a suitable model with the right level of complexity is obtained. The simulation should be run over a 3-month (12-week) period of five 8-hour workdays per week. A schematic flowchart of the manufacturing process is shown in Figure 9.35.

After the cabinets are completed and inspected, the finished cabinets leave the factory. It is noteworthy that each workstation can handle only one item at a time. Workstations 1 through 3 cannot store items. Similarly, there is no room to store items before Workstation 4, but after it, there is room to store two assembled cabinets. At Workstation 5, there is ample space to store cabinets before the workstation, but after it, there is only room to store at most two painted cabinets. At the inspection station, there is ample space to store cabinets both before and after the station. Each cabinet that is made requires one unit each of five different components/raw materials delivered to the inbound storage area. Workstation 1 requires one unit each of raw materials 1 and 2, Workstation 2 requires one unit of raw material 3, and Workstation 3 requires one unit each of raw materials 4 and 5. Table 9.18 specifies the estimated processing times in Workstations 1 through 5. Table 9.19 shows collected inspection time data that has not yet been analyzed. This needs to be done to build a valid model of the process.

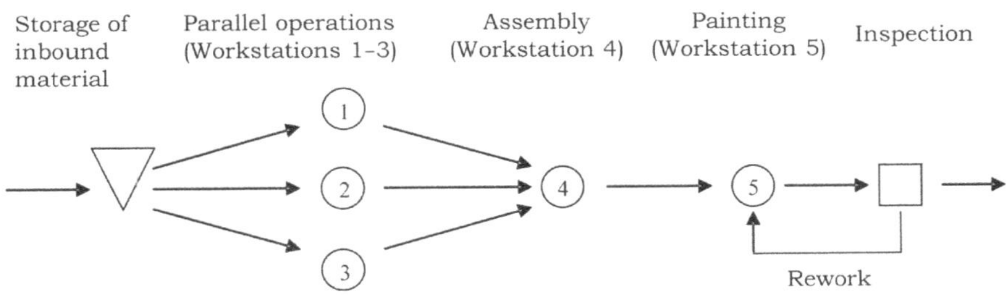

FIGURE 9.35
Flowchart of the LeedSim manufacturing process.

TABLE 9.18

Estimated Processing and Inspection Times

Processing Unit	Processing Time Distribution	Parameter Values (Hours)
Workstation 1	Triangular	Max = 3; Min = 1.5; Most likely = 2
Workstation 2	Triangular	Max = 3; Min = 1.5; Most likely = 2
Workstation 3	Triangular	Max = 3; Min = 1.5; Most likely = 2
Workstation 4	Triangular	Max = 4; Min = 2; Most likely = 3
Workstation 5	Triangular	Max = 6; Min = 3; Most likely = 4

TABLE 9.19

Observed Inspection Time Data in Minutes

No.	Inspection Time	No.	Inspection Time	No.	Inspection Time	No.	Inspection Time
1	0.994	31	0.218	61	0.343	91	1.532
2	3.084	32	1.498	62	2.129	92	2.561
3	0.169	33	2.433	63	0.756	93	6.198
4	7.078	34	1.491	64	0.991	94	1.663
5	2.440	35	0.088	65	1.001	95	0.984
6	5.546	36	0.502	66	2.070	96	0.183
7	0.201	37	2.324	67	3.216	97	1.385
8	1.185	38	0.458	68	2.037	98	0.212
9	5.308	39	1.474	69	5.358	99	0.757
10	0.989	40	1.180	70	0.024	100	1.291
11	0.590	41	0.307	71	2.397	101	0.063
12	8.476	42	5.252	72	4.718	102	3.571
13	3.676	43	6.797	73	1.478	103	7.869
14	0.504	44	2.461	74	1.089	104	0.233
15	0.016	45	0.418	75	12.196	105	0.661
16	1.392	46	0.699	76	0.109	106	0.697
17	0.552	47	0.293	77	4.355	107	4.937
18	2.059	48	4.245	78	1.158	108	0.045
19	2.858	49	1.594	79	0.003	109	1.239
20	5.982	50	0.733	80	0.137	110	0.357
21	2.337	51	0.389	81	0.293	111	1.143
22	2.426	52	1.088	82	0.193	112	3.068
23	0.252	53	1.457	83	1.263	113	0.548
24	0.290	54	0.206	84	2.249	114	3.460
25	5.139	55	0.755	85	0.689	115	1.271
26	1.727	56	1.786	86	2.376	116	3.401
27	3.859	57	0.510	87	0.729	117	3.082
28	3.356	58	3.400	88	1.408	118	0.357
29	0.884	59	1.690	89	5.199	119	2.098
30	1.992	60	2.186	90	3.286	120	0.728

The performance measure that LeedSim is most interested in is the number of cabinets produced in a 3-month period. However, they also want to keep track of the following:

- The WIP levels (measured in units of raw material)—the total as well as at different parts of the workshop (the mean and the standard deviation for a single run, and the mean with 95 percent confidence intervals in the case of multiple simulation runs)
- The inbound storage levels of the raw materials (the maximum, the mean, and the standard deviation for a single run, and the mean with 95 percent confidence intervals in the case of multiple simulation runs)
- The cycle time, measured from the instant a component arrives at the storage area until a finished cabinet leaves the factory (the mean and the standard deviation for a single run and the mean with 95 percent confidence intervals in the case of multiple simulation runs). (*Hint*: Note that all components for a particular cabinet have the same cycle time)
- The use of workstations and equipment such as the forklift truck (the mean with 95 percent confidence intervals in the case of multiple simulation runs)

Questions

1. The raw material arrives by truck once every week on Monday morning. Every shipment contains 15 units each of the five necessary components. The internal logistics within the factory is such that the transportation times for the incomplete cabinets can be neglected. However, to transport the fully assembled cabinet to and from the paint shop, a special type of forklift truck is needed. The transportation time from the assembly line to the paint shop is exponentially distributed with a mean of 45 minutes. The transportation time between the paint shop and the inspection station at the loading dock is normally distributed with a mean of 60 minutes and a standard deviation of 15 minutes. After each delivery, the forklift truck always returns to the strategically located parking area to get new instructions. The travel time for the forklift truck, without load, between the parking area and each of the workstations is negligible. Transportation of painted cabinets is prioritized. This means that whenever a forklift truck is available for a new assignment (at its parking area), and there are unpainted and painted cabinets awaiting transport, the latter will be transported first. Currently, there is one forklift truck available at the factory. For the first model, assume that all painted cabinets pass inspection, so no rework occurs.
    a. Analyze the input data for the inspection times and fit a suitable distribution. Build the model and run the simulation once with random seed = 5. How many cabinets are being produced? How is the WIP situation? What does a plot over the storage inventory levels tell us? Where is the bottleneck?

   b. Run the simulation 30 times with different random seeds. How many cabinets are being produced on average? What is the standard deviation? How is the WIP situation? Where is the bottleneck? (*Hint*: Use the Statistics block in the Value library to collect data and analyze it efficiently. For the number of units produced, connect a Mean & Variance block to the Exit block, and choose the dialog options Calculate for Multiple Simulations, and Number of Inputs-1.)

2. In reality, only 75 percent of the painted cabinets pass the inspection. If a cabinet fails the inspection, it needs to be transported back to the paint shop to be repainted. The transportation time is the same as in the opposite direction (normally distributed with a mean of 60 minutes and a standard deviation of 15 minutes). The transportation of painted cabinets that failed inspection, back to the paint shop, has a higher priority than any other transportation assignment. The forklift truck will always go back to the parking area after a completed mission. When arriving at the paint shop, repainting has higher priority than the ordinary paint jobs and follows an exponential distribution with a mean of 2 hours. Inspecting the reworked cabinets is no different from inspecting non-reworked cabinets. How does the introduction of these features affect the performance measures?
   a. Run the simulation once with random seed = 5. How many cabinets are being produced? How is the WIP situation? Where is the bottleneck?
   b. Run the simulation 30 times with different random seeds. How many cabinets are being produced on average? What is the standard deviation? How is the WIP situation? Where is the bottleneck?
3. Based on your understanding of the process, suggest a few design changes and try them out. What is your recommendation to LeedSim regarding how to improve their operations?
4. In this model, the collection of statistics data starts at time zero, which means that the system is empty when the simulation starts. It would be more accurate to run the system for a warm-up period, say 1 week, before starting to collect data. Implement this and see the difference. Does it change your conclusions?

9.8.3 Case 3: Redesign of a Credit Applications Process

The management of a mortgage company has decided with limited information that the company can save money if it reduces its staff. Before downsizing, management asks you to model the credit application process to provide reassurance that service will not be severely affected by the reduction in staff.

The mortgage company currently employs three loan agents, two of whom perform an initial review of credit applications and the third performs a second review of the applications that fail the initial review. The second review is performed as an attempt to correct the deficiencies by contacting the originating party. The process has the following characteristics:

- Approximately four to eight credit applications (and most likely six) arrive every hour.
- It takes 12–16 minutes to complete the first review.
- About 20 percent of the applications fail the first review.
- It takes 25–35 minutes to complete the second review.
- About 50 percent of the applications fail the second review.

Your task is to compare the performance of the current process with the performance of the process using two loan agents. For the downsized process, management wants the two remaining loan agents to work in either of the two reviewing steps; that is, the loan agents are not assigned to the first or the second review step, but rather they are to perform initial or second reviews as needed.

Questions

1. Create a simulation model of the current process. Use the triangular distribution to model the arrivals of credit applications and the uniform distribution for the reviewing times.
2. Simulate the process for 5 working days (40 hours) and collect the following data: use of loan agents, waiting time, and cycle time.
3. Modify the model to simulate the downsized process. Repeat Question 2 for the new model. Compare the performance of the two processes by analyzing the data collected during the simulation runs.

9.8.4 Case 4: Redesigning the Adoption Process in a Humane Society

The purpose of this project is to redesign the pet adoption process of a Humane Society. One of the main goals of the project is the development of a simulation model of the process. The adoption department of the Humane Society would like to use this model as a tool for evaluating the effect of proposed changes to the current pet adoption process. Management considers that the model significantly reduces the risks associated with this redesign project because a number of what-if scenarios can be tested before implementing any changes. Furthermore, management believes that the model can help them obtain buy-in from the employees directly involved in the process.

The modeling process consists of the following steps:

1. Flowcharting and analysis of the current process
2. Simulation modeling and validation
3. Performance analysis of the current process
4. Discussion of different options for redesigning the current process
5. Development of several scenarios
6. Modeling and testing of scenarios
7. Selection of final proposed process

A task force is created to understand the current process. After several meetings of the task force, the current process is summarized as follows.

Patrons arrive by car at the Humane Society and look for a place to park. The arrival rate is about 10 cars per hour, and the interarrival times follow an exponential distribution. If a patron finds the parking lot full, he or she leaves. If the parking lot is not full, the patron parks the car and enters the building (between 2 and 5 uniformly distributed minutes). Patrons then walk through the kennels (between 10 and 45 uniformly distributed minutes) and may decide to leave if they don't find a suitable pet. About 15 percent of patrons leave at this point. If a patron finds a suitable pet, then he or she finds out what to do next to start the adoption process (between 1 and 5 uniformly distributed minutes).

After receiving the instructions on how to proceed, the patrons count the number of people in the waiting area. (This includes people filling out the sign-in form and people waiting for a counselor.) If a patron finds fewer than 10 people in the waiting area, he or she lines up to sign in. If 10 people or more are in the waiting area, the patron leaves with a 70 percent probability. The time it takes a patron to fill out the entrance form is normally distributed with a mean of 5 minutes and a standard deviation of 1 minute, denoted Normal(5, 1) minutes. After signing in, patrons wait to be called to the counter to meet with a counselor. Counseling takes between 10 and 60 minutes, and 20 percent of the time, patrons also must talk to a supervisor after counseling, an activity that takes Normal(15, 2) minutes. After finishing counseling (or talking to the supervisor), patrons decide to stay and continue with the adoption process, or they leave. About 35 percent of the patrons leave at this stage of the adoption process.

If a patron decides to stay, he or she must wait for a kennel technician. After a brief conversation with a kennel technician (between 2 and 5 uniformly distributed minutes), the patron and the technician visit the animals and explore options (between 5 and 45 uniformly distributed minutes). After visiting the animals, about 15 percent of the patrons decide to leave. Those who stay work with the technician to check on holds, which takes Normal(15, 3) minutes. This is necessary because some of the animals are held for people who have visited them at the Humane Society and currently are considering adoption. If there are holds (10 percent of the time), the patron is asked to return later. If there are no holds, the patron receives instructions from the kennel technician, Normal(10, 2) minutes, and walks (between 1 and 2 uniformly distributed minutes) to meet with a counselor.

A counselor and the patron then fill out the adoption contract (between 10 and 45 uniformly distributed minutes). After the contract is signed, the health of the animal is checked (5 exponentially distributed minutes). After the health check is complete, the patron walks to the car with his or her pet and leaves (between 2 and 5 uniformly distributed minutes).

Note: A processing time given as a range of numbers is assumed to follow a uniform distribution. For example, if the time is between 10 and 45 minutes, the actual time is assumed to follow a uniform distribution with a minimum value of 10 minutes and a maximum value of 45 minutes. Also, a processing time of Normal(15, 3) minutes means that the actual time follows a normal distribution with a mean of 15 minutes and a standard deviation of 3 minutes.

The current process operates 10 hours per day and uses the following resources:

- Parking spaces
- Eight counselors
- One supervisor
- Five kennel technicians
- One veterinarian

It is assumed that the process is empty when the doors open every morning. Also, the doors are closed after 10 hours, but personnel stay until the last patron leaves the building. Working time beyond 10 hours is considered overtime.

Part I

1. Draw a flowchart of the current process.
2. Create a simulation model of the current process.
3. Run 30 days of operation (using 55 for the random seed number).
4. Analyze the performance of the system according to cycle time, resource utilization, ratio of number of adoptions per number of arrivals, and daily overtime.

The report for Part I of this project consists of an executive summary with the objectives and main findings. The supporting materials should include a flowchart of the current process, a printout of the simulation model (i.e., an annotated task network), and the following charts and tables:

- A frequency distribution of the cycle times observed during the 30-day simulation.
- A table of daily resource utilization with five columns, one for each resource, and 30 rows, one for each day. Three additional rows should contain the minimum, the average, and the maximum utilization for each resource type.
- A table with 30 rows, one for each day, and two columns, one for the daily ratio of the number of adoptions per number of arrivals and one for the daily overtime. Four additional rows should contain the minimum, the average, and the maximum overtime and the adoption ratio.

Part II

1. Discuss different options for redesigning the current process. For example, consider eliminating activities or performing some activities in parallel.
2. Develop a redesign scenario.
3. Model a redesigned process.
4. Predict the performance of the redesigned process in terms of cycle time, resource utilization, the ratio of adoptions to patron arrivals, and daily overtime.

The report for Part II of this project consists of an executive summary with objectives, methodology, main findings, and recommendations. The supporting materials should be the same as in the report for Part I.

9.8.5 Case 5: Performance Analysis and Improvement of an Internet Ordering Process

The management of a software company wants to study the performance of the company's web order processing. The interarrival times of orders are exponentially distributed with a mean of 7 minutes. Orders are placed through the company's webpage, using an electronic ordering form, or via e-mail. On arrival, a clerk looks for the buyer's name in the company's database. The time required to look for a name in the database is uniformly distributed between 20 and 45 seconds. If the buyer is not in the database, the clerk enters the buyer's information, which includes name, address, phone number, and e-mail address. The time required to enter the buyer's information in the database is uniformly distributed between 10 and 30 seconds. Approximately 60 percent of the time, the buyer's name is not in the database.

Some orders are for an upgrade of the software, and the others are from first-time buyers. For all practical purposes, it takes no time to figure out whether an order is for an upgrade or from a first-time buyer. Approximately 30 percent of the orders are for upgrades. If the order is for an upgrade, then the clerk simply enters a code in the electronic purchase order (PO), created when the buyer's name was entered or found in the database. (This code is later e-mailed to the customer, so he or she can download an upgrade from the company's website.) Entering the code requires an exponentially distributed time with a mean of 2 minutes because the clerk needs to verify the customer's current software version and platform. Once the code is entered, the electronic PO goes to accounting.

When the order is from a first-time buyer, the clerk checks whether the buyer wants the CD version with printed documentation or whether he or she prefers to download the software from the company's website. This requires an exponentially distributed time with a mean of 1 minute because sometimes this information has been misplaced. About 70 percent of the buyers prefer the CD version. When the CD version is preferred, the clerk needs to retrieve it from the storage room. This activity requires a normally distributed time with a mean of 5 minutes and a standard deviation of 1 minute. The clerk then prepares the software for shipping, which takes between 3 and 6 minutes (uniform distribution). If the buyer prefers to download the software, the clerk enters an appropriate code in the electronic PO. Entering the code requires an exponentially distributed time with a mean of 1 minute because a computer program sometimes is slow at generating a license for each customer.

Purchase orders for upgrades and first-time buyers go to accounting after a clerk has either entered a code for downloading or prepared the CD version for shipping. Accounting personnel charge the purchase to a credit card (exponential distribution with a mean of 2 minutes) and prepare the invoice. Data on invoice preparation times has been collected and is available in Table 9.20. Finally, the accounting personnel mail the software or e-mail the access code with the invoice. This activity requires a uniformly distributed time between 45 and 90 seconds.

Currently, the company employs two people for this process: one clerk for the initial processing and one person in charge of the accounting. However, management is considering adding one person to the process and would like to use simulation to determine where to add this new employee to obtain the maximum customer service benefit.

TABLE 9.20

Observed Invoice Preparation Times in Minutes

No.	Invoice Preparation	No.	Invoice Preparation	No.	Invoice Preparation	No.	Invoice Preparation
1	0.987	31	1.037	61	0.875	91	0.721
2	0.881	32	1.103	62	0.763	92	1.132
3	1.022	33	0.965	63	1.283	93	0.705
4	0.799	34	1.199	64	0.856	94	1.211
5	0.986	35	0.989	65	1.091	95	0.831
6	1.053	36	1.055	66	1.028	96	0.866
7	0.493	37	0.916	67	1.163	97	1.243
8	0.883	38	0.611	68	0.972	98	0.431
9	1.028	39	1.044	69	1.082	99	1.193
10	1.057	40	0.775	70	0.578	100	1.193
11	0.793	41	0.975	71	0.717	101	1.220
12	1.068	42	0.699	72	1.473	102	0.954
13	0.836	43	0.948	73	1.418	103	1.050
14	0.984	44	1.072	74	1.119	104	0.747
15	1.092	45	0.929	75	1.102	105	1.270
16	1.006	46	0.752	76	1.208	106	0.754
17	1.113	47	1.041	77	0.970	107	0.807
18	0.914	48	0.878	78	0.757	108	0.890
19	0.877	49	1.180	79	1.189	109	0.771
20	1.196	50	1.326	80	1.183	110	1.099
21	1.139	51	1.064	81	1.300	111	1.114
22	0.937	52	0.803	82	1.280	112	0.982
23	0.640	53	1.035	83	1.023	113	1.235
24	1.347	54	1.073	84	0.852	114	1.152
25	0.961	55	1.011	85	1.148	115	0.708
26	1.004	56	1.300	86	0.972	116	0.745
27	1.068	57	1.333	87	0.747	117	0.938
28	1.170	58	1.043	88	1.031	118	1.384
29	1.406	59	0.810	89	1.190	119	1.033
30	1.288	60	1.141	90	1.017	120	1.378

Questions

1. The first task is to understand this process, so you should develop a flowchart. This chart should be the first exhibit in your written report.
2. Using the flowchart as a guideline, develop a simulation model of this process. To build a valid model, the available data on how long it takes to prepare an invoice needs to be analyzed. More precisely, a suitable distribution should be fitted to the data. Because only two resource types are in this process (the clerks and the accounting personnel), your model should have only two queues.

3. Set the random seed value to 34 in the Simulation Setup of the Run menu. Run the model for 15 working days and collect the waiting time at the queues, cycle time, resource use, and WIP. (A working day consists of 8 hours.)
4. Discuss the performance of the current process based on the collected data. Include the following exhibits to support your arguments: queue statistics, line graphs of resource use, a histogram of cycle times, and the WIP value at the end of the 15 days.
5. Identify the bottleneck and add the new employee to the bottleneck. Compare the use of clerks and accounting personnel before and after adding the new employee. Also, compare the frequency distribution of cycle times before and after adding the new employee. Include line graphs for the use of resources after adding the new employee and a histogram of cycle times.

Appendix 9A: Hypothesis Testing, Confidence Intervals, and Statistical Tables

9A.1 Goodness-of-Fit Tests (Section 9.2.1)

In this section, the Chi-square test and the Kolmogorov–Smirnov test are described and illustrated by numerical examples.

9A.1.1 The Chi-Square Test

The Chi-square test is probably the most commonly used goodness-of-fit test. Its name is related to the use of the Chi-square distribution to test the significance of the statistic that measures the differences between the frequency distribution from the sample data and the theoretical distribution being tested. The procedure starts with building a histogram. The lower and upper bounds for each bin in the histogram can be used to calculate the probability that the random variable x takes on a value within a given bin. Suppose the lower bound of the ith bin is l_i, and the upper bound is u_i. Also, suppose that the theoretical distribution under consideration is the exponential distribution with a mean of $1/\mu$, a PDF denoted $f(x)$, and a CDF denoted $F(x)$. Then, the probability associated with the ith bin, p_i, is calculated as $F(u_i) - F(l_i)$ or, equivalently, as the area under $f(x)$ over the interval $l_i \leq x \leq u_i$ as shown in Figure 9.36.

The probability values for the most common theoretical probability distribution functions are available in tables. These tables can be found in most statistics books, and a few are included in Section 9A.4. Alternatively, probabilities can be computed easily using software such as Microsoft Excel.

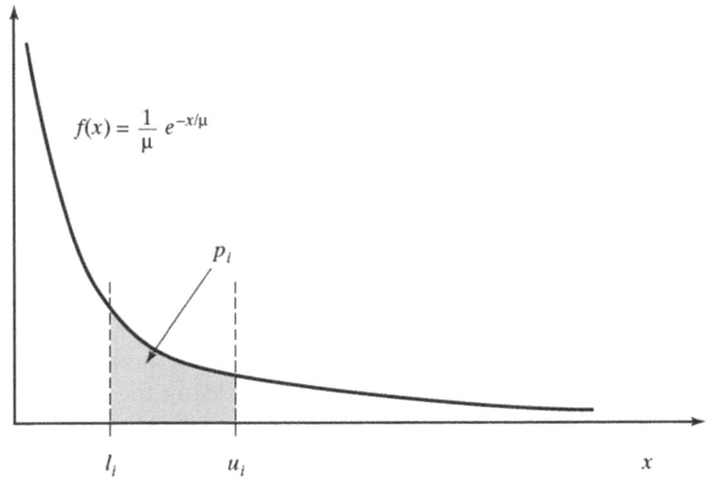

FIGURE 9.36
Probability calculation of an exponential distribution.

Suppose the lower bound of a bin is in Cell A1 of a spreadsheet, and the upper bound is in Cell A2. Also, suppose the PDF being tested is the exponential distribution with a mean of $1/\lambda$ (where λ is the estimated arrival rate of jobs to a given process). Then, the probability that the random variable X takes on a value within the range [A1, A2] defined by a particular bin can be found using the following Excel expression:

$$= \text{EXPON.DIST}(\text{A2}, \lambda, \text{TRUE}) - \text{EXPON.DIST}(\text{A1}, \lambda, \text{TRUE})$$

The form of the Excel function that returns the probability values for the exponential distribution is

$$= \text{EXPON.DIST}(x\text{-value, rate, TRUE or FALSE})$$

The first argument is a value of the random variable x. The second argument is the rate (e.g., λ for arrivals or μ for service) of the exponential distribution under consideration. The third argument indicates whether the function should return the distribution's density value (FALSE) or the cumulative probability (TRUE) associated with the given value of x. The aforementioned expression means that the cumulative probability value up to the lower bound (given in Cell A1 of the spreadsheet) is subtracted from the cumulative probability value up to the upper bound (given in Cell A2 of the spreadsheet). The statistical functions in Excel include the calculation of probability values for several well-known theoretical distributions in addition to the exponential, as indicated in Table 9.21.

TABLE 9.21

Excel Functions for Some Well-Known Theoretical Probability Distribution Functions

Theoretical Probability Distribution	Excel Function	Argument Description
Beta	BETA.DIST(x, α, β, *TRUE*, A)	α and β = shape parameters A = optional minimum value TRUE = returns the cumulative probability value
Exponential	EXPON.DIST(x, λ, TRUE)	λ = mean rate value TRUE = returns the cumulative probability value
Gamma	GAMMA.DIST(x, α, β, TRUE)	α = shape and β = scale TRUE = returns the cumulative probability value
Normal	NORM.DIST(x, μ, σ, TRUE)	μ = mean σ = standard deviation TRUE = returns the cumulative probability value
Weibull	WEIBULL.DIST(x, α, β, TRUE)	α = shape and β = scale TRUE = returns the cumulative probability value

The Excel functions in Table 9.21 return the cumulative probability value up to the value of x. Note that the default range for the beta distribution is from 0 to 1, but the distribution can be relocated and its range can be adjusted with the appropriate specification of the values for A and B. After the probability values p_i for each bin i have been calculated, the test statistic is computed in the following way:

$$\chi^2 = \sum_{i=1}^{N} \frac{(O_i - np_i)^2}{np_i}$$

where

n is the total number of observations in the sample

O_i is the number of observations in bin i

N is the number of bins

The product np_i represents the expected frequency for bin i

The statistic compares the theoretical expected frequency with the actual frequency, and the deviation is squared. The value of the χ^2 statistic is compared with the value of the Chi-square distribution with $N - k - 1$ degrees of freedom, where k is the number of estimated parameters in the theoretical distribution being tested (using $N-1$ degrees of freedom is sometimes suggested as a conservative estimate; see Law and Kelton, 2000). For example, the exponential distribution has one parameter, the mean, and the normal distribution has two parameters, the mean and the standard deviation, that may need estimation. The test assumes that the parameters are estimated using the observations in the sample. The hypothesis that the sample data come from the candidate theoretical distribution cannot be rejected if the value of the test statistic does not exceed the Chi-square value for a given level of significance. Chi-square values for the most common significance levels (0.01 and 0.05) can be found in Section 9A.4. Spreadsheet software such as Excel also can be used to find Chi-square values. The comparison for a significance level of 5 percent can be done as follows:

Do not reject the hypothesis if the test statistic $\chi^2 \leq \text{CHISQ.INV.RT}\,(0.05, N-k-1)$

The Chi-square test requires that the expected frequency in each bin exceeds 5; that is, the test requires $np_i > 5$ for all bins. If the requirement is not satisfied for a given bin, the bin must be combined with an adjacent bin. The value of N represents the number of bins after any merging has been done to meet the specified requirement.

Clearly, the results of the Chi-square test depend on the histogram and the number and size of the bins used. Because the initial histogram is often built intuitively, the results may vary depending on the person performing the test. In a sense, the histogram must capture the "true" grouping of the data for the test to be reliable. Regardless of this shortcoming, the Chi-square goodness-of-fit test is popular and often gives reliable results. The precision (or strength) of the test is improved if the bins are chosen so that the probability of finding an observation in a given bin is the same for all bins. A common approach is to reconstruct the histogram accordingly before performing the Chi-square test. It should be noted that other tests exist (such as the Kolmogorov–Smirnov test) that are applied directly to the sample data without the need for constructing a histogram first. At the same time, some of these tests are better suited for continuous distributions and tend to be unreliable for discrete data. The applicability of a test depends on each situation. The Chi-square test is illustrated by the following example.

Example 9.14

Consider the interarrival time data in Example 9.1. The shape of the histogram, along with the fact that the sample mean is almost equal to the sample standard deviation, indicates that the exponential distribution might be a good fit. The exponential distribution is fully characterized by the mean value (because the standard deviation is equal to the mean). The sample mean of 104.55 is used as an estimate of the true but unknown population mean. Table 9.22 shows the calculations necessary to perform the Chi-square test on the interarrival time data.

TABLE 9.22
Chi-Square Test Calculations

Bin	Observed Frequency (O_i)	Expected Frequency (np_i)	$\frac{(O_i - np_i)^2}{np_i}$
0–59	23	25.88	0.32
60–119	20	14.58	2.01
120–179	7	8.21	0.18
180–239	3	4.63	0.02
240–299	4	2.61	
300–359	2	1.47	
360–∞	1	0.83	
Totals	60		2.54

The expected frequency in Bins 4 to 7 is less than 5, so these bins are combined, and the statistic is calculated using the sum of the observed frequencies compared with the sum of the expected frequencies in the combined bins. The total number of valid bins is then reduced to four, and the degrees of freedom for the associated Chi-square distribution are $N - k - 1 = 4 - 1 - 1 = 2$. The level of significance for the test is set at 5 percent. Because $\chi^2 = 2.54$ is less than the Chi-square value CHISQ.INV.RT(0.05,2) = 5.99, the hypothesis that the sample data come from the exponential distribution with a mean of 104.55 cannot be rejected. In other words, the test indicates that the exponential distribution provides a reasonable fit for the field data. ■

9A.1.2 The Kolmogorov–Smirnov Test

The Kolmogorov–Smirnov (KS) test is another frequently used goodness-of-fit test. One advantage of this test over the Chi-square test is that it does not require a histogram of the field data. A second advantage of the KS test is that it gives fairly reliable results even when dealing with small data samples. On the other hand, the major disadvantage of the KS test is that it applies only to continuous distributions.

In theory, all the parameters of the candidate distribution should be known to apply the KS test (as originally proposed). This is another major disadvantage of this test because the parameter values typically are not known, and analysts can only hope to estimate them using field data. This limitation has been overcome in a modified version of the test, which allows the estimation of the parameters using the field data. Unfortunately, this modified form of the KS test is reliable only when testing the goodness of fit of the normal, exponential, or Weibull distribution. However, in practice, the KS test is applied to other continuous distributions and even discrete data. The results of these tests tend to reject the hypothesis more often than desired, eliminating some PDFs from consideration when in fact they could be a good fit for the field data.

The KS test starts with an empirical cumulative distribution of the field data. This empirical distribution is compared with the CDF of a theoretical PDF. Suppose that $x_1, \ldots, x_n$ are the values of the sample data ordered from the smallest to the largest. The empirical probability distribution has the following form:

$$F_n(x) = \frac{Number\ of\ x_i \leq x}{n}$$

Then, the function is such that $F_n(x_i) = i/n$ for $I = 1, \ldots, n$. The test is based on measuring the largest absolute deviation between the theoretical and empirical cumulative probability distribution functions for every given value of x. The calculated deviation is compared with a tabulated KS value (see Table 9.23) to determine whether or not the deviation can be due to randomness; in this

TABLE 9.23

Kolmogorov–Smirnov Critical Values

Degrees of Freedom (n)	$D_{0.10}$	$D_{0.05}$	$D_{0.01}$
1	0.950	0.975	0.995
2	0.776	0.842	0.929
3	0.642	0.708	0.828
4	0.564	0.624	0.733
5	0.510	0.565	0.669
6	0.470	0.521	0.618
7	0.438	0.486	0.577
8	0.411	0.457	0.543
9	0.388	0.432	0.514
10	0.368	0.410	0.490
11	0.352	0.391	0.468
12	0.338	0.375	0.450
13	0.325	0.361	0.433
14	0.314	0.349	0.418
15	0.304	0.338	0.404
16	0.295	0.328	0.392
17	0.286	0.318	0.381
18	0.278	0.309	0.371
19	0.272	0.301	0.363
20	0.264	0.294	0.356
25	0.240	0.270	0.320
30	0.220	0.240	0.290
35	0.210	0.230	0.270
Over 35	$\frac{1.22}{\sqrt{n}}$	$\frac{1.36}{\sqrt{n}}$	$\frac{1.63}{\sqrt{n}}$

case, one would not reject the hypothesis that the sample data come from the candidate distribution. The test consists of the following steps.

1. Order the sample data from smallest to largest value.
2. Compute D^+ and D^- using the theoretical cumulative distribution function $\hat{F}(x)$.

$$D^+ = \max_{1 \le i \le n}\{D_i^+\} = \max_{1 \le i \le n}\left\{\frac{i}{n} - \hat{F}(x_i)\right\}$$

$$D^- = \max_{1 \le i \le n}\{D_i^-\} = \max_{1 \le i \le n}\left\{\hat{F}(x_i) - \frac{i-1}{n}\right\}$$

3. Calculate $D = \max(D^-, D^+)$.
4. Find the KS value for the specified level of significance and the sample size n.
5. If the critical KS value is greater than or equal to D, then the hypothesis that the field data come from the theoretical distribution cannot be rejected.

The following example illustrates the application of the KS test.

Example 9.15

This example uses the KS test to determine whether the exponential distribution is a good fit for 10 interarrival times (in minutes) collected in a service operation: 3.10, 0.20, 12.1, 1.4, 0.05, 7, 10.9, 13.7, 5.3, and 9.1. Table 9.24 shows the calculations needed to compute the value of the D statistic.

Note that the x_i values in Table 9.24 are ordered from the smallest to the largest, that the average interarrival time calculated from the sample

TABLE 9.24

Calculations for KS Test

i	x_i	$\hat{F}(x_i)$	i/n	Di^+	Di^-
1	0.05	0.0079	0.1	0.0921	0.0079
2	0.20	0.0313	0.2	0.1687	–0.0687
3	1.40	0.1997	0.3	0.1003	–0.0003
4	3.10	0.3894	0.4	0.0106	0.0894
5	5.30	0.5697	0.5	–0.0697	0.1697
6	7.00	0.6717	0.6	–0.0717	0.1717
7	9.10	0.7649	0.7	–0.0649	0.1649
8	10.90	0.8235	0.8	–0.0235	0.1235
9	12.10	0.8542	0.9	0.0458	0.0542
10	13.70	0.8869	1	0.1131	–0.0131
			Max:	0.1687	0.1717

data is 6.285, and that the theoretical CDF values for the corresponding exponential distribution are obtained from the expression:

$$\hat{F}(x_i) = 1 - e^{-x_i/6.285}$$

Because $D^+ = 0.1687$ and $D^- = 0.1717$, then $D = 0.1717$. The critical KS value for a significance level of 0.05 and a sample size of 10 is 0.410 (see Table 9.23.) The KS test indicates that the hypothesis that the underlying probability distribution of the field data is exponential with a mean of 6.285 minutes cannot be rejected. ■

Although no precise recipes can be followed to choose the best goodness-of-fit test for a given situation, experts agree that for large samples (i.e., samples with more than 30 observations) of discrete values, the Chi-square test is more appropriate than the KS test. For smaller samples of a continuous random variable, the KS test is recommended. However, the KS test has been applied successfully to discrete probability distributions, so it cannot be ruled out when dealing with discrete data.

9A.2 Confidence Interval for a Population Proportion (Section 9.5.3)

Some output variables from a process simulation are proportions: for example, the proportions of customers serviced on time, the percentage of jobs going through a specialized operation, the percentage of jobs requiring rework, and so on. In general, a proportion is a value representing the percentage of one type of outcome (often referred to as *success*) in a number of trials.

The true but unknown probability of success is denoted by p. Therefore, the probability of failure is $1 - p$. The resulting random variable takes on two values: zero with probability $1 - p$ and 1 with probability p. A random variable of this kind follows a Bernoulli distribution, for which the mean value is p and the variance is $p(1 - p)$. In a sample of n observations, the ratio of the number of successes to the total number of trials, $\bar{p}$, is a point estimate of p. The distribution of the $\bar{p}$ statistic is approximately normal for n larger than 10 and both $np > 5$ and $n(1 - p) > 5$. A $(1 - \alpha)$ percent confidence interval for p can be constructed from the following probability statement:

$$P\left\{-Z_{\alpha/2} \le \frac{\bar{p} - p}{\sqrt{p(1-p)/n}} \le Z_{\alpha/2}\right\} = 1 - \alpha$$

Note that this probability statement requires the knowledge of p to be able to calculate the standard deviation. Because p is the parameter to be estimated, $\bar{p}$ can be used to approximate p for the calculation of the standard deviation when the sample size is sufficiently large. Substituting p with $\bar{p}$ in the denominator and rearranging the elements of the probability statement gives the following confidence interval for a proportion:

$$\bar{p} - Z_{\alpha/2}\sqrt{\frac{\bar{p}(1-\bar{p})}{n}} \le p \le \bar{p} + Z_{\alpha/2}\sqrt{\frac{\bar{p}(1-\bar{p})}{n}}$$

Example 9.16

Consider the cycle time data in Table 9.7. Suppose the proportion p of interest is the percentage of jobs with cycle times greater than 7. Suppose also that the goal is to construct a 99 percent confidence interval for p. Table 9.7 shows 28 observations with cycle times greater than 7; therefore, $\bar{p} = 0.56$. The resulting confidence interval is as follows:

$$0.56 - 2.576\sqrt{\frac{0.56(1-0.56)}{50}} \le p \le 0.56 + 2.576\sqrt{\frac{0.56(1-0.56)}{50}}$$

$$0.379 \le p \le 0.741$$

Note that the interval width is reduced as the number of observations in the sample is increased, just as in the case of the confidence interval for a population mean. The value of 2.576 in the aforementioned calculation is the Z value (obtained from Excel) that yields a cumulative probability of 0.995 in a standard normal distribution. ■

9A.3 Hypothesis Testing (Section 9.5.5)

A hypothesis states a certain relationship for an unknown population parameter that might or might not be true. This relationship is known as the *null hypothesis* (denoted by H_0). Because the test is based on estimated values, it is possible to commit errors. When the null hypothesis is true and the test rejects it, it is said to have committed a Type I error. A Type II error occurs when the test fails to reject a false null hypothesis. The level of significance of the test is denoted by α and is interpreted as the maximum risk associated with committing a Type I error. The confidence level of the test is $1 - \alpha$.

Suppose $\bar{x}$ is the mean and s is the standard deviation estimated from a sample of n observations. Also, suppose that we would like to test the following hypothesis:

$$H_0: \mu = a$$

$$H_A: \mu \ne a$$

Then, the null hypothesis is rejected (with a significance level of α) if the following relationship does not hold:

$$-Z_{\alpha/2} \le \frac{\bar{x} - a}{s/\sqrt{n}} \le Z_{\alpha/2}$$

TABLE 9.25

One-Tail Hypothesis Test on One Mean

Hypothesis test	Reject null hypothesis if ...
$H_0: \mu \geq a$ $H_A: \mu < a$	$\frac{\bar{x} - a}{s/\sqrt{n}} < -Z_\alpha$
$H_0: \mu \leq a$ $H_A: \mu > a$	$\frac{\bar{x} - a}{s/\sqrt{n}} > Z_\alpha$

This is a so-called *two-tail test* on a population mean. The two related one-tail tests are shown in Table 9.25.

In addition to testing whether a population mean is equal to, greater than, or less than a given value, it is sometimes desirable to be able to compare the difference between two means. This is particularly useful in the context of process design, where it might be necessary to show that a new design is better than an existing design in terms of a key output variable. Suppose μ_1 is the population mean of a new process design and that its value is estimated with $\bar{x}_1$. Suppose also that μ_2 is the population mean of the existing process design, and its value is estimated with $\bar{x}_2$. Then, the following two-tail test can be formulated:

$$H_0 : \mu_1 - \mu_2 = a$$
$$H_A : \mu_1 - \mu_2 \neq a$$

The null hypothesis is rejected (with a significance level of α) if the following relationship does not hold:

$$-Z_{\alpha/2} \leq \frac{\bar{x}_1 - \bar{x}_2 - a}{\sqrt{\frac{s_1}{n_1} + \frac{s_2}{n_2}}} \leq Z_{\alpha/2}$$

where n_1 and n_2 are the sample sizes for each population, respectively. Also, s_1 and s_2 are the sample standard deviations for the populations. One-tail tests similar to those in Table 9.25 can be formulated for the difference between two means, as shown in Table 9.26.

TABLE 9.26

One-Tail Hypothesis Test on the Differences of Two Means

Hypothesis test	Reject null hypothesis if ...
$H_0: \mu_1 - \mu_2 \geq a$ $H_A: \mu_1 - \mu_2 < a$	$\frac{\bar{x}_1 - \bar{x}_2 - a}{\sqrt{\frac{s_1^2}{n_1} + \frac{s_2^2}{n_2}}} < -Z_\alpha$
$H_0: \mu_1 - \mu_2 \leq a$ $H_A: \mu_1 - \mu_2 > a$	$\frac{\bar{x}_1 - \bar{x}_2 - a}{\sqrt{\frac{s_1^2}{n_1} + \frac{s_2^2}{n_2}}} > Z_\alpha$

Example 9.17

Consider the cycle time data in Table 9.7. A new design has been proposed for the process, and a computer simulation model has been built. The cycle times in Table 9.27 were obtained from running the simulation model of the proposed process design.

TABLE 9.27

Cycle Times for a Proposed New Process Design

2.7	10.5	5.7	9.5	3.3	8.8	7.1	11.8	5.5	7.6
5.8	5.5	4.0	2.8	5.1	6.1	3.5	2.8	11.3	6.1
8.9	11.5	2.2	7.2	7.7	7.9	5.5	5.6	4.2	7.7
9.5	2.7	6.1	7.8	4.2	8.8	3.9	11.0	8.3	8.6
11.3	2.2	8.8	7.5	2.0	3.9	9.8	2.9	8.2	7.9

Does the mean cycle time in the proposed process represent an improvement over the mean cycle time of the existing process? To find out, one can test the null hypothesis $\mu_1 - \mu_2 \geq 0$, where μ_1 is the true but unknown mean cycle time for the existing process and μ_2 is the true but unknown mean cycle time for the proposed process. Compared with the expressions in Table 9.26, this corresponds to a one-tail test with $a = 0$. Assume that a significance level of 5 percent will be used for this test.

Using the sample data, calculate $\bar{x}_1 = 7.418$, $\bar{x}_2 = 6.586$, $s_1^2 = 2.450$, and $s_2^2 = 7.981$.

The observed value of the test statistic is then calculated as follows:

$$Z = \frac{7.418 - 6.586 - 0}{\sqrt{\dfrac{2.450}{50} + \dfrac{7.981}{50}}} \approx 1.82$$

From the normal distribution table in Section 9A.4, one can obtain $Z_{0.05} = 1.645$. Because $Z > Z_{0.05}$, the null hypothesis can be rejected. This conclusion is reached at the confidence level of 95 percent. ■

The Data Analysis Toolpack of Microsoft Excel provides a tool for performing hypothesis testing directly on an electronic spreadsheet. Assume that the data in Table 9.7 is copied onto an Excel spreadsheet in the range A1:A50. Also, assume that the data in Table 9.27 are copied onto the same Excel spreadsheet in the range B1:B50. Select the Data Analysis item in the Tools menu under the Data tab in the Excel menu bar. Then, select *t*-Test: Two-Sample Assuming Unequal Variances. Note that at this point, one cannot choose the *Z* test because this test assumes that the variances are known. The estimates of the variance could be entered in the corresponding text boxes; however, it is preferable to let Excel calculate the variances and use the *t* statistic. (Recall that the values of the *t* distribution with more than 30 degrees of freedom are similar to the values of the standard normal distribution.) Figure 9.37 shows the completed dialog for the selected *t*-test. Table 9.28 shows the results of the *t*-test.

TABLE 9.28

Results Obtained from Running the Excel Function *t*-Test: Two-Sample Assuming Unequal Variances

Characteristics	Variable 1	Variable 2
Mean	7.418	6.586
Variance	2.450	7.981
Observations	50	50
Hypothesized mean difference	0	
Degrees of freedom	76	
t statistic	1.822	
$P(T \leq t)$ one-tail	0.036	
t Critical one-tail	1.665	
$P(T \leq t)$ two-tail	0.072	
t Critical two-tail	1.992	

The *t* statistic of 1.822 is identical to the *Z* statistic that was used before. The *t* critical one-tail value of 1.665 is similar to $Z_{0.05} = 1.645$. The one-tail test shows the rejection of the null hypothesis. It also shows that the null hypothesis would not be rejected at a significance level of 3.6 percent (see the $P(T \leq t)$ one-tail value in Table 9.28). The two-tail test shows that the null hypothesis $\mu_1 - \mu_2 = 0$ cannot be rejected because the *t* statistic (1.822) falls within the range (–1.992, 1.992). The two-tail null hypothesis would be rejected at a level of significance of 7.2 percent (as specified by the two-tail *p*-value in Table 9.28).

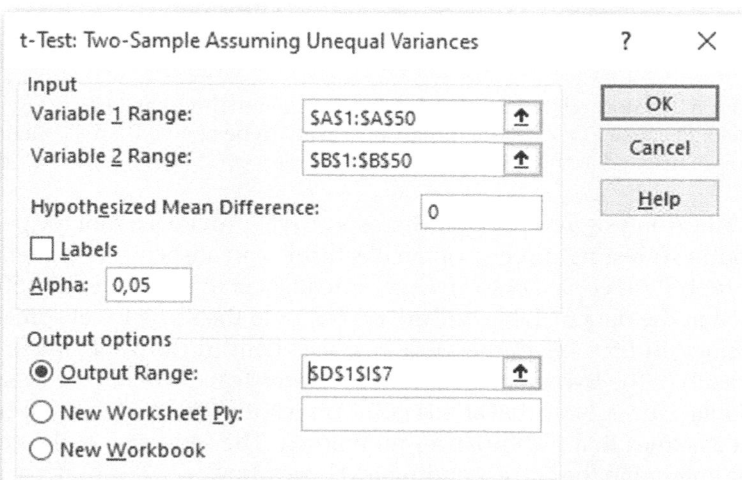

FIGURE 9.37

The Excel input dialog for the function *t-Test: Two-Sample Assuming Unequal Variances* used for investigating the difference between two population means.

9A.4 Statistical Tables

Statistical tables for the Chi-square distribution, the standardized normal distribution, and the Student *t* distribution are provided in Tables 9.29 through 9.31.

TABLE 9.29

Chi-Square Distribution Table

n	$\chi^2_{0.995}$	$\chi^2_{0.99}$	$\chi^2_{0.975}$	$\chi^2_{0.95}$	$\chi^2_{0.90}$
1	7.88	6.63	5.02	3.84	2.71
2	10.60	9.21	7.38	5.99	4.61
3	12.84	11.34	9.35	7.81	6.25
4	14.86	13.28	11.14	9.49	7.78
5	16.75	15.09	12.83	11.07	9.24
6	18.55	16.81	14.45	12.59	10.64
7	20.28	18.48	16.01	14.07	12.02
8	21.95	20.09	17.53	15.51	13.36
9	23.59	21.67	19.02	16.92	14.68
10	25.19	23.21	20.48	18.31	15.99
11	26.76	24.73	21.92	19.68	17.28
12	28.30	26.22	23.34	21.03	18.55
13	29.82	27.69	24.74	22.36	19.81
14	31.32	29.14	26.12	23.68	21.06
15	32.80	30.58	27.49	25.00	22.31
16	34.27	32.00	28.85	26.30	23.54
17	35.72	33.41	30.19	27.59	24.77
18	37.16	34.81	31.53	28.87	25.99
19	38.58	36.19	32.85	30.14	27.20
20	40.00	37.57	34.17	31.41	28.41
21	41.40	38.93	35.48	32.67	29.62
22	42.80	40.29	36.78	33.92	30.81
23	44.18	41.64	38.08	35.17	32.01
24	45.56	42.98	39.36	36.42	33.20
25	46.93	44.31	40.65	37.65	34.38
26	48.29	45.64	41.92	38.89	35.56
27	49.65	46.96	43.19	40.11	36.74
28	50.99	48.28	44.46	41.34	37.92
29	52.34	49.59	45.72	42.56	39.09
30	53.67	50.89	46.98	43.77	40.26
40	66.77	63.69	59.34	55.76	51.81
50	79.49	76.15	71.42	67.50	63.17
60	91.95	88.38	83.30	79.08	74.40
70	104.21	100.43	95.02	90.53	85.53
80	116.32	112.33	106.63	101.88	96.58
90	128.30	124.12	118.14	113.15	107.57
100	140.17	135.81	129.56	124.34	118.50

TABLE 9.30

Distribution Table for the Standardized Normal Distribution (Mean = 0 and Standard Deviation = 1)

Z	**0.00**	**0.01**	**0.02**	**0.03**	**0.04**	**0.05**	**0.06**	**0.07**	**0.08**	**0.09**
0.0	0.5000	0.5040	0.5080	0.5120	0.5160	0.5199	0.5239	0.5279	0.5319	0.5359
0.1	0.5398	0.5438	0.5478	0.5517	0.5557	0.5596	0.5636	0.5675	0.5714	0.5753
0.2	0.5793	0.5832	0.5871	0.5910	0.5948	0.5987	0.6026	0.6064	0.6103	0.6141
0.3	0.6179	0.6217	0.6255	0.6293	0.6331	0.6368	0.6406	0.6443	0.6480	0.6517
0.4	0.6554	0.6591	0.6628	0.6664	0.6700	0.6736	0.6772	0.6808	0.6844	0.6879
0.5	0.6915	0.6950	0.6985	0.7019	0.7054	0.7088	0.7123	0.7157	0.7190	0.7224
0.6	0.7257	0.7291	0.7324	0.7357	0.7389	0.7422	0.7454	0.7486	0.7517	0.7549
0.7	0.7580	0.7611	0.7642	0.7673	0.7704	0.7734	0.7764	0.7794	0.7823	0.7852
0.8	0.7881	0.7910	0.7939	0.7967	0.7995	0.8023	0.8051	0.8078	0.8106	0.8133
0.9	0.8159	0.8186	0.8212	0.8238	0.8264	0.8289	0.8315	0.8340	0.8365	0.8389
1.0	0.8413	0.8438	0.8461	0.8485	0.8508	0.8531	0.8554	0.8577	0.8599	0.8621
1.1	0.8643	0.8665	0.8686	0.8708	0.8729	0.8749	0.8770	0.8790	0.8810	0.8830
1.2	0.8849	0.8869	0.8888	0.8907	0.8925	0.8944	0.8962	0.8980	0.8997	0.9015
1.3	0.9032	0.9049	0.9066	0.9082	0.9099	0.9115	0.9131	0.9147	0.9162	0.9177
1.4	0.9192	0.9207	0.9222	0.9236	0.9251	0.9265	0.9279	0.9292	0.9306	0.9319
1.5	0.9332	0.9345	0.9357	0.9370	0.9382	0.9394	0.9406	0.9418	0.9429	0.9441
1.6	0.9452	0.9463	0.9474	0.9484	0.9495	0.9505	0.9515	0.9525	0.9535	0.9545
1.7	0.9554	0.9564	0.9573	0.9582	0.9591	0.9599	0.9608	0.9616	0.9625	0.9633
1.8	0.9641	0.9649	0.9656	0.9664	0.9671	0.9678	0.9686	0.9693	0.9699	0.9706
1.9	0.9713	0.9719	0.9726	0.9732	0.9738	0.9744	0.9750	0.9756	0.9761	0.9767
2.0	0.9772	0.9778	0.9783	0.9788	0.9793	0.9798	0.9803	0.9808	0.9812	0.9817
2.1	0.9821	0.9826	0.9830	0.9834	0.9838	0.9842	0.9846	0.9850	0.9854	0.9857
2.2	0.9861	0.9864	0.9868	0.9871	0.9875	0.9878	0.9881	0.9884	0.9887	0.9890
2.3	0.9893	0.9896	0.9898	0.9901	0.9904	0.9906	0.9909	0.9911	0.9913	0.9916
2.4	0.9918	0.9920	0.9922	0.9925	0.9927	0.9929	0.9931	0.9932	0.9934	0.9936
2.5	0.9938	0.9940	0.9941	0.9943	0.9945	0.9946	0.9948	0.9949	0.9951	0.9952
2.6	0.9953	0.9955	0.9956	0.9957	0.9959	0.9960	0.9961	0.9962	0.9963	0.9964
2.7	0.9965	0.9966	0.9967	0.9968	0.9969	0.9970	0.9971	0.9972	0.9973	0.9974
2.8	0.9974	0.9975	0.9976	0.9977	0.9977	0.9978	0.9979	0.9979	0.9980	0.9981
2.9	0.9981	0.9982	0.9982	0.9983	0.9984	0.9984	0.9985	0.9985	0.9986	0.9986
3.0	0.9987	0.9987	0.9987	0.9988	0.9988	0.9989	0.9989	0.9989	0.9990	0.9990
3.1	0.9990	0.9991	0.9991	0.9991	0.9992	0.9992	0.9992	0.9992	0.9993	0.9993
3.2	0.9993	0.9993	0.9994	0.9994	0.9994	0.9994	0.9994	0.9995	0.9995	0.9995
3.3	0.9995	0.9995	0.9995	0.9996	0.9996	0.9996	0.9996	0.9996	0.9996	0.9997
3.4	0.9997	0.9997	0.9997	0.9997	0.9997	0.9997	0.9997	0.9997	0.9997	0.9998
3.5	0.9998	0.9998	0.9998	0.9998	0.9998	0.9998	0.9998	0.9998	0.9998	0.9998

TABLE 9.31

Student *t* Distribution Table

n	$t_{0.995}$	$t_{0.99}$	$t_{0.975}$	$t_{0.95}$	$t_{0.90}$
1	63.656	31.821	12.706	6.314	3.078
2	9.925	6.965	4.303	2.920	1.886
3	5.841	4.541	3.182	2.353	1.638
4	4.604	3.747	2.776	2.132	1.533
5	4.032	3.365	2.571	2.015	1.476
6	3.707	3.143	2.447	1.943	1.440
7	3.499	2.998	2.365	1.895	1.415
8	3.355	2.896	2.306	1.860	1.397
9	3.250	2.821	2.262	1.833	1.383
10	3.169	2.764	2.228	1.812	1.372
11	3.106	2.718	2.201	1.796	1.363
12	3.055	2.681	2.179	1.782	1.356
13	3.012	2.650	2.160	1.771	1.350
14	2.977	2.624	2.145	1.761	1.345
15	2.947	2.602	2.131	1.753	1.341
16	2.921	2.583	2.120	1.746	1.337
17	2.898	2.567	2.110	1.740	1.333
18	2.878	2.552	2.101	1.734	1.330
19	2.861	2.539	2.093	1.729	1.328
20	2.845	2.528	2.086	1.725	1.325
21	2.831	2.518	2.080	1.721	1.323
22	2.819	2.508	2.074	1.717	1.321
23	2.807	2.500	2.069	1.714	1.319
24	2.797	2.492	2.064	1.711	1.318
25	2.787	2.485	2.060	1.708	1.316
26	2.779	2.479	2.056	1.706	1.315
27	2.771	2.473	2.052	1.703	1.314
28	2.763	2.467	2.048	1.701	1.313
29	2.756	2.462	2.045	1.699	1.311
30	2.750	2.457	2.042	1.697	1.310
∞	2.576	2.326	1.960	1.645	1.282

Exercises

1. Use the Chi-square goodness-of-fit test and a significance level of 0.1 to test the set of interarrival times in Table 9.32 for possible fit to the exponential distribution.

TABLE 9.32

Interarrival Times for Exercise 9.1

1.9	5.3	1.1	4.7	17.8	44.3
9.0	18.9	114.2	47.3	47.1	11.2
60.1	38.6	107.6	56.4	10.0	31.4
58.6	5.5	11.8	62.4	24.3	0.9
44.5	115.8	2.0	50.3	21.1	2.6

2. Use the Chi-square goodness-of-fit test and a significance level of 0.05 to test the set of service times in Table 9.33 for possible fit to the normal distribution.

TABLE 9.33

Service Times for Exercise 9.2

8.9	13.6	21.6	20.8	19.9	20.6
21.0	27.7	10.0	27.7	13.1	25.5
22.7	33.9	14.5	29.0	9.9	26.8
18.7	17.3	14.1	29.0	14.2	24.3
31.2	25.3	27.3	23.4	18.4	20.0

3. An analyst is interested in determining whether the Weibull distribution with parameters $\alpha = 6$ and $\beta = 10$ is a good fit to the data set in Table 9.34. Use Microsoft Excel to perform a Chi-square goodness-of-fit test to help the analyst make a decision.

TABLE 9.34

Data Set for Exercise 9.3

11.14	11.11	8.24	7.52	9.53	10.85
6.10	7.13	9.83	12.23	9.97	8.05
11.18	7.62	10.41	9.33	7.41	7.48
6.81	9.03	9.09	8.62	10.23	9.87
8.46	10.87	8.52	9.74	9.60	9.40
10.30	10.95	8.09	11.45	10.45	6.99
9.74	7.66	8.30	6.11	5.08	12.38
11.77	9.52	9.52	6.59	10.32	4.19
5.69	11.09	11.31	9.18	8.19	9.07
8.51	9.39	10.00	10.10	9.10	12.00

4. Use the Kolmogorov–Smirnov test to examine on the significance level of 0.01 whether the observations in Table 9.35 are random numbers uniformly distributed between 0 and 1.

TABLE 9.35

Data Set for Exercise 9.4

0.385	0.855	0.309	0.597	0.713	0.660
0.137	0.396	0.238	0.657	0.583	0.194

5. Apply the Kolmogorov–Smirnov test to the data in Exercise 9.1.
6. Apply the Kolmogorov–Smirnov test to the data in Exercise 9.2.
7. Consider the data in Exercise 9.4. Perform the runs test to determine with a 95 percent confidence level whether or not the numbers in the sequence (ordered by rows) are independent.
8. The data set in Table 9.36 consists of cycle times of jobs in a given process. Perform the runs test to determine with a 99 percent confidence level whether or not the cycle times (ordered by columns) are independent.

TABLE 9.36

Cycle Times for Exercise 9.8

10.76	18.73	10.55	8.17	17.42	15.22
8.54	13.09	13.53	14.30	12.32	10.75
12.52	25.10	18.23	14.71	13.23	8.22
10.19	20.39	12.80	36.53	14.14	11.34
13.35	13.88	11.06	29.41	14.78	10.97

9. Starting with a seed of 3461, generate 10 four-digit random numbers using the mid-square method.
10. Use the linear congruential method with parameters $Z_0 = 79$, $a = 56214$, $c = 17$, and $m = 999$ to generate 10 three-digit random numbers.
11. Perform the following tests on the numbers generated in Exercise 9.10:
    a. Test for randomness using Kolmogorov–Smirnov with a significance level of 0.05.
    b. Test for independence using the runs test with a 0.05 level of significance.
12. Use the data in Exercise 9.4 to generate random variates from the discrete probability distribution function in Table 9.37.

TABLE 9.37

Probability Distribution for Exercise 9.12

Value	Probability
5	0.12
6	0.28
7	0.27
8	0.33

13. Use the data in Exercise 9.4 to generate random variates from an exponential distribution with a mean of 25.
14. The data set in Table 9.38 represents the WIP at the end of a day for a 40-day simulation of a given business process. Find a 99 percent confidence interval for the mean WIP.

TABLE 9.38

WIP Data for Exercise 9.14

26	31	27	29	27
28	28	27	28	26
28	27	31	32	24
25	25	29	28	28
27	26	23	30	25
31	26	25	27	30
29	25	30	30	27
27	26	25	23	27

15. An analyst has suggested a new design for the process in Exercise 9.14. A simulation of the proposed process yielded the WIP values in Table 9.39. Test the hypothesis that the new process has a mean WIP that is less than the average of the old process. Assume a 5 percent significance level.

TABLE 9.39

WIP Data for Exercise 9.15

25	21	21	24	27
24	23	25	26	24
25	26	29	24	23
26	24	26	24	28
28	17	21	26	26
26	27	26	22	23
25	29	25	19	20
27	22	22	25	25

16. With a preliminary simulation of 10 days of operation of an emergency room, an analyst found the results shown in Table 9.40 in terms of the average number of patients waiting. Suppose the analyst would like to estimate the average number of patients waiting with a 95 percent confidence interval whose total width is four patients. Calculate the number of times the model must be run so as to be able to construct the desired confidence interval.

TABLE 9.40

Data for Exercise 9.16

Day (Run Number)	Average No. of Patients	Day (Run Number)	Average No. of Patients
1	23	6	19
2	45	7	32
3	12	8	8
4	21	9	14
5	31	10	5

17. The management team of a major ski resort wants to simulate the rental program during the high season. The resort has two types of customers: people who pay cash and people who use a credit card. The interarrival times of these customers is exponentially distributed with means of 130 and 60 seconds, respectively. After arrival, customers must fill out a waiver form. The time taken to fill out the waiver form is normally distributed with a mean of 60 and a standard deviation of 20 seconds.[4]

    After filling out the form, customers get in line to pay. The time to pay depends on the method of payment. The time required to pay in cash is uniformly distributed between 30 and 45 seconds; the time required to pay by credit card is uniformly distributed between 90 and 120 seconds.

    Three types of equipment are rented: boots, skis, and poles. Not everyone rents all three types of equipment because some people already have some of the equipment. After paying, 80 percent of the people rent boots, 10 percent go directly to skis, and the rest only rent poles. At the boot counter, an employee takes a normally distributed time with a mean of 90 and a standard deviation of 5 seconds to obtain the boots for each customer. The time to try the boots is uniformly distributed between 60 and 240 seconds. About 20 percent of the people need a different size of boot; the rest go to get their skis.

    At the ski rental counter, everyone waits until a resort employee is free to obtain the right size of ski, which takes a uniformly distributed time between 45 and 75 seconds. Twenty percent of these people need their bindings adjusted. The binding-adjustment process is exponentially distributed with a mean of 180 seconds. (The binding-adjustment process is considered separately in terms of staffing.) Seventy percent of the people go on to get poles; the rest leave. Ninety percent of the people who get their bindings adjusted go on to get poles, and the rest leave. At the station where the poles are rented, service is normally distributed with a mean of 60 and a standard deviation of 10 seconds.

a. Assume that one employee is located in each service station (one at the cash register, one at the boots counter, one at the skis counter, and one at the poles). Simulate this process for 60 days and collect the daily average cycle time. Set a 95 percent confidence interval with the data collected during the simulation.
b. Test the hypothesis (using a 5 percent significance level) that a difference in the mean cycle time results when one employee is added to the boots counter.
c. Test the hypothesis (using a 5 percent significance level) that a difference in the mean cycle time results when one employee is added to the ski counter.
d. Test the hypothesis (using a 5 percent significance level) that a difference in the mean cycle time results when two employees are added, one to the boots counter and one to the ski counter.

Notes

1. Adapted from G. Hansen. 1994. *Automating business process reengineering*. Upper Saddle River, NJ: Prentice Hall.
2. Adapted from A. Pristker. 1984. *Introduction to simulation and SLAM II*. New York: Halsted Press.
3. Adapted from R. Anupindi, S. Chopra, S. D. Deshmukh, J. A. Van Mieghem, and E. Zemel. 1999. *Managing business process flows*. Upper Saddle River, NJ: Prentice Hall.
4. Adapted from Khoshnevis (1994).

References

ExtendSim User Reference. 2024. www.ExtendSim.com, ANDRITZ Inc., 13560 Morris Road, Suite 1250, Alpharetta, GA 30004 USA 770.640.2500, Info.ExtendSim@Andritz.com.

Khoshnevis, B. 1994. *Discrete systems simulation*. New York: McGraw-Hill.

Law, A.M., and Kelton, W.D. 2000. *Simulation modeling and analysis*. 3rd ed. Boston: McGraw-Hill.

10

Prescriptive Analytics for Process Performance Optimization

After a simulation model has been developed to represent an existing or proposed business process, one might want to find the configuration that is best (according to some performance measure) among a set of possible choices. For simple processes, finding the best configuration might be easy. For instance, in the drive-through example of Section 7.5, three scenarios were considered, and the main measure of performance was total revenue during the lunch hours. To estimate the revenue, it was assumed that a simulation model of the drive-through was run to simulate 30 days of operation under each scenario. This trial-and-error approach works well in this situation because only a handful of scenarios are being considered.

When processes are complex, and the configuration depends on several strategic choices, the trial-and-error approach can be applied with only limited success. Consider, for example, a supermarket with nine cash registers. Suppose the manager wants to find the optimal staffing level in a 10-hour period; that is, the manager would like to determine how many cash registers to open during each period to optimize the checkout process according to a specified measure of service. At the same time, the manager might be faced with budget constraints that prevent overstaffing the system. With a simulation model of the supermarket and a quantifiable definition of the performance criterion, an optimizer (such as the ExtendSim Optimizer) can be invoked to search for the best configuration (as determined by a set of values for input factors to the simulation model). Optimizers are designed to search for configurations that meet the user's constraints while maximizing or minimizing the value of an objective function. The number of possible configurations in the supermarket case is large ($9^{10} \approx 3.5$ billion), rendering the enumeration of all possible combinations impractical.

This chapter is devoted to exploring how to determine the main drivers of process performance and how process performance can be optimized when a simulation model is available. The use of the ExtendSim Optimizer, which can search for optimal values of the input parameters of simulation models, is introduced.

DOI: 10.1201/9781032617237-10

10.1 Identifying the Main Drivers of Process Performance

Design of experiments (DoE) sometimes called multifactor testing (MFT) has a long tradition in the manufacturing process as a tool to increase the understanding of a product or process. It has provided a means to achieve breakthrough improvements in product quality and process efficiency. DoE enables the simultaneous investigation of the effects of several system or process variables, leading to an increased understanding of the system. DoE is a direct replacement for the traditional one-factor-at-a-time or trial-and-error approach to experimentation. DoE has been widely applied to various problems in manufacturing, but little attention has been paid to its applications in the context of business processes.

One of the difficulties in applying designed experiments in service and transactional businesses is that it is often difficult or impossible to physically experiment with the system under study. For example, suppose that a hospital would like to improve service operations in its emergency department (ED). The measure of performance may be patient waiting time and the main contributing factors may be the number of personnel on duty, the mix of skills in the on-duty personnel, the number of treatment rooms, the types of treatment and diagnostic equipment available, the physical layout of the ED, and the sequencing procedure that determines the order in which arriving patients are processed. Clearly, some of these factors should have an effect on patient throughput and hence waiting times. However, varying these factors in a designed experiment would be impractical and in most instances impossible. This situation is encountered in many improvement projects involving service and transactional operations (Antony, et al. 2011).

The usual approach to determine the factors, and their interactions, that have the largest influence on process performance in these situations is to build a computer model of the process and then apply designed experiments to the model. If the model is built properly and validated, then results from the experiment conducted on the model can be transferred to the actual process. As we saw in Chapter 8, business processes are well-suited for transaction-based models, such as discrete-event-simulation models, where probability distributions are used to model random elements. As we have seen, these random components produce an output response (i.e., a measure of performance) that is a random variable. Consequently, the full range of standard experimental design methods, including factorial and fractional factorial designs and response surface designs, can be applied to these models.

Some potential issues associated with applying to DoE to computer simulations include sample size and simulation run length determination, dealing with multiple measures of performance and non-linearity of the output, duration of the warm-up period, and a potentially large number of input factors. For a relatively small number of input factors and levels of these factors, experimental design can provide valuable insights into the performance of a process.

10.1.1 Factorial Design for Simulation Models

Simulation models often have several input factors, and determining which ones have a significant impact on performance measures (responses) of interest can be a difficult task. The common approach of changing one factor at a time is statistically inefficient and, more importantly, is very often just incorrect because, for many models, responses are impacted by the interaction of input factors (Law 2014). In DoE for simulation models, an "experiment" refers to the execution of the computer simulation model. In DoE terminology, the input parameters and structural assumptions composing a model are called factors, and output performance measures are called responses. Factors can be quantitative or qualitative (also called categorical). Quantitative factors naturally assume numerical values (e.g., the number of people in a team), while qualitative factors represent structural assumptions that are not easily quantified (e.g., a queue discipline that can be first-in, first-out or shortest-job first). The major goal of experimental design in simulation is to determine which factors have the greatest effect on a response and to do so with the least amount of simulating. This is called factor screening and is typically performed using 2^k factorial designs.

Suppose that a simulation model of a business process has $k \geq 2$ factors and that we want to estimate how each factor affects a single response (i.e., the measure of performance). We also want to determine if the factors interact with each other, i.e., whether the effect of one factor on the response depends on the levels of the others. One way of doing this within the DoE framework requires that we choose two levels for each factor and perform simulation runs for each of the 2^k possible factor-level combinations, which are called design points.

The two levels in a 2^k factorial design, which should be chosen in consultation with subject-matter experts, should be far enough apart to see a difference in the response while avoiding combinations that do not make sense in the context. The design, which is limited to two levels per factor, assumes that the response is approximately linear (or at least monotonic) over the range of the factor. If the response is nonmonotonic over the range, then the results could be misleading. For instance, they could indicate that a factor has no effect on the response when in fact it does. (There are methods for testing the linearity assumption that are beyond the scope of this introductory discussion.)

Table 10.1 shows a tabular representation of a 2^k factorial design for $k = 3$. This is called the design matrix. The R_1 to R_8 values represent the response when running the simulation with each combination of the factor levels. A minus sign is associated with one level of a factor and a plus sign with the

TABLE 10.1

Design Matrix for a 2^3 Factorial Design

Factor Combination (Design Point)	Factor 1	Factor 2	Factor 3	Response
1	–	–	–	R_1
2	+	–	–	R_2
3	–	+	–	R_3
4	+	+	–	R_4
5	–	–	+	R_5
6	+	–	+	R_6
7	–	+	+	R_7
8	+	+	+	R_8

other. For example, R_4 is the response resulting from running the simulation with factors 1 and 2 at their respective "+" levels and factor 3 at its "–" level. The design matrix facilitates the calculation of the factor effects and interactions.

The main effect of factor j, denoted by e_j, is the difference between the average response when the factor j is at its "+" level and the average response when it is at its "–" level. For the 2^3 factorial design of Table 10.1, the main effect of factor 1 is thus:

$$e_1 = \frac{R_2 + R_4 + R_6 + R_8}{4} - \frac{R_1 + R_3 + R_5 + R_7}{4}$$

This can be rewritten as:

$$e_1 = \frac{-R_1 + R_2 - R_3 + R_4 - R_5 + R_6 - R_7 + R_8}{4}$$

Therefore, the calculation of the main effect of any factor is simply the sum of all the response values using the corresponding sign and dividing by 4.

The main effects measure the average change in the response due to a change in an individual factor. It could be, however, that the effect of one factor depends in some way on the level of some other factor, in which case the factors are said to interact. A measure of the interaction is the difference between the average response when both factors are at the same level and the average response when they are at opposite levels. The easiest way of calculating the interaction effect is by creating a new column in the design matrix that consists of multiplying the signs of the factors involved in the interaction. Table 10.2 shows the column corresponding to the interaction of factors 1 and 2.

The interaction effect of factors 1 and 2 is then given by:

$$e_{12} = \frac{R_1 - R_2 - R_3 + R_4 + R_5 - R_6 - R_7 + R_8}{4}$$

TABLE 10.2

Signs for the Interaction of Factors 1 and 2

Factor Combination (Design Point)	Factor 1	Factor 2	Interaction of 1 and 2	Response
1	–	–	+	R_1
2	+	–	–	R_2
3	–	+	–	R_3
4	+	+	+	R_4
5	–	–	+	R_5
6	+	–	–	R_6
7	–	+	–	R_7
8	+	+	+	R_8

Higher level interactions are difficult to interpret but they can be calculated in the same way, that is, by creating a column of signs with the product of the signs corresponding to the factors in the interaction.

10.1.2 Illustrative Example of Design of Experiments

Consider a technical support process for which the main measure of performance is customer rating, whose maximum value is 100. A model has been built to simulate the arrivals of requests with a complexity that varies. The main resource in the model is a pool of agents that handle the requests. Performance is affected by three input factors with two levels each. The first factor refers to the level of technology available to the agents (low or high). The second factor is associated with the level of training that the agents have received (basic or advanced). And the third factor refers to the scope of the requests that the pool of agents is asked to handle (normal or extended).

A 2^3 factorial DoE is used to identify the main effects of the three factors in the model. It is also desired to identify interaction effects between factor pairs. Table 10.3 shows the design matrix for this model.

TABLE 10.3

Design Matrix for Technical Support Process Model

Design Point	Technology	Training	Scope	Ratings
1	Low	Basic	Normal	80
2	High	Basic	Normal	85
3	Low	Advanced	Normal	76
4	High	Advanced	Normal	90
5	Low	Basic	Extended	72
6	High	Basic	Extended	73
7	Low	Advanced	Extended	72
8	High	Advanced	Extended	86
Main effect	34	14	–28	

TABLE 10.4

Data for Interaction Plots

Technology vs. Training		
	Basic Training	Advanced Training
Low Technology	76	74
High Technology	79	88
Scope vs. Technology		
	Low Technology	High Technology
Normal Scope	78	87.5
Extended Scope	72	79.5
Training vs. Scope		
	Normal Scope	Extended Scope
Basic Training	78	87.5
Advanced Training	72	79.5

The last row of Table 10.3 shows the main effects for each factor. Technology has the largest positive impact on ratings. The experiment predicts an average increase of 34 points in ratings when transitioning from low to high technology. The experiment also shows that expanding the scope of the support that the agent pool is provided will drop ratings by 28 points.

The interaction effects are better understood using plots. An interaction plot for two factors consists of lines showing the change in the average response caused by changes in the levels of the factors. Table 10.4 shows the values needed for the interaction plots.

Each value in Table 10.4 is the average of two ratings. For instance, the (Low Technology, Basic Training) cell has a value of 76, which is the average of the ratings corresponding to design points for which technology is "Low" and training is "Basic," that is, design points 1 and 5 in Table 10.3. The average rating is (80 + 72)/2 = 76. All other values in Table 10.4 are obtained in a similar way. Figure 10.1 shows the interaction plots.

Figure 10.1a shows the interaction between Technology and Training. The crossing of the lines indicates that these two factors interact. For both levels of training, the move from low to high technology has a positive effect on the ratings. However, the increase in ratings for high technology is more modest when the training of the agents is basic than when the agents have advanced training. The interaction occurs because at the low technology level basic training results in better ratings (although slightly) than advanced training, while the opposite is true at the high technology level (with a larger difference in ratings).

Figure 10.1b shows no interaction between Scope and Technology. At both technology levels, the move from normal to an extended scope results in a similar decrease in ratings. Finally, Figure 10.1c shows that

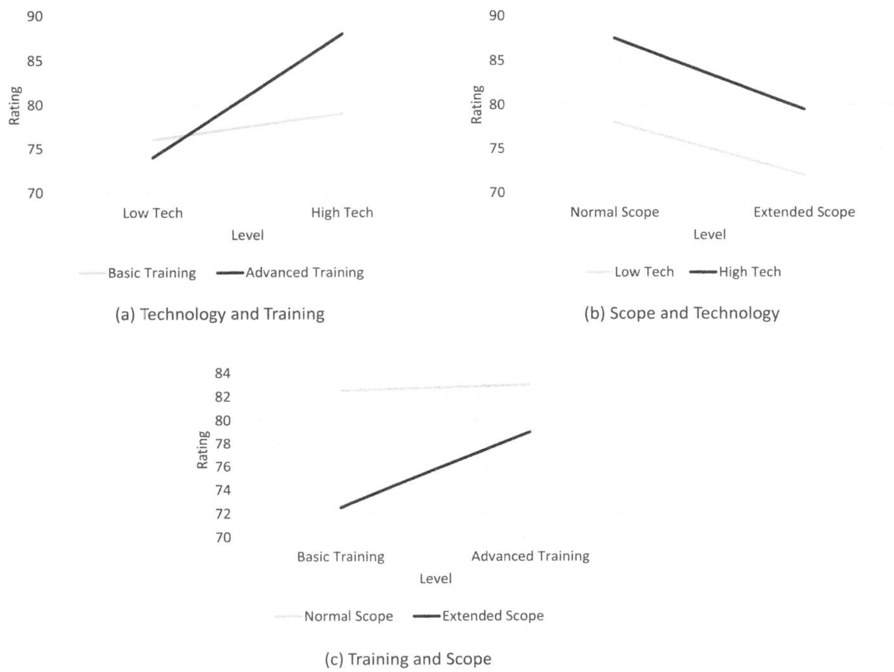

FIGURE 10.1
Interaction plots for the factorial design of the technical support process.

training can help increase ratings for both scope levels. Training seems to be particularly helpful in increasing ratings when the process operates at an extended scope. Figure 10.1c shows no interaction between the factors.

10.2 Business Process Optimization

There are essentially two levels of process optimization: structural and operational. Structural optimization refers to finding the best design for the process, while operational optimization deals with selecting the best configuration of the design. The design determines how the process operates and what resources and technology are employed and, thus, the determining logic behind the flow of work. Clearly, one would like to determine the best or optimal design to improve how business is conducted. The effectiveness and efficiency of a design, however, depend not only on how the process is structured but also on how it is operated. Selecting one technology over another is a design choice that creates a particular structure, while choosing the policies to run the chosen technology is an operational decision. Ideally, the analyst would like to explore multiple combinations of structural and operational choices to choose the best.

Depending on the complexity of the process, creating a simulation model for which structural changes to the process could be made by direct changes to input parameters may be a very difficult task. For instance, suppose that the analyst is considering a shared database system that would allow some activities to be performed in parallel instead of in series. Typically, two simulation models would be needed to test the differences between a design with the database in place and one without it. When only two choices are being considered, the optimization process is simple. However, if the number of combinations is large, due to the availability of several choices with several alternatives, then creating a model for each combination becomes impractical. In such a situation, a single simulation model with input parameters representing structural choices would be ideal. However, this is where the modeling could become too complex, or the available software could be incapable of supporting it.

The most common form of process optimization, and the focus of this chapter, relates to operational decisions. The modeling for this type of optimization does not require that the analyst applies any significant additional effort because the input parameters are common elements in most simulations. For instance, a number of operational decisions are associated with choosing the right level of resources to be applied to the process at the right time. Simulation software is equipped with simple ways of changing resource levels throughout the simulation, making it straightforward to create optimization models based on these types of decisions. Changing the input values and, therefore, the configuration of the process becomes a trivial exercise that simulation software operationalizes within automated systems that search for the best.

To illustrate the difference between structural and operational optimization, consider the drive-through example of Section 7.5. The current design shown in the flowchart of Figure 7.3 includes a single drive-through with three resources and their corresponding queues. Each resource deals with one order at a time, and the flow of cars and orders is determined by the design choices (i.e., a single drive-through with a single menu board, kitchen, and pickup window). Operationally, this design could be optimized by changing resource levels, as is done in the first two scenarios considered in Section 7.5. In the first one, the queue space was eliminated, and in the second one, the queue space was increased. These changes are very simple to implement in simulation software, including ExtendSim. There is an element that models the queue, whose limit could be changed to explore the performance of the system under these two scenarios. The third scenario, however, involves a structural change. Recall that the analyst is considering the adoption of a new technology that reduces the time to place an order. Because this change involves the reduction of the average processing time of a single activity, the change in the model is not difficult to implement. In fact, in this case, the structural change could be "parameterized" by creating an input to the simulation model that sets the average processing speed at the menu board, which, in turn, indicates the technology that is being considered. Other structural changes are not as easy to implement. For instance, suppose that the analyst would like to test a scenario where a second drive-through is added to the process. There are two separate lines, and customers arriving at the fast-food restaurant must choose one. The orders are now placed and picked up at two different locations, but they are made in a single kitchen. Clearly, this structural change cannot be simply implemented by changing an input parameter in the simulation model that was built to study the single-drive-through design.

Operational decisions are not limited to changing the availability of resources in a process. Improving the performance of a business process by selecting the most qualified available personnel for each activity, without changing the structure of the business process or hiring new employees, is a form of operational optimization that has received less attention than others in the operations management literature (Kamrani et al. 2012). This approach is appropriate for scenarios in which a fairly large number of employees within an organization are able to perform a set of activities. For instance, consider military personnel who are appointed to positions in a command center or a group of software engineers collaborating in the design of a software system. The activities in these processes require various levels of expertise, and the level of qualification of the available personnel varies. Hence, the employee performance may vary according to the assigned activity, making the outcome of the process depend heavily on the assignment of personnel to activities.

The problem of optimally assigning tasks to workers, known as the *assignment problem,* is not new to the operations management literature, and the first solution approach was introduced more than 50 years ago. However, the classical assignment problem and its variants assume a static system where the benefit of assigning one person to an activity is known with certainty and does not relate to the assignments. A process view of this problem considers that activities are interconnected by a process and its corresponding operational rules. The process includes decision points that determine the path of the workflow. Some activities may be repeated several times, or different courses of action may be taken, resulting in alternative activities to be performed. A business process often involves uncertainty, which must be addressed and incorporated into the analysis adequately. This makes the assignment of personnel to a business process a more complex form of the classical assignment problem.

Within a simulation environment, however, this problem may be modeled in such a way that each employee possesses a set of attributes that are translated into a score that measures the benefits of assigning them to particular activities. At the same time, the cost of the assignment could be approximated, which would result in a trade-off that the optimizer must resolve. The value of assigning an employee to a particular activity could translate into, for instance, fewer errors (i.e., a decrease in the probability that the activity must be performed again) or increased speed. An even more complex view of this problem would include the modeling of employees working as teams and the possible interactions (negative or positive) among team members.

10.3 The Role of Simulation Optimization in Business Process Management

A growing number of business process management software vendors and consultants are offering simulation capabilities to extend their modeling functions and enhance their analytical proficiencies. Detailed simulation models connected are sometimes referred to as Digital Twins, offering a digital copy of the real process. Simulation is positioned as a means to evaluate the impact of process changes and new processes in a model environment through the creation of "what-if" scenarios. Simulation is promoted as enabling the examination and testing of decisions prior to actually making them in the "real" environment. Since simulation approximates reality, it enables the forecasting of process performance under assumptions of uncertainty and variability. This section[1] explores how new approaches are significantly expanding the power of simulation for business process management.

The need for optimization of simulation models arises when the process analyst wants to find a set of model specifications (i.e., input parameters and/or structural assumptions) that leads to optimal performance. On the one hand, the range of parameter values and the number of parameter combinations are too large for analysts to simulate all possible scenarios, so they need a way to guide the search for good solutions. Suppose a simulation model depends on only two input parameters. If each parameter has 10 possible values, trying each combination requires 100 simulations (10^2 alternatives). If each simulation were short (e.g., 2 seconds), then the entire process would be completed in approximately 3 minutes of computer time. However, if instead of two variables, the model depends on six, then trying all combinations would require 1 million simulations (10^6 alternatives), or approximately 23 days of computer time. A complete enumeration could easily take weeks, months, or even years to carry out.

On the other hand, without simulation, many real-world problems are too complex to be modeled by mathematical formulations, which are at the core of pure optimization methods. This creates a conundrum; pure optimization models alone are incapable of capturing all the complexities and dynamics of the system, so one must resort to simulation, which cannot easily find the best solutions. Simulation optimization resolves this conundrum by combining both methods.

The merging of optimization and simulation technologies has seen remarkable growth since the beginning of the 2000s. The simulation community was at the beginning reluctant to use optimization tools. Optimization models were thought to over-simplify the real problem, and it was not always clear why a certain solution was the best. However, a vast body of research in the area of metaheuristics, coupled with improved statistical methods of analysis, reduced this resistance considerably. In 1986, Prof. Fred Glover coined the term *metaheuristics* to describe a master strategy that guides and modifies other heuristics to produce solutions beyond those that are normally generated in a quest for local optimality. The heuristics guided by such a meta-strategy may be high-level procedures or may embody nothing more than a description of available moves for transforming one solution into another together with an associated evaluation rule.

Metaheuristic research has enabled the development of powerful algorithms to guide a series of simulations to produce high-quality solutions in the absence of tractable mathematical structures. Furthermore, techniques have been developed to allow optimization procedures to compare different solutions in terms of quality in the presence of uncertainty (as is the case when the quality of a solution is measured by running a simulation model). Nearly every commercial discrete-event simulation software package, including ExtendSim, contains an optimization module—with OptQuest[2] being one of the most popular ones—that performs some sort of search for optimal values of input parameters.

The optimization of simulation models deals with the situation in which the analyst would like to find which of possibly many sets of model specifications (i.e., input parameters and/or structural assumptions) leads to optimal performance. As mentioned in Section 10.1, in the area of DoE, the input parameters and structural assumptions associated with a simulation model are called *factors*. The output performance measures are called *responses*. For instance, a simulation model of a hospital may include factors such as number of nurses, number of doctors, number of beds, and available equipment (e.g., X-ray machines). The responses may be cycle times, waiting times, and resource utilization.

In the world of optimization, the factors become *decision variables*, and the responses are used to model an *objective function* and *constraints*. Whereas the goal of experimental design is to find out which factors have the greatest effect on a response, optimization seeks the combination of factor levels that minimizes or maximizes a response (subject to constraints imposed on factors and/or responses). In the hospital example, an optimization model may seek to minimize waiting times by manipulating the number of nurses, doctors, and equipment while restricting capital investment and operational costs as well as maintaining a minimum utilization level of all resources. A model for this optimization problem would consist of decision variables associated with labor and equipment as well as a performance measure based on waiting times obtained from running the simulation of the hospital. The constraints are formulated with both decision variables and responses (i.e., utilization of resources).

With simulation optimization, when changes are proposed to business processes to improve performance, the projected improvements can be not only simulated but also optimized. Changes may entail adding, deleting, and modifying activities, processing times, required resources, schedules, work rates, skill levels, and budgets. Performance measures may include throughput, costs, inventories, cycle times, resource and capital utilization, start-up times, cash flow, and waste. In the context of business process management and improvement, simulation can be thought of as a way to understand and communicate the uncertainty related to making the changes, while optimization provides a way to manage that uncertainty.

Four approaches account for most of the academic literature in simulation optimization. They are (1) stochastic approximation (gradient-based approaches); (2) (sequential) response surface methodology; (3) random search; and (4) sample path optimization (also known as *stochastic counterpart*). However, none of these approaches has been used to develop optimization for commercial simulation software, mainly because these "methods generally require a considerable amount of technical sophistication on the part of the user, and they often require a substantial amount of computer time as well" (Andradóttir, 1998).

Leading commercial simulation software employs metaheuristics as the methodology of choice to provide optimization capabilities to its users. Like other developments in the operations research/computer science interface (e.g., those associated with solving large combinatorial optimization problems), commercial implementations of simulation optimization procedures have only become practical with the exponential increase of computational power and the advance in metaheuristic research. The metaheuristic approach to simulation optimization is based on viewing the simulation model as a black box function evaluator.

Figure 10.2 shows the black box approach to simulation optimization favored by procedures based on metaheuristic methodology. In this approach, the metaheuristic optimizer chooses a set of values for the input

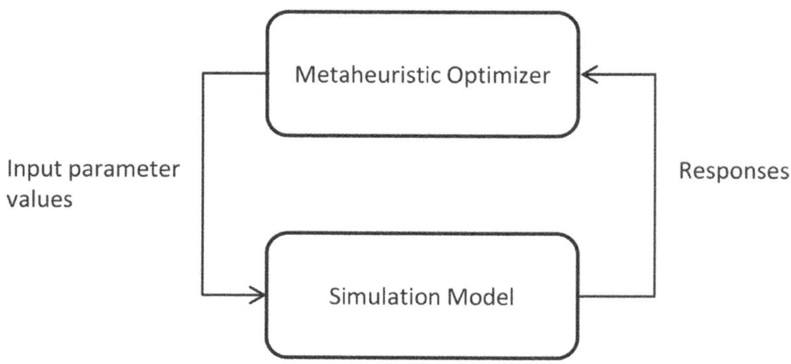

FIGURE 10.2
Black box approach to simulation optimization.

parameters (i.e., factors or decision variables) and uses the responses generated by the simulation model to make decisions regarding the selection of the next trial solution.

Most of the optimization engines embedded in commercial simulation software are based on evolutionary approaches. Evolutionary approaches search the solution space by building and then evolving a population of solutions. The evolution is achieved by mechanisms that create new trial solutions out of the combination of two or more solutions that are in the current population. The transformation of a single solution into a new trial solution is also considered in these approaches. Examples of evolutionary optimization approaches are genetic algorithms and scatter search. The latter, in conjunction with a memory-based approach called *tabu search*, is used in OptQuest.

In the context of simulation optimization, a simulation model can be thought of as a "mechanism that turns input parameters into output performance measures" (Law and Kelton, 2000). In other words, the simulation model is a function (whose explicit form is unknown) that evaluates the merit of a set of specifications, typically represented as a set of values. Viewing a simulation model as a function has motivated a family of approaches to optimize simulations based on response surfaces and metamodels.

A response surface is a numerical representation of the function that the simulation model represents. A response surface is built by recording the responses obtained from running the simulation model over a list of specified values for the input factors. A metamodel is an algebraic model of the simulation. A metamodel approximates the response surface, and, therefore, optimizers use it instead of the simulation model to estimate performance. Metamodels could be built with simple techniques like standard linear regression or with more sophisticated machine learning/AI techniques like neural networks and kriging. Once a metamodel is built, in principle, appropriate deterministic optimization procedures can be applied to obtain an estimate of the optimum.

An important feature in simulation optimization software is the ability to specify constraints. Constraints define the feasibility of trial solutions. Constraints may be specified as mathematical expressions or as statements based on logic. In the context of simulation optimization, constraints may be formulated with input factors or responses. If the constraints in a simulation optimization model depend only on input parameters, then a new trial solution can be checked for feasibility before running the simulation. An infeasible trial solution may be discarded or may be mapped to a feasible one when its feasibility depends only on constraints formulated with input parameters. OptQuest, for instance, has a mechanism to map infeasible solutions of this type into feasible ones (Laguna 2011). On the other hand, if constraints depend on responses, then the feasibility of a solution is not known before running the simulation.

The following two examples[3] provide additional context on how companies are using simulation optimization technology.

Example 10.1: Outpatient Appointment Scheduling

Indiana University Health Arnett Hospital (UHAH), a full-service acute care hospital and a multispecialty clinic, faced poor statistics because the number of no-show patients (those who do not show up for their scheduled appointments) rose dramatically to 30 percent. This was primarily caused by clinic schedules driven by the individual preferences of the medical staff, which led to increased variations in scheduling rules. To eliminate the problem, UHAH wanted to develop a scheduling system that would take into consideration metrics related to the clinic, doctors, and patients. The system was developed with the goal of optimizing doctors' schedules and decreasing the number of no-shows. The multiobjective optimization problem included

- Increasing physician efficiency
- Increasing facility utilization
- Decreasing physician overtime
- Decreasing waiting time for patients

To address the challenge of improving appointment scheduling, a discrete-event simulation model was developed. The model simulated the patients' appointment process and further checkups. Patients in the model were divided into five groups:

1. Patients requesting same-day appointments
2. New patients of high priority
3. Re-check of high priority
4. New patients of low priority
5. Re-check of low priority

A major distinction in the model is that high-priority patients were those with insurance, as opposed to those of low priority. The model allowed the adjustment of critical non-controllable parameters (such as no-show rates) to test different scenarios. It also exposed controllable parameters (such as doctors' availability and appointment slots per day) to be used for optimization purposes. Simulation output included metrics such as staff utilization, amount of overtime, distribution of patients among staff, and patient waiting time. The simulation output allowed UHAH to see how schedule policies affected the clinic's process flow and provided insight to make better-informed staff management decisions. ■

Example 10.2: Call Center Management

The United Services Automobile Association (USAA), a Fortune 500 group of companies, owns large call centers with highly complex infrastructures. The USAA representatives wanted to model the call center framework to optimize the headcount and the scheduling and routing of calls by using aggregated data. These steps were aimed at improving call center overall utilization and customer satisfaction rates as well as lowering the abandonment rate. Call center simulation models are quite widespread. However, because of deficiencies in the approaches used, these models are often neglected. Some of the deficiencies are

- Low granularity in terms of time and groups of people
- Missing effects of abandonment behavior and multi-skilled call center representatives
- Over-simplistic routing strategies
- Inability to take into account call and sales representative attributes, and attribute-based routing

USAA contracted the development of a simulation model with a high level of detail on calls, call center representatives and their skills, routing, and abandonments. Simulation and optimization provided insights that formed the basis for improvements in the call center working process. As a result, customer service indexes rose significantly due to reduced waiting times, which, combined with a decrease in the abandonment rate, produced increased revenues. Changes in the working process resulted in cuts in hiring and training costs. ■

10.4 Simulation Optimization with ExtendSim

Traditional search methods work well when finding local solutions around a given starting point or when model data are precisely known. However, these methods fail when searching for global solutions to real-world problems that contain a significant amount of uncertainty, as in the case of most business processes. Important developments in optimization have produced efficient search methods, such as those based on evolutionary computation, capable of finding optimal solutions to complex problems involving elements of uncertainty.

The ExtendSim Optimizer incorporates metaheuristics to guide its search algorithm toward improved solutions. As mentioned in the previous section, metaheuristics is the term used to describe a family of optimization approaches that includes genetic algorithms, simulated annealing, tabu search, scatter search, and others (including hybrids). The particular approach embedded in the ExtendSim optimizer uses the evolutionary computation paradigm (see Appendix 10A for details). Because this technique does not use the hill-climbing approach of simple optimizers, it typically does not get trapped in local optimal solutions, and it does not get thrown off course by noisy (uncertain) model data.

After an optimization model is formulated, the optimizer runs the simulation to evaluate the output for different sets of decision variable values. The optimizer uses the outputs of the current and previous runs to determine a new set of values for the decision variables to evaluate. This iterative process successively generates new sets of values and runs the simulation to measure their performance (in terms of maximizing profits or minimizing costs, for example). Not all of these values improve the objective, but over time, this search provides a trajectory to the best solutions. The search process continues until the optimizer reaches some termination criteria.

When optimizing the performance of a business process, analysts are faced with many difficult decisions; such decisions include staffing or operational policies and might involve thousands or millions of potential alternatives. As mentioned previously, considering and evaluating each of the alternatives would be impractical or perhaps impossible. A simulation model can provide valuable assistance in analyzing decisions and finding good solutions. Simulation models capture the most important features of a problem and present them in a form that is easy to interpret. Models often provide insights that intuition alone cannot provide. To optimize a simulation model, one must formulate an optimization model, which is a model that seeks to maximize or minimize some quantity, such as profit or cost. Optimization models consist of three major elements: decision variables, constraints, and an objective.

- *Decision variables*: Quantities that the decision-maker can control; for example, the number of people to employ, the number of dollars to allocate among different resources, or the operational rule to use from a set of alternatives. Decision variables are selected from the dialogs of blocks such as the Resource Pool block. If a decision is important in the context of a given process, this decision must be linked to a value that appears in the dialog of a block used in the simulation model. The optimization model can be formulated in terms of the decision variables that are selected (by cloning the variable, as explained later). The values of the decision variables selected change with each trial simulation until the best value for each is found during the optimization run.
- *Constraints*: Relationships among decision variables that restrict the values of the decision variables. For example, a constraint might ensure that the total amount of money allocated among various resources cannot exceed a specified amount or that at least a specified minimum number of nurses are assigned to the emergency room (ER) of a hospital at a given time. Constraints restrict the possible values for the decision variables. For example, if the total monthly budget for a process with two resources (labeled *A* and *B*) is $50,000, the following constraint could be added to an optimization model:

$$2000A + 1000B \leq 50,000$$

Here, it is assumed that each *A* resource costs $2000 per month, and each *B* resource costs $1000 per month. It also is assumed that *A* and *B* are initial values in separate Resource Pool blocks and that they have been cloned to be added to the Optimizer block. Given this constraint, combinations of values for the two resource variables whose sum is $50,000 or less are considered during the optimization run.

Not all optimization models need constraints; however, those that do must deal with the distinction between a feasible and an infeasible solution. A feasible solution is one that satisfies all constraints. An optimization model is infeasible when no combination of values of the decision variables can satisfy the set of constraints. Note that a solution (i.e., a single set of values for the decision variables) can be infeasible if it fails to satisfy the problem constraints, but this doesn't imply that the problem or model itself is infeasible. For example, suppose that in a business process with two resource types, *A* and *B*, one insists on finding an optimal configuration that meets the following constraints:

$$A + B \leq 4$$

$$A + B \geq 5$$

Clearly, no combination will make the sum of the units of *A* and the units of *B* no more than 4 and at the same time greater than or equal to 5. Infeasible models can be made feasible by fixing the inconsistencies of the relationships modeled by the constraints. For a constrained problem, the optimal solution is the best solution that satisfies all constraints.

Objective: A mathematical representation of the optimization model's objective, such as maximizing profit or minimizing cost, in terms of the decision variables. Each optimization model has one objective function that represents the model's goal mathematically. The optimizer's job is to find the optimal value of the objective by selecting different values for the decision variables. When model data are uncertain and can only be described using probability distributions, the objective itself will have some probability distribution for any set of decision variables. This probability distribution can be found by performing statistical analysis on the values of the output variable of interest. For example, a plotter can be used to create a frequency distribution of cycle time values if these values are used to calculate the objective function.

A conceptualization of an optimization model is shown in Figure 10.3. The solution to an optimization model provides a set of values for the decision variables that optimizes (maximizes or minimizes) the associated objective (e.g., a measure of process performance). If the world were simple and the future were predictable, all data in an optimization model would be constant (making the model deterministic), and techniques such as linear and nonlinear programming could be used to find optimal solutions.

However, a deterministic optimization model can't capture all the relevant intricacies of a practical decision environment. When model data are

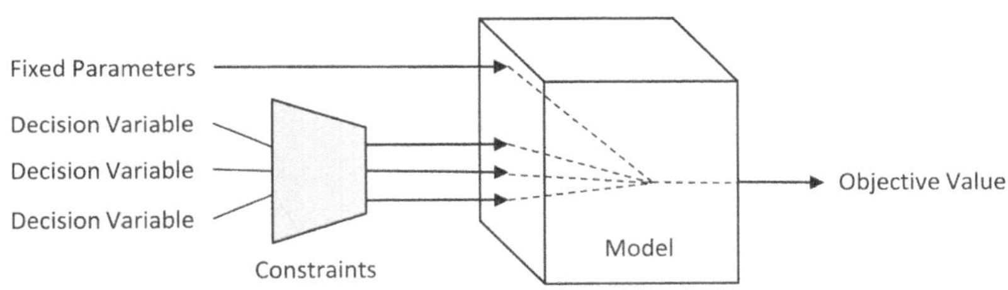

FIGURE 10.3
Conceptualization of an optimization model.

uncertain and can only be described probabilistically, the objective will be a random variable that follows an unknown probability distribution for any chosen set of decision variables. An optimization model with uncertainty has several additional elements, including the following:

- *Assumptions*: Capture the uncertainty of model data using probability distributions. Assumptions are modeled primarily by choosing appropriate probability distributions in Input Random Number or Create blocks.
- *Output Variables*: Capture values of interest such as waiting time, cycle time, and resource utilization. Output variables typically are represented by frequency distributions.
- *Output Statistics*: Summary values of an output variable, such as the mean, standard deviation, or variance. The optimization can be controlled by maximizing, minimizing, or restricting output statistics, such as the average waiting time or the maximum queue length.
- *Requirements*: Relationships among output variables. Requirements are constraints that involve output variables. The ExtendSim Optimizer does not allow the user to set requirements, but other optimizers such as OptQuest do. A typical requirement may involve limiting waiting time in a cost-minimization model.

Figure 10.4 shows a conceptualization of an optimization model with uncertainty. This figure shows that instead of constant data, the assumptions turn the calculation of the objective function into a stochastic evaluation. For example, if average processing times are used to calculate the cycle time of a process, the result is a deterministic model that uses constant values to estimate performance. However, if probability distributions are used to estimate processing times, then the cycle time calculation becomes a stochastic evaluation performed with a simulator. The steps for adding optimization to an Extend simulation are as follows:

1. Open an existing simulation model or develop one to optimize its performance by changing the values of some decision variables.
2. Add an Optimizer block to the model by clicking on the Optimizer item of the Model Analysis submenu of the Value library.

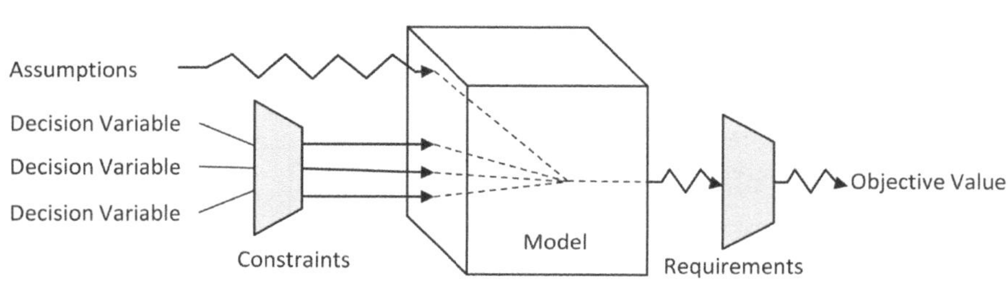

FIGURE 10.4
Conceptualization of an optimization model with uncertainty.

3. On paper, define a cost or profit equation (also called the *objective function*) to optimize.
4. Determine which variables the objective function needs and "clone-drop" them onto the Optimizer. Clone-drop is achieved by dragging the needed variables, with the clone tool, from the block where they reside to the Optimizer block.
5. Open the Optimizer block dialog by double-clicking on the Optimizer block. Set the limits for the chosen decision variables in the Optimizer's Variables table (within the Objectives tab).
6. Write the profit or cost equation in the Optimizer's dialog. Note that the profit or cost equation must be written using the variables that have been added to the Optimizer block.
7. Add appropriate constraint equations in the Constraints tab.
8. In the Run Parameters tab, set the optimizer defaults for either a random or a nonrandom model.
9. Click on the Results tab and then click on the New Run button.

The following tutorial illustrates these steps.

10.4.1 Tutorial: Process Optimization with ExtendSim

A business process consists of 10 activities as illustrated in the flowchart in Figure 10.5. Four resource types perform these activities: clerks, agents, technicians, and supervisors. As shown in Figure 10.5, some activities require more than one resource type; for example, a technician and a supervisor perform Activity F. Jobs arrive at a rate of one every 10 minutes with interarrival times following an exponential distribution. The activity times are uncertain and approximately follow the distributions shown in Table 10.5. The process manager would like to minimize cycle time but would not like to add resources in such a way that they would become underutilized. Therefore, the process

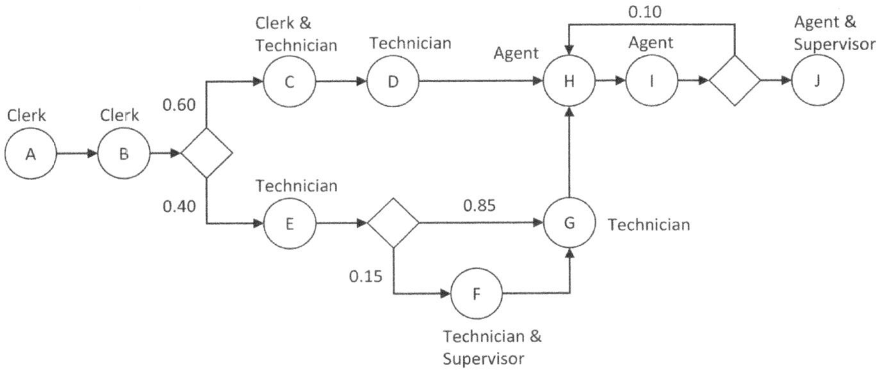

FIGURE 10.5
Business process with multiple resource types.

TABLE 10.5

Activity Times for the Process in Figure 10.5

Activity	Distribution of Processing Times
A	Uniform between 1 and 4 minutes
B	Exponential with a mean of 6 minutes
C	Normal with a mean of 7 and standard deviation of 2 minutes
D	Uniform between 3 and 8 minutes
E	Exponential with a mean of 7 minutes
F	Normal with a mean of 15 and standard deviation of 2 minutes
G	Uniform between 2 and 10 minutes
H	Uniform between 2 and 4 minutes
I	Exponential with a mean of 8 minutes
J	Triangular with parameter values of 10, 17, and 21 minutes

manager also would like to set limits on the number of resources of each type and an overall limit on the total number of resources. The manager's optimization problem can be summarized as follows.

Minimize average cycle time

Subject to Clerks + Technicians + Supervisors + Agents ≤ 12

≤ Clerks ≤ 10

≤ Technicians ≤ 10

≤ Supervisors ≤ 5

≤ Agents ≤ 10

Figure 10.6 shows an ExtendSim simulation model of the process depicted in Figure 10.5. The model uses four Resource Pool blocks, one for each labor pool in the problem. There are five queues in the model because the supervisors are employed in two different parts of the process for Activities F and G. All queues are first-in-first-out, and Resource Pool Release blocks are used to release laborers after completion of their assigned activities. The timing attribute *CycleTime* is attached to all items generated in the Create block. The attribute is then used by the Information block to calculate cycle time statistics.

For the purpose of this tutorial, assume that the simulation model has been built already and that optimization is to be added. The first step consists of adding the Optimizer block to the model. This block—which appears to the right of the Resource Pool blocks in Figure 10.6 with the label "Minimize Cycle Time"—is in the Model Analysis submenu of the Value library. After the block has been inserted in the simulation model, the optimization problem needs to be modeled. The optimization model consists of four decision variables, the objective function, and one constraint. Therefore, the next step is to add the decision variables to the Optimizer block. The variables are added with the Clone Cursor. ExtendSim has four types of cursors in the

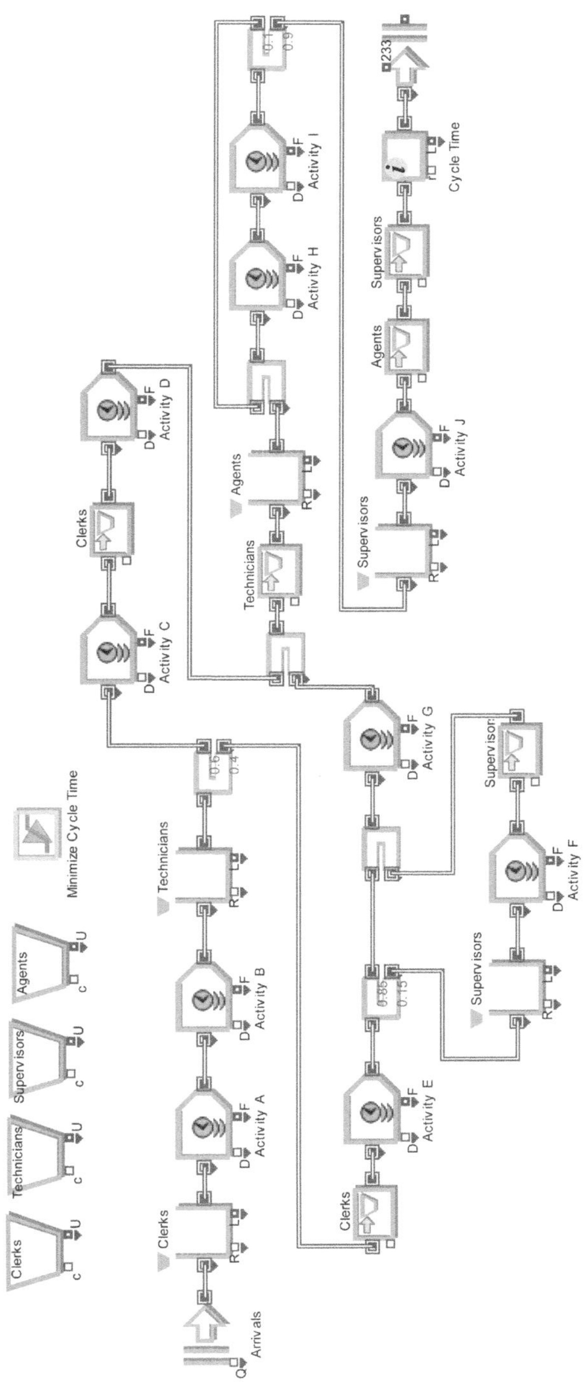

FIGURE 10.6
ExtendSim model of the process in Figure 10.4 with an optimizer block.

Cursor toolbar: Block/Text Cursor (an arrow), Graphics Cursor (an arrow with a box below), Clone Cursor (a black arrow overlapping a white arrow), and All Objects Cursor (an arrow with an infinity sign below).

The Clone Cursor clones dialog items and allows the cloned items to be placed in other parts of the model. For example, the utilization rate in the Results tab of a Resource Pool block can be cloned and placed next to the block. When the model is run, changes in the utilization rate of the labor pool can be seen without opening the block dialog. The Clone tool is used to create a copy of the initial value for each resource pool in the model. The dialog of the Resource Pool that models the clerks is shown in Figure 10.7.

To clone the initial value of the available clerks and add it as a decision variable in the Optimizer block, select the Clone Cursor and click-point the Initial number item to the Optimizer block. Enter "Clerks" in response to the dialog that opens to ask for a variable name. Do the cloning three more

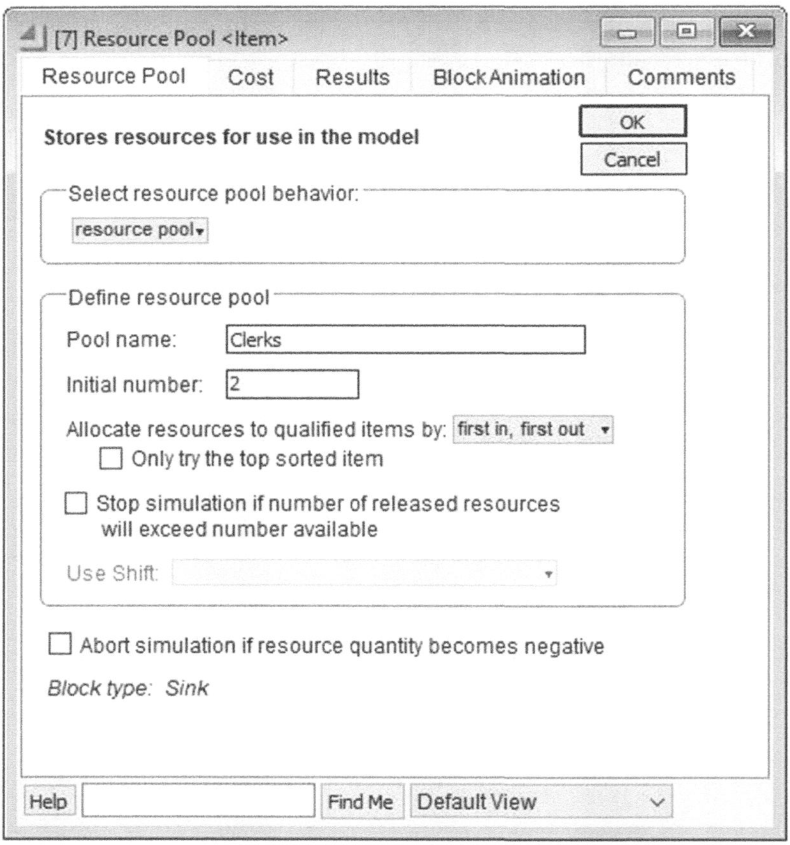

FIGURE 10.7
Resource pool block dialog for clerks.

times to add the initial value of technicians, supervisors, and agents as decision variables in the Optimizer block. Next, add the average cycle time to the Optimizer block to be able to formulate the objective function. The average cycle time is calculated in the Statistics tab of the Information block. (See dialog in Figure 10.8.) To add the average cycle time to the Optimizer block, select the Clone Cursor and click-point the Average within the cycle time section of the Statistics tab in the Information block. Enter "CycleTime" in response to the dialog that opens to ask for a variable name.

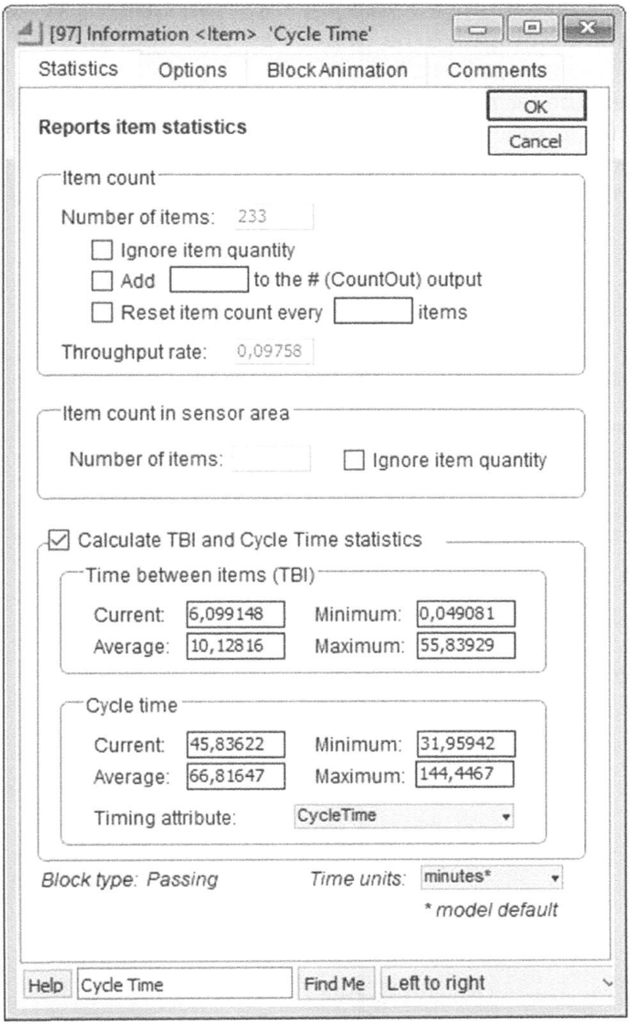

FIGURE 10.8
Statistics tab of the information block dialog.

Next, add the bounds on the decision variables and formulate the objective function. This is done in the Objectives tabs of the Optimizer block dialog. The bounds are added in the Minimum Limit and Maximum Limit columns. No bounds are associated with the average cycle time because this variable will hold an output value from the simulation model and not the value of a decision variable. The Equation Variable column shows the variable names entered when the variables were added to the Optimizer. However, the names can be changed in this dialog if desired. These names cannot contain spaces or special characters, but they can contain underscores. The objective function has the form MinCost = CycleTime to indicate the desire to minimize the value of average cycle time. The completed dialog is shown in Figure 10.9.

To complete the formulation, the constraint that restricts the total number of workers to no more than 12 needs to be added. Constraints are added in the Constraints tab of the Optimizer dialog. ExtendSim has two types of constraints: constraints on individual variables and global constraints.

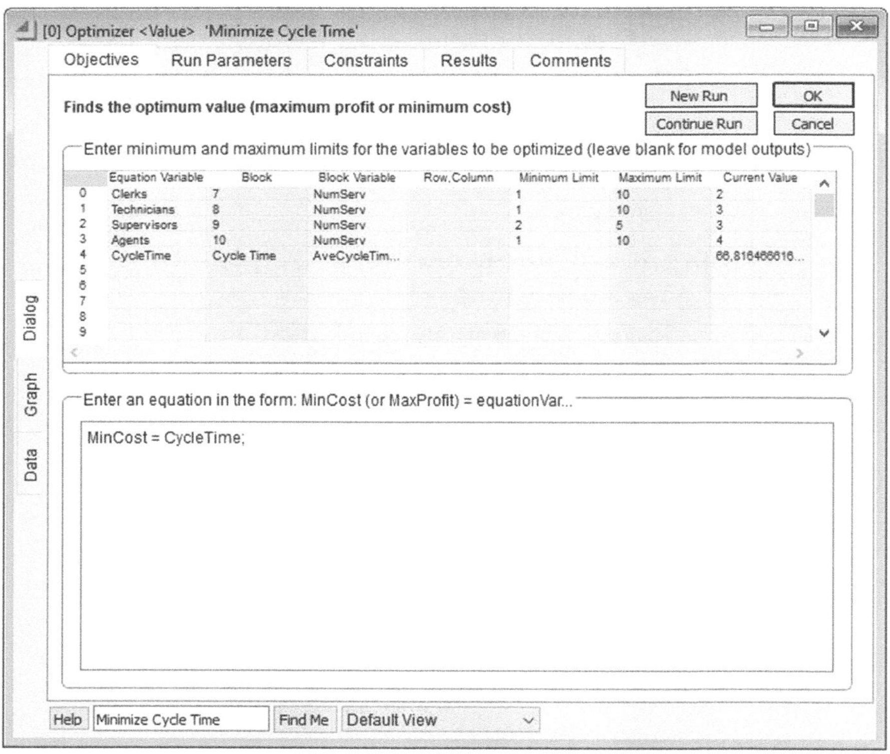

FIGURE 10.9
Objectives tab of the optimizer block dialog.

A global constraint is a relationship that involves more than one decision variable. Therefore, the constraint that limits the number of total workers in the process must be formulated as a global constraint. Figure 10.10 shows the Constraints tab of the Optimizer block. As shown in this figure, global constraints are modeled using the "reject system." In the example, ExtendSim is instructed to reject solutions for which the sum of the variables Clerks, Technicians, Supervisors, and Agents exceeds 12. The optimization model is now completed, and the search for the optimal values for the decision variables can begin. The search is controlled by a set of parameter values, which can be adjusted in the Run Parameters tab of the Optimizer block. ExtendSim provides a fair amount of flexibility in terms of setting these values for particular optimization problems, and the task of choosing the right setting could be difficult. Fortunately, Extend includes two default parameter settings for models with random elements (such as those used for business process design). The parameter settings are labeled Quicker Defaults and Better Defaults in the Random Model section of the Run Parameters tab.

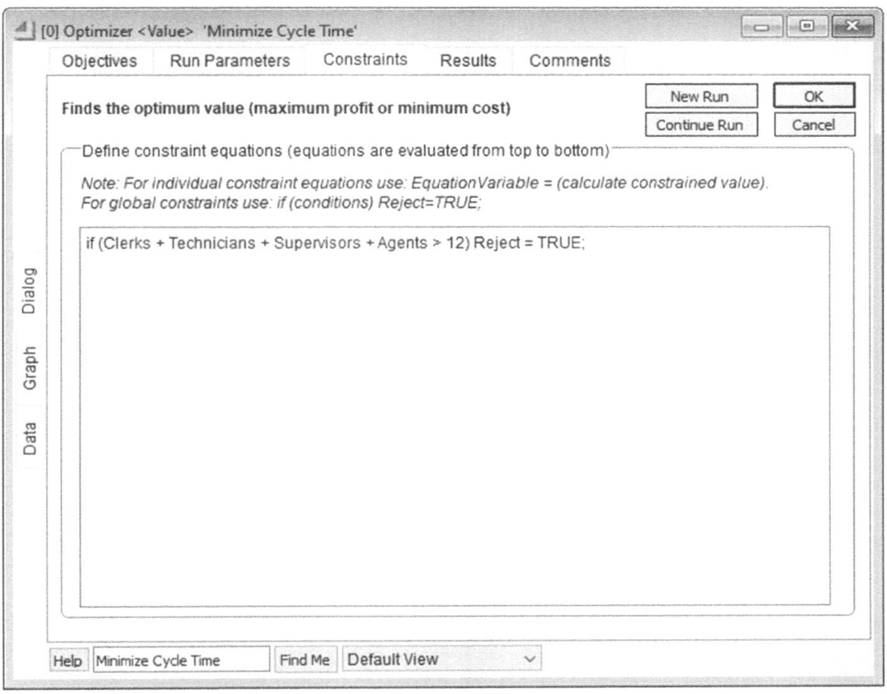

FIGURE 10.10
Global constraint to model limit in the total number of workers.

The Quicker Defaults are meant for optimization runs where reasonably good solutions are desired within a short amount of computer time. For better solutions, Better Defaults is used; however, the computational time increases. After selecting either the Quicker or the Better defaults, click on the New Run button to begin the search. During the search, Extend displays the Results tab and the Optimization Value plot. The Results tab (Figure 10.11) shows the population of solutions ordered from best to worst. These results were found using the Quicker Defaults.

The solutions in Figure 10.11 show that very similar results are obtained with solutions that employ either two clerks and three technicians or three clerks and two technicians. There seems to be a consensus that the best solutions require three supervisors and four agents. The last solution in the table is interesting because it achieves an average cycle time of just under 62 minutes with 11 employees. The process manager should pay attention to this solution and perhaps perform additional tests and simulation runs

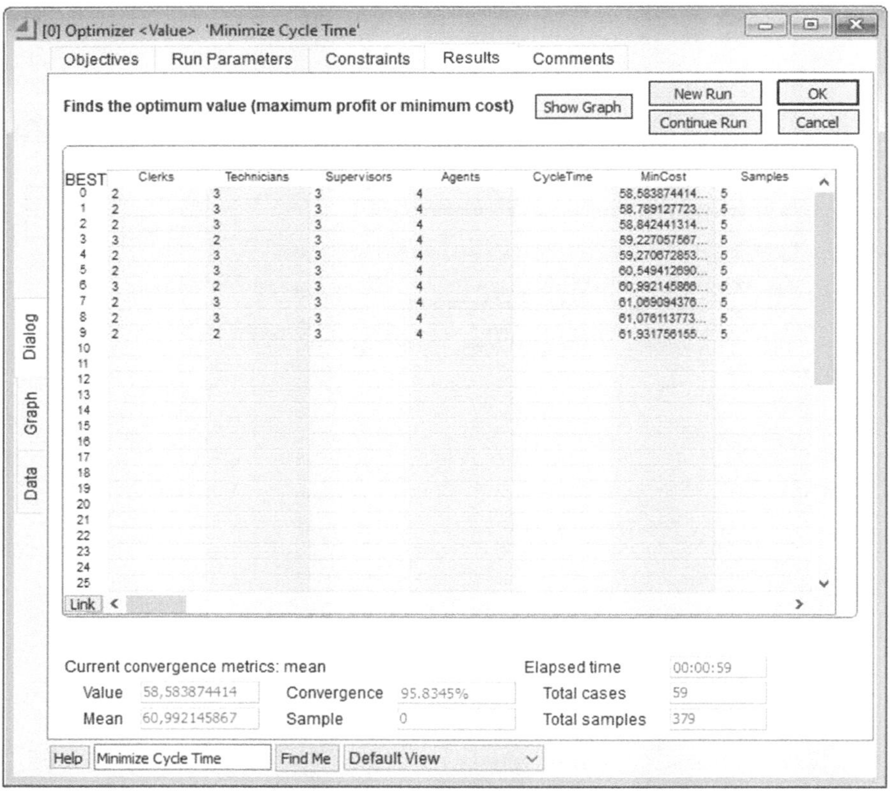

FIGURE 10.11
Results tab showing best solutions.

to determine whether the cycle times achieved with a total of 11 workers are acceptable for the purpose of reaching customer service goals. Several samples of longer simulation runs would allow the detection of significant differences between the two staffing levels.

The Optimization Value plot (see Figure 10.12) shows the search trajectory; that is, the objective function value of the best solution found during the current optimization run. Because the problem is to minimize the average cycle time, the trajectory shows a downward trend as the optimization search progresses. This plot also shows the convergence rate, which is a value between 0 percent and 100 percent that indicates the similarity of the objective function values of the best and worst solutions in the population. The Optimization Value plot in Figure 10.12 has two scales on the y-axis. The first scale (on the left) refers to the objective function value; that is, the average cycle time in this case. The second scale (on the right) refers to the convergence. The plot shows that the population converged (with a convergence value of 95.83 percent) after 59 generations.

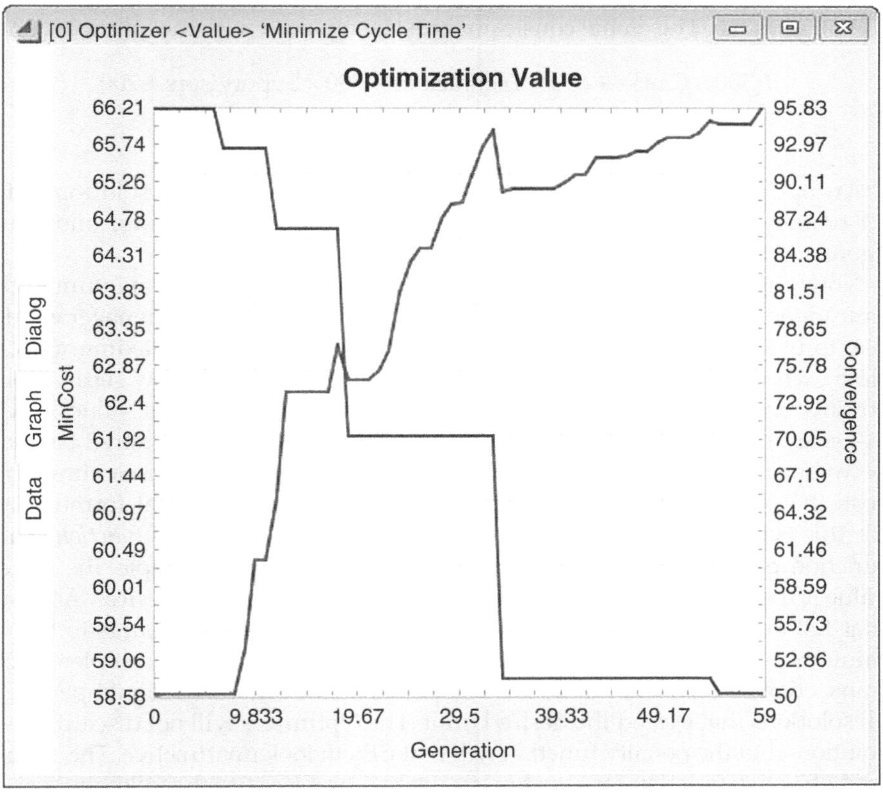

FIGURE 10.12
Optimization value plot.

10.4.2 Alternative Optimization Models

In the tutorial of the previous section, it was assumed that the process manager was interested in minimizing the average cycle time subject to some limits in the number of individual resources and an overall limit on the total amount of resources. This is just one optimization model among several possible variations. For example, the manager might wish to minimize the average cycle time subject to a budget constraint. In this case, the global constraint would limit the amount of money instead of the number of people. Because the simulation model of this example is set up for 1 working week, it would be possible to try to optimize performance subject to a weekly labor cost budget. Suppose the budget is \$8500, and the weekly salaries are \$500 for a clerk, \$650 for a technician, \$700 for an agent, and \$790 for a supervisor. The resulting constraint would be

$$500 \times \text{Clerks} + 650 \times \text{Technicians} + 790 \times \text{Supervisors} + 700 \times \text{Agent} \leq 8500$$

To enforce this constraint, open the Optimizer block dialog and click on the Constraints tab. The global constraint must be replaced with the following one:

$$\text{if}(500 \times \text{Clerks} + 650 \times \text{Technicians} + 790 \times \text{Supervisors} + 700$$
$$\times \text{Agents} > 8500)\text{Reject} = \text{TRUE};$$

An optimization run with the new constraint results in a solution with 13 employees (four clerks, two technicians, three supervisors, and four agents) and an estimated average cycle time of 58.5 minutes.

Now, suppose the manager is concerned with limiting the maximum cycle time (denoted by *MaxCycleTime*) at a minimum cost. Ideally, the manager would like to use a requirement that rejects solutions for which the maximum cycle time exceeds a desired limit. However, ExtendSim does not allow setting constraints on output variables—that is, variables that hold output values, such as cycle time or resource utilization. Therefore, to achieve the desired balance between customer service (modeled as a limit on the maximum cycle time) and cost, the objective function needs to be reformulated. A typical formulation for this type of problem consists of creating a so-called *penalty function*. This function penalizes deviations from a target value. In the example, the target value is the limit on the maximum cycle time that the manager desires. Assume that 120 minutes is such a limit. In other words, the manager wants to find a minimum-cost staffing mix of the process that is predicted to complete each transaction in 120 minutes or less. The penalty function, then, should penalize all solutions that exceed the desired limit. The Optimizer will not discard these solutions, but the penalty function will make them look unattractive. The penalized objective function for a cost-minimization problem has the following form:

$$\text{Minimize Cost} + \text{Penalty Function}$$

The magnitude of a penalty function should be such that the Optimizer would prefer a solution for which the penalty is zero. A possible penalty function for the current example might be the following:

Penalty Function = Number of minutes that *MaxCycleTime* exceeds 120 minutes

The problem with such a penalty function is that the cost is in units that are two orders of magnitude larger than the time units. That is, the cost is in the thousands, while the cycle time is in the tens. A sensible solution, then, is to use a multiplier for the penalty function in such a way that the deviations from the target have a weight that is significantly larger than the cost to force the optimizer to focus on solutions that meet the desired constraint.

Alternatively, the objective function can be formulated with an "if" statement that is similar to the one used to impose global constraints. The idea is to assign a large cost to solutions that do not meet the requirement. The cost should be larger than any cost of a feasible solution. In this case, the largest possible labor cost occurs when the maximum limits of all Resource Pools are chosen:

$$500 \times 10 + 650 \times 10 + 790 \times 5 + 700 \times 10 = 22,450$$

Therefore, solutions that do not meet the requirement could be assigned a cost larger than 22,450, making the optimizer prefer any feasible solution to any solution for which the requirement is not satisfied. The Clone Cursor is used to add the maximum cycle time variable (*MaxCycleTime*) to the Optimizer block. This output variable is found in the Statistics tab of the Information block. Once the variable has been added to the Optimizer block, the objective function can be reformulated:

```
if(MaxCycleTime>120)
MinCost=22500;
else
MinCost=500*Clerks+650*Technicians+790*Supervisors+700*Agents;
```

The value of 22,500 is large enough that any solution with a maximum cycle time exceeding the target limit of 120 minutes will have an objective function value (i.e., the value of MinCost) that is larger than the total labor cost of any feasible solution. ExtendSim, through its ModL programming language, offers a fair amount of flexibility to formulate a wide range of objective functions to implement quite complex optimization models.[4]

An optimization run of this model (with the Quicker Defaults) results in a process configuration with three clerks, three technicians, three supervisors, and five agents for a total weekly cost of $9320. To confirm the validity of these results, a simulation of 50 runs is performed. Before this validation runs, the Statistics block from the Value Library is added to the model. This block accumulates results from multiple runs and computes statistics, including confidence intervals. In the Statistics tab, the "mixed blocks (clone drop)" option is selected in the "Block type to report." Then, the Clone Cursor is used to drag and drop the maximum cycle time and the average cycle time from the Information block. Finally, the simulation is changed to 50 runs (in the "Simulation Setup ..." of the Run menu). After the simulation is performed, the option "confidence interval (compressed)" is chosen. The results show that with the suggested staffing levels, the 95 percent confidence interval for the average cycle time is 54.29 ± 0.47 minutes, and the interval corresponding to the maximum cycle time is 114.1 ± 3.59 minutes. Note that the interval for the maximum cycle time does not contain the 120-minute limit, and therefore it can be concluded, with a high level of confidence, that the solution found with the Optimizer will meet the specified requirement.

10.5 Optimization of Process Simulation Models

The best way to appreciate the benefits of simulation optimization in the area of business process modeling and design is to see it in action. This section describes two applications of simulation optimization in business process improvement. The following examples have been adapted from April et al. (2006).

10.5.1 Configuring a Hospital Emergency Room Process

Consider the operation of an ER in a hospital. Figure 10.13 shows a high-level view of the overall process. The process begins when a patient arrives through the doors of the ER and ends when a patient is either released from the ER or admitted into the hospital for further treatment. On arrival, patients sign in, are assessed in terms of their condition, and are transferred to an ER room. Depending on their condition, patients must then go through the registration process and through the treatment process before being released or admitted into the hospital.

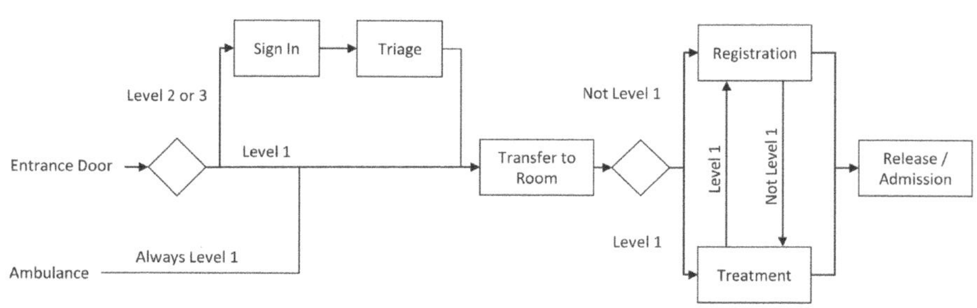

FIGURE 10.13
High-level view of ER process.

Arriving patients are classified into different levels according to their condition, with Level 1 patients being more critical than Levels 2 and 3. Level 1 patients are taken to an ER room immediately upon arrival. Once in the room, they undergo their treatment. Finally, they complete the registration process before being either released or admitted into the hospital for further treatment. Level 2 and Level 3 patients must first sign in with an administrative clerk. After they have signed in, their condition is assessed by a triage nurse, and then they are taken to an ER room. Once in the room, Levels 2 and 3 patients must first complete their registration; then, they go on to receive their treatment, and finally, they are either released or admitted into the hospital for further treatment. The treatment process consists of the following activities:

- A secondary assessment performed by a nurse and a physician
- Laboratory tests, if necessary, performed by a patient care technician (PCT)
- The treatment itself, performed by a nurse and a physician

The registration process consists of the following activities:

- A data collection activity performed by an administrative clerk
- An additional data collection activity performed by an administrative clerk if the patient has worker's compensation insurance
- A printing of the patient's medical chart for future reference, performed by an administrative clerk

On average, 90 percent of all patients are released from the ER, while the remaining 10 percent are admitted into the hospital for further treatment. The final release/hospital admission process consists of the following activities:

- In the case of release, either a nurse or a PCT fills out the release papers (whoever is available first).
- In the case of admission into the hospital, an administrative clerk fills out the patient's admission papers. The patient must then wait for a hospital bed to become available. The time until a bed is available is handled by an empirical probability distribution. Finally, the patient is transferred to the hospital bed.

The ER has the following resources:

- Nurses
- Physicians
- PCTs
- Administrative clerks
- ER rooms

In addition, the ER has one triage nurse and one charge nurse at all times. Due to cost and layout considerations, hospital administrators have determined that the staffing level must not exceed seven nurses, three physicians, four PCTs, and four administrative clerks. Furthermore, the ER has 20 rooms available; however, using fewer rooms would be beneficial, since other departments in the hospital could use the additional space more profitably. The hospital wants to find a configuration of resources that minimizes the total asset cost. The asset cost includes the staff's hourly wages and the fixed cost of each ER room used. Management wants to make sure that, on average, Level 1 patients do not spend more than 2.4 hours in the ER. This can be formulated as an optimization problem as follows:

Minimize the Expected Total Asset Cost

Subject to the following constraints:

- Average Level 1 cycle time is less than or equal to 2.4 hours
- Number of Nurses is greater than or equal to 1 and less than or equal to 7
- Number of Physicians is greater than or equal to 1 and less than or equal to 3
- Number of PCTs is greater than or equal to 1 and less than or equal to 4
- Number of Administrative Clerks is greater than or equal to 1 and less than or equal to 4
- Number of ER Rooms is greater than or equal to 1 and less than or equal to 20

This is a relatively simple problem in terms of size with five decision variables (one for each labor pool) and six constraints. However, the total number of possible solutions is $7 \times 3 \times 4 \times 4 \times 20 = 6720$. Although not all of these solutions are feasible with respect to the first constraint, the feasibility of the solution is not known until it is simulated and the cycle time is determined. In other words, the first constraint is what is referred to as a *requirement* in Section 10.3. Simulating all possibilities would have required days of computer time.

Instead of examining all possible solutions, this problem was approached with the OptQuest optimizer and the SIMPROCESS[5] simulation software. As a base case, it was decided to use the upper resource limits provided by hospital administrators to obtain a reasonably good initial solution. This configuration yielded an Expected Total Asset Cost of $36,840 for 5 days of operation and a Level 1 patient cycle time of 1.91 hours. Once the problem was set up in OptQuest, the optimizer performed 100 iterations (experiments) of five simulation runs per iteration (each run simulated 5 days of the ER operation). After 21 iterations, OptQuest identified a solution with four nurses, two physicians, three PCTs, three administrative clerks, and 12 ER rooms. The expected cost of this solution is $25,250 (a 31 percent improvement over the base case) with an average Level 1 patient cycle time of 2.17 hours. The time to run all 100 iterations was approximately 28 minutes.

After obtaining this solution, the analysts redesigned some features of the current model to improve the cycle time of Level 1 patients. In the proposed model, it was assumed that Level 1 patients could go through the treatment process and the registration process in parallel. That is, the assumption was that, while the patient is undergoing treatment, the registration process is being done by a surrogate or whoever is accompanying the patient. If the patient's condition is very critical, then someone else can provide the registration data; however, if the patient's condition allows it, then the patient can provide the registration data during treatment. In the process view of Figure 10.13, the change amounts to the elimination of the arrow from Treatment to Registration.

If the current resource level is used under the new process, the asset cost estimation does not change, but the expected Level 1 patient cycle time is reduced from 2.17 to 1.98 hours, a 12 percent improvement. The optimization of the new model results in a solution with four nurses, two physicians, two PCTs, two administrative clerks, and nine ER rooms. This configuration yields an Expected Total Asset Cost of $24,574 and an average Level 1 patient cycle time of 1.94 hours. In this example, simulation optimization played a critical role in identifying the configurations that are predicted to yield the best performance of a current and a redesigned process. These results allow hospital management to make informed decisions with respect to process changes that will improve asset utilization and service.

10.5.2 Staffing Levels for a Personal Insurance Claims Process

A personal claims department in an insurance company handles claims made by their clients. Claims arrive according to a Poisson process with a mean interarrival time of 5 minutes. Figure 10.14 depicts the personal claims process as a service system map (SSM). Recall that an SSM is an extension of traditional flowcharting where horizontal bands are used to organize activities in terms of the different players in a process (see Chapter 4).

The first band corresponds to work done by a claims handler located at the client's local service center. On arrival of a claim, the claims handler determines whether the client has a valid policy. If the policy is not valid (5 percent of all cases), then the case is terminated; otherwise (95 percent of all cases), the claims handler enters the appropriate information in the system. In the second band, an assessor located at the service center receives the information from the claims handler. The assessor first determines whether the claim is covered by the client's policy. If the claim is not covered (5 percent of all cases), the case is terminated; otherwise (95 percent of all cases), the assessor approves the preliminary estimate of the damage. If the damage exceeds $2000 (35 percent of all cases), the claim is sent to an assessor

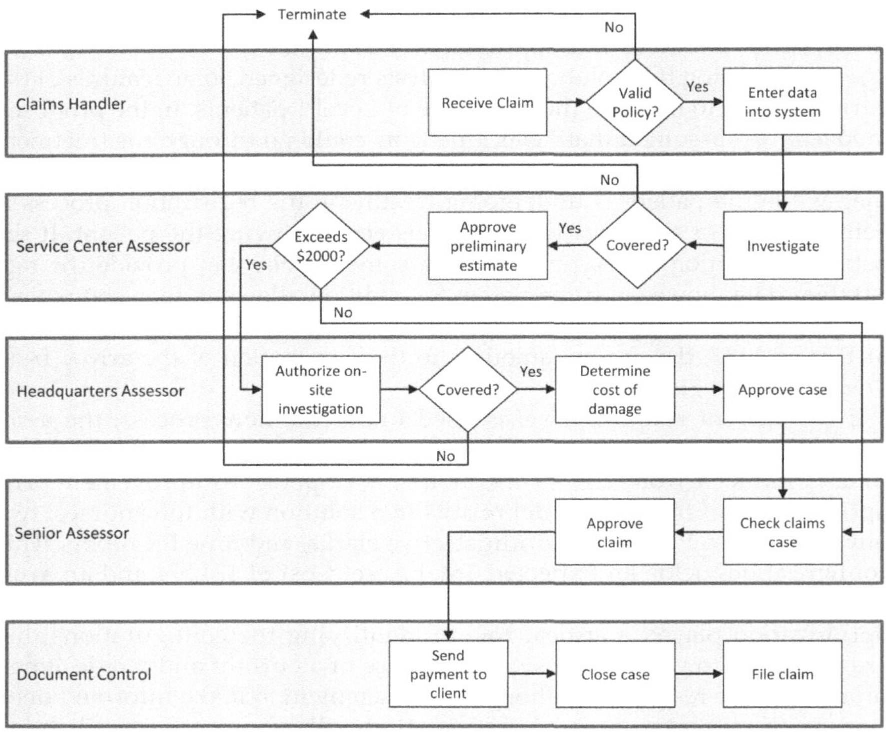

FIGURE 10.14
Service system map of the personal claims process.

at headquarters for approval; otherwise (65 percent of all cases), it is sent directly to a senior assessor.

The third band corresponds to the assessor at headquarters. The assessor first authorizes the on-site investigation of the accident. If the investigation determines that the incident is not covered by the client's policy (2 percent of all cases), then the case is terminated; otherwise (98 percent of all cases), a final price is determined, and the case is approved.

In the fourth band, the senior assessor receives the claim, checks it, completes it, and provides the final approval. Once the claim is approved, it is sent to documentation control. Documentation control, in the fifth band, is in charge of processing the payment to the client, closing the case, and finally filing the claim. The optimization problem in this example is to find staffing levels for each of the five resource types to minimize headcount while keeping average throughput above 1500 claims during 4 weeks and a maximum of 20 people in each resource pool.

A what-if analysis of all the possible solutions to this problem would require 3.2 million (i.e., 20^5) experiments. Although not all scenarios are

TABLE 10.6

Best Solutions for the Claims Process Found with Simul8 and OptQuest

Solution	Claims Handler	Service Center	Headquarters	Senior Assessor	Doc. Control	Throughput	Cycle Time	Headcount
1	9	17	17	15	16	1568	639	74
2	9	17	17	14	16	1564	658	73
3	9	17	16	15	16	1567	646	72
4	9	18	12	15	11	1622	503	65
5	9	18	11	15	11	1621	510	64

feasible, because a number of them do not satisfy the throughput requirement, the feasibility of a scenario is not known before the simulation run. The process was simulated with Simul8[6] and optimized with OptQuest. The optimization process consisted of 100 iterations of five simulation runs each.

Since some of the solutions obtained from the optimization are relatively close in terms of throughput and cycle time, a closer examination of a reduced set of solutions was performed to assess the precision of the results. Table 10.6 shows the best five solutions found with OptQuest by conducting an experiment of 20 simulation runs. The table allows the analysis of the trade-offs between headcount and throughput or cycle time to decide which configuration best aligns with service levels and process goals. For example, Solutions 1, 2, and 3 are statistically the same. Solutions 4 and 5 are significantly better than 1, 2, and 3 in terms of headcount, throughput, and cycle time, so the focus should be on these solutions. To select the best, an additional experiment with 60 simulation runs was performed to obtain confidence intervals around the throughput and cycle times. These intervals overlapped, indicating that both solutions are essentially the same, statistically speaking. Therefore, the fourth solution seems more attractive in terms of reducing the headcount by one.

10.6 Summary

This chapter introduced the notion of optimizing simulations. In recent years, this has been a relevant and exciting research topic. Coupling metaheuristic optimizers and simulation modeling has resulted in powerful tools for designing effective business processes.

After an introduction to metaheuristic optimization, a tutorial was presented that shows how to set up optimization models using the Optimizer block within ExtendSim. The Clone Cursor is needed to create a connection between the variables in the simulation model and the variables used for optimization in the Optimizer. The Optimizer is fairly simple to use; however, one must keep in mind that any metaheuristic optimizer can converge to a suboptimal solution, particularly if it is not run long enough. ExtendSim recommends that at least 100 generations are used, and the model should be run several times with convergence to the same or near solution. Finally, longer confirmation runs should be performed with the proposed solution before the solution is implemented.

Appendix 10A: Evolutionary Computation

The idea of applying the biological principle of natural evolution to artificial systems was introduced by John Holland in the early 1970s. Usually grouped under the term *evolutionary algorithms* or *evolutionary computation,* one finds the domains of genetic algorithms, evolution strategies, evolutionary programming, genetic programming, and scatter search. Evolutionary algorithms have been applied successfully to numerous problems from different domains, including optimization, machine learning, economics, social systems, ecology, population genetics, studies of evolution, and learning.

An evolutionary algorithm is an iterative procedure that consists of a constant-size population of individuals, each one represented by a finite string of symbols, known as the *chromosome* (or *genome*), encoding a possible solution in a given problem space. This space, referred to as the *search space,* comprises all possible solutions to the problem at hand. Generally speaking, genetic algorithms are applied to spaces that are too large to be searched exhaustively (such as those in combinatorial optimization).

The standard evolutionary algorithms proceed as follows. An initial population of individuals is generated at random or heuristically. At every evolutionary step, known as a *generation,* the individuals in the current population are decoded and evaluated according to some predefined quality criterion, referred to as the *fitness* or *fitness function.* To form a new population (the next generation), individuals are selected according to their fitness. Many selection procedures are currently in use, one of the simplest being Holland's original fitness-proportionate selection, in which individuals are selected with a probability proportional to their relative fitness. This ensures that the expected number of times an individual is chosen is approximately proportional to its relative performance in the population. Thus, high-fitness ("good") individuals stand a better chance of "reproducing," and low-fitness ones are more likely to disappear.

Genetically inspired operators are used to introduce new individuals into the population (i.e., to generate new points in the search space). The best known of such operators are crossover and mutation. Crossover is performed with a given probability between two selected individuals, called *parents,* by exchanging parts of their genomes (encoding) to form two new individuals called *offspring;* in its simplest form, substrings are exchanged after a randomly selected crossover point. This operator tends to enable the evolutionary process to move toward promising regions of the search space. The mutation operator is introduced to prevent premature convergence to local optima by randomly sampling new points in the search space. Mutation entails flipping bits at random with some (small) probability. Evolutionary algorithms are stochastic iterative processes that are not guaranteed to converge to the optimal solution; the termination condition can be specified as some fixed, maximum number of generations or as the attainment of an acceptable fitness level for the best individual.

As noted, the problem with optimization algorithms based on evolutionary computation stems from the inability to tell when the optimal solution has been found or even whether the optimal solution has been simulated. A good approach within ExtendSim is to allow the optimization to continue for an "adequate" amount of time and check to see whether the population of solutions converges. After that, the user could try the optimization procedure several times to make sure the answers from several optimization runs agree or are close.

Exercises

1. An operations analyst developed a simulation model for the design of a warehouse. She wants to use the model to study two levels of four design factors, as shown in Table 10.7. The average cycle time (ACT) for all stock keeping units (SKUs) is the criterion used to assess the performance of the simulation experiments. ACT (in hours) is calculated as the average time difference between receiving and shipping SKUs in a simulation run.

TABLE 10.7

Design Matrix for Exercise 10.1

Design Point	Size	Layout	Operational Policies	Manpower	ACT
1	Small (−)	Traditional (−)	Random (−)	Low (−)	10.92
2	Large (+)	Traditional (−)	Random (−)	Low (−)	11.02
3	Small (−)	Fishbone (+)	Random (−)	Low (−)	10.82
4	Large (+)	Fishbone (+)	Random (−)	Low (−)	11.71
5	Small (−)	Traditional (−)	Standard (+)	Low (−)	9.29
6	Large (+)	Traditional (−)	Standard (+)	Low (−)	9.07
7	Small (−)	Fishbone (+)	Standard (+)	Low (−)	8.59
8	Large (+)	Fishbone (+)	Standard (+)	Low (−)	9.19
9	Small (−)	Traditional (−)	Random (−)	High (+)	10.08
10	Large (+)	Traditional (−)	Random (−)	High (+)	9.91
11	Small (−)	Fishbone (+)	Random (−)	High (+)	9.58
12	Large (+)	Fishbone (+)	Random (−)	High (+)	10.27
13	Small (−)	Traditional (−)	Standard (+)	High (+)	7.9
14	Large (+)	Traditional (−)	Standard (+)	High (+)	8.4
15	Small (−)	Fishbone (+)	Standard (+)	High (+)	8.14
16	Large (+)	Fishbone (+)	Standard (+)	High (+)	8.28

a. Use a factorial design to calculate the main effects of each input factor for the warehouse simulation model. What factor has the largest effect in reducing the average cycle time? What factor has the largest effect in increasing the average cycle time? (Note that there are four factors in this design and therefore the calculation of the average effect is the sum of the responses (with the corresponding sign) divided by 8.)

b. Determine whether the factors identified in part (a) interact with each other.

2. Consider the process of insuring a property described in Exercise 12 of Chapter 8.

a. Develop a simulation model of this process. Assume that work in process (WIP) at the end of each day becomes the beginning WIP for the next day.

b. Assuming 8-hour workdays, add data collection to estimate average cycle time, WIP, and throughput. Run a sufficiently long simulation to verify that Little's law (see Section 5.1.4) correctly establishes the relationship among average cycle time, WIP, and throughput.

c. Using average values, estimate the capacity utilization of this process. Compare your estimated capacity utilization with the results from the simulation model.

d. If one unit of resource (of any type) could be added to the process, where would you add it to decrease average cycle time, decrease average WIP, and increase average throughput? Did the change make a (statistically) significant impact on the performance of the process?

3. The insurance company in Exercise 10.2 would like to configure the process to optimize performance. The company would like to find the best labor pool capacities to minimize average cycle time.

a. Set up a simulation optimization model to find the number of clerks, underwriters, raters, and policy writers to minimize the average cycle time without exceeding 24 employees in total. Use simulation runs of 10 days during the optimization process and a range of 1–10 employees in each labor pool.

b. The company has determined that reducing the average cycle time is not as important as being able to minimize the proportion of requests that require more than 6 hours to complete. Change the simulation optimization model of Part (a) to find the number of clerks, underwriters, raters, and policy writers to minimize the proportion of requests that are completed in more than 6 hours without exceeding 24 employees in total. Use simulation runs of 10 days during the optimization process.

c. Suppose that the insurance company would like to minimize the total labor cost while maintaining a desired level of service (e.g., limiting the proportion of requests with cycle times exceeding 6 hours to 10 percent). Modify the simulation model to solve this optimization problem. *Hint*: The proportion of cycle times that exceed 6 hours is a simulation output. ExtendSim does not include the capability of setting constraints on output variables. Therefore, an objective function that penalizes infeasible solutions is necessary to handle this situation (see Section 10.4.2).

Simulation Optimization Projects

This chapter concludes with four simulation optimization projects. The goals of the projects are to create a simulation model of the process and to apply the Optimizer block to optimize key input parameters to the model.

Project 1: Emergency Room Staffing[7]

An emergency department (ED) is open 24 hours a day and receives an average of 145 patients daily. Besides its internal capacity, the ED shares resources with other hospital services, such as X-rays, scanners, clinical laboratories, and pharmacy. Patients arriving at the ED follow a process as depicted in Figure 10.15. The process begins when a patient arrives through the doors of the ED and ends when a patient is either released from the ED or admitted into the hospital for further treatment. There are two types of arrivals: walk-in patients, who are required to see a receptionist, and ambulance patients, who bypass the receptionist and enter the examination room directly. The acuity of the patient's illness is assessed by a doctor in the examination room. Also in the examination room, doctors will decide whether the patient needs further tests, such as X-rays and clinical lab tests, performed by a patient care lab technician. Patients are classified as critical (Category 1) or noncritical according to their condition. After an assessment has been performed by the doctor, the noncritical patients are further classified into two categories. Category 2 patients are asked to wait for a minor treatment, which is performed by a nurse in the treatment room. Category 3 patients receive their medication and are released from the hospital. Each critical patient is assigned to a bed in the emergency room, where he/she receives complete treatment and stays under close observation. The treatment services in the emergency room are provided by a nurse and a doctor; the doctor is called from the examination room when needed. Finally, critical patients are either released or admitted into the hospital for further treatment. Patients who arrive at the hospital in an ambulance are considered

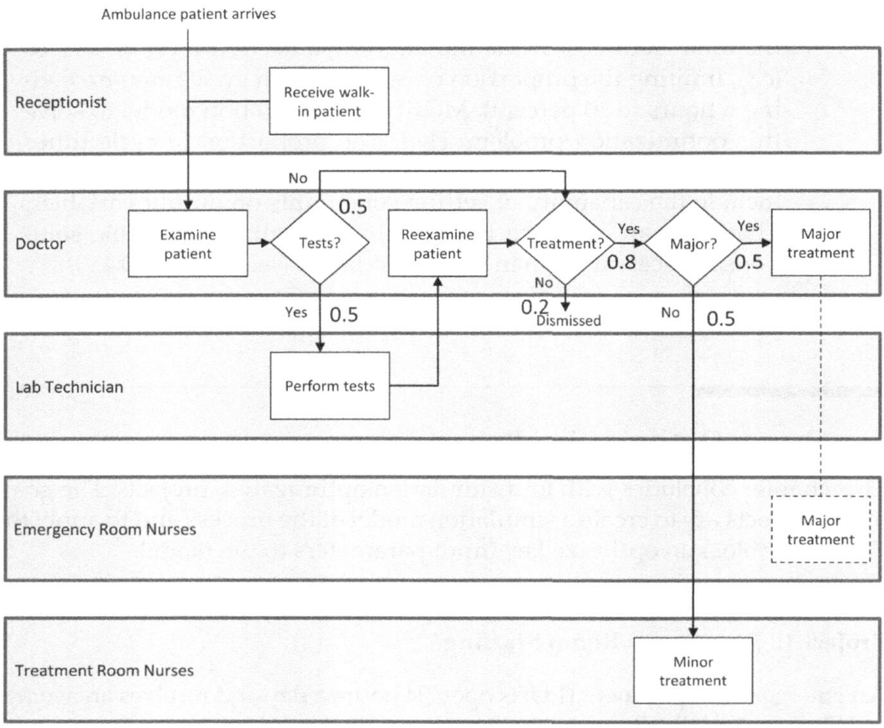

FIGURE 10.15
Service system map of emergency department.

critical patients (Category 1) and are rushed immediately to the emergency room. It is observed that 88 percent of all patients are released from the emergency room, while the remaining 12 percent are admitted into the hospital for further treatment. The ED has the following resources:

- Receptionists
- Doctors
- Lab technicians
- Treatment room nurses
- Emergency room nurses

Due to layout considerations, hospital administrators have determined that the staffing level must not exceed three receptionists, six doctors, three lab technicians, three treatment room nurses, and eight emergency room nurses. The hospital would like to find the configuration of resources that maximizes patient throughput (patients dismissed per unit of time) subject to budget constraint and a constraint imposed on the average waiting time in the system for patients of Category 1.

TABLE 10.8

Mean Interarrival Times (in Minutes) for Walk-in Patients

0:00	2:00	4:00	6:00	8:00	10:00	12:00	14:00	16:00	18:00	20:00	22:00
11.43	15.79	20.00	12.50	8.57	7.27	6.67	7.74	8.57	7.50	9.23	18.43

A comprehensive survey has been carried out at the ED to collect data on the arrival process, the service times at the examination room, the service times at the treatment room, and the total turnaround time in the ED. After observing the process for 3 weeks and collecting additional data from interviewed doctors, nurses, and hospital personnel in charge of each of these activities, the results of these efforts were used to determine the best theoretical distribution to represent each stage of the process under study. It was determined that the interarrival times of walk-in patients follow an exponential distribution with a mean value that depends on the time of the day, as shown in Table 10.8. The interarrival times of ambulance patients follow an exponential distribution with a mean value of 30 minutes. The routing probabilities of patients at each stage inside the system are shown in Figure 10.15. The service time distributions and the values of their parameters are shown in Table 10.9. Due to the hospital request for information privacy, especially in budget data details, budget data were recoded in terms of budget units (BU). The cost units for the staff are as follows: 0.4 BU per receptionist, 1.2 BU per doctor, 0.5 BU per lab technician, and 0.3 BU per treatment or emergency nurse.

1. Build an ExtendSim model that simulates the operation of the emergency room department and that collects cycle time information for Category 1 patients. Set up the model to simulate 100 days (in the Simulation Setup dialog under the Run Menu). Use trial and error to find the minimum staffing levels that will keep the average cycle time for Category 1 patients under 3.5 hours. *Hint*: Click on the "ExtendSim Help ..." item in the Help menu to access the User Guide. Search for "Random intervals with dynamic parameters." This section of the User Guide explains how to combine the Create block and the Lookup Table block to generate items for which the mean interarrival time changes during the simulation.

TABLE 10.9

Activity Times for Emergency Department

Activity	Distribution of Processing Times
Reception	Uniform between 5 and 10 minutes
Lab tests	Triangular with parameter values of 10, 20, and 30 minutes
Examination	Uniform between 10 and 20 minutes
Reexamination	Uniform between 7 and 12 minutes
Minor treatment	Uniform between 20 and 30 minutes
Major treatment	Uniform between 60 and 120 minutes

2. Add the Optimizer block to the simulation model. Use the Clone Cursor to add the initial value of each of the Resource Pool blocks to the Optimizer block. This creates five decision variables in the Optimizer block. Also add the average cycle time for Category 1 patients to the Optimizer block. Set up an optimization model that minimizes the average cycle time for Category 1 patients while restricting the labor budget to 10 BU. Set up the simulation for 30 days and the Optimizer to run one sample per case (Run Parameters tab).
3. Compare the two solutions found in Parts 1 and 2 using simulation runs of 100 days. The comparison should be made in terms of cost, average and maximum cycle time for Category 1 patients, and utilization of all resources.

Project 2: Call Center Configuration[8]

Consider a telephone call center that handles three types of calls: (a) technical support, (b) sales, and (c) order-status checking. The time between arrivals of calls follows an exponential distribution with a mean of 1 minute. Calling customers choose one of the three options with equal probability (i.e., each with probability 1/3). A customer selecting (a) must also select one of three products (1, 2, or 3) for which to obtain technical support. Technical support for these products is requested with equal probability. The support is provided by technical personnel who specialize in each product: Tech1, Tech2, and Tech3 for products 1, 2, and 3, respectively. There is also a pool of TechAll technicians who are capable of providing support for any of the products. However, these technicians are only called to help when no specialized technician is able to attend an incoming call. For instance, if there are two Tech1 technicians, and both are busy when a new call for Product 1 support comes in, then a TechAll technician (if available) takes care of the new call. The time required to provide technical support is independent of the product and follows an exponential distribution with a mean of 3 minutes.

Customers selecting (b) are routed to a sales representative or put on hold if all representatives are busy attending other calls. The time for the sales activity follows a triangular distribution with a minimum value of 2 minutes, a most likely value of 5 minutes, and a maximum value of 10 minutes. For option (c), a computerized system handles requests for checking the status of an order, and it requires an exponentially distributed time with a mean of 2 minutes. There is no practical limit on the number of calls that the computerized system can handle at a time. However, after receiving this computerized information, there is a probability of 1/2 that the customer asks to speak with a sales representative. Then, the system routes the call to a sales representative, but if they are all busy, the customer is put on hold and is given a lower priority than customers that chose (b) when entering the process. Customers calling when there are no phone lines available get a busy signal and cannot enter the system.

The telephone lines are open from 8:00 a.m. to 6:00 p.m., so no incoming calls are accepted after 6:00 p.m. However, all calls that are already in the system are completed even after 6:00 p.m. Each simulation run starts and ends with an empty system. The manager of the call center wants to restrict the number of rejected calls (i.e., those calls that find a busy signal) to no more than 30 per day. This represents about 5 percent of the expected number of calls in any given day. The manager would like to know how to configure the center to meet this requirement at a minimum cost. The configuration consists of setting the number of phone lines and the staff levels. The labor costs per day are $100 per specialized technician, $150 per sales representative, and $200 per TechAll technician.

1. Develop an ExtendSim simulation of the call center. A run of the model should represent a single day. Use a Shift block to limit the arrival of calls to between 8:00 a.m. and 6:00 p.m. The arrivals should occur for 10 hours, but the simulation should continue to run until all calls complete the process. Also, the queues for technical support should be linked to two resource pools, one for the corresponding specialized technicians (Tech1, Tech2, or Tech3) and one for the TechAll technicians. The first resource pool in the list should be the one for the specialized technicians and the second one for the TechAll technicians. The "Take from any resource pool" option should be selected. A priority queue should be used for the calls waiting for a sales representative. Set up the model to operate with 10 phone lines, two specialized technicians of each type, one TechAll technician, and two sales representatives. Execute 30 runs and estimate the percentage of lost calls. (*Hint*: Use a Statistics block and the Clone Cursor to collect data on the number of calls exiting the system as "lost" and the number of calls exiting the system as "completed".)
2. Add an Optimizer block to the simulation model to find the configuration with the minimum cost that meets the requirement that the number of lost calls on a day is less than or equal to 30. Set the limits for phone lines to a minimum of 1 and a maximum of 20, and for specialized technicians and sales representatives to a minimum of 1 and a maximum of 5. The limits on TechAll technicians should be set to a minimum of 0 and a maximum of 3. Use the "if-statement" described in Section 10.4.2 to enforce the lost-call requirement. Run the optimization with the Quicker Defaults for a random model. What is the best configuration that the optimizer finds? Use the configuration for a 30-run simulation and confirm that the predicted number of lost calls does not exceed 30 per day.
3. Suppose that the manager wants to find a configuration that takes into account a customer service cost associated with putting calls on hold. In particular, the manager wants to impose a cost of $1 per

minute that a caller must wait to be served. Add this cost to the objective function and rerun the optimization. What is the new configuration? What are the differences between the configuration in Part 2 and this new configuration in terms of resources and waiting times? (*Hint*: In the Options tab of the Create block, check the "Define item costs" and set the waiting cost to $1/minute. In the Options tab of all the Queue blocks, check the "Calculate waiting costs" box. Add the total cost from the Results tab of the Queue blocks as output variables in the Optimizer block. Add these costs to the objective function.)

Project 3: Loan Application Process

The following description is for a loan application approval process. Loan amounts tend to be large; therefore, some require several days to be reviewed and approved. When a loan is approved, the loan company begins to collect interest. The company's profit is based on the interest it collects, so it is imperative that processing time be reduced to a minimum.

1. Loan applications are sent into a main office by loan officers at a random rate of five per day. It is not possible to predict exactly how many will arrive every day, but the interarrival time is believed to be exponentially distributed. Since a business day consists of 8 hours, on average, the time between arrivals is 1.6 hours. A clerk checks the loan applications for completeness. This check takes about 48 minutes (a time that is also exponentially distributed). After the application passes the initial check, it is forwarded, using interoffice mail, to a processing team. Interoffice mail takes 0.5 days 50 percent of the time, 1 day 25 percent of the time, 1.5 days 12.5 percent of the time, and 2 days 12.5 percent of the time. The time required to process an application by a single person follows a triangular distribution with a minimum of 1 day, a most likely time of 1.5 days, and a maximum of 2 days. Use average processing times and capacity analysis (see Section 5.2.1) to determine the number of clerks and processing team members that will result in a theoretical process capacity of at least five applications per day. Build an ExtendSim model of the process with the resources given by the capacity analysis calculation. Use days as the global time unit in the Simulation Setup—also adjust the Non-Calendar date definitions to 8 hours per day—and run the model for 1000 days. Note that this is a nonterminating process. Plot the utilization of both resources and determine whether the process is stable.
2. Consider now that loan applications are broken into five types (A, B, C, D, and E) based on their dollar value. The percentages of each type are shown in Table 10.5 along with their corresponding processing time distribution (actual dollar values are not important

at this point, and time is given in 8-hour days). Three labor pools (1, 2, and 3) for processing loan applications are formed under the current process. The application types that can be handled by each pool are shown in Table 10.10. Use capacity analysis to determine the minimum staffing levels (i.e., number of employees in each labor pool) necessary to make the process stable. Confirm that these levels are appropriate by modifying and running the simulation model of Part 1.

3. Members of the processing team may find that some applications have not been filled in correctly or are missing some information. Historically, 95 percent of Type 1 applications are correct, 92 percent of Type 2 are correct, 90 percent of Type 3 are correct, 85 percent of Type 4 are correct, and 80 percent of Type 5 are correct. When an application is missing information or has incorrect information, it is sent back to the originator, a process that takes 1 day 25 percent of the time, 2 days 50 percent of the time, and 3 days 25 percent of the time. The number of working days includes the time required to send the application back to the field office, the time required for the loan officer to correct the application, and the time required to send it into the main office directly to the processing team. Do the staffing levels for Types 1, 2, and 3 of Part 2 need to be changed for the process to be stable? Note that the rework occurs after the processing activity is performed and that the activity must be performed again once the application is returned by the field officer.
4. A committee has proposed the use of an expert workflow automation system that will eliminate all problems with applications (i.e., it will eliminate the rework of Part 3); implement electronic data interchange (EDI) and therefore eliminate the interoffice mail; and reduce processing time for Type 1 applications by 30 percent, Type 2 applications by 33 percent, Type 3 applications by 40 percent, Type 4 applications by 60 percent, and Type 5 applications by 75 percent. In addition, the initial check will not be needed, and applications will go directly to the processing team. The clerk will stay with the company but will perform other duties. Top

TABLE 10.10

Loan Application Data

Loan	Labor Pool	Percentage	Processing Time (Days)
A	1	5	Exponential (3.5)
B	2	15	Exponential (2.75)
C	3	20	Triangular (2.5, 3.25, 4)
D	3	25	Triangular (2, 2.375, 2.75)
E	3	35	Triangular (1, 1.5, 2)

management has concluded that productivity must be factored into the labor cost calculations and has decided that labor cost per day should be calculated as follows:

Labor cost = (number of workers × daily rate)/utilization

Labor rates for loan processors depend on the loan type they process. Rates are as follows: Type 1 loan processors receive $350 per day, Type 2 receive $200 per day, and Type 3 receive $155 per day. How much should management be willing to pay per day for the new technology, given the opportunities for labor cost reductions? Compare the average cycle time of the proposed process with the technology and the process in Part 3. Is there a significant reduction in average cycle time?

Project 4: Process with Multiple Job Types and Deadlines

A process is designed to handle three types of jobs. Jobs arrive about every 10 minutes, with interarrival times following an exponential distribution. About 10 percent of the jobs are Type 1, about 30 percent of the jobs are Type 2, and about 60 percent of the jobs are Type 3. The process consists of 10 activities (labeled A to J), and each job is routed according to its type, as shown in Table 10.11. The activity times depend on each job type. The probability distribution of times and the resource (employee) type needed to perform each activity are fixed for all job types. Table 10.12 summarizes the time and resource data associated with each activity.

Currently, there are five employees of each type, that is, Types I, II, and III, and they are paid $1250 per week, $750 per week, and $500 per week, respectively. A cost is associated with cycle times greater than 80 minutes. In particular, if the cycle time for a Type 1 job is greater than 80 minutes, a $300 late fee must be paid. The late fees for Type 2 and 3 jobs are $200 and $10, respectively.

1. Create an ExtendSim simulation of this process that will run for 5 days of 10 working hours. This is a nonterminating process; that is, the work that is not completed at the end of a day remains in the system for the beginning of the next day. Use the "_Item priority" property in the Set block to set the job type: _Item priority = 1 for

TABLE 10.11

Routing of Jobs for Project 4.

Job	Routing
1	A, C, E, G, H, and I
2	B, C, D, E, G, H, and J
3	A, B, D, F, H, I, and J

TABLE 10.12

Process Time Information

Activity	Resource Type	Distribution	Parameters
A	I	Exponential	17
B	II	Normal	(8, 1)
C	I	Lognormal	(10, 3)
D	III	Gamma	(7, 2)
E	II	Uniform	(6, 9)
F	III	Exponential	12
G	III	Exponential	9
H	I	Uniform	(8, 20)
I	III	Gamma	(10, 3)
J	I	Normal	(10, 1)

Exponential: Mean
Normal: (Mean, Standard deviation)
Lognormal: (Mean, Standard deviation) with Location = 0
Gamma: (Scale, Shape) with Location = 0
Uniform: (Min, Max)

job-type 1, _Item priority = 2 for job-type 2, and _Item priority = 3 for job-type 3. Perform 50 runs to estimate the average late fee cost. (*Hint*: Use the Equation(I) block to calculate the late fees. The cycle time can be calculated as the difference between the system variable "CurrentTime" and the Timing Attribute set by the Create block. The output variable (i.e., the late fee) can be connected to a Mean & Variance block to calculate the average.)

2. Considering that the payroll for this process cannot exceed $6250 per day, use the Optimizer block to set up an optimization model to find the staffing levels that minimize the average late fees.
3. An analyst has suggested that a priority system should be used for this process. Specifically, the analyst suggests changing the priority rule in the queues from first-in-first-out to a rule where Type 1 jobs have highest priority, followed by Type 2 jobs and, finally, Type 3 jobs. Change the model to use the suggested priority rule and predict the average late fees with the staffing levels found in Part 2. Would this change improve the performance of the process in terms of late fees? Is there a better configuration of staff that takes advantage of the priority rule to reduce the late fees but maintains the same limit in total labor costs? (*Hint*: If the job types are set with the priority attribute as described in Part 1, then select "priority" in the "Allocate resources to qualify items by" in the Resource Pool block.)

Notes

1. Adapted from April et al. (2006).
2. OptQuest is a trademark of OptTek Systems Inc., Boulder, CO (www.opttek.com).
3. Adapted from www.anylogic.com/outpatient-appointment-scheduling-using-discrete-event-simulation-modeling/
4. A complete list of mathematical functions that can be used within ExtendSim can be found in the ExtendSim Help option under the Help menu. The function list starts on page 1043 (Bookmark: ModL – Reference – ModL functions – Math functions).
5. SIMPROCESS is a registered trademark of CACI (http://simprocess.com).
6. Simul8 is a registered trademark of Simul8 Corp. (www.simul8.com).
7. Adapted from Ahmed and Alkhamis (2009).
8. Adapted from Kleijnen et al. (2008).

References

Ahmed, M.A., and Alkhamis, T.M. 2009. Simulation optimization for an emergency department healthcare unit in Kuwait. *European Journal of Operational Research 198*: 936–942.

Andradóttir, S. 1998. *A review of simulation optimization techniques.* In *Proceedings of the 1998 Winter Simulation Conference,* Washington, DC, editors D.J. Medeiros, E.F. Watson, J.S. Carson, and M.S. Manivannan, pp. 151–158.

Antony, J., Coleman, S., Montgomery, D.C., Anderson, M.J., and Silvestrini, R.T. 2011. Design of experiments for non-manufacturing processes: Benefits, challenges and some examples. *Proceedings of the Institution of Mechanical Engineers, Part B: Journal of Engineering Manufacture 225*(11): 2078–2087.

April, J., Better, M., Glover, F., Kelly, J., and Laguna, M. 2006. Enhancing business process management with simulation optimization. In *Proceedings of the 2006 Winter Simulation Conference,* Monterey, CA, editors L.F. Perrone, F.P. Wieland, J. Liu, B.G. Lawson, D.M. Nicol, and R.M. Fujimoto, pp. 642–649.

Kamrani, F., Ayani, R., and Moradi, F. 2012. A framework for simulation-based optimization of business process models. *Simulation 88*(7): 852–869.

Kleijnen, J.P.C., van Beers, W., and van Nieuwenhuyse, I. 2008. Constrained optimization in expensive simulation: Novel approach. *European Journal of Operational Research 202*(1): 164–174.

Law, A.M. 2014. A tutorial on design of experiments for simulation modeling. In Proceedings of the 2014 Winter Simulation Conference, editors A. Tolk, S. Y. Diallo, I. O. Ryzhov, L. Yilmaz, S. Buckley, and J. A. Miller, pp. 66–80.

Law, A., and Kelton, D. 2000. *Simulation modeling and analysis.* 3rd ed. New York: McGraw-Hill.

Laguna, M. 2011. *OptQuest: Optimization of complex systems.* White paper, OptTek Systems Inc., www.opttek.com/white-papers (accessed: May 4, 2018).

11

Business Process Analytics

Business analytics has been defined as "the extensive use of data, statistical and quantitative analysis, explanatory and predictive models, and fact-based management to drive decisions and actions" (Davenport and Harris 2007). Complementary to this is the application of technology with the goal of exploiting data to understand and analyze business performance. Activities related to turning data into useful information to improve performance have been discussed under the umbrella of *business intelligence*; however, it has been argued that this area is in fact part of the larger field of business analytics, which centers on the following key organizational needs (Dietrich et al. 2010):

- *Information access.* This is the foundation of business analytics, which promotes informed and collaborative decision-making across the organization. Its main goal is to ensure that decision-makers understand how their area of business is doing so that they can make informed decisions.
- *Insight.* This relates to gaining a deeper understanding of why things are happening; for example, gaining a full view of customers (transaction history, segmentation, sentiment and opinion, etc.) to make better decisions and enable profitable growth.
- *Foresight.* This refers to leveraging the past to predict future outcomes so as to make decisions that will move the organization closer to achieving its objectives and meeting requirements.
- *Business agility.* This refers to driving real-time decision analysis and optimization in both people-centric and process/automated-centric processes.
- *Strategic alignment.* This emphasizes the aligning of the organization from strategy to execution by enabling enterprise and operational visibility as well as documenting the preferences, priorities, objectives, and requirements that drive decision-making.

There are three main categories of business analytics:

- *Descriptive analytics.* A set of tools that use data to understand and analyze business performance
- *Predictive analytics.* The extensive use of data and mathematical techniques to formulate explanatory and predictive models of business performance that uncover the inherent relationship among inputs and outputs/outcomes

DOI: 10.1201/9781032617237-11

- *Prescriptive analytics.* Mathematical modeling to determine effective alternative actions or decisions for a complex set of objectives, requirements, and constraints, with the goal of improving business performance

Collectively, these categories show that analytics facilitates (1) the achievement of business objectives through the reporting of data for trend analysis, (2) the creation of predictive models for forecasting, and (3) the optimization of business processes for enhanced performance. Dietrich et al. (2010) expand on the roles of each of the categories of business analytics.

Descriptive analytics is the most commonly used and most well understood type of analytics, which consists of categorizing, characterizing, consolidating, and classifying data. Descriptive analytics includes dashboards, reports (e.g., budget, sales, revenue, and costs), and various types of queries. Tools for descriptive analytics typically provide mechanisms for interfacing with enterprise data sources. These tools may include report generation and distribution as well as data visualization. Descriptive analytics facilitates both an understanding of the past and monitoring of the events occurring in real time.

Insight into what is happening now or has happened in the past can be useful in making decisions about the future, but descriptive analytics relies on the human review of data. In other words, the tools within this category of analytics provide significant insight into business performance and enable users to better monitor and manage their business processes. However, they do not contain techniques that estimate what might happen in the future or models to suggest decisions on what should be done next. Descriptive analytics often serves as a first step in the successful application of predictive or prescriptive analytics. Organizations that effectively use descriptive analytics typically have reached an agreement about what happened in the past and, consequently, are able to focus on the present and the future.

Predictive analytics uses the understanding of the past to make predictions about the future. Predictive analytics is applied both in real time to affect the operational process (for instance, real-time retention actions via chat messages or real-time identification of suspicious transactions) and in batch (e.g., by targeting new customers via websites or direct mail). Predictions are made by examining historical data to detect patterns and uncover relationships. For example, a particular type of insurance claim that falls into a category (pattern) that has proved troublesome in the past might be flagged for closer investigation. Predictive analytics can be classified into the following six categories:

- Data mining—What data are correlated with other data?
- Pattern recognition and alerts—When should action be taken to correct or adjust a process or a resource?

- Simulation—What could happen?
- Forecasting—What happens if these trends continue?
- Root cause analysis—Why did something happen?
- Predictive modeling—What will happen next if?

Predictive analysis applies advanced techniques to examine scenarios and helps to detect hidden patterns in large quantities of data to project future events. It uses techniques that segment and group data (e.g., transactions, individuals, or events) into coherent sets to predict behavior and detect trends. Forecasting is often applied to predict workload, which is then translated into resource (including human) requirements. In simulation, a sufficiently accurate model of the system is created to estimate or make predictions about future behavior under a variety of scenarios. Predictive modeling techniques can also be used to examine data to evaluate hypotheses. If each data point (or observation) consists of multiple attributes, then it may be useful to understand whether some combinations of a subset of attributes are predictive of a combination of other attributes. For example, insurance claims may be examined to validate the hypothesis that age, gender, and zip code can predict the likelihood of an auto insurance claim.

Once the past is understood, and predictions can be made about what might happen in the future, it is then important to identify the best response or action that takes into account both the goals of the organization and its limited resources. This is the area of *prescriptive analytics*. Improvements in the speed and memory size of computers, as well as significant progress in the performance of the underlying mathematical algorithms, have made the application of prescriptive analytics a reality. Many of the problems associated with prescriptive analytics are complex and involve too many choices or alternatives for a human decision-maker to consider effectively and within a reasonable time frame. Embedding problem-solving capabilities in decision support systems was impractical when the combination of computing power and solution methods was not able to provide timely results. Nowadays, however, results can be obtained in real time to support operational as well as tactical and strategic decisions.

Prescriptive analytics, based on mathematical optimization, is used to model a system of potential decisions, the interactions among these decisions, and the factors or constraints limiting the choices associated with the decisions. In addition to the models, the area of prescriptive analytics includes the development of algorithms that are designed to search for the best set of decisions that meet the constraints of the problem. Optimization is used extensively in many industries in applications ranging from strategic planning to operational issues. Applications include those where uncertainty in the data must be considered. Combining predictive and prescriptive analytics is an effective formula to deal with these situations. The simulation optimization approach described in Chapter 10 for optimizing the performance of business processes is a good example of combining predictive (simulation) and prescriptive (optimization) analytics.

11.1 Competing on Analytics[1]

Business today is inundated with data and tools to process data, enabling organizations to compete on analytics. In the market, there are many firms offering similar products and using comparable technologies, making business processes one of the last remaining points of differentiation. Firms competing on analytics squeeze every drop of value from those processes. Like other companies, analytics competitors know what products their customers want; however, they also know what prices those customers will pay, how many items each will buy in a lifetime, and what will make them buy more. Like other companies, they know compensation costs and turnover rates, but they are also able to evaluate how personnel contribute to or detract from the bottom line, and how salary levels relate to the performance of key individuals. Like other companies, they know when inventories are running low, but they can also predict problems with demand and supply chains to minimize inventory costs while operating at a high service level.

Analytics competitors coordinate their efforts as part of an overarching strategy championed by top leadership and pushed down to decision-makers at every level of the organization. Employees hired for their expertise with numbers or trained to recognize their importance search for evidence by applying appropriate quantitative tools. As a result, they make good decisions, big and small, all the time. Competing on analytics requires a significant investment in technology to gather and process data, the human resources with the right quantitative training, the establishment of a fact-based management culture, and the formulation of companywide strategies to implement and maintain an analytics focus that recognizes the difference between "thinking" that something is true and "knowing" it.

Example 11.1: Total Hotel Optimization at Marriott International

Marriott International has perfected to a science its system for establishing the optimal price for guest rooms (the key analytics process in hotels, known as *revenue management*). Through its Total Hotel Optimization program, Marriott has expanded its quantitative expertise to areas such as conference facilities and catering and has made related tools available over the Internet to property revenue managers and hotel owners. It has developed systems to optimize offerings to loyal customers and assess the likelihood of those customers defecting to competitors. It has given local revenue managers the power to override the system's recommendations when certain local factors can't be predicted (such as the large number of Hurricane Katrina evacuees arriving in Houston). The company has even created a revenue opportunity model, which computes actual revenues as a percentage of the optimal rates that could have been charged. That figure grew from 83 percent to 91 percent as Marriott's revenue-management analytics was adopted throughout the entire organization. Marriott is recognized as a best-practice firm for maximizing the revenue of perishable inventory. ■

Any company can generate simple descriptive statistics about aspects of its business. For instance, most companies know their average revenue per employee or average order size. Analytics competitors look well beyond basic statistics and use predictive modeling to identify things such as their most profitable customers, those with the greatest profit potential, and even the ones most likely to cancel their accounts. They pool data generated in-house and data acquired from outside sources (which they analyze more deeply than do their less statistically savvy competitors) for a comprehensive understanding of their customers. Prescriptive analytics is part of their toolbox, for instance, to optimize their shipment routes and supply chains or to find robust system configurations—through simulation optimization—that mitigate the impact of uncertainty and unexpected events. Analytics competitors establish prices in real time to maximize benefits from each of their customer transactions. They create complex models of how their operational costs relate to their financial performance.

Analytics competitors understand that most business functions can be improved with sophisticated quantitative techniques. These organizations are not looking for one killer application but rather for multiple applications supporting many parts of the business. These programs operate not just under a common label but also under common leadership and with common technology and tools. The proliferation of user-developed spreadsheets—of which research has shown that between 20 percent and 40 percent contain errors—and databases inevitably leads to multiple versions of key indicators within an organization. This is why analytics competitors have centralized groups that ensure that critical data and other resources are well managed and that facilitate the sharing of data among different parts of the organization without the impediments of inconsistent formats, definitions, and standards.

A companywide embrace of analytics impels changes in culture, processes, behavior, and skills for many employees. Like any major transition, it requires leadership from executives at the very top who fully subscribe to fact-based management approaches. The chief executive officer (CEO) is the ideal principal advocate. CEOs leading the analytics charge require both an appreciation of and a familiarity with quantitative approaches. A background in statistics isn't necessary, but those leaders must understand the principles behind various quantitative methods as well as the methods' limitations.

Example 11.2: UPS Transformation to Analytics

UPS embodies the evolution from targeted analytics user to comprehensive analytics competitor. Although in 2005 the company was among the world's most rigorous practitioners of operations research and industrial engineering, its capabilities were narrowly focused. As the role of analytics expanded, UPS employed its statistical skills on data associated with the movement of packages and to anticipate and influence the actions of people—assessing the likelihood of customer attrition and identifying sources of problems. The UPS Customer Intelligence Group, for example, was able to accurately predict customer defections by examining usage patterns and complaints. When the data point to a potential defector, a salesperson contacts that customer to review and resolve the problem, dramatically reducing the loss of accounts. UPS continues to launch initiatives that will move it closer to a full-bore analytics competitor. ■

The idea of competing on analytics is not entirely new. A few organizations—primarily within financial services and particularly in financial investment and trading businesses—have competed on this basis for decades. The trading of stocks, bonds, currencies, and commodities has long been driven by analytics. What is new is the spreading of analytical competition to a variety of other industries—from consumer finance, to retailing, to travel and entertainment, to consumer goods—and within companies, from individual business units to an enterprise-wide perspective. Even professional sports teams—which are part of one of the most traditionally intuitive industries—are moving in this direction.

Example 11.3 Analytics in Professional Sports

The New England Patriots football team uses data and analytical models extensively, both on and off the field. In-depth analytics helps the team select its players, stay below the salary cap, decide whether to punt or "go for it" on fourth down, and try for one point or two after a touchdown. Both its coaches and players are renowned for their extensive study of game films and statistics, and Head Coach Bill Belichick peruses articles by academic economists on statistical probabilities of football outcomes. Off the field, the team uses detailed analytics to assess and improve the "total fan experience." At every home game, for instance, 20–25 people have specific assignments to make quantitative measurements of the stadium's food, parking, personnel, bathroom cleanliness, and other factors. External vendors of services are monitored for contract renewal and have incentives to improve their performance. Thanks to Michael Lewis's best-selling book *Moneyball*, which demonstrated the power of statistics in professional baseball, the Oakland A's are almost as famous for their geeky number crunching as they are for their athletic prowess.

Analytical competition is not only taking root in US sports. Some soccer teams in Europe also have begun to employ similar techniques. AC Milan, one of the leading teams in Europe, uses predictive models to prevent player injuries by analyzing physiological, orthopedic, and mechanical data from a variety of sources. Bolton, an English soccer team, is known for its manager's use of extensive data to evaluate player performance and team strategies. ■

Analytical cultures and processes benefit not only professional sports teams but also any business that can harness extensive data, complex statistical processing, and fact-based decision-making. The gaming firm Harrah's, for example, has chosen to compete on analytics for customer loyalty and service rather than on building the mega-casinos in which its competitors have invested. Amazon.com uses extensive analytics to predict what products will be successful and to optimize the efficiency of its supply chain. At the mutual fund company Dreyfus, analysis of customer information defined segmentation that helped reduce fund attrition from 22 percent to 7 percent a year. Capital One conducts more than 30,000 experiments a year with different interest rates, incentives, direct-mail packaging, and other variables. Its goal is to maximize the likelihood that potential customers will sign up for credit cards and that they will not default. Progressive employs similar experiments using widely available insurance industry data. The company defines narrow groups, or cells, of customers: for example, motorcycle riders aged 30 and above, with college education, credit scores over a certain level, and no accidents. For each cell, the company performs a regression analysis to identify factors that most closely correlate with the losses that the group engenders. It then sets prices for the cells, which should enable the company to earn a profit across a portfolio of customer groups, and uses simulation software to test the financial implications of those hypotheses. With this approach, Progressive can profitably insure customers in traditionally high-risk categories. Seven areas where the application of analytics has large potential benefits are shown in Table 11.1.

The most proficient analytics practitioners include customers and vendors in the process. Wal-Mart, for instance, insists that suppliers use its Retail Link system to monitor product movement by store, to plan promotions and layouts within stores, and to reduce stock-outs. E&J Gallo provides distributors with data and analysis on retailers' costs and pricing, enabling them to calculate the per-bottle profitability for each of Gallo's 95 wines. The distributors, in turn, use that information to help retailers optimize their mixes while persuading them to add shelf space for Gallo products. Procter & Gamble offers data and analysis to its retail customers, as part of a program called Joint Value Creation, and to its suppliers to help improve responsiveness and reduce costs. Hospital supplier Owens & Minor furnishes similar services, enabling customers and suppliers to access and analyze their buying and

TABLE 11.1

Common Areas of Application of Analytics

Area	Application	Examples
Supply chain	Simulate and optimize supply chain flows; reduce inventory and stock-outs	Dell, Wal-Mart, Amazon
Customer selection, loyalty, and service	Identify customers with the greatest profit potential; increase the likelihood that they will want the product or service offering; retain their loyalty	Harrah's, Capital One, Barclays
Pricing	Identify the price that will maximize yield or profit	Progressive, Marriott
Human capital	Select the best employees for particular tasks or jobs, at particular compensation levels	New England Patriots, Oakland A's, Boston Red Sox
Product and service quality	Detect quality problems early and minimize them	Honda, Intel
Financial performance	Better understand the drivers of financial performance and the effects of nonfinancial factors	MCI, Verizon
Research and development	Improve quality, efficacy, and where applicable, safety of products and services	Novartis, Amazon, Yahoo

selling data, track ordering patterns in search of consolidation opportunities, and move off-contract purchases to group contracts that include products distributed by Owens & Minor and its competitors. For example, Owens & Minor might show a hospital chain's executives how much money they could save by consolidating purchases across multiple locations or help them see the trade-offs between increasing delivery frequency and carrying inventory.

In sum, the key attributes of companies that are recognized leaders in using analytical techniques for performance improvements are

- Strong top-leadership support, with one or more senior executives who are intensely advocating analytics and fact-based decision-making
- Widespread use of not just descriptive statistics but predictive modeling and sophisticated prescriptive techniques
- Substantial use of analytics approaches across multiple business functions or processes
- Movement toward an enterprise-level approach (which might include customers and vendors) to managing analytical tools, data, and organizational skills and capabilities

It is reasonable to believe that business analytics will grow in acceptance as the availability of both the necessary data and the analytical tools continues to increase. Analytics is positioned to help businesses compete by keeping their best customers, spending less on logistics and inventory, developing new products and services in a timely way, and hiring capable employees.

11.2 Business Process Management Systems

Business process management (BPM) is a holistic approach to managing an organization to align customer needs with corporate strategies. Business process analytics (BPA) is the set of quantitative tools that support BPM, and business process management systems (BPMSs) are the technologies to implement BPA. Business process analytics simply refers to the application of analytics (descriptive, predictive, and prescriptive) to business processes. Analytics tools are data intensive, and therefore, information technology (IT) plays a crucial role in the successful application of these quantitative techniques in support of BPM initiatives. Because the process management literature is confusing when it comes to distinguishing between a philosophy and its implementation, these definitions are meant to make a distinction among the management approach (BPM), the associated quantitative models, algorithms, and tools (analytics), and the IT (BPMSs) to support their implementation.

BPMS[2] is a class of software that allows organizations to devise process-centric IT solutions. Process-centric means that BPMS solutions are able to integrate people, systems, and data. Organizations that use BPMS to accomplish IT-enabled business process realize the following benefits:

- Close involvement of business analysts in designing IT-enabled business processes
- Integration of people and systems that participate in business processes
- Business processes simulation functionality
- Monitoring, controlling, and improvement of business processes in real time
- Changing existing business processes in real time without an elaborate conversion effort

One of the major advantages of BPMS over traditional IT-enabled business process improvement efforts is that BPMS brings IT closer to the business process owners. IT solutions are typically conceived from a collection of functional specifications. A gap between what the business wants and what IT implements is created when the specifications do not capture the business requirements appropriately. BPMS is able to bridge the gap by allowing business process owners to be directly involved in designing the IT solution. BPMSs typically include a graphical *process designer* tool that enables the design of processes by process owners and business analysts. The tool automatically generates computer code that sometimes can be deployed without IT development help. To help the business process owners and business analysts in the process design, BPMSs include process simulation and modeling functionality. This means that business process owners or analysts can design business processes and run the process designs in simulation mode (as discussed in Chapters 7 through 10). In other words, BMPSs enable the application of predictive and prescriptive analytics.

Simulation is only a part of BPMSs; in fact, they play a large role as the supervisory systems that oversee the business processes once they have been implemented. The supervisory aspect of BPMS provides the ability to monitor, control, and improve business processes. Because BPMSs oversee all the steps, whether manual or automated, in the business process, they can provide valuable process information and, in that role, serve as the performance monitor for the processes. Process owners can obtain statistics such as average cycle time per transaction, the wait time before a process task is performed by human participants, and cost data. This is the BMPS support for implementing descriptive analytics.

BPMS gives organizations the ability to implement real-time process improvement without the extensive process conversion effort. The original business processes already exist in the business process designer. This eliminates the need to gather current process information. When process deficiencies have been identified (for instance, a bottleneck), business process owners or analysts can incorporate improvements to the process using the business process designer. After the improved business process solution is implemented, BPMS allows any work that started on the original process to finish using the original process and any new work to be performed using the improved process. In essence, the system allows both the original and improved processes to coexist until all work from the original process is finished. Using BPMS, process improvement could be made without disruption to process output. This is an important benefit to continuous process improvement.

In sum, BPM technology allows the business analyst to collaborate more closely with IT people in implementing projects. The various tools BPMSs offer provide a new paradigm for how solutions can be implemented. Organizations are no longer tied to the business processes ingrained in their business applications. With automatic workflow generation and Web portal capabilities, workflow can be easily deployed across multiple applications, thus integrating people into the business processes. These technological innovations enable technology to better fulfill the ideals of BPM and the principles of BPA (i.e., fact-based management). Processes can be managed in a process framework, and decisions can be based on quantitative analysis. This framework allows organizations to design, execute, monitor, measure, and enhance their processes.

11.2.1 Business Rules

Related to the implementation of a BPMS is the notion of *business rules.* A business rule is a statement that defines or constrains some aspect of the business. It is intended to assert business structure or to control or influence the behavior of the business. Business rules are typically atomic; that is, they cannot be broken down further. They have existed since companies started automating business processes using software applications. Traditionally, they have been buried deep inside the application in some procedural programming language. As business rules matured, business analysts tried to control and manage these rules without directly having to deal with the IT aspects of the system. Rules are virtually everywhere in an organization; for example, a bank may have the rule to deny a loan to a customer if his or her annual income is less than $10,000. Business rules may be categorized as follows (Pant 2009):

- *Business policies.* These are rules associated with the general business policies of a company, for example, loan approval policies.
- *Constraining.* These are the rules that businesses have to keep in mind and work within their scope while going about their operations. Rules associated with regulatory requirements will fall under this category.
- *Computation.* These are the rules associated with decisions involving any calculations, for example, discounting rules or premium adjustments.
- *Reasoning capabilities.* These are the rules that apply logic and inference to a course of actions based on multiple criteria, for example, rules associated with the up-sell or cross-sell of products and services to customers based on their profile.
- *Allocation rules.* There are some rules that are applicable in terms of determining the course of action for the process based on information from the previous tasks. They also include rules that manage the receiving, assignment, routing, and tracking of work.

A business rule is primarily divided into the following four blocks:

- *Definition of terms.* This provides a vocabulary for expressing rules. For example, customer, car, or claims acts as the category for the rules.
- *Facts.* These are used to relate terms in definitions with each other. For example, a customer may file a claim.
- *Constraints.* They limit or control how an organization wants to gather, use, and update data. For example, for opening an account, a customer's passport details or social security details are required.
- *Inference.* This applies to logical assertions such as "if X, then Y" to a fact and infers new facts. For example, if there is a single account validation rule (if an applicant is a defaulter, then the applicant is high-risk), and it is known that Harry (the applicant) has defaulted earlier on his payments for other bank services, it can be inferred that Harry is a high-risk customer.

Rules may exist within a computer application with the goal of implementing decision logic. For instance, a bank may purchase an application to assess risk, and the computer code used to implement the assessment model may include a number of rules. This system will work fine as long as changes are not needed, and as long as it is executed independently of other systems, thus avoiding external interactions of the rules. The reality, however, is that systems interact, and rules may have to be changed over time. Therefore, rules that are embedded within the complexity of the application code and that spread across multiple systems make it extremely difficult for the organization to introduce changes quickly and without creating a domino effect across systems. Modern BPMSs drive the decision logic used during the execution of the process from a central repository of rules. The rules are centrally stored and managed, creating a flexible environment to implement changes and reuse.

For the loan approval example shown earlier, the business rule would traditionally be embedded in the application code as follows:

```
public boolean checkAnnualIncome(Customer name)
{
boolean declineLoan = false;
int income = name.getincome();
if(income < 10000)
{
declineLoan = true;
}
return declineLoan;
}
```

The rule, as coded, is difficult for the business analyst to understand and change. With the need for organizations to be agile, waiting for days or even weeks before a small change can be implemented by IT is impractical and unproductive. This is why BPMSs provide functionality that allows business analysts to define and control their own rules, thus enabling them to make changes easily and quickly.

11.2.2 Monitor and Control

The lifecycle of business process excellence consists of the design, implementation, execution, and monitoring and control of processes (Kirchmer 2011). The business process reengineering movement that started in the 1990s emphasized the designed phase of the process lifecycle. It wasn't until a decade later that the application of technology and quantitative analysis became an important part of monitoring and controlling the execution of business processes. As mentioned in Section 11.2, BPMSs record many types of events that occur during process executions, including the start and completion time of each activity, input and output data, resources involved in the completion of activities, and even failures. The monitor and control functionality of BPMSs is made possible by the addition of *process intelligence* (PI), which is defined as the set of integrated tools that support business and IT analysts in managing process execution quality (Grigori et al. 2004). These tools are standard in most modern BPMSs.

Process logs (also called *event logs*) are the main input to PI tools. PI tools are designed to extract knowledge from process logs stored in data warehouses, where the logs have been cleaned and aggregated. The discovery might lead to identifying high- or low-quality process executions to find explanations of why they occurred as well as predicting future behavior. Data visualization tools play an important role in monitor and control activities. Dashboards and mashups are two common process intelligence tools related to these activities. *Dashboards* are visual displays that show the status or "health" of a business process via numeric and graphical key performance indicators.[3] *Mashups* refer to the integration of information from different sources by using editors that allow the remix of data without programming. Mashup in music means blending existing content to create something new. The technology behind mashups enables analysts to combine and aggregate process information from multiple sources to create and manage their own dashboards.

Business activity monitoring (BAM), an additional PI tool, is designed to determine abnormal events in the process and issue alerts when predefined conditions occur. In other words, BAM is software for the real-time monitoring of business processes. An example of a BAM alert is a credit card that has been charged multiple times in a short period of time. This might trigger an alert to the fraud prevention department. Another example of a BAM alert occurs when inventory of a particular product is below a specified level at retail stores. This alert might trigger an automatic process to place a purchase order with the vendor to deliver merchandise at specified points of the supply chain. The implementation of BAM does not necessarily have to involve BPMSs. However, BPMS engines and the business process designer contain the design and runtime tools that facilitate the implementation of BAM solutions.

11.2.3 Process Mining

Process mining is a set of techniques that are useful to analyze existing processes based on how they are actually being executed. Process mining operates on process logs that contain data associated with the order in which events took place. This is why these analytics tools are embedded in BMPSs. There are three classes of process mining techniques:

- *Discovery*. A set of procedures designed to construct a model of a process from data associated with the actual execution of the process.[4] In other words, without an *a priori* model, information is gathered about the process as it takes place, and an explicit process model is extracted. The model of the actual process should be consistent with the observed dynamic behavior.

- *Conformance analysis*. In this case, there is a process model that is compared with the event log, and discrepancies between the log and the model are analyzed.
- *Enhancement*. A discovered or existing model is enhanced by the analysis of information related to performance, cost, or social structures.

The principles of discovery—the main process mining technique—may be illustrated with the following example from van der Aalst and Weijters (2004). Consider the process log in Table 11.2. This log contains information about five cases (i.e., process instances). The log shows that Tasks *a*, *b*, *c*, and *d* have been executed for Cases 1 to 4. For Case 5, only two different tasks (*e* and *f*) were executed. This log can be represented as a set of traces or sequences in the following form:

$$L=\left[\langle a,b,c,d\rangle^2,\langle a,c,b,d\rangle^2,\langle e,f\rangle\right]$$

Cases 1 and 3 follow the first sequence, hence the superscript two. Cases 2 and 4 follow the second sequence. Case 5 is represented by the last sequence. A simple event log *L* is the main input to process discovery algorithms, which are defined as functions that map *L* onto a workflow network (WF-net). The α-algorithm is an example of a mapping function for process discovery (van der Aalst, 2016). This algorithm was one of the first process discovery

TABLE 11.2

Process Log

Case	Activity
1	a
2	a
3	a
3	b
1	b
1	c
2	c
4	a
2	b
2	d
5	e
4	c
1	d
3	c
3	d
4	b
5	f
4	d

algorithms to be able to handle concurrency. The basic α-algorithm is capable of discovering a large class of WF-nets, but it does have some limitations. In particular, it might generate WF-nets that are unnecessarily complex. In addition, it does not handle short loops (i.e., loops of length one or two) correctly. Some extensions to the original algorithm have been created that are able to correct these problems.

The quality of the model produced by a process discovery algorithm depends not only on the algorithm but also on the completeness of the event log. It cannot be assumed that the event log will contain all traces. Therefore, the model must be able to extrapolate. The issues of overfitting and underfitting that are typical of any data mining technique are also present in process discovery. Therefore, an event log is typically split into a training log and a test log. The training log is used to discover the process model, and the test log is used to evaluate the model. Van der Aalst (2016) proposes the following four dimensions to measure the quality of a model resulting from a process discovery:

- *Fitness*. This refers to the ability of the model to replay the traces in the event log. A model has a perfect fitness if it is capable of replaying all traces in the log.
- *Simplicity*. The simplest model that can explain the behavior in the log is considered the best. A measure of simplicity is the number of nodes and arcs in the resulting process graph.
- *Generalization*. Since it is not possible for a log to contain all possible traces, a model must be able to generalize. An overfitted model is one that is not able to generalize.
- *Precision*. A precise model does not allow behavior that is not possible. An underfitted model results in over-generalization; that is, it allows more behavior when there are no indications in the log that suggest this additional behavior. A model that is not precise allows behavior that is very different from what can be observed in the log.

The α-algorithm is based on establishing the relationships between all the activities in a log L:

- $a \rightarrow b$: indicates that b immediately follows a, and a never immediately follows b
- $a \# b$: indicates that neither b ever immediately follows a nor a ever immediately follows b
- $a \| b$: indicates that both b immediately follows a and a immediately follows b

TABLE 11.3

Footprint for Log in Table 11.2

	a	b	c	d	e	f
a		→	→	#	#	#
b			‖	→	#	#
c				→	#	#
d					#	#
e						→

Because precisely one of these relations holds for any pair of activities, a footprint of a log *L* can be captured in a matrix. Table 11.3 shows the footprint for the event log in Table 11.2. If Task *b* is executed, then Task *c* is also executed. However, Task *c* does not always follow Task *b*. For example, in the second trace of the log *L* above, Task *c* is executed before Task *b*. Based on the information shown in Table 11.2 and summarized as log *L*, with the corresponding footprint shown in Table 11.3, the process model in Figure 11.1 may be constructed. It is assumed that the cases are representative and the set of observed events is sufficiently large to characterize the behaviors of the process. The model includes an arc for each pair of activities labeled with an arrow (→) in the footprint of Table 11.3. Activities *b* and *c* are shown in parallel, as indicated in the footprint (‖). Note that *b* and *c* appear in both orders, immediately following each other in the event log. The process starts with either Task *a* or Task *e*. Task *f* always follows Task *e*, and Task *d* always follows Tasks *b* and *c*.

Table 11.2 contains the minimum information that is required to extract a process model. However, process logs generated by BPMSs often include timestamps and case attributes that can be used to extract additional causality information. For this example, process discovery is relatively simple. However, large processes represent a challenge because of the large number of possible combinations generated by the presence of alternative and parallel routes. In addition, paths with low probability of occurring are difficult to

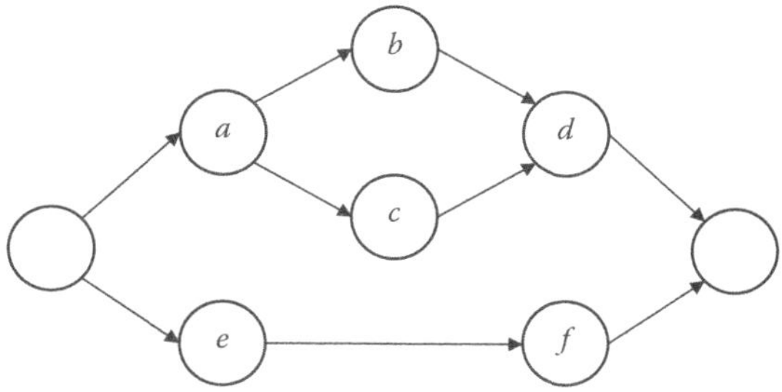

FIGURE 11.1
Process model discovered from the event log in Table 11.2.

detect. Extensions of the α algorithm have been suggested to remedy some of its shortcomings. For instance, the *heuristics miner* takes into consideration the number of times that each relationship between activities occurs (e.g., the number of times that a immediately followed b in the event log). These counts are used to build a *dependency graph* that shows a numerical relationship between pairs of events. The larger the values, the stronger the dependency. The heuristics are designed to find reliable causality relations even if the event log contains noise.

The α algorithm and the heuristics miner are just two of a variety of process mining procedures. A wealth of information about process mining, which is beyond the scope of this book, can be found in www.processmining.org. This website includes process-mining tools embedded in ProM, the process-mining framework, as well as information associated with Will van der Aalst's seminal book on process mining, titled *Process Mining: Data Science in Action.*

11.3 Machine Learning

Machine learning and data mining are closely related and use similar techniques to extract knowledge from data. Data mining's primary goal is to support decision-making by discovering patterns and relationships from large databases. Machine learning, on the other hand, focuses on algorithms and models that can make predictions with the goal of automating decision-making. Data mining, also called *knowledge discovery in databases,* involves uncovering patterns that may not be initially evident in the data. The goals of the data mining process depend on the intended use of the outcomes and can be classified into two types: *verification,* where the process is limited to verifying the analyst's hypothesis, and *discovery,* where the system independently finds new patterns and present them to a user in a human-understandable form (Wegener and Rüping 2010). Machine learning is centered on predictive modeling and making accurate predictions on new, unseen data.

Data mining and machine learning have become an integral part of successful businesses. Their application ranges from determining where to advertise, to what products to recommend, to how to detect fraud. These applications have important implications for the organization's bottom line. The redesign of business processes is often triggered by discoveries associated with extracting knowledge from data. For example, a data exploration project may discover that a particular advertisement channel performs very well for a certain group of customers, triggering a change in the advertisement process. Likewise, a fraud detection procedure based on mining may discover that certain characteristics of a payment transaction are associated with a high fraud rate, triggering a change in the payment process.

Machine learning may be applied to configure rules in processes for which customer classification is critical. In a loan application process, applicants may be classified as low-risk, regular, or high-risk by the application of data mining techniques that cluster items based on attributes. Statistical methods are used on historical data, and rules emerge that make the decisions in a loan application process fact-based.

11.3.1 Support Vector Machines

A Support Vector Machine (SVM) is a supervised machine learning algorithm used for classification and regression tasks. The primary goal of a SVM in classification is to find a hyperplane that best separates different classes in the feature space. In a two-dimensional space, this hyperplane is a line, and in higher dimensions, it becomes a hyperplane. The key concept in SVM is to maximize the margin, which is the distance between the hyperplane and the nearest data points from each class (called support vectors). Support vectors are the data points that lie closest to the decision boundary (hyperplane). They play a crucial role in determining the position and orientation of the hyperplane. A simple linear SVM classifier works by making a straight line between two classes. That means all the data points on one side of the line will represent a category and the data points on the other side of the line will be put into a different category. This means there can be an infinite number of lines from which to choose. SVM chooses the line that separates the data and is as far as possible from the closet data points.

Consider, for instance, the loan application rejection rule in the previous section and assume that the goal is to find a business rule that goes beyond the simple consideration of salary as the only key attribute for making an initial decision. A BPMS can extract the data in Table 11.4 for the purpose of analysis and the creation of an effective business rule. The table contains 20 historical observations of loans that are classified into two groups: (1) paid and (2) defaulted. The credit rating (a value between 300 and 999) and the salary (in thousands) were recorded at the time that the loan application was submitted. With this data, a business analyst would like to find a rule to reject applications without engaging in any additional review. The problem is therefore to find a classification rule that has a large probability of classifying applicants correctly. For problems with two groups and real-valued attributes (i.e., the values on which the classification is based), *linear regression* is a simple way of implementing a linear SVM to find such a rule.

The application of linear regression in this example consists of constructing an equation of the following form:

$$y = ax + b$$

where

x represents the set of attributes (i.e., credit rating and salary)

y is the classification score

TABLE 11.4

Attributes and Classification of 20 Loan Applications

Observation	Credit Score	Salary	Group
1	670.5	36.1	1
2	648.0	42.0	1
3	637.5	30.8	1
4	603.0	42.6	1
5	628.5	37.1	1
6	589.5	33.8	1
7	579.0	40.3	1
8	571.5	28.5	1
9	568.5	44.7	1
10	552.0	42.6	1
11	537.0	35.2	1
12	585.0	33.1	2
13	562.5	36.4	2
14	541.5	30.8	2
15	534.0	27.2	2
16	516.0	39.0	2
17	492.0	35.2	2
18	471.0	29.4	2
19	442.5	37.9	2
20	451.5	33.9	2

The linear regression model finds the values for the attribute coefficients (i.e., the values of a) and the intercept (i.e., the value of b). Table 11.5 shows the classification scores (i.e., the values of y) found using the linear regression functionality in Microsoft Excel. The classification score for the first observation (Cell E2) is found with the following Excel formula:

$$=\text{TREND}(\$D\$2{:}\$D\$21,\$B\$2{:}\$C\$21,B2{:}C2)$$

The same formula is applied to Cells E3 to E21. The cutoff value (Cell E22) is used to classify new observations, and it is the value between the average classification score for Group 1 and the average classification score for Group 2. In Excel, this is calculated as follows:

$$=\frac{(\text{AVERAGEIF}(\$D\$2{:}\$D\$21,\text{"}=1\text{"},\$E\$2{:}\$E\$21)+\text{AVERAGEIF}(\$D\$2{:}\$D\$21,\text{"}=2\text{"},\$E\$2{:}\$E\$21))}{2}$$

The predicted group for each of the observations is found simply by comparing the classification score with the cutoff value. If the classification score

TABLE 11.5

Spreadsheet to Formulate a Business Rule Using Linear Regression

	A	B	C	D	E	F
1	Observation	Credit Rating	Salary	Group	Classification Score	Predicted Group
2	1	670.5	36.1	1	0.855	1
3	2	648.0	42.0	1	0.813	1
4	3	637.5	30.8	1	1.173	1
5	4	603.0	42.6	1	1.034	1
6	5	628.5	37.1	1	1.049	1
7	6	589.5	33.8	1	1.345	1
8	7	579.0	40.3	1	1.223	1
9	8	571.5	28.5	1	1.584	2
10	9	568.5	44.7	1	1.159	1
11	10	552.0	42.6	1	1.303	1
12	11	537.0	35.2	1	1.583	2
13	12	585.0	33.1	2	1.387	1
14	13	562.5	36.4	2	1.416	1
15	14	541.5	30.8	2	1.679	2
16	15	534.0	27.2	2	1.817	2
17	16	516.0	39.0	2	1.591	2
18	17	492.0	35.2	2	1.821	2
19	18	471.0	29.4	2	2.089	2
20	19	442.5	37.9	2	2.008	2
21	20	451.5	33.9	2	2.070	2
22				Cutoff value	1.479	

is less than or equal to the cutoff value, then the observation is classified as belonging to Group 1. Otherwise, the observation is classified as belonging to Group 2. The Excel formula corresponding to Cell F2 is

$$= \text{IF}(\text{E2} <= \$\text{E}\$22,1,2)$$

This classification model, based on a simple linear SVM, is not perfect, but no classification procedure is. As shown in Figure 11.2 and in column F of Table 11.5, observations 8, 11, 12, and 13 are misclassified. Therefore, the accuracy of the classification is 16/20 or 80 percent.

A business rule can now be formulated: "deny a loan if the classification score is more than 1.479." Suppose that two applications are submitted with

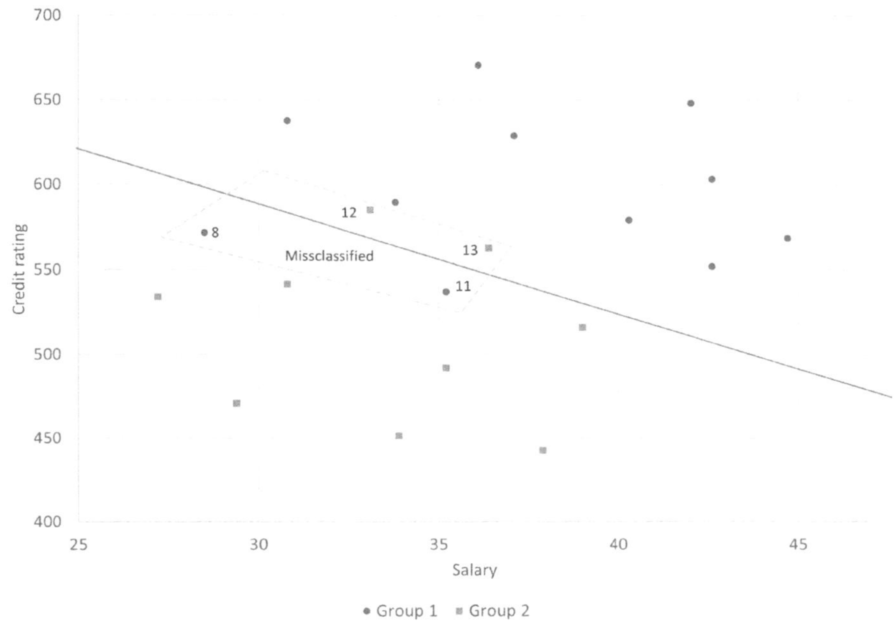

FIGURE 11.2
Linear SVM example.

the information shown in Table 11.6. The classification scores are calculated with the regression model, and a group prediction is found. The results are that Applicant A will be given further consideration, while Applicant B will be denied.

11.3.2 *k*-Nearest Neighbor Classifier

Often, the attributes on which the formulation of a business rule is based belong to categorical data instead of real values. Categorical data consist of indicators that specify a category or label (e.g., male and female or low, medium, and

TABLE 11.6
Prediction for Two New Loan Applicants

Applicant	Credit Rating	Salary	Classification Score	Predicted Group
A	607.5	33.4	1.261	1
B	442.5	34.4	2.104	2

high). A general and simple method for classification that can be applied to both real and categorical data is known as the *k-nearest neighbor classifier* (*k*-NN). This method classifies new observations by measuring the proximity to observations for which their classification is known. In contrast to other methods, such as the one based on linear regression presented earlier, *k*-NN does not build a model (e.g., a linear equation). Instead, it just identifies the observations that are nearest in the attribute space and then assigns the most common classification in the set to the new observation. The neighbors are found using a measure of distance between pairs of observations. Let $d(x,y)$ represent the distance between point x and y in a dataset, where each point consists of n attribute values. Then, two common measures of distance are:

- **Euclidean distance.** This is the most used distance measure, and it is limited to real-valued vectors. The distance measures a straight line between the two points.

$$d(x,y)=\sqrt{\sum_{i=1}^{n}(x_i-y_i)^2}$$

- **Manhattan distance.** The distance between two points is the sum of the absolute differences of their Cartesian coordinates. That is, the distance between two points is measured along axes at right angles. It is also referred to as taxicab distance or city block distance as it is commonly visualized with a grid, illustrating how one might navigate from one address to another via city streets.

$$d(x,y)=\sum_{i=1}^{n}|x_i-y_i|$$

When all attribute values are binary, the Manhattan distance is equivalent to the Hamming distance.

Suppose that instead of using the credit rating and the salary to classify a loan applicant, the attributes in Table 11.7 are employed. These categorical data must be transformed into numerical data before the application of *k*-NN. We use the following transformation:

1. Assign integer values to each category (e.g., 0, 1, and 2 for attributes with three categories)
2. For each attribute, calculate the average and the standard deviation
3. Normalize values under each attribute

In the example of Table 11.7, the transformation may start by assigning the following values for each attribute:

Age: 0 = young, 1 = middle, and 2 = old
Job: 0 = no and 1 = yes
Home owner: 0 = no and 1 = yes
Credit rating: 0 = fair, 1 = good, and 2 = excellent

TABLE 11.7

Categorical Data for Loan Applications

Observation	Age	Job?	Homeowner?	Credit rating	Paid?
1	young	yes	no	fair	no
2	young	no	no	good	no
3	young	yes	no	good	yes
4	young	yes	yes	fair	yes
5	young	no	no	fair	no
6	middle	no	yes	fair	no
7	middle	no	no	good	no
8	middle	yes	yes	good	yes
9	middle	no	yes	excellent	yes
10	middle	no	yes	good	no
11	old	no	yes	excellent	yes
12	old	no	yes	good	no
13	old	yes	no	good	yes
14	old	yes	no	excellent	yes
15	old	no	no	fair	no

The average and the sample standard deviation[5] corresponding to each attribute in the sample are given in Table 11.8. These values are used to perform the normalization of the data as indicated in the third step of the transformation procedure.

$$\text{Normalized value} = \frac{(\text{Value} - \text{Average})}{\text{Standard deviation}}$$

The normalized values are

Age: –1.183 = young, 0 = middle and 1.183 = old

Job: –0.789 = no and 1.183 = yes

Homeowner: –0.904 = no and 1.033 = yes

Credit rating: –1.166 = fair, 0.179 = good, and 1.525 = excellent

The result of this normalization is that for the new dataset, the average of each column (that is, each attribute) is 1, and the standard deviation is 0. The observations are now ready for the application of *k*-NN. The *k*-NN score for a new

TABLE 11.8

Average and Sample Standard Deviation for the Attribute Values in Table 11.7

Statistic	Age	Job?	Home owner?	Credit rating
Average	1.000	0.400	0.467	0.867
Std. Dev.	0.845	0.507	0.516	0.743

applicant is the average of the "Paid?" values for the k nearest observations in the sample. (Note that the "Paid?" values are not normalized and are simply transformed from "no" and "yes" as shown in Table 11.5 to 0 and 1.) A value of $k = 3$ and the Manhattan distance will be used to determine the category of a new applicant. For the Manhattan distance, x_i and y_i are the normalized values of the ith attribute of observations x and y, respectively. Once the d values are calculated, the k nearest neighbors are determined, and the k-NN score is evaluated. The business rule is to reject the loan application if the k-NN score is less than 0.5.

Suppose that a young applicant who is currently employed, is not a homeowner, and has excellent credit applies. Would the application be immediately rejected? The normalized attribute values for this applicant are −1.183, 1.183, −0.904, and 1.525, corresponding to age, job, homeowner status, and credit rating, respectively. The distances between this applicant and all the observations in the sample of Table 11.7 are shown in Table 11.9. The distances in the table are ordered from smallest to largest. For instance, the distance between the new applicant and Observation 3 is calculated as

$$d = |-1.183 + 1.183| + |1.183 - 1.183| + |-0.904 + 0.904| + |1.525 - 0.179| = 1.345$$

According to the distances shown in Table 11.9, the three nearest neighbors to the new applicant are Observations 3, 14, and 1. The average "Paid?" value for these observations is $(1 + 1 + 0)/3 = 0.667$, which is greater than 0.5. This means that the new application will not be rejected immediately and will be reviewed further. When using the k-NN procedure for classification

TABLE 11.9

Distances from the New Applicant to the Observations in the Sample

Observation	Distance	Paid?
3	1.345	1
14	2.366	1
1	2.691	0
2	3.318	0
13	3.712	1
8	4.465	1
7	4.501	0
4	4.627	1
5	4.663	0
9	5.092	1
11	6.275	1
10	6.437	0
15	7.029	0
12	7.620	0
6	7.783	0

problems with two groups—such as those that arise in situations where a "yes or no" answer is required—it is convenient to use an odd k value to avoid a potential tie, where exactly half of the nearest observations belong to one group and half belong to the other group.

Analytics competitors configure business rules employing quantitative tools, such as the techniques shown earlier.[6] The result of BPMSs incorporating additional analytics tools (e.g., machine learning) as well as business rules management functionality is that the quality of the decisions in automated business processes tends to increase.

11.3.3 Neural Networks

Neural networks are a class of machine learning models inspired by the structure and functioning of the human brain. They are a fundamental component of deep learning, a subset of machine learning that involves training models on large amounts of data.

A neural network consists of interconnected nodes, also known as neurons or artificial neurons, organized into layers. As shown in Figure 11.3, these layers are typically categorized into three types:

- **Input Layer**. The first layer of the neural network is responsible for receiving input data. Each node in the input layer represents a feature or attribute of the input data.
- **Hidden Layers**. These are intermediate layers between the input and output layers. These layers process the input data through a series of weighted connections and apply activation functions to introduce non-linearities. The presence of multiple hidden layers gives rise to the term "deep" in deep learning.
- **Output Layer**. The final layer produces the output of the neural network. The number of nodes in the output layer depends on the type of task the neural network is designed for (e.g., binary classification, multi-class classification, or prediction).

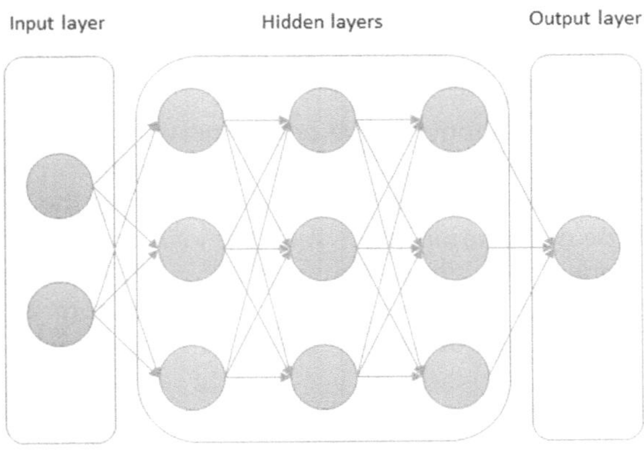

FIGURE 11.3
Structure of a neural network.

The connections between nodes, often represented by weights, determine the strength of the relationships between neurons (i.e., the nodes in the network). During the training process, the neural network learns to adjust these weights based on the input data and the desired output. The classical neural network training involved a process called backpropagation, which iteratively updates the weights to minimize the difference between the predicted output and the actual output.

The development of a neural network involves four key concepts:

- **Activation Function.** Each node (called a neuron) in a neural network typically applies an activation function to its weighted input. Common activation functions include sigmoid, hyperbolic tangent, and rectified linear units. Activation functions introduce non-linearities, enabling neural networks to learn complex relationships in the data.
- **Loss Function.** The loss function measures the difference between the predicted output and the actual output. The goal during training is to minimize this loss, and the choice of the loss function depends on the specific task (e.g., mean squared error for prediction or cross-entropy for classification).
- **Optimization**. Optimization algorithms, such as gradient descent, are used to minimize the loss function by adjusting the weights of the neural network. Variants like stochastic gradient descent and adaptive learning rate methods are commonly employed.
- **Architecture**. The architecture of a neural network refers to its structure, including the number of layers, the number of nodes in each layer, and the connectivity pattern. Different architectures are suited for different types of tasks.

Neural networks have demonstrated remarkable success in various domains, especially when dealing with large and complex datasets. Deep learning, powered by neural networks, has significantly advanced the state of the art in tasks such as image recognition, natural language processing, and speech recognition.

We illustrate the development of a neural network with the loan application categorical data in Table 11.7. This is a classification problem with two groups, one that paid the loan and one that did not pay the loan. We assign a value of 1 to those observations for which the loan was paid and a value of 0 to the observations for which the loan was not paid. We use the same transformation from categorical to numerical values as before, to create the four input variables (x) and the output variable (y) in our model:

$x_1 = 0, 1, 2$ (Age = young, middle, old)

$x_2 = 0, 1$ (Job = no, yes)

$x_3 = 0, 1$ (Homeowner = no, yes)
$x_4 = 0, 1, 2$ (Credit rating = fair good and excellent)
$y = 0, 1$ (Paid = no, yes)

There is a weight value associated with each arc in the network. These weights are used to create a signal entering the node in the next layer. All the signals are combined into a single value z that the activation function of the node uses. Figure 11.4 shows the four nodes in the input layer connecting with a node in a hidden layer. The weights are represented by w_1 to w_4 and the combined signal entering the hidden is z. The combined signal is the sum product of the input values and the weights. The activation function f is the so-called sigmoid function and depends on the value of z.

Because our neural network will have a single hidden layer with three nodes, we will need three sets of four weights for a total of 12 weights between the input layer and the hidden layer. We also need three weights for the arcs going from the hidden layer to the output layer. For classification models, the customary loss function l is the so-called *cross-entropy* function. The form of the cross-entropy function for binary classification models is:

$$l = -\frac{1}{n}\sum_{i=1}^{n}\left(y_i \log\left(\hat{y}_i\right) + \left(1 - y_i\right)\log\left(1 - \hat{y}_i\right)\right)$$

In this function, y_i is the output for the *ith* observation a dataset with n observations and $\hat{y}_i$ is the prediction of the model for that observation. The prediction is the value of the activation function corresponding to the output node. Figure 11.5 is a screen capture of the Excel worksheet that models our neural network.

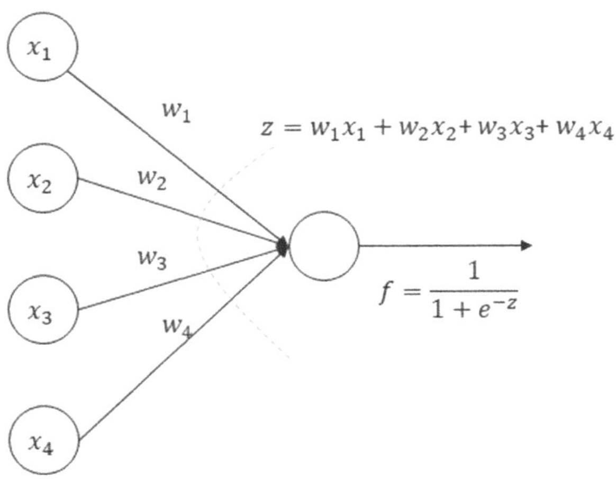

FIGURE 11.4
Activation function for the hidden layer.

	A	B	C	D	E	F	G	H	I	J	K	L	M	N	O
1	Obs	x1	x2	x3	x4	z1	z2	z3	f(z1)	f(z2)	f(z3)	z out	y hat	y	Loss
2	1	0	1	0	0	2.2003	-10.6216	-5.0732	0.9003	0.0000	0.0062	-4.6843	0.0092	0	-0.0040
3	2	0	0	0	1	-18.3033	-0.0528	9.2551	0.0000	0.4868	0.9999	-2.0757	0.1115	0	-0.0513
4	3	0	1	0	1	-16.1031	-10.6744	4.1819	0.0000	0.0000	0.9850	7.5454	0.9995	1	-0.0002
5	4	0	1	1	0	22.2003	-11.8110	7.5353	1.0000	0.0000	0.9995	2.4012	0.9169	1	-0.0376
6	5	0	0	0	0	0.0000	0.0000	0.0000	0.0000	0.0000	0.0000	0.0000	0.0000	0	0.0000
7	6	1	0	1	0	23.4206	-1.0292	21.3891	1.0000	0.2632	1.0000	-2.8592	0.0542	0	-0.0242
8	7	1	0	0	1	-14.8827	0.1073	18.0357	0.0000	0.5268	1.0000	-2.8751	0.0534	0	-0.0238
9	8	1	1	1	1	7.3175	-11.7036	25.5711	0.9993	0.0000	1.0000	2.4088	0.9175	1	-0.0373
10	9	1	0	1	2	-13.1861	-1.1348	39.8994	0.0000	0.2433	1.0000	2.7957	0.9424	1	-0.0257
11	10	1	0	1	1	5.1173	-1.0820	30.6443	0.9940	0.2531	1.0000	-2.6256	0.0675	0	-0.0303
12	11	2	0	1	2	-9.7655	-0.9747	48.6800	0.0001	0.2739	1.0000	2.1819	0.8986	1	-0.0464
13	12	2	0	1	1	8.5379	-0.9219	39.4249	0.9998	0.2846	1.0000	-3.2849	0.0361	0	-0.0159
14	13	2	1	0	1	-9.2619	-10.3541	21.7432	0.0001	0.0000	1.0000	7.6599	0.9995	1	-0.0002
15	14	2	1	0	2	-27.5652	-10.4069	30.9983	0.0000	0.0000	1.0000	7.6605	0.9995	1	-0.0002
16	15	2	0	0	0	6.8412	0.3203	17.5612	0.9989	0.5794	1.0000	-9.1767	0.0001	0	0.0000
17	New	0	1	0	1	-16.1031	-10.6744	4.1819	0.0000	0.0000	0.9850	7.5454	0.9995	Average	0.0198

FIGURE 11.5
Excel model of the neural network for the loan application data.

Our neural network has four nodes in the input layer, one for each attribute (i.e., age, job status, homeowner status, and credit rating). We use three nodes in a single hidden layer and a single node in the output layer. The input values are in columns B to E of the spreadsheet. Columns F, G, and H have the combined signals entering the hidden nodes. The Excel function in cell F2 that calculates the combined signal entering the first hidden node for the first observation is:

$$= \text{SUMPRODUCT}(\text{B2}:\text{E2}, \$\text{B}\$21:\$\text{E}\$21)$$

The weights corresponding to the arcs entering the first node in the hidden layer are in the B21 to E21 range, as shown in Figure 11.6. The SUMPRODUCT function can be copied down to the remaining cells in the column. The functions for G2 and H2 are:

$$= \text{SUMPRODUCT}(\text{B2}:\text{E2}, \$\text{B}\$22:\$\text{E}\$22)$$

$$= \text{SUMPRODUCT}(\text{B2}:\text{E2}, \$\text{B}\$23:\$\text{E}\$23)$$

Note that the references to the table of weights change to row 22 and row 23 for the second and third nodes in the hidden layer, respectively. Columns I, J, and K calculate the value of the activation function. Therefore, the function for the first node in the hidden layer and the first observation (i.e., cell I2) is:

$$= \text{IF}\left(\text{F2} = 0, 0, 1/\left(1 + \text{EXP}(-\text{F2})\right)\right)$$

We use an IF statement because the data contains an observation (number 5) for which all the input values are zero. This results in a combined signal $z = 0$. The activation (sigmoid) function for such an observation is 0.5 regardless of the weights, since EXP(0) = 1. The three signals leaving the nodes in

	A	B	C	D	E
19	Weights				
20		x1	x2	x3	x4
21	Hidden 1	3.421	2.200	20.000	–18.303
22	Hidden 2	0.160	–10.622	–1.189	–0.053
23	Hidden 3	8.781	–5.073	12.609	9.255
24					
25		f(z1)	f(z2)	f(z3)	
26	Output	–5.25559	–20	7.661075	

FIGURE 11.6
Weights for the neural network model in Figure 11.5.

the hidden layer are combined using the weights in cells B26, C26, and D26. Therefore, the combined signal entering the output node for the first observation (i.e., cell L2) is calculated with the following function:

$$= \text{SUMPRODUCT}(\text{I2}:\text{K2},\$\text{B}\$26:\$\text{D}\$26)$$

The value of the activation function for the output node is the predicted value for the observation (i.e., $\hat{y}$). The predicted values are in column M and the function for the first observation is:

$$= \text{IF}\left(\text{L2} = 0, 0, 1/\left(1 + \text{EXP}(-\text{L2})\right)\right)$$

We again screen for combined signal values that are equal to zero. Once we have calculated all the predictions, we can determine the loss associated with each observation. For the first observation, the loss function is:

$$= \text{N2} * \text{LOG}(\text{M2} + 0.0001) + (1 - \text{N2}) * \text{LOG}(1 - \text{M2} + 0.0001)$$

We add 0.0001 to the LOG calculations for cases when $\hat{y} = 0$ or 1 because LOG(0) is undefined.

The last piece of the development of a neural network is the optimization of the weights. The weight values shown in Figure 11.6 are the best values found using the Evolutionary solver in Excel. The use of a metaheuristic optimizer such as the Evolutionary solver is very common in practice. The optimization model consists of minimizing the average loss cell (O17) by changing the values of the weights (cells B21:E23 and B26:D26). These values are continuous in the range between –20 and 20. The Evolutionary solver and the GRG Non-linear solver with random restarts require that the variables are bounded. These bounds can be found by trial and error, as we did here.

The optimized weights result in a very small average loss value. In this case, the rounded $\hat{y}$ prediction values match the data output. In practice, the weight values are found with a so-called training dataset. The training set contains data that is representative. Then, the trained neural network is used on additional data for which the output values are known. This set is called the test set. This testing allows the analyst to have a better idea of the magnitude of the error that could result from using the trained network in practice.

In our example, we could use our trained neural network to determine whether the young applicant in Section 11.3.2 who is currently employed, is not a homeowner, and has excellent credit would be immediately rejected. The input values are $x_1 = 0, x_2 = 1, x_3 = 0$, and $x_4 = 1$. As shown in row 17 of Figure 11.5, our neural network predicts $\hat{y} = 0.9995$. Therefore, the model suggests that this new application is not rejected immediately and is reviewed further.

11.3.4 Classification Problems in Business Processes

Classification problems are common in business processes, and they play an important role in decision-making and operational efficiency. As we saw in the previous sections, in a classification problem the goal is to categorize input data into predefined classes or categories based on certain features or attributes. Businesses often encounter various scenarios where classification is essential. Here are some common examples:

- **Credit scoring.** Financial institutions use classification models to assess the creditworthiness of applicants. The models classify applicants into different risk categories, aiding in the decision to approve or deny credit.
- **Customer churn.** Businesses use classification to predict whether a customer is likely to churn (stop using a service or product). This allows proactive retention strategies to be implemented for high-risk customers. Telecommunications companies have been found to use machine learning classification models for churn prediction.
- **Customer segmentation.** Businesses use classification to categorize customers into different segments based on their purchasing behavior, demographics, or other relevant features. This helps in targeted marketing and personalized customer experiences.
- **Employee attrition**. Human resources departments use classification models to predict the likelihood of employees leaving the organization. This allows proactive measures to retain valuable talent.
- **Fraud detection.** Classification models are employed to detect fraudulent activities in transactions, whether in banking, insurance, or e-commerce. Unusual patterns or anomalies can be classified as potentially fraudulent. In the case of an anomaly detection problem, the machine learning model is trained to predict a particular class (e.g., normal or anomalous), and the classification model is then used to classify new data.
- **Product categorization**. E-commerce companies use classification to automatically categorize products into relevant categories. This streamlines inventory management and improves the user experience in online shopping platforms.

These examples illustrate the versatility of classification in addressing a wide range of business challenges. Machine learning algorithms, such as decision trees, support vector machines, and neural networks, are commonly applied to develop classification models for these purposes. The ability to automate decision-making based on data patterns is an asset for businesses seeking efficiency and accuracy.

Discussion Questions and Exercises

1. *Business intelligence and business process management.*[7] Nearly all business processes involve complex value-based decision points, such as loan approvals or up-sell opportunities. Complex decision-making may require information that often is not delivered within the workflow, and decision-makers must spend time gathering the information required to support the decision. Surveys show that users, on average, spend 20 hours per week gathering and analyzing information. This ad hoc activity may be informal or formal, ranging from accessing and querying a database to soliciting advice from coworkers. Whatever the approach, this information-gathering step adds processing time and may create bottlenecks while taking workers away from more productive uses of their time and talent. This leads to inefficient and ineffective processes. Organizations that have invested in business intelligence (BI) solutions to spread the culture of analytics in an organization must make sure that the right information is available to the decision-makers in their business processes. The merging of BI and BPM incorporates the right analytic reports from a BI system into a process workflow. The result is that when the workflow gets routed to a decision-maker, it is accompanied by relevant and timely information that supports the decision. Give an example of how a business process may benefit from the merger of BI and BPM in the following industries:
   a. Banking
   b. Healthcare
2. An *if-then* statement is the simplest form of the inference block of a business rule. A set of "if-then" statements is such that the "if" is the condition of the rule, and the "then" is the action. Typically, rules that are based on multiple if-then statements are evaluated sequentially, and each "true" condition triggers some action. Therefore, an if-then statement set may result in more than one action. Consider an online vendor that gives discounts, matches a competitor's price, or offers free shipping to loyal customers depending on the customer loyalty status and the size of the order.
   a. Create an inference block (i.e., a set of if-then statements) for a rule that provides different discounts (5 percent, 10 percent, and 15 percent) to customers according to their loyalty status (bronze, silver, gold).
   b. Enhance the rule to provide free shipping to any loyal customer for orders of more than $100.
   c. Gold customers are given a "best price" guarantee, which means that the price they pay is never higher than the competitors' price. Explain how a business rule could be used to implement this policy.

3. The manager of a commercial loan department for a bank wants to develop a rule to make loan application decisions.[8] The manager believes that liquidity, profitability, and activity are the three key indicators of performance that are helpful in deciding on a company's loan application. Liquidity is measured as the ratio of current assets to current liabilities. Profitability is measured as the ratio of net profit to sales. Activity is measured as the ratio of sales to fixed assets. Table 11.10 shows a sample of 20 loans that the bank has made in recent years. The loans are classified into two groups: (1) those that were acceptable and (2) those that should have been rejected.
   a. Use the linear SVM approach to develop a classification rule that can be used to accept or reject new loan applications. State the classification rule.
   b. Suppose that five loan applications are received from companies with the key indicators shown in Table 11.11. Use the classification rule to determine which companies should be considered for a loan.

TABLE 11.10

Key Indicator Data for 20 Recent Loans

Loan	Group	Liquidity	Profitability	Activity
1	1	0.65	0.31	1.75
2	1	0.64	0.29	1.50
3	1	0.62	0.23	1.45
4	1	0.77	0.27	1.20
5	1	0.70	0.28	1.95
6	1	0.85	0.32	1.65
7	1	0.65	0.26	1.79
8	1	0.77	0.29	1.88
9	1	0.64	0.30	1.99
10	1	0.67	0.32	1.84
11	2	0.82	0.25	1.77
12	2	0.65	0.34	1.42
13	2	0.68	0.27	1.91
14	2	0.75	0.22	1.88
15	2	0.85	0.25	1.60
16	2	0.67	0.25	1.34
17	2	0.87	0.22	1.65
18	2	0.82	0.29	1.37
19	2	0.82	0.30	1.46
20	2	0.85	0.29	1.44

TABLE 11.11

Key Indicators for Five New Loan Applications

Company	Liquidity	Profitability	Activity
A	0.78	0.27	1.58
B	0.91	0.23	1.67
C	0.68	0.33	1.43
D	0.78	0.23	1.23
E	0.67	0.26	1.78

4. Apply the *k*-NN classifier to the data in Exercise 3. Use the classifier with $k = 3$ to make decisions with respect to the five loan applications in Part (b) of Exercise 3. Are the recommendations from the *k*-NN classification rule the same as those found with the application of the classification model based on linear SVM?
5. A telecommunications company is concerned with the number of customers leaving for a competitor. The company wants to understand who is leaving and identify current customers that may be at risk. The company has identified six attributes that may be used to build a model capable of detecting customers who are at a high risk of leaving (churn). Table 11.12 shows data associated with 40 customers.
   a. Transform the categorical data in Table 11.12 to numerical data using the following values.

      Gender (0 = female, 1 = male)

      Dependents (0 = no, 1 = yes)

      Tenure (0 = short, 1 = long)

      Multiple phone lines (0 = no, 1 = yes)

      Internet service (0 = no, 1 = basic, 2 = fast)

      TV service (0 = no, 1 = yes)

      How many of the customers left the company? Out of the customers that churned, what proportion of them were female?
   b. Use Excel to develop a neural network with four neurons in a single hidden layer to predict customer churn. Use the Evolutionary solver to optimize the weight values. What is the average loss? How many customers are misclassified by the optimized neural network?
   c. The company wants to know whether a male customer with no dependents, a long tenure with the company, with multiple lines, a fast internet connection, and no TV service is at risk of leaving. Use the neural network to classify this customer. Is he likely to leave?

TABLE 11.12
Customer Data from Telecommunication Company

Customer	Gender	Dependents	Tenure	Multiple lines	Internet service	TV service	Churn
1	male	yes	long	yes	fast	yes	no
2	male	yes	long	no	basic	no	no
3	female	no	long	no	fast	yes	yes
4	female	no	short	yes	fast	yes	yes
5	female	no	short	no	fast	no	yes
6	female	no	short	no	fast	no	yes
7	male	no	short	no	basic	no	yes
8	female	yes	long	no	basic	yes	yes
9	male	yes	short	no	basic	no	yes
10	female	no	short	yes	fast	no	yes
11	female	no	long	no	no	no	no
12	female	no	short	yes	fast	yes	yes
13	male	no	long	yes	fast	yes	no
14	female	no	long	yes	fast	yes	yes
15	male	no	long	no	basic	no	no
16	female	no	short	no	fast	yes	yes
17	female	no	short	no	basic	no	no
18	female	no	short	yes	fast	no	yes
19	male	no	long	no	no	no	no
20	male	no	short	no	no	no	no
21	female	no	long	no	basic	no	no
22	female	yes	long	no	basic	no	no
23	female	no	short	no	basic	no	no
24	male	yes	long	yes	fast	yes	yes
25	male	yes	long	no	basic	no	no
26	male	yes	long	no	basic	no	no
27	female	yes	long	yes	fast	yes	no
28	male	no	short	no	fast	no	yes
29	female	yes	short	no	fast	yes	yes
30	female	no	short	yes	fast	yes	yes
31	male	no	long	no	basic	no	no
32	male	no	long	yes	fast	yes	yes
33	male	no	long	yes	basic	yes	no
34	male	no	long	no	fast	yes	no
35	male	no	short	no	fast	yes	no
36	female	no	long	no	fast	yes	yes
37	female	yes	short	no	basic	no	yes
38	female	no	long	no	fast	no	no
39	female	no	long	yes	fast	no	no
40	male	no	short	no	no	no	yes

Notes

1. This section is based on the articles by Davenport (2006) and Davenport et al. (2005).
2. The discussion on BPMSs is based on the book by Chang (2006).
3. Financial or nonfinancial metrics that quantify business process performance, for example, cycle time.
4. Automated process discovery is another term used to describe these process mining techniques.
5. In the calculation of the sample standard deviation, the denominator is the number of observations minus 1. STDEV.S is the Excel function that implements this calculation for a range of numbers in the spreadsheet.
6. The combination of rules and analytics is sometimes referred to as *enterprise decision management*.
7. Adapted from "Using business process management and business analytics together for smarter work," IBM Software, WebSphere.
8. Adapted from Ragsdale (2011).

References

Chang, J.F. 2006. *Business process management systems*. New York: Auerbach Publications.

Davenport, T.H. 2006. Competing on analytics. *Harvard Business Review 84*(1): 98–107.

Davenport, T.H., Cohen, D., and Jacobson, A. 2005. Competing on analytics. Babson Executive Education, Working Knowledge Research Center.

Davenport, T.H., and Harris, J.G. 2007. *Competing on analytics: The new science of winning*. Boston, MA: Harvard Business School.

Grigori, D., Casati, F., Castellanos, M., Dayal, U., Sayal, M., and Shan, M.C. 2004. Business process intelligence. *Computers in Industry 53*: 321–343.

Kirchmer, M. 2011. *High performance through process excellence*. Berlin, Germany: Springer-Verlag.

Dietrich, B., Johnson, C., and Dziekan, C. 2010. The analytics journey. Analytics, December 5, 2010, https://doi.org/10.1287/LYTX.2010.06.01.

Pant, K. 2009. *Business rules Management, BPM, and SOA*. UK: Packt Publishing. Available at www.packtpub.com/article/business-rules-management-bpm-and-soa

Ragsdale, C.T. 2011. *Managerial decision modeling*. Canada: South-Western Cengage Learning.

van der Aalst, W.M.P. 2016. Process discovery: An introduction. In: *Process mining*. Springer, Berlin, Heidelberg, ISBN 978-3-662-49850-7

van der Aalst, W.M.P., and Weijters, A.J.M.M. 2004. Process mining: A research agenda. *Computers in Industry 53*: 231–244.

Wegener, D., and Rüping, S. 2010. On Integrating Data Mining into Business Processes. In: Abramowicz, W., Tolksdorf, R. (eds) Business Information Systems. BIS 2010. *Lecture Notes in Business Information Processing*, vol 47. Springer, Berlin, Heidelberg. https://doi.org/10.1007/978-3-642-12814-1_16

Epilogue

In the first three chapters, this book examined the issues of process design from a conceptual, high-level view. Chapter 4 began the move from a high-level to a low-level, detailed examination of processes and flows. This detailed examination led readers through deterministic models for cycle time and capacity analysis in Chapter 5, and stochastic queuing models in Chapter 6. Then, Chapter 7 introduced the notion of building computer simulations for modeling business processes. Chapters 8 and 9 moved from theory to practice in the realm of process simulation with the introduction and fairly extensive use of the ExtendSim software. Chapter 10 climbed to a higher-level view of a process by treating simulation models as black boxes. In the black-box view, the internal details of the simulation model were not of concern; instead, the focus was on finding effective values for input parameters, where effectiveness was measured by retrieving relevant output from the simulation model. Finally, Chapter 11 looked at processes from both a white-box and a black-box perspective while embracing analytics as an approach to business process management (BPM). Two particular analytics tools for processes were explained in some detail: process mining and machine learning. In process mining, the focus is on establishing how existing processes operate. Machine learning, on the other hand, enhances the operation of a process by creating models that can automate decision-making. The techniques can be used in a variety of business contexts that involve classification.

The main goal of this book is to provide a balanced approach to BPM by not only discussing the managerial implications of the approach but also presenting details of the key supporting analytical tools. The quantitative approaches are not meant to be the answer to all problems related to process performance. As discussed in Chapter 7, computer simulation is a very powerful tool; however, it is not appropriate in all situations. The intent has not been to build an argument for the application of quantitative tools and technology, but rather to make the reader familiar with a spectrum of techniques—from the simplistic to the somewhat sophisticated—to support BPM initiatives.

The BPM community is very active both inside and outside academia. Organizations such as the Institute for Operations Research and the Management Sciences (informs.org), the Production and Operations Management Society (poms.org), and the Institute of Industrial and Systems Engineers (iise.org) are excellent resources in areas related to analytics and business process excellence within the academic community. Likewise, in the practitioner's world, organizations such as the Association of Business Process Management Professionals (abpmp.org) have a wealth of information about the latest developments in the practice of BPM.

Index

Note: **Bold** page numbers refer to **tables** and *Italic* page numbers refer to *figures*

D

H

I

Q

For Product Safety Concerns and Information please contact our EU representative GPSR@taylorandfrancis.com Taylor & Francis Verlag GmbH, Kaufingerstraße 24, 80331 München, Germany